Additive Manufacturing

Additive Manufacturing

Alloy Design and Process Innovations

Special Issue Editors

Prashanth Konda Gokuldoss
Zhi Wang

MDPI • Basel • Beijing • Wuhan • Barcelona • Belgrade • Manchester • Tokyo • Cluj • Tianjin

Special Issue Editors

Prashanth Konda Gokuldoss
Head of Additive Manufacturing
Laboratory, Department of
Mechanical and Industrial
Engineering, Tallinn University
of Technology
Estonia

Zhi Wang
National Engineering Research
Center of Near-net-shape
Forming for Metallic Materials,
South China University of
Technology
China

Editorial Office
MDPI
St. Alban-Anlage 66
4052 Basel, Switzerland

This is a reprint of articles from the Special Issue published online in the open access journal *Materials* (ISSN 1996-1944) (available at: https://www.mdpi.com/journal/materials/special_issues/ AMADPI).

For citation purposes, cite each article independently as indicated on the article page online and as indicated below:

LastName, A.A.; LastName, B.B.; LastName, C.C. Article Title. *Journal Name* **Year**, *Article Number*, Page Range.

Volume 2
ISBN 978-3-03928-352-1 (Pbk)
ISBN 978-3-03928-353-8 (PDF)

Volume 1-2
ISBN 978-3-03928-416-0 (Pbk)
ISBN 978-3-03928-417-7 (PDF)

Cover image courtesy of Prasanth Konda Gokuldoss

Contents

About the Special Issue Editors

Prashanth Konda Gokuldoss (Prof.) is Head of the Additive Manufacturing Laboratory and Professor of Additive Manufacturing in the Department of Mechanical and Industrial Engineering, Tallinn University of Technology, Tallinn, Estonia. In addition, he is a guest scientist at the Erich Schmid Institute of Materials Science, Austrian Academy of Science, Leoben, Austria, and an Adjunct Professor at the department of CBCMT, School of Engineering, Vellore Institute of Technology, Vellore, India. was a awarded a PhD by the Technical University Dresden, Germany (2014), and conducted postdoctoral research at the Leibniz Institute of Solid State and Materials Research (IFW) Dresden, Germany. He also worked as an R&D Engineer (Sandvik, Sweden), Senior Scientist (Erich Schmid Institute of Materials Science, Austrian Academy of Science, Leoben, Austria), and Associate Professor (Norwegian University of Science and Technology, Gjøvik, Norway) before taking a Full Professorship at the Tallinn University of Technology, Tallinn, Estonia. His present research is focused on but not limited to additive manufacturing (alloy, process and product development), fabrication of meta-stable materials, powder metallurgy, light materials, solidification and biomaterials. He has published over 125 peer-reviewed journal papers and has an h-index of 32 (google scholar). A multiple award winner, he actively collaborates with/visits China, India, the USA, Austria, Poland, Norway, Germany, Spain, Taiwan, South Korea and Iran.

Zhi Wang (Prof.) is a Full Professor at the National Engineering Research Center of Near-net-shape Forming for Metallic Materials, School of Mechanical and Automotive Engineering, South China University of Technology, Guangzhou, China. He received his PhD from Technical University Dresden, Germany (2014), and did his postdoctoral research at the WPI-Advanced Institute for Materials Research, Tohoku University, Japan. He research revolved around powder metallurgy, additive manufacturing, Al and its alloys, metal matrix composites, solidification, and meta-stable materials. He is an active reviewer for more than 20 Elsevier journals and has published more than 65 articles in international peer reviewed journals. He is actively collaborating with groups around the world, including in Estonia, China, India, Germany, Austria, Australia, Poland and the USA.

Preface to "Additive Manufacturing"

Additive manufacturing (AM) is revolutionizing the manufacturing sector, particularly in terms of the production of metallic components with added functionality, including complex or intricate geometries, conformal cooling channels, and highly customized parts with small production cycles. In addition, material saving and weight reduction of the components in automobile and aerospace sectors, helping reduce fuel consumption, and promoting a green environment fall into this category. The AM field is undergoing rapid development with improvements and innovations taking place at all times. However, the AM field will face several challenges before it can be adapted completely to an industrial environment. The challenges include the process capabilities and material aspects, including the microstructure formation and properties, and the process cycle. It is, therefore, necessary to devote attention to the research and development activities in the field of AM, and to promote the industrialization process. This includes the production of powder, the properties of powder, AM process developments, alloy systems used for the AM process, and post-processing of the components. In view of the growing importance of additive manufacturing, this book addresses key issues related to emerging science and technology in this area (especially alloy design for additive manufacturing and process innovations). Detailed and informative articles are presented in this book, related to different additive manufacturing processes like selective laser melting (laser-based powder bed fusion process), electron beam melting, direct metal laser sintering, laser classing, ultrasonic consolidation, and wire arc additive manufacturing. Of key importance in the area of materials science are the end properties and the correlation with their microstructure. Accordingly, the articles presented critically discuss the effects of microstructural features, such as porosity, forming defects, and heat treatment induced effects on mechanical properties, with a focus on applications in the aerospace, automobile, defense and aerospace sectors. Overall, the cutting edge information presented in this book is of significant importance to researchers in the field of additive manufacturing.

Prashanth Konda Gokuldoss, Zhi Wang
Special Issue Editors

Editorial

Additive Manufacturing: Alloy Design and Process Innovations

Konda Gokuldoss Prashanth [1,2,3,*] and Zhi Wang [4]

[1] Department of Mechanical and Industrial Engineering, Tallinn University of Technology, Ehitajate Tee 5, 19086 Tallinn, Estonia

[2] Erich Schmid Institute of Materials Science, Austrian Academy of Science, Jahnstrasse 12, A-8700 Leoben, Austria

[3] CBCMT, School of Mechanical Engineering, Vellore Institute of Technology, Vellore 632014, Tamil Nadu, India

[4] School of Mechanical and Automotive Engineering, South China University of Technology, Guangzhou 510640, China; wangzhi@scut.edu.cn

* Correspondence: kgprashanth@gmail.com; Tel.: +372-5452-5540

Received: 22 January 2020; Accepted: 22 January 2020; Published: 23 January 2020

Additive Manufacturing (AM) is an emerging manufacturing technique of immense engineering and scientific importance and is also regarded as the technique of the future. AM can fabricate any kind of materials including metals, polymers, ceramics, composites, etc. It also offers several advantages, like added functionality, offering intricacy for free, near-net-shape fabrication with minimal or no post-processing, shorter lead time, etc. AM has been used in several industrial sectors like aerospace, automobile, oil refinery, marine, construction, food industry, jewelry, etc. However, the several shortcomings in the field of AM are (a) alloy development that suits the AM processes (b) pre-mature failure of materials, even though improved properties are observed (c) process development and innovation (d) structure-property correlation, etc.

Accordingly, the present Special Issue (book) focuses on the two main aspects: alloy design and process development and innovation. Alloy design and development that suits the process conditions and the process is requisite, and the conventionally designed alloys for powder metallurgy/casting process are unusable. In addition, process development is happening at a rapid pace in the field of AM, which also warrants attention. Overall, 45 papers were published under this Special Issue with the following themes:

Selective Laser Melting/Laser-based powder bed fusion process of materials—23 papers including processing of Al-based alloys, Fe-based alloys, Ni-based alloys, Ti-based alloys and Zr-based alloys:

Direct Metal Laser Sintering—1 paper

Laser Cladding—3 papers

Electron Beam Melting—2 papers

Hybrid Manufacturing (Additive + Subtractive Manufacturing)—1 paper

Wire Arc Additive Manufacturing—3 papers

Fusion Deposition Modeling—2 papers

Ultrasonic Consolidation—1 paper

Miscellaneous fields—9 papers

The outcome of the Special Issue suggests that research is thriving in the field of AM, especially alloy design and process innovations. The present Special Issue is particularly interesting because it covers a wider range of AM processes and materials and gives an overview of research in this field, including alloy design and development, process development, microstructure-property correlation, simulation on melt pool dynamics, etc.

Finally, we would like to thank all the contributing authors for their excellent contribution to this Special Issue, to the reviewers for constructively improving the quality of the Special Issue and to the

Materials **2020**, *13*, 542

Materials staff for giving us the opportunity to host this Special Issue and to publish the articles in a timely manner and make this Special Issue a great success.

Conflicts of Interest: Authors declare no conflict of interest.

Article

A Comprehensive Approach to Powder Feedstock Characterization for Powder Bed Fusion Additive Manufacturing: A Case Study on AlSi7Mg

Jose Alberto Muñiz-Lerma [1], Amy Nommeots-Nomm [1], Kristian Edmund Waters [2] and Mathieu Brochu [1,*]

1   REGAL Aluminum Research Center, Department of Mining and Materials Engineering, McGill University, Montreal, QC H3A 0C5, Canada; jose.muniz@mcgill.ca (J.A.M.-L.); amy.nomm@mcgill.ca (A.N.-N.)
2   Department of Mining and Materials Engineering, McGill University, Montreal, QC H3A 0C5, Canada; kristian.waters@mcgill.ca
*   Correspondence: mathieu.brochu@mcgill.ca; Tel.: +1-514-398-2354

Received: 8 November 2018; Accepted: 25 November 2018; Published: 27 November 2018

Abstract: In powder bed fusion additive manufacturing, the powder feedstock quality is of paramount importance; as the process relies on thin layers of powder being spread and selectively melted to manufacture 3D metallic components. Conventional powder quality assessments for additive manufacturing are limited to particle morphology, particle size distribution, apparent density and flowability. However, recent studies are highlighting that these techniques may not be the most appropriate. The problem is exacerbated when studying aluminium powders as their complex cohesive behaviors dictate their flowability. The current study compares the properties of three different AlSi7Mg powders, and aims to obtain insights about the minimum required properties for acceptable powder feedstock. In addition to conventional powder characterization assessments, the powder spread density, moisture sorption, surface energy, work of cohesion, and powder rheology, were studied. This work has shown that the presence of fine particles intensifies the pick-up of moisture increasing the total particle surface energy as well as the inter-particle cohesion. This effect hinders powder flow and hence, the spreading of uniform layers needed for optimum printing. When spherical particles larger than 48 µm with a narrow particle distribution are present, the moisture sorption as well as the surface energy and cohesion characteristics are decreased enhancing powder spreadability. This result suggest that by manipulating particle distribution, size and morphology, challenging powder feedstock such as Al, can be optimized for powder bed fusion additive manufacturing.

Keywords: additive manufacturing; metal powders; powder flowability; powder properties; aluminum; water absorption

---

1. Introduction

Laser powder bed fusion (LPBF), an additive manufacturing (AM) process that allows the fabrication of complex geometries, is creating opportunities to fabricate innovative parts with enhanced functionality [1]. Aluminum alloys have shown to be of great interest in the AM sector acting as an economical method to manufacture parts that cannot be produced through traditional casting or forging routes. However, aluminum powders are amongst the most difficult powders to process via LPBF due to the lack of repeatability associated with the quality of powder feedstock [2].

The powder feedstock quality can be assessed by its extrinsic and intrinsic properties. Extrinsic properties refer to the relationship between powder morphology, including particle size distribution and shape, to the final part performance. While intrinsic properties refer to the influence of the powder composition and microstructure on the final quality of the additively manufactured part [3].

In terms of extrinsic properties, it is well accepted that spherical powder particles positively contribute towards powder flowability and higher particle packing densities in the powder bed [4]. Spherical morphologies are also typically associated with low levels of particle-particle friction which results in enhanced powder flow. Another important extrinsic property is the apparent density of the powder feedstock; some reports in literature have suggested that this factor has the most significant effect on the final density of the manufactured parts [5,6]. Apparent density is directly influenced by a variety of factors such as particle size distribution (PSD), particle shape, inter-particle friction, surface chemistry, and agglomeration. However, among these factors, the particle shape, size and distribution are suggested to be the most influential [7]. Secondarily to this, it has recently been reported that even with powder feedstock optimisation, particle segregation can occur within a powder bed affecting the local apparent density [8,9] and the mechanical properties of the produced part.

For LPBF, suppliers typically provide powders in the size range of 15–45 μm. However, selecting an appropriate powder size and size distribution is a key factor as even within this range, competitive powder interactions can occur. Theoretically, the presence of fine particles in the powder feedstock (smaller than the recoating layer thickness) would be beneficial. As this would lead to an increase in the powder density which would improve the surface roughness and reduce the number of defects in the printed parts [10,11]. Smaller particle sizes are also known to have increased laser energy absorption, thus improving their processing. The advantages gained from fine particles are contradicted by the significant disadvantages associated with their use, such as their high tendency to agglomerate [12], and their large surface area to mass ratio. The agglomeration of fine powders are more likely to promote pore and void formation during processing due to the irregular shapes and variable sizes of the agglomerates formed [13]. Additionally, agglomerated powders are known to increase the reflectivity of the powder bed resulting in less energy absorption during the manufacturing process [12,14]. The discussed factors makes the spreading of uniform powder layers more challenging thus compromising the process stability since inhomogeneous powder bed densities give rise to non-uniform heating increasing the risk of defect formation [12,13,15].

Intrinsic properties are those which relate to the powder composition and microstructure of the resulting part. Here, issues arise due to powder stability. Metallic powder composition during storage, handling and building (including re-use), can have significant changes over time due to oxygen and moisture exposure. This phenomenon is significantly influenced by the surface-to-volume ratio of the powder. Both oxygen and moisture contamination are known to produce reaction products that can change and degrade the microstructure of the particles, resulting in an increased particle porosity and ultimately reducing the mechanical properties such as ductility and toughness of 3D printed parts [2].

Some factors affect both the extrinsic and intrinsic properties. One of these factors is the likelihood of elemental evaporation of alloying elements during the LBPF process. Smaller powder particles are more susceptible to this phenomenon due to their high laser absorptivity [15,16]. As discussed there are a wide number of aspects which need to be considered when looking to optimize suitable powder feedstock for the LPBF process. Reaching the perfect balance in terms of particle size distribution, contamination and printability is not a trivial process.

Recently, a new atomization technique has been developed that produces metallic powders with high sphericity, narrow PSD and uniform microstructure [17]. The powders formed are expected to provide superior properties compared to typical gas atomized powders for AM applications, leading to a higher density and improved mechanical properties in the final part.

The purpose of this paper is to compare the powder properties of various AlSi7Mg powders obtained from different production methods. Different characterization techniques have been employed to obtain information regarding the apparent density, powder flow, spread density, moisture sorption, surface energy and work of cohesion of the studied powders. The conclusions contribute to the growing knowledge on powder characterization methods and optimum powder feedstock properties for AM processing.

2. Materials and Methods

Three powders produced using the gas atomization principle were used in the current study. The powders are termed A, B, and C, and are all based on the A356-A357 systems. Scanning electron microscopy (SEM) was used to characterize the morphology of the powders using a Hitachi SU3500 Scanning Electron Microscope, Tokyo, Japan. The powder PSD was obtained by a Horiba laser particle size analyser (Model LA-920, Kyoto, Japan). The log-normal slope parameter method was used to calculate the dimensionless PSD width (S_w) which represents the breadth of the PSD [18]. In this method, the standard deviations of the cumulative distribution were plotted against the logarithmic particle size resulting in a linear plot. The slope of the linear plot represents the S_w while the intersection with the x-axis is equivalent to the median particle diameter, D_{50}. If the particle size is very narrow, high values of S_w are expected.

Powder flowability was measured using Hall and Carney funnels (QPI-HFM1800SS, Qualtech Products Industry, Denver, CO, USA) according to the ASTM standards [19,20]. Apparent density was assessed using the Hall and Carney funnels, as well as the Arnold meter according to the ASTM standards B212, B417, and B703, respectively [21–23]. The latter method, not commonly studied for AM characterization, is representative of a gravity fed powder delivery mechanism. Theoretical densities used for the calculation of the apparent density were 2.70 g/cm^3 for powders A and B and 2.68 g/cm^3 for powder C.

Powder spread density was studied using a single layer of powder with a thickness of ~100 μm CT scanned by a Zeiss Xradia 520 Versa 3D X-ray microscope, Oberkochen, Germany. A bespoke testing holder made of stainless steel 316L was printed using a Renishaw AM250 (Gloucestershire, UK) to simulate the build plate. The testing holder was then stuck onto a cylindrical aluminum mounting rod. Once the testing holder was fixed in the CT holder, the powder from the batch of interest was spread over the simulated built plate region using the doctor blade technique and covered by a centrifuge tube to limit movement of the powders during testing. Each tomography scan was collected using the HE2 filter with sample rotation of 360°, voltage of 60 KV, current of 82 μA, and an exposure time of 18 s. A magnification of 4× objective was selected to provide a 3D isotropic voxel size of 1 μm across the scanned volume. After scanning, the acquired micro-computed tomography images were segmented using the open source image-processing platform imageJ Fiji (ImageJ 1.52b, National Institute of Health, Bethesda, MD, USA) [24] and reconstructed using the Dragonfly software version 3.1.0.319 from Object Research Systems (ORS, Montreal, QC, Canada) [25]. Finally, the total particle volume as a percentage occupied within the scanned area was considered as the powder spread density.

In order to study the particle segregation within a single 50 μm thick layer, the surface of an Al built plate was patterned by rastering the laser of a Renishaw AM400 to simulate the surface roughness created in a part during printing. The powder was then spread over the Al plate using the Renishaw AM400 spreading system, Gloucestershire, UK). Once the powder layer was deposited over the plate, representative samples were collected from the top and bottom sections for PSD analysis by SEM to determine if powder segregation occurred.

The moisture sorption behavior of the studied powders was evaluated using the gravimetric dynamic vapor sorption (DVS) technique (DVS Intrinsic Plus, from Surface Measurement Systems, London, UK). The DVS Intrinsic apparatus measures mass change (±0.1 μg) under controlled temperature and humidity. Prior to testing, samples were individually dried under vacuum at 200 °C. Dried powder samples (~100 mg) were loaded into an aluminum pan and placed into a chamber kept at 25 °C, and allowed to reach equilibrium, i.e., until the change in mass as a function of time was less than 0.002% per minute. Testing started by reaching 0% relative humidity (RH) and soaked for 12 h. Then, the RH was increased in steps of 10% until 80% RH was reached. Each RH step was held until equilibrium. Once 80% RH had been reached, the RH was ramped down to 0% in steps of 10 mirroring the ramp up.

To evaluate the effect of humidity on the flowability of the tested powders, a multiple cycle flowability test using the Carney funnel was performed. The test consisted on drying 50 g of each powder at 200 °C under vacuum for a period of 2 h. Immediately after drying, Carney flow tests were carried out continuously using the same powder at a 40% RH environment until a total of 30 measurements were obtained.

The surface energy, defined as the energy per unit area required to create a new interface and described as the sum of the dispersive component (London dispersion forces) and the specific component (dipole-dipole, induced dipole, and hydrogen bonding interactions), was determined using inverse gas chromatography (IGC) [26,27]. Details of the method can be found in ref. [28]. Pre-salinized glass columns (300 mm length, 4 mm inner diameter) were filled with the different powders (A, B, and C), and the ends plugged with salinized glass wool. The columns were then placed in a surface energy analyser (IGC-SEA, from Surface Measurement Systems, London, UK), with helium flowing through at 10 mL·min^{-1} as an inert carrier gas. Methane was used for dead volume corrections. All measurements were carried out at 30 °C and 0% RH. In order to determine the specific surface area of the samples, octane was passed through each column at set molecular amounts, and an isotherm was produced. From this isotherm, the BET equations were used to determine the specific surface area. To determine the dispersive component of the surface energy, alkane probes with increasing chain length from heptane to nonane (HPLC grade, Sigma Aldrich, St. Louis, MO, USA), were passed through the sample at a set (fractional) surface coverages from 0.1 to 0.3. Sufficient time was allowed between injections as to allow for the probe to pass completely through the column. For the specific surface energy component, dichloromethane and toluene (both HPLC grade, Sigma Aldrich, St. Louis, MO, USA) were used as the polar probes.

Two different techniques, static and dynamic angles of repose, were also used to assess the powders cohesiveness. The static angle of repose consists on freely flowing powders through a funnel to form a characteristic conical heap onto a horizontal plate. The angle developed between the surface of the conical heap and the plate represents the static angle of repose. For this purpose, 50 g of as-received powders were dispensed through a Carney funnel onto the center of a plate. The angle of repose was then measured between the surface of the powder heap and the plate. This procedure was repeated three times for each powder and the average values are reported.

The dynamic angle of repose measurement was obtained using the granular material flow analyzer, Granudrum® (Granutools™, Awans, Belgium). The analysis consists of a transparent drum that is filled with 50 cm^3 of powder and rotated at an angular velocity to induce powder flow. The rotating drum is backlight and a camera is used to capture images of the avalanche at different times. Angular velocities ranging from 2–20 rpm were used. For each angular velocity, 50 images of the drum separated by 1000 ms were acquired. The interface location between the air and powder was automatically detected and the average position as well as the deviations around this average position were automatically computed for each velocity. The dynamic angle of repose was determined from the center of the flow and the deviations from the interface was directly related to the cohesion inside the drum and denominated as the cohesion index [29]. The process was repeated three times and the average values are presented.

3. Results

3.1. Powder Morphology and Particle Size Distribution

The powder morphology corresponding to powders A, B, C imaged by SEM are shown in Figure 1a–f. Spherical particles with regular morphologies, smooth surfaces, and a limited number of satellites were observed for powders A and B. In comparison, the particles present in powder C were more varied in their nature with irregular surfaces; the smaller particles were spherical, but the larger particles were distorted with satellites present.

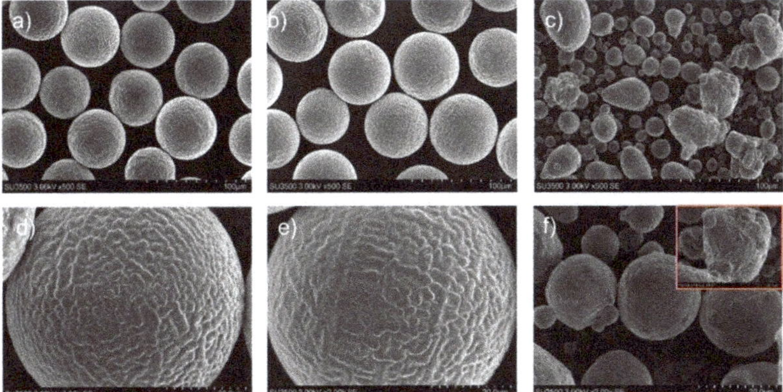

Figure 1. Representative micrographs of the powders tested within this study: (**a**) powder A, (**b**) powder B, (**c**) powder C, and corresponding high magnification images of (**d**) powder A, (**e**) powder B, (**f**) small particles of powder C. The red square in image (**f**), corresponds to the high magnification image of large particles encountered on image (**c**).

The PSD expressed in terms of the "log-normal slope" parameter method is a way to easily compare and identify in a single plot the S_w and the median value of the PSD. A graphical representation of the cumulative particle size against the standard deviations used to determine the S_w of powders A, B, and C is depicted in Figure 2. It was clearly seen that powder C presents the lowest gradient with finer modal particles when it was compared with powders A and B. The corresponding PSD D_{10}, D_{50}, D_{90}, and the dimensionless S_w values are summarised in Table 1. Powders A and B present a narrow PSD distribution with S_w values of 10.6 and 10.8 respectively, while powder C shows a S_w of 4.2 indicating a wide PSD.

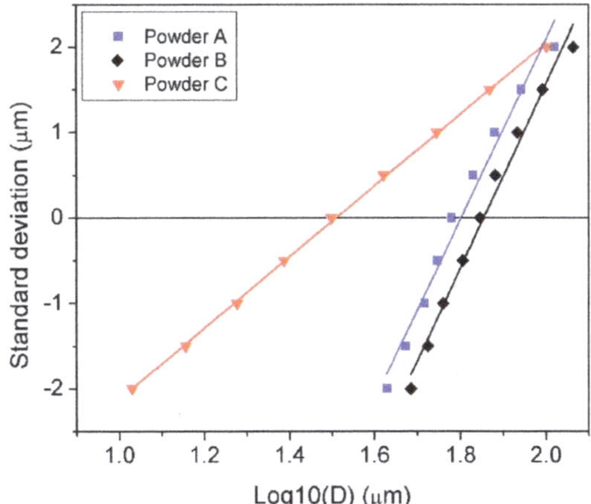

Figure 2. Graphical representation of the cumulative logarithmic particle size versus the standard deviation of powders A, B, and C.

Table 1. Particle size distribution of the tested powders A, B, and C.

Powder	D_{10} (µm)	D_{50} (µm)	D_{90} (µm)	S_w
Powder A	48	63	83	10.6
Powder B	54	70	91	10.8
Powder C	14	31	58	4.2

3.2. Density and Flow Characteristics

Apparent density and flow characteristics are reported to provide useful insight into how the powders will pack in a loose state, which simulates the powder layer spreading during AM processing where no compression or tapping forces are applied [30]. Table 2 presents the flow characteristics of powders A, B, and C, obtained by the Hall and Carney funnels. The measured flow times under these characterization methods indicate no statistical difference between powders A and B. In contrast, powder C showed a lack of flow through the Hall funnel while for the Carney funnel, the measured flow corresponds to 15.3 ± 0.4 s for 50 g of powder. The results obtained by both techniques, indicate that powders A and B exhibit better flow behaviors than powder C.

Table 2. Summarises the powder flowability of powders A, B, and C determined by Hall and Carney funnels.

Powder	Hall Flow (s/50 g)	Carney Flow (s/50 g)
Powder A	33.0 ± 0.4	6.1 ± 0.1
Powder B	32.7 ± 0.7	6.1 ± 0.1
Powder C	No flow	15.3 ± 0.4

Table 3 summarizes the apparent density values obtained by three different traditional techniques recommended for powder metallurgy, i.e., Hall funnel, Carney funnel, and Arnold meter. Powder A presents the highest apparent density followed by powder B and powder C. It is known that the hall funnel is unable to quantify the apparent density of non-flowing powders. However, alternative techniques used in the powder metallurgy field such as the Carney funnel and Arnold meter are able to provide consistent values with a difference of approximately 1% between each technique.

Table 3. Apparent density of powders A, B, and C estimated by the Hall and Carney funnels as well as the Arnold meter.

Powder	Hall Apparent Density (%)	Carney Apparent Density (%)	Arnold Apparent Density (%)
Powder A	54.8 ± 0.5	55.5 ± 0.1	56.3 ± 0.1
Powder B	53.3 ± 0.5	54.8 ± 0.1	55.8 ± 0.1
Powder C	-	51.6 ± 0.4	52.7 ± 0.2

3.3. Spread Density

The laser-powder interactions during LPBF processing influence the melt pool dynamics. Ideally, during powder spreading, a homogeneous powder bed with high packing density would be preferred. In order to evaluate the spread density, X-ray micro-computed tomography. This technique provides a 3D visualization of particles, which allows the determination of the internal porosity as well as the packing density of powders contained in a powder bed, commonly known as spread density. Figure 3a–c depicts the 2D and 3D reconstructed images of powders A, B, and C, respectively. Trapped gas was evident in powders A and B with a volume fraction of 0.7 and 0.2% respectively while no evidence of porosity was found in powder C. Trapped gas within the powder feedstock is an important quality parameter since it has been demonstrated that during AM processing, trapped gas can be released into the molten metal leading to rounded residual porosity in as-built parts [31].

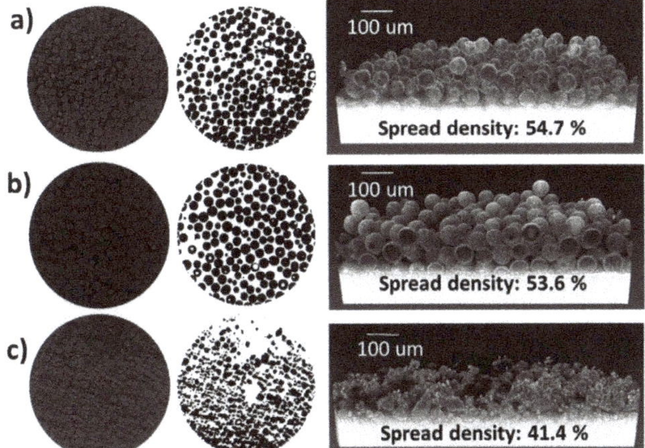

Figure 3. 2D and 3D reconstructed images of (**a**) powder A, (**b**) powder B, and (**c**) powder C used to determine the spread density.

The spread density obtained from powders A and B was consistent with the apparent density measurements obtained by the Hall, Carney, and Arnold meter methods with 54.7 and 53.6%, respectively. However, the apparent density obtained for powder C of 41.4% using micro-CT differs significantly from the traditional methods used.

3.4. Particle Segregation in A Powder Bed

Little attention has been paid to particle segregation occurring during the powder recoating process. Differences in local PSD within a powder bed can have the potential to produce local variations of powder bed density during manufacturing and process instability in terms of melt pool signature [15]. In order to study the particle segregation during powder coating, a layer of 50 μm of powder C was spread over a previously laser rastered Al plate and the PSD was then analyzed at different locations over the plate. Figure 4, depicts the PSD of samples from powder C collected from the top and bottom sections of the Al plate. The top section corresponds to a position close to the beginning of the powder spreading while the bottom section corresponds to a position at the end of powder spreading. Differences in PSD between the top and bottom section of the powder bed were observed. Large particles with a D_{50} of 13.0 μm segregated in the region where spreading starts while smaller particles with a D_{50} of 9.8 μm, segregate at the end of the powder bed.

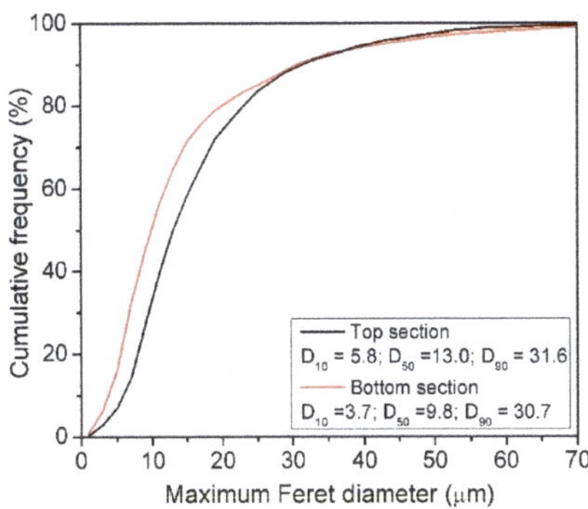

Figure 4. Particle segregation in a powder bed from the top and bottom sections of the building plate.

3.5. Water Vapor Adsorption Characteristics

The water vapor adsorption test results of powders A, B, and C determined by means of DVS are shown in Figure 5a–c. In order to see the effect of PSD on the moisture adsorption, powder C was sieved to obtain a PSD of D_{10} = 48 µm, D_{50} = 65 µm, and D_{90} = 92 µm as to be comparable to powders A and B. The DVS isotherm for the sieved powder is presented in Figure 5d. Powder C has a significantly larger degree of vapor sorption compared to powders A and B. By eliminating the fine particles present in powder C, the degree of vapor sorption for the sieved powder C was reduced.

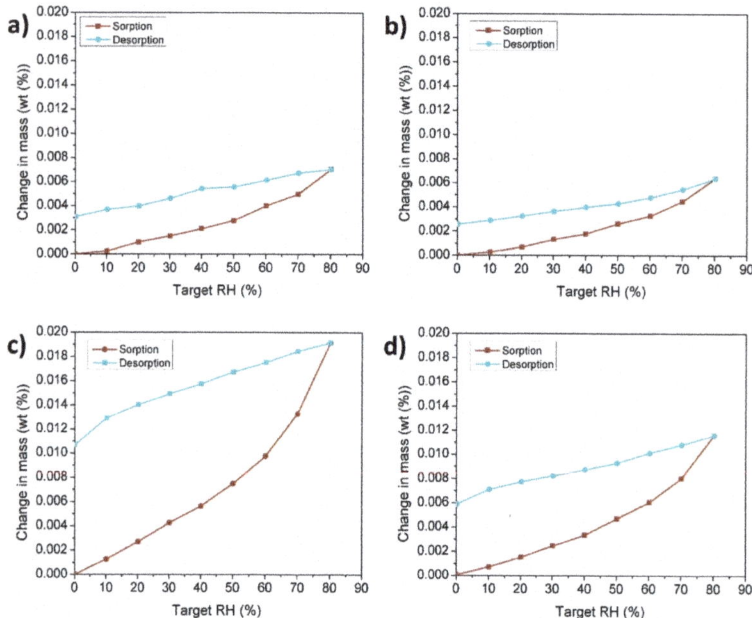

Figure 5. Dynamic vapor sorption (DVS) isotherms for (**a**) powder A, (**b**) powder B, (**c**) powder C, and (**d**) sieved powder C.

3.6. Multiple Cycle Flowability Test

In order to evaluate the effect of humidity on powder flow, a multiple cycle flowability test was carried out at a RH of 40% using 50 g of dried powders. Figure 6 shows the Carney flow corresponding to powders A, B, and C as a function of measurement sequence. No significant variation in flow was observed between powders A and B which remained stable over time with an average flow of 6 s. However, powder C showed a strong variation in flow within the first 5 min after removal from the furnace. During this time, the flow time increased from 16 s to approximately 22 s. The flow time then subsequently decreased continuously until reaching a flow time of 14 s after 30 consecutive measurements. The vapor sorption characteristics of the tested powders depicted by the previous DVS analysis (Figure 5) plays an important role in the powder flowability. It is well known that the surface of aluminum powders reacts with oxygen during powder production to form a passivation alumina layer. During powder handling, the passivated aluminum powders can absorb humidity and form a hydrated aluminum oxide [32,33]. The high surface area present in powder C due to the greater number of fine particles, increases the amount of adsorbed water which decreases the powder flow and/or contributes to the development of powder agglomerates [34]. After the first 5 min, it is believed that the observed increment in flowability may be related to the breaking up of the agglomerates due to the shearing strength produced by continuously flowing the same powder.

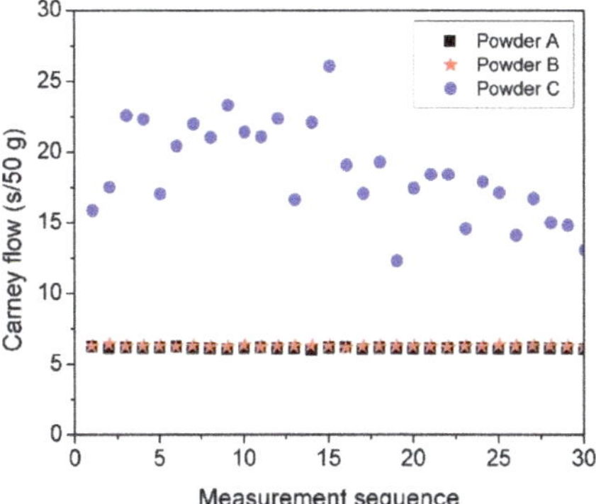

Figure 6. Carney flow variations of powders A, B, and C after oven drying and exposure to RH 40%.

3.7. Particle Cohesiveness

Figure 7a–c, shows the surface energy as a function of surface coverage for powders A, B, and C, respectively. For all powders, the dispersive component of the surface energy was the most dominant. It can be seen that powder C displayed a much greater degree of surface energy heterogeneity, as indicated by the decrease in surface energy (dispersive and specific, leading to a decrease in total) with increasing surface coverage. The surface energy of powder C was also much greater than the other two powders at low surface coverage, which was likely to be due to the finer particle range present having a higher energy. Powders A and B have a higher degree of surface energy homogeneity, with the general trend of powder A having a greater surface energy across the entire surface.

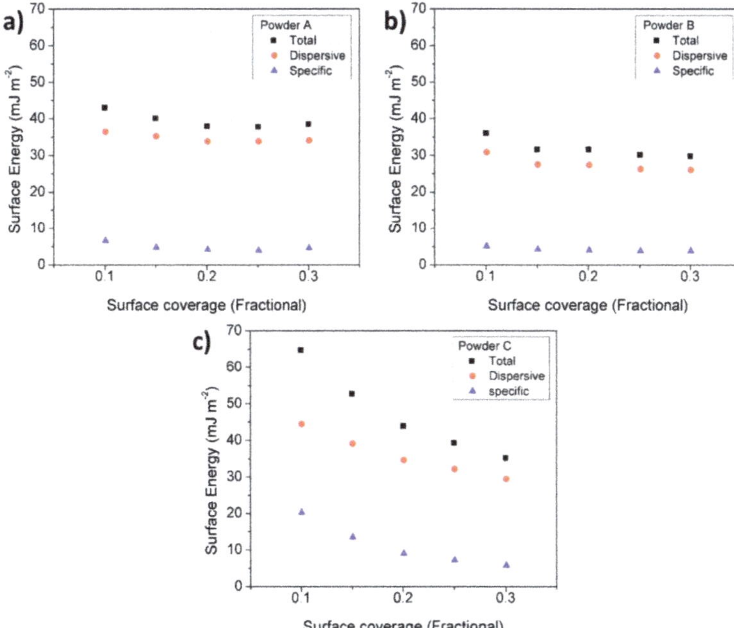

Figure 7. Surface energy of (**a**) powder A, (**b**) powder B, and (**c**) powder C.

The work of cohesion gives an indication of the natural affinity of the powders for agglomeration. Figure 8 shows the Work of Cohesion (Wco) of the powders A, B, and C. It can be seen that at the lowest surface coverage, powder C had a much greater Wco, which is most likely dominated by the fine particles present. Thus suggest that the presence of fine powders favors cohesion giving rise to powder agglomeration [34]. From this result, it would indicate that powder C would have the lowest flowability.

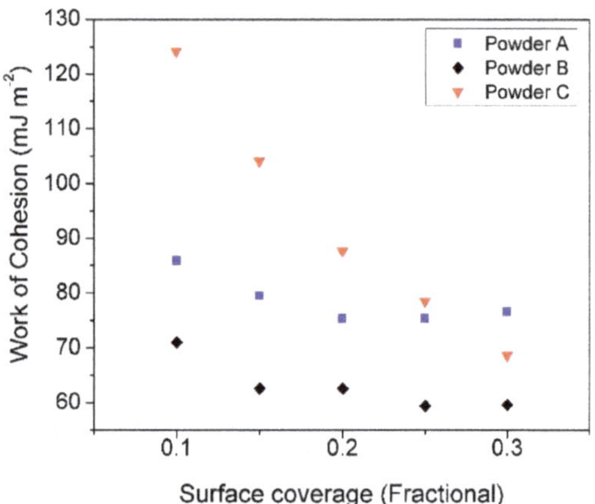

Figure 8. Comparison between the work of cohesion between powders A, B, and C.

3.8. Static and Dynamic Angle of Repose

During AM powder layer spreading, the powder is typically deposited in two steps: (1) the powder is piled up by gravity in front of a re-coating blade, and (2) the powder is spread over the powder bed by the horizontal movement of the re-coater. The first recoating step, may be properly represented by static flow tests while the second step, which is subjected to shear stresses due to the horizontal movement of the re-coater, might be better represented by studying the dynamic angle of repose. The static flow behavior for each powder characterized by the conical angle of repose is summarized in Table 4. It can be seen that the powder containing spherical and coarser particles, powder B, presents the lowest static angle of repose followed by powder A. In contrast, powder C which contains the largest amount of fine and irregular particles showed the highest static angle of repose.

Table 4. Static angle of repose of particles A, B, and C.

Powder	Static Angle of Repose (°)
Powder A	30 ± 4
Powder B	26 ± 3
Powder C	39 ± 4

Figure 9a presents the evolution of the dynamic angle of repose as a function of the rotating speed for powders A, B, and C. Dynamic powder flow measurements clearly show reproducible differences between the tested powders up to a rotating speed of 14 rpm. Below this speed, the spherical and coarser powders (A and B) show the lowest dynamic angles while the fine and non-spherical powders (C) present the highest angles of repose. At rotating speeds higher than 14 rpm, no statistical difference was observed. Specifically, powders A and B present a typical shear thinning behavior due to the slipping layers passing each other at rotating speeds between 2 and 6 rpm with a minimum dynamic angle of 26 ° and 30 °, respectively. At rotating speeds higher than 6 rpm, there was a transition from shear thinning to shear thickening behavior in both powders associated to the breakdown of layers and the formation of large aggregates [35]. The maximum angle of repose was obtained at 20 rpm for powders A and B with values of 37° and 36°, respectively. In contrast, powder C presents the highest dynamic angle of repose at the lower rotating speeds, i.e., between 2 and 6 rpm, with a maximum value of 38°. After 6 rpm, powder C shows a slight shear thinning behavior decreasing its angle of repose to an average minimum of 35° between 14 and 20 rpm.

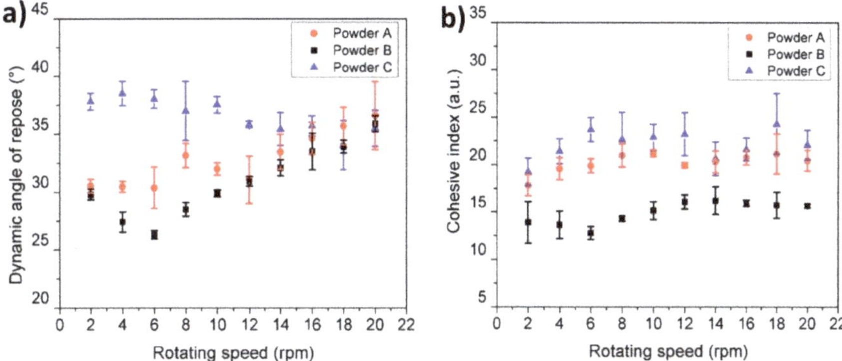

Figure 9. (**a**) Dynamic angle of repose and (**b**) cohesive index determined for powders A, B, and C.

Quantification of the cohesion which occurs in the powders during drum rotation, can be carried out by studying the cohesive index which is determined from the fluctuations of the avalanche interface [29]. Higher values of cohesive index represent higher cohesion while a lower cohesion index represents lower cohesion between particles. Figure 9b depicts the evolution of the cohesive index as a function of the rotating speed for the tested powders. The coarser and spherical powder B, produced the lowest cohesion index whereas the finest and non-spherical powder C, presents the highest cohesion index. It was clearly seen that the particle cohesion of the Al powders was affected by the surface properties and PSD. The high surface energy and water adsorption characteristics measured in powder C compared with powders A and B, evidently increased the particle cohesiveness. Additionally, as the particles become smaller, the gravitational force contribution become negligible compared to the cohesive forces causing the powders to agglomerate and to reduce their flow [36,37].

4. Discussion

Ensuring process reliability and quality of additive manufactured components is one of the key challenges for widespread adoption of AM technologies. In LPBF, factors associated to laser optics, process parameters, and powder feedstock quality, have a direct effect on the process repeatability and quality of manufactured parts. Among these factors, powder feedstock quality is of paramount importance since changes in powder characteristics influence the laser-powder interactions resulting in process defects [38]. Baitimerov et al. [39], has shown the complex relationship between powder feedstock quality and printed part porosity. In their work powders with equivalent chemical compositions but different flow and apparent density characteristics, produced parts with relative densities between 94.4 ± 2.3% and 99.4 ± 0.3% using the same processing conditions. Thus, it has been demonstrated that flow and apparent density are important characteristics to consider due to the fact that homogeneous and highly packed powder layers are typically desired to enhance laser absorption [15].

Powder flow and apparent density are not independent properties, they are extrinsically related to other powder characteristics such as morphology, PSD, porosity, and surface chemistry. The particularities of powders A and B such as mono-sized PSD, high sphericity, good flowability and lack of satellite particles, produced a homogeneous powder bed with high apparent density similar to the one obtained by traditional powder characterization methods. However, the presence of large amounts of small and irregular particles such as the ones present in powder C, decreases the powder flow resulting in an inhomogeneous powder bed with low apparent density which is not comparable with traditional test results.

One of the main factors that influence the apparent density of a powder bed, is the presence of inter-particle forces which lead to cohesion [37]. It has been reported that non-cohesive and mostly spherical particles provide high apparent densities [40]. However, when cohesion between particles is present, agglomerates are likely to form giving rise to lower apparent densities [41,42]. The formation of agglomerates in the form of cage-like structures during packing have been observed by means of discrete element simulations when cohesive forces are taken into account. Such structures produce low apparent densities and are generated due to the presence of particles with high surface energy and cohesive forces typically larger than gravitational forces present [40].

The cohesiveness of the particles studied based on the measure of total surface energy, was evaluated by means of IGC. The total surface energy and work of cohesion measured in powders A, B, and C indicated that the PSD plays an important role in the total surface energy of these powders. Hence contributing to their cohesion behavior and consequently their flow. Powder C has a wide size distribution, with high surface energy at lower surface coverages, which was most likely due to the fine particles dominating. As the surface coverage increased, the lower energy sites of the fine particles adsorb the probe molecules, as do the larger particles in the system. It can be seen in Figure 7a that as the surface coverage increased the total surface energy of powder C tends towards that of the other two powders which contain the least quantity of fine particles.

The cohesiveness of the particles has also been determined using the rotating drum technique which provides a cohesion index based on the variation of the dynamic angle of repose. The results show that powder C presents the largest dynamic angle of repose and cohesive index followed by powder A and B. These results are in good agreement with the work of cohesion obtained by IGC. Additionally, this technique provides useful information regarding the flow of the powders at different shear rates. In the tested powders, shear thinning and shear thickening behaviors are observed depending on the rotation speed (shear rate). This observation provides insights into the critical powder spreading speed required to obtain the best flow characteristics for specific powders. This approach to obtain a value of 'spreading speed' cannot be obtained by traditional static flow measurements which only provide qualitative insight to the powder flow.

The presence of liquid bridges has been reported to increase the cohesion between particles due to the coalescence of the bridges leading to poor flow and low apparent densities [43]. In the present study, water adsorption tests have been carried out to investigate the water adsorption characteristics of powders A, B, and C. From the results, it was evident that a wide PSD with a large amount of fine particles such as powder C, led to higher amounts of adsorbed water. By sieving powder C to a PSD similar to that of powders A and B, the water adsorption behavior tends towards those of powders A and B. Fine particles would be more likely to adsorb a higher degree of water vapor relative to their mass due to the higher specific surface area. Multiple powder flow measurements carried out at constant RH of 40% indicate that humidity has an important effect on powder flowability depending on the PSD. Powder C, which adsorbed the highest amount of water, shows a significant change in flowability while powders A and B remained constant over time.

Powder segregation occurring within the powder bed has a direct effect on the local apparent density. Discrete element method simulation results of the powder spreading process during AM processing [8,9], have shown evidence of particle segregation. Specifically, fine particles are segregated at the beginning of the spreading process while coarser particles are deposited at the end. In the present study, the cohesive powder C was tested to observe if particle segregation was likely to occur during the spreading process. In the present study, fine particles are segregated at the end of the powder bed while larger particles are present at the beginning. The simulation results provided in the literature contradict the ones observed experimentally within the present study. However, it is important to note that the discrete element simulations were carried out assuming free flowing spherical particles where the effect of surface energy and particle–particle interactions such as cohesion, were not considered. The flow behavior observed in non-cohesive and cohesive granular material differs significantly [44]. For the case of non-cohesive particles, axial segregation of larger particles is observed while in cohesive powders a lower degree of segregation is present [45]. Assuming that the shear strength during recoating between the coater and the particles is higher than the cohesion force of the agglomerates, the segregation observed in the present study could be explained by the breaking up of agglomerates during powder spreading resulting in the deposition of finer particles at the end of the powder bed.

5. Conclusions

Traditional and non-traditional quantitative powder characterization techniques were used to evaluate the properties of different AlSi7Mg powders. The traditional characterization techniques are able to provide quantitative measurements of the powder performance in terms of flowability and apparent density. However, if the powders do not flow through the Hall and Carney funnels, no metrics can be obtained. In contrast to the traditional techniques, non-traditional techniques, such as CT scanning, dynamic vapor sorption, inverse gas chromatography, and the rotating drum technique, provide quantitative metrics of properties for free and non-free flowing powders. Additionally, they can be used to gain insight into characterisation parameters such as flow and apparent density. The results of this study have suggested that the presence of a PSD with a high number of fine particles facilitates water absorption and powder cohesion due to high surface energy. Such characteristics may hinder the spreading of uniform powder layers, which may lead to defects in AM parts. In contrast, when a narrow PSD and particles larger than 48 μm are present, water absorption and powder cohesion was decreased improving powder flow and apparent density.

Author Contributions: Conceptualization, J.A.M.-L. and M.B.; Data curation, J.A.M.-L.; Formal analysis, J.A.M.-L.; Funding acquisition, M.B.; Investigation, J.A.M.-L. and A.N.-N.; Methodology, J.A.M.-L.; Project administration, M.B.; Resources, K.E.W. and M.B.; Supervision, M.B.; Validation, J.A.M.-L., K.E.W. and M.B.; Visualization, J.A.M.-L. and A.N.-N.; Writing—Original draft, J.A.M.-L.; Writing—Review & editing, A.N.-N.

Funding: This research was funded by the Natural Science and Engineering Research Council (NSERC) for the support of this project through grant EGP 521786-17.

Acknowledgments: The authors acknowledge Rui Tahara for assistance with X-ray microscopy scans and a CFI Innovation grant (33122) awarded to Jackie Vogel and the integrated Quantitative Biology initiative.

Conflicts of Interest: The authors declare no conflict of interest.

References

1. Rüßmann, M.; Lorenz, M.; Gerbert, P.; Waldner, M.; Justus, J.; Engel, P.; Harnisch, M. *Industry 4.0: The Future of Productivity and Growth in Manufacturing Industries*; Boston Consulting Group: Boston, MA, USA, 2015; Volume 9.
2. Ding, Y.; Muñiz-Lerma, J.A.; Trask, M.; Chou, S.; Walker, A.; Brochu, M. Microstructure and mechanical property considerations in additive manufacturing of aluminum alloys. *MRS Bull.* **2016**, *41*, 745–751. [CrossRef]
3. Olakanmi, E.O.; Cochrane, R.F.; Dalgarno, K.W. A review on selective laser sintering/melting (SLS/SLM) of aluminium alloy powders: Processing, microstructure, and properties. *Prog. Mater. Sci.* **2015**, *74*, 401–477. [CrossRef]
4. Strondl, A.; Lyckfeldt, O.; Brodin, H.; Ackelid, U. Characterization and control of powder properties for additive manufacturing. *Jom* **2015**, *67*, 549–554. [CrossRef]
5. German, R.M. *Particle Packing Characteristics*; Metal Powder Industries Federation: Princeton, NJ, USA, 1989.
6. Olakanmi, E. Selective laser sintering/melting (SLS/SLM) of pure Al, Al-Mg, and Al-Si powders: Effect of processing conditions and powder properties. *J. Mater. Process. Technol.* **2013**, *213*, 1387–1405. [CrossRef]

7. Carson, J.W.; Pittenger, B.H. Bulk properties of powders. In *ASM Handbook, Volume 7: Powder Metal Technologies and Applications*; Lee, P.W.T., Iacocca, R., German, R.M., Ferguson, B.L., Eisen, W.B., Moyer, K., Madan, D., Sanderow, H., Eds.; ASM International: Almere, The Netherlands, 1998; pp. 287–301.

8. Mindt, H.W.; Megahed, M.; Lavery, N.P.; Holmes, M.A.; Brown, S.G.R. Powder bed layer characteristics: The overseen first-order process input. *Metall. Mater. Trans. A* **2016**, *47*, 3811–3822. [CrossRef]

9. Zielinski, J.; Vervoort, S.; Mindt, H.-W.; Megahed, M. Influence of powder bed characteristics on material quality in additive manufacturing. *BHM Berg Hüttenmänn. Monatshefte* **2017**, *162*, 192–198. [CrossRef]

10. Spierings, A.B.; Herres, N.; Levy, G. Influence of the particle size distribution on surface quality and mechanical properties in AM steel parts. *Rapid Prototyp. J.* **2011**, *17*, 195–202. [CrossRef]

11. Liu, B.; Wildman, R.; Tuck, C.; Ashcroft, I.; Hague, R. Investigation the effect of particle size distribution on processing parameters optimisation in selective laser melting process. *Addit. Manuf. Res. Group Loughb. Univ.* **2011**, *84*, 227–238.

12. Simchi, A. The role of particle size on the laser sintering of iron powder. *Met. Mater. Trans. B* **2004**, *35*, 937–948. [CrossRef]

13. Parteli, E.J.R.; Pöschel, T. Particle-based simulation of powder application in additive manufacturing. *Powder Technol.* **2016**, *288*, 96–102. [CrossRef]

14. Yoshimi, K.; Hirozumi, A.; Satoshi, Y.; Hisashi, S.; Soya, T. Phenomenological studies in laser cladding. Part I. Time-resolved measurements of the absorptivity of metal powder. *Jpn. J. Appl. Phys.* **1993**, *32*, 205.

15. Boley, C.D.; Khairallah, S.A.; Rubenchik, A.M. Calculation of laser absorption by metal powders in additive manufacturing. *Appl. Opt.* **2015**, *54*, 2477–2482. [CrossRef] [PubMed]

16. Aboulkhair, N.T.; Everitt, N.M.; Ashcroft, I.; Tuck, C. Reducing porosity in AlSi10Mg parts processed by selective laser melting. *Addit. Manuf.* **2014**, *1*, 77–86. [CrossRef]

17. Equispheres. Available online: https://equispheres.com/powder/ (accessed on 28 August 2018).

18. German, R.M.; Park, S.J. *Mathematical Relations in Particulate Materials Processing: Ceramics, Powder Metals, Cermets, Carbides, Hard Materials, and Minerals*; Wiley: Hoboken, NJ, USA, 2008; p. 419.

19. ASTM B213-17. *Standard Test Methods for Flow Rate of Metal Powders Using the Hall Flowmeter Funnel*; ASTM International: West Conshohocken, PA, USA, 2017.

20. ASTM B964-16. *Standard Test Methods for Flow Rate of Metal Powders Using the Carney Funnel*; ASTM International: West Conshohocken, PA, USA, 2016.

21. ASTM B212-17. *Standard Test Method for Apparent Density of Free-Flowing Metal Powders Using the Hall Flowmeter Funnel*; ASTM International: West Conshohocken, PA, USA, 2017.

22. ASTM B417-13. *Standard Test Method for Apparent Density of Non-Free-Flowing Metal Powders Using the Carney Funnel*; ASTM International: West Conshohocken, PA, USA, 2013.

23. ASTM B703-10. *Standard Test Method for Apparent Density of Metal Powders and Related Compounds Using the Arnold Meter*; ASTM International: West Conshohocken, PA, USA, 2010.

24. Schindelin, J.; Arganda-Carreras, I.; Frise, E.; Kaynig, V.; Longair, M.; Pietzsch, T.; Preibisch, S.; Rueden, C.; Saalfeld, S.; Schmid, B. Fiji: An open-source platform for biological-image analysis. *Nat. Methods* **2012**, *9*, 676–682. [CrossRef] [PubMed]

25. Object Research Systems (ORS) Inc., Montreal, Canada. Dragonfly 3.1 (Computer Software). 2016. Available online: http://www.theobjects.com/dragonfly (accessed on 7 November 2018).

26. Johnson, K.; Kendall, K.; Roberts, A. Surface energy and the contact of elastic solids. *Proc. R. Soc. Lond. A Math. Phys. Eng. Sci.* **1971**, *324*, 301–313. [CrossRef]

27. Voelkel, A.; Strzemiecka, B.; Adamska, K.; Milczewska, K. Inverse gas chromatography as a source of physiochemical data. *J. Chromatogr. A* **2009**, *1216*, 1551–1566. [CrossRef] [PubMed]

28. Mohammadi-Jam, S.; Waters, K.E. Inverse gas chromatography applications: A review. *Adv. Colloid Interface Sci.* **2014**, *212*, 21–44. [CrossRef] [PubMed]

29. Lumay, G.; Boschini, F.; Traina, K.; Bontempi, S.; Remy, J.C.; Cloots, R.; Vandewalle, N. Measuring the flowing properties of powders and grains. *Powder Technol.* **2012**, *224*, 19–27. [CrossRef]

30. Spierings, A.B.; Voegtlin, M.; Bauer, T.; Wegener, K. Powder flowability characterisation methodology for powder-bed-based metal additive manufacturing. *Prog. Addit. Manuf.* **2016**, *1*, 9–20. [CrossRef]

31. Cunningham, R.; Nicolas, A.; Madsen, J.; Fodran, E.; Anagnostou, E.; Sangid, M.D.; Rollett, A.D. Analyzing the effects of powder and post-processing on porosity and properties of electron beam melted Ti-6Al-4V. *Mater. Res. Lett.* **2017**, *5*, 516–525. [CrossRef]

32. Pethrick, R.A.; Hayward, D.; Jeffrey, K.; Affrossman, S.; Wilford, P. Investigation of the hydration and dehydration of aluminium oxide-hydroxide using high frequency dielectric measurements between 300 kHz-3 GHz. *J. Mater. Sci.* **1996**, *31*, 2623–2629. [CrossRef]

33. Litvintsev, A.I.; Arbuzova, L.A. Kinetics of degassing of aluminum powders. *Sov. Powder Met. Met. Ceram.* **1967**, *6*, 1–10. [CrossRef]

34. Pietsch, W. *Agglomeration Processes: Phenomena, Technologies, Equipment*; John Wiley & Sons: Hoboken, NJ, USA, 2002.

35. Hao, T. Analogous viscosity equations of granular powders based on Eyring's rate process theory and free volume concept. *RSC Adv.* **2015**, *5*, 95318–95333. [CrossRef]

36. de Ryck, A.; Condotta, R.; Dodds, J.A. Shape of a cohesive granular heap. *Powder Technol.* **2005**, *157*, 72–78. [CrossRef]

37. Lumay, G.; Fiscina, J.; Ludewig, F.; Vandewalle, N. Influence of cohesive forces on the macroscopic properties of granular assemblies. *AIP Conf. Proc.* **2013**, *1542*, 995–998.

38. Foster, B.; Reutzel, E.; Nassar, A.; Hall, B.; Brown, S.; Dickman, C. Optical, layerwise monitoring of powder bed fusion. In Proceedings of the 6th International Solid Freeform Fabrication Symposium, Austin, TX, USA, 10–12 August 2015.

39. Baitimerov, R.; Lykov, P.; Zherebtsov, D.; Radionova, L.; Shultc, A.; Prashanth, K. Influence of powder characteristics on processability of AlSi12 alloy fabricated by selective laser melting. *Materials* **2018**, *11*, 742. [CrossRef] [PubMed]

40. Deng, X.L.; Davé, R.N. Dynamic simulation of particle packing influenced by size, aspect ratio and surface energy. *Granul. Matter* **2013**, *15*, 401–415. [CrossRef]

41. Yu, A.B.; Bridgwater, J.; Burbidge, A. On the modelling of the packing of fine particles. *Powder Technol.* **1997**, *92*, 185–194. [CrossRef]

42. Parteli, E.J.; Schmidt, J.; Blümel, C.; Wirth, K.-E.; Peukert, W.; Pöschel, T. Attractive particle interaction forces and packing density of fine glass powders. *Sci. Rep.* **2014**, *4*, 6227. [CrossRef] [PubMed]

43. Fiscina, J.E.; Lumay, G.; Ludewig, F.; Vandewalle, N. Compaction dynamics of wet granular assemblies. *Phys. Rev. Lett.* **2010**, *105*, 048001. [CrossRef] [PubMed]

44. Harnby, N. Chapter 5—The mixing of cohesive powders. In *Mixing in the Process Industries*, 2nd ed.; Butterworth-Heinemann: Oxford, UK, 1992; pp. 79–98.

45. Chaudhuri, B.; Mehrotra, A.; Muzzio, F.J.; Tomassone, M.S. Cohesive effects in powder mixing in a tumbling blender. *Powder Technol.* **2006**, *165*, 105–114. [CrossRef]

 *materials*

Article

Characterization of Composite Powder Feedstock from Powder Bed Fusion Additive Manufacturing Perspective

Eskandar Fereiduni *, Ali Ghasemi and Mohamed Elbestawi *

Department of Mechanical Engineering, McMaster University, Hamilton, ON L8S 4L7, Canada;
ghasemia@mcmaster.ca
* Correspondence: fereidue@mcmaster.ca (E.F.); elbestaw@mcmaster.ca (M.E.)

Received: 18 October 2019; Accepted: 5 November 2019; Published: 7 November 2019

Abstract: This research aims at evaluating the characteristics of the 5 wt.% B_4C/Ti-6Al-4V composite powder feedstock prepared by two different categories of mechanical mixing for powder bed fusion (PBF) additive manufacturing (AM) of metal matrix composites (MMCs). Microstructural features, particle size, size distribution, sphericity, conditioned bulk density and flow behavior of the developed powders were examined. The flowability of the regularly mixed powders was significantly lower than that of the Ti-6Al-4V powder. However, the flowability of the ball-milled systems was a significant function of the milling time. The decrease in the flowability of the 2 h ball-milled powder compared to the Ti-6Al-4V powder was attributed to the mechanical interlocking and the entangling caused by the B_4C particles fully decorating the Ti-6Al-4V particles. Although the flattened/irregular shape of powder particles in the 6 h milled system acted to reduce the flowability, the overall surface area reduction led to higher flowability than that for the 2 h milling case. Regardless of the mixing method, incorporation of B_4C particles into the system decreased the apparent density of the Ti-6Al-4V powder. The composite powder obtained by 2 h of ball milling was suggested as the best possible condition, meeting the requirements of PBF–AM processes.

Keywords: additive manufacturing; powder bed fusion; selective laser melting; regular mixing; ball milling; flowability; Ti-6Al-4V

---

1. Introduction

Metal matrix composites (MMCs) are outstanding materials bilaterally benefitting from the properties of at least two constituents: the metal matrix (usually an alloy), and the reinforcement (in general, an oxide, an intermetallic compound, a carbide or a nitride) [1–3]. Incorporation of reinforcements into the metallic matrix is generally associated with the improvement in hardness, specific strength, wear resistance, fracture toughness, and stiffness compared to the monolithic counterparts [4–7]. Owing to their desired structural and functional properties, MMCs have found their application in numerous technological fields including automotive, aerospace and biomedical industries [8,9].

Several conventional manufacturing processes already exist for incorporating reinforcements into the metallic matrices to produce a wide variety of MMCs [10–14]. However, it is rather challenging to fabricate geometrically complex parts using these processes. Additive manufacturing (AM) is regarded as a major revolution in the manufacturing technology and competes with conventional manufacturing processes in many aspects, including, but not limited to, the design freedom, fabrication cost and time, accuracy and the part quality [15,16].

Powder bed fusion (PBF) refers to the AM processes in which an object is manufactured layer-by-layer from a batch of loose powder using a mobile heat source. During this process, a thin layer of powder is deposited on the building platform by the recoater. The powder flow behavior during the recoating process plays a critical role on the uniformity, surface roughness and the thickness of the deposited powder layer and consequently, the dimensional accuracy of the final part [17,18]. On the other hand, the powder bed packing density directly affects the density and mechanical properties of the additively manufactured components [19,20]. Therefore, the flowability and the packing density of the powders need to be investigated prior to the AM process to ensure the soundness of the final parts [21–24].

The desired properties of MMCs are achieved when the reinforcements are homogeneously distributed throughout the matrix with a strong reinforcement/matrix interfacial bonding [25,26]. Conventional MMC fabrication methods generally yield inhomogeneous microstructures, making it rather difficult to fully exploit the strengthening potentials of reinforcements. Therefore, it is of crucial importance to develop new fabrication routes providing a more homogenous distribution of reinforcements within the matrix [16]. The noticeably localized melt pool and the extremely high solidification rates associated with the PBF–AM technology can lead to MMC structures which are much more homogenous than the conventionally processed parts [27]. However, there are still some challenges with the PBF–AM processing of MMCs for achieving highly uniform microstructures due to the following reasons [3,28]:

i.   In nano-composites, reinforcing particles tend to agglomerate and form coarsened clusters in the matrix due to the presence of van der Waals attraction forces among them.
ii.   A large difference between the densities of the reinforcing particles and the liquid matrix encourages the non-uniform distribution of reinforcements in the microstructure.
iii.   The convection flows (i.e., Marangoni effect) induced in the melt pool may not be sufficient to disperse the reinforcing particles throughout the system.

Therefore, when it comes to the PBF–AM of MMCs, particular emphasis should be placed on pre-processing of the composite powder feedstock in order to achieve parts with homogenous microstructures and consequently, uniform mechanical properties. Development of a suitable composite powder feedstock with a uniform distribution of reinforcing particles is the first step to mitigate the mentioned challenges.

Since the feedstock of MMCs is not commercially available for AM processing [29], several techniques have been employed in recent years to prepare these powders. The mechanical routes such as regular mixing [30–32] and ball milling [33–39], and non-mechanical methods including gas atomization of a pre-alloyed system [40,41], agent-assisted deposition [42,43] and electrodeposition [44,45] are among these methods. Compared to the mechanical mixing routes, which have attracted a great deal of attention in recent years, the non-mechanical methods have been rarely adopted to prepare composite powder feedstocks for AM purposes. Although the non-mechanical methods can probably produce composite powder systems with a better flowability and a more uniform distribution state of powder constituents compared to the mechanical routes, they are much more complex and expensive. Due to their low cost as well as their applicability to many powder systems, the mechanical routes are the most frequently used methods in powder feedstock preparation for the PBF–AM of MMCs [29,46].

Incorporation of the guest powder particles (e.g., ceramic particles) into the host powder particles (usually metallic particles) leads to the production of composite powder feedstocks with different characteristics from the individual constituents. The particle size, particle size distribution, and distribution state of the guest powder particles are among these features which dictate the apparent density, flowability, and laser absorptivity of the composite powder [15,16]. The laser absorptivity influences the heat absorption, and the melt pool size [44,47,48], while the flowability and the apparent packing density of the powder play crucial roles in layer thickness (dimensional accuracy) and density of the final part [17,49–52]. A literature review of the AM processing of MMCs reveals that the mixing of powders, in most cases, has been performed to achieve a distribution of free (non-attached) guest particles throughout the mixed powder system [30–39,53–55]. On the other hand, preserving the spherical shape of metallic powder particles has been considered in some of these studies [36,53,55,56]. However, the effect of powder preparation routes and process variables on the composite powder features affecting the quality of AM processed parts is still unclear and needs in-depth analysis and characterization from the AM perspective to obtain high-quality MMCs.

The present research aims to study the characteristics of the mechanically mixed 5 wt.% B_4C/Ti-6Al-4V (Ti64) composite powder feedstock from the PBF–AM viewpoint. For this purpose, the regular mixing and the ball-milling methods were employed with different mixing times to prepare a wide variety of mixed powder systems. The effects of the mixing method and the mixing time on the size, size distribution, sphericity, shape, distribution state of guest particles, phase formation, plastic deformation, apparent density, and flow behavior of the prepared composite powder systems were studied. Moreover, the mechanisms involved in the flow behavior of the developed feedstocks were identified, and the best possible powder feedstock meeting the requirements of PBF–AM processing of MMCs was proposed.

2. Materials and Methods

2.1. Powder Preparation

The powders used in this research were Ti64 (15–45 µm) and B_4C (1–3 µm) named as "host" and "guest" powders, respectively. The nominal chemical composition of these powders is provided in Table 1. For preparing the composite powder feedstock, the regular mixing and the ball-milling processes were employed, which are schematically shown in Figure 1. In both cases, the mixed powder systems contained 5 wt.% B_4C (guest), and the powders were mixed under a protective argon gas atmosphere to avoid oxidation. The mixing process was performed using a high-performance planetary Pulverisette 6 machine at a fixed rotational speed of 200 rpm and mixing times in the range of 1 to 6 h. The regularly mixed and the ball-milled samples have been marked as R(1–6) and B(1–6), respectively, depending on their mixing time. In the regular mixing, the powders were mixed without balls. However, stainless steel balls with a diameter of 10 mm were added to the system in the ball-milling process. The ball-to-powder ratio was selected to be 5:1, and every 30 min of milling was followed by a 15 min pause in order to avoid a temperature increase during the process [57]. For each composite powder system, the mass was measured after the mixing to find the material loss.

Table 1. Nominal chemical composition of the starting Ti-6Al-4V and B_4C powders.

Powder	Elements (wt.%)								
Ti-6Al-4V (Host)	Ti	Al	V	Fe	O	N	C	H	B
	Bal.	6.3	4.0	0.03	0.1	0.01	0.01	<0.1	-
B_4C (Guest)	B_4C	Al	V	Fe	O	N	Free C	H	Free B
	Bal.	<0.001	-	<0.001	<0.04	<0.001	3	-	4

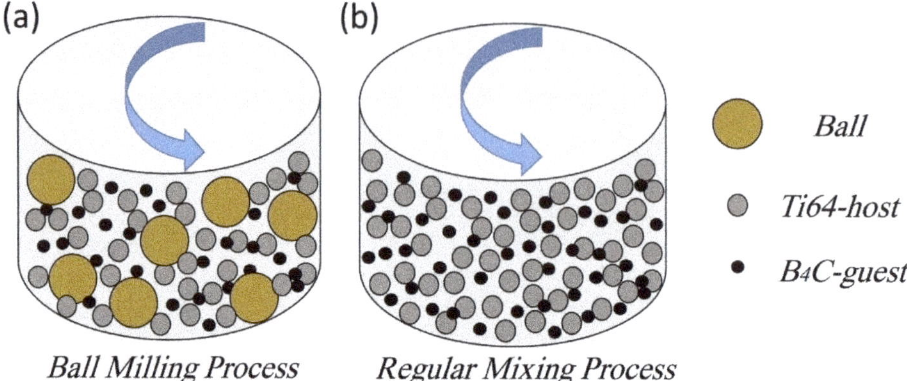

Figure 1. Schematic illustration of: (**a**) ball milling and (**b**) regular mixing processes.

2.2. Powder Characterization

2.2.1. Microstructure and XRD Analysis

The morphology of the starting Ti64 (host) and B_4C (guest) powders, as well as the composite powder systems, were studied using Vega Tescan scanning electron microscopy (SEM) operating at an accelerating voltage of 20 kV. The X-ray diffraction (XRD) analysis was employed to study the effect of the mixing method and the mixing time on the phase formation and the plastic deformation of the developed composite powder feedstocks. This analysis was performed at ambient temperature over a wide range of $2\theta = 20°–80°$ using a PANalytical X'Pert powder X-ray diffractometer (Westborough, MA, USA, Cu Kα target, operating at the voltage and the current of 45 kV and 35 mA, respectively, with a step size of 0.0167°) equipped with an X-ray monochromator. In order to have a better understanding of the microstructural features, the starting Ti64 and composite powders were also sectioned and characterized using SEM.

2.2.2. Particle Size, Size Distribution and Sphericity

A Retsch Camsizer X2 machine (Haan, Germany) was used to measure the particle size, size distribution and sphericity of the Ti64 and the developed 5wt.% B_4C/Ti64 composite powder systems fabricated through the regular mixing and ball-milling methods. This equipment utilizes a high-resolution dual-camera system to characterize fine and agglomerating powders ranging from 800 nm to 8 mm in diameter [58]. The reported results are the average of three measurements. The sphericity of the powder particles was calculated based on the equation suggested by the ISO 9276-6 standard [59]:

$$Sphericity = \frac{4\pi A}{P^2} \tag{1}$$

where P and A are the measured circumference and area covered by a particle projection, respectively. Given the fact that the sphericity of an ideal sphere is unity, deviation from the ideal spherical shape results in lower sphericities.

2.2.3. Flow Characteristics

The flow behavior of powder particles could be characterized using different techniques such as ring shear cell tester, Hausner ratio (HR), angle of repose (AOR)/Hall flowmeter, avalanche angle and Freeman Technology 4 (FT4) powder rheometer [60–62]. When deciding to choose one of these techniques, characteristics of both the technique and the process need to be considered since the selected flowability measurement technique should be as close to the employed process as possible. In recent years, the FT4 Powder Rheometer has emerged as a unique technology to measure the flow behavior of the powder, whilst the powder is in motion. In this device, a precision 'blade' is rotated and moved downwards through a fixed mass of the powder bed to establish a flow pattern. The work required to drive the rotating impeller a certain distance into the powder bed yields the flow energy. Due to its dynamic nature, this technique is capable of differentiating the flowability of powders that exhibit similar behavior under other flow measurement techniques [63]. The FT4 powder rheometer technique provides measurement of several parameters related to the process performance of powders. The interaction of the precision blade with the powder in this technique resembles that of recoater/powder in the PBF–AM process. It is worth noting that the control over the speed of the blade provided by this technique enables the characterization of powder sensitivity to the changes in flow rate. This unique feature also facilitates analysis of the powder flowability for different PBF–AM machines with different recoater speeds.

The flow characteristics of the host powder, as well as the developed composite powder feedstocks in the present study, were measured using an FT4 powder rheometer (Freeman technology, Tewkesbury, UK). In order to study the powder rheological properties using this technique, the standard "Stability and Variable Flow Rate (SVFR)" method was employed, which consists of a stable tip speed zone with seven test cycles followed by a variable tip speed zone having four test cycles. During each test cycle, the precision blade rotated downwards and upwards through the fixed mass of powder to establish a flow pattern, where the powder resistance to the blade yielded the flow properties. During the stability part, the blade operated with a tip speed of -100 mm/s (anti-clockwise) and a helix angle of $5°$. For the variable flow rate zone, the tip speed varied as -100, -70, -40 and -10 mm/s for the test cycles eight, nine, ten and eleven, respectively, with the same helix angle of $5°$. The upward speed remained constant at 40 mm/s, and the helix angle was $-5°$ upwards throughout the experiment. It is noteworthy that before all tests, a conditioning cycle was performed that involves the downward and then upward movement of the blade into the powder bed to gently slice the powder and provide a reproducible, uniform and low-stress packing state, allowing an objective comparison of samples. For each sample in this study, three tests were run to ensure consistency of the results. The flow characteristics of the samples were studied by analyzing the variation in the flow energy as well as by measuring the basic flowability and specific energy (SE). In addition, the conditioned bulk density (CBD) of the samples was examined to study the powder bed density.

The basic flowability, defined as the ability of the powder to flow when forced, is qualitatively measured as the basic flow energy (BFE). The BFE represents the energy required for the rotation of the blade for the seventh test cycle (BFE $= E_{\text{test 7, down}}$) of the stable tip speed part.

The SE shows the energy required to establish a particular flow pattern in a precise volume of conditioned powder and is defined as the average energy of the upward blade rotation for the 7th and 8th test cycles divided by the mass of the remaining powder in the vessel (Equation (2)). By gently lifting the powder, the upward motion of the blade generates a low stress and unconfined flow mode in the powder.

$$\text{SE} = \frac{\frac{E_{\text{test 7,up}} + E_{\text{test 8,up}}}{2}}{m_{\text{split}}} \tag{2}$$

where m_{split} is the mass of the powder after the excess powder is removed.

The CBD describes the packing state or the density of the powder in its reference state. In order to measure the CBD, each powder system was gently filled in a 25 mL volume splitting cylindrical vessel with a diameter of 25 mm. The conditioning process was performed with a conditioning blade, which slices the powder bed to remove the excess air and create a uniform powder bed with a low-stress packing state. After conditioning, the vessel was split in order to remove the excess mass of powder, so that the remaining powder had a volume of 25 mL. For each sample, three measurements were performed, and the average value was reported as the CBD, based on Equation (3):

$$\text{CBD} = m_{\text{split}} / v_{\text{split}} \tag{3}$$

In which v_{split} signifies the volume of the powder (the vessel volume), after the excess powder is removed.

3. Results and Discussions

3.1. XRD Analysis: Plastic Deformation and Phase Formation

Figure 2a presents the XRD patterns of the starting host and guest powders, as well as the developed composite powder systems subjected to the regular mixing and ball milling for different mixing times. The diffraction peaks of Ti64 in the R6 system had almost the same position and intensity as that of the starting Ti64. However, those for the B2 and the B6 systems exhibited the Ti64 diffraction peaks with a decreased intensity and increased width. The observed phenomenon was more pronounced for the B6 sample than for the B2 system. Moreover, a close examination of the Ti64 peaks in the ball-milled systems revealed that the severe plastic deformation induced in the ball milling process led to the shift in the peaks' position due to the structural changes such as the crystallite refinement and the accumulation of micro-strain [64,65]. The lattice micro-strain of the Ti64 constituent in the composite powder systems was determined using the standard Williamson–Hall analysis as follows [66]:

$$\beta \cos\theta = \left[\frac{k\lambda}{t}\right] + 4\varepsilon \sin\theta \tag{4}$$

where k is the shape factor (0.9), λ represents the wavelength of the X-ray (1.5406), θ signifies the diffraction angle, t is the effective crystallite size, β is the full width at the half maximum of the XRD peak, and ε is the micro-strain. By constructing a linear plot of ($\beta \cos\theta$) against ($4\sin\theta$), the slope gives the strain (ε). According to the micro-strain results provided in Figure 2b, the B2 and B6 systems showed increased lattice strain compared to the regularly mixed feedstocks. Also, longer milling times led to higher lattice strain (B6 compared to B2) due to the higher levels of plastic deformation imparted to the powder particles.

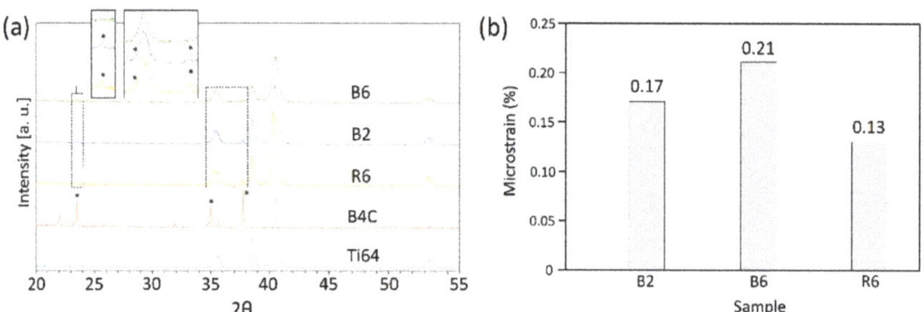

Figure 2. (a) XRD patterns of starting Ti-6Al-4V (Ti64) and B_4C powders, as well as composite powder systems prepared by regular mixing and ball milling, and (b) The microstrain of Ti64 constituent in the developed composite powder systems derived from the XRD patterns.

Referring to Figure 2a, all the peaks obtained for the composite powder feedstocks corresponded to those for the Ti64 and B_4C powders due to two probable scenarios: (i) no in-situ reaction has been activated in the system during the applied range of mixing times, or (ii) if formed, the amount of in-situ synthesized phases is below the detection limit of the XRD analysis. As indicated in Figure 2a, the intensity of the B_4C peaks in the R6 composite powder system decreased compared to those for the starting B_4C powder. When employing the ball-milling method, the intensity of these weak peaks further decreased for the B2 sample and finally, disappeared in the B6 system.

3.2. Microstructural Characterization

Figure 3 presents SEM micrographs of the starting host Ti64 and the guest B$_4$C powders used in this research. As observed in Figure 3a,b, the starting host powder particles had an almost spherical morphology and a very smooth surface, which are both characteristics of the gas-atomized Ti64 powders [17]. However, B$_4$C particles exhibited an irregular morphology (Figure 3c). Despite their smooth surface, some of the Ti64 particles have satellites. These satellites are formed when the finer solidified particles stick to the molten or semi-molten surface of the coarser ones as a result of the in-flight collisions before the solidification of the coarser molten droplets [67].

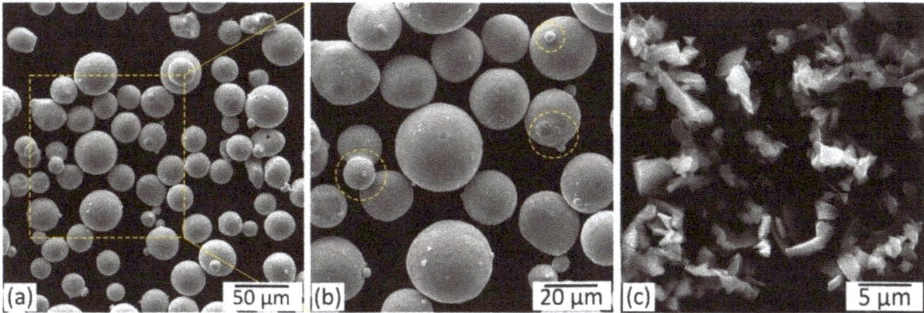

Figure 3. SEM micrographs of starting: (**a**,**b**) Ti-6Al-4V (host) and (**c**) B$_4$C (guest) powder particles. (**b**) Higher-magnification micrographs of the selected region in (**a**). The circles in (**b**) indicate the satellites.

The micrographs of the regularly mixed and ball-milled composite powders subjected to various mixing times are provided in Figures 4 and 5, respectively.

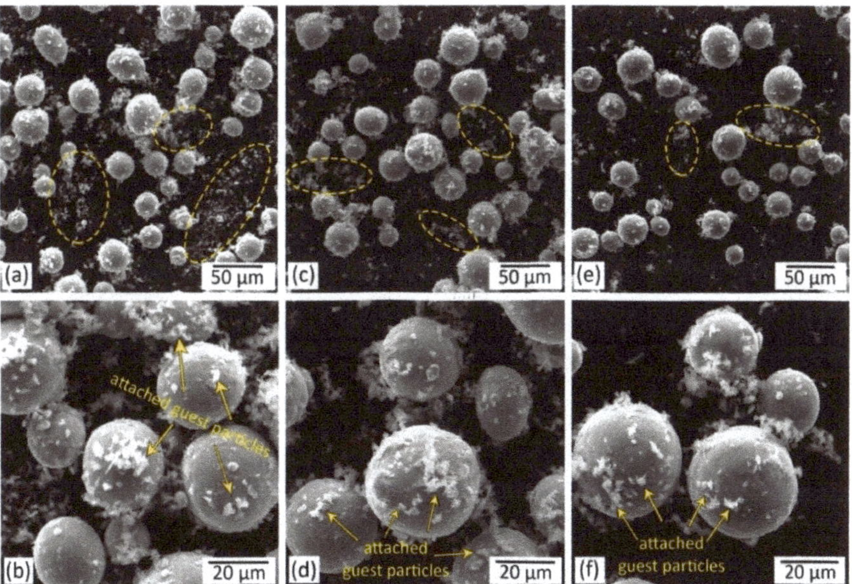

Figure 4. SEM micrographs of 5 wt.% B$_4$C/Ti-6Al-4V mixed powder systems subjected to regular mixing for: (**a**,**b**) 1 h (R1); (**c**,**d**) 3 h (R3); and (**e**,**f**) 6 h (R6). The ellipses in (**a**,**c**,**e**) indicate the free B$_4$C (guest) particles that are not attached to the Ti-6Al-4V (host) particles and are agglomerated.

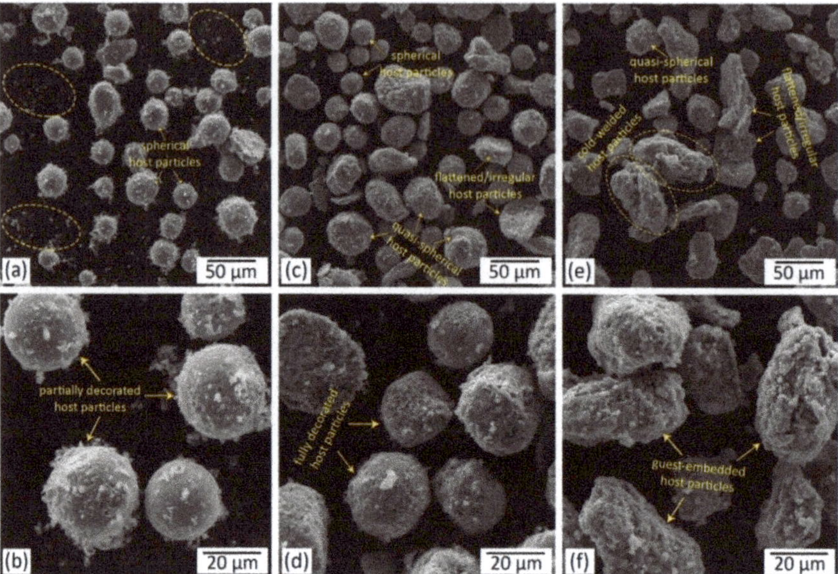

Figure 5. SEM micrographs of 5 wt.% B$_4$C/Ti-6Al-4V mixed powder systems subjected to ball milling for: (**a,b**) 1 h (B1); (**c,d**) 3 h (B3); and (**e,f**) 6 h (B6). The ellipses in (**a**) show the free B$_4$C (guest) particles that are not attached to the Ti-6Al-4V (host) particles.

The cross-sectional SEM micrographs of the starting Ti64 powder, as well as the B2 and the B6 powder systems, are presented in Figure 6.

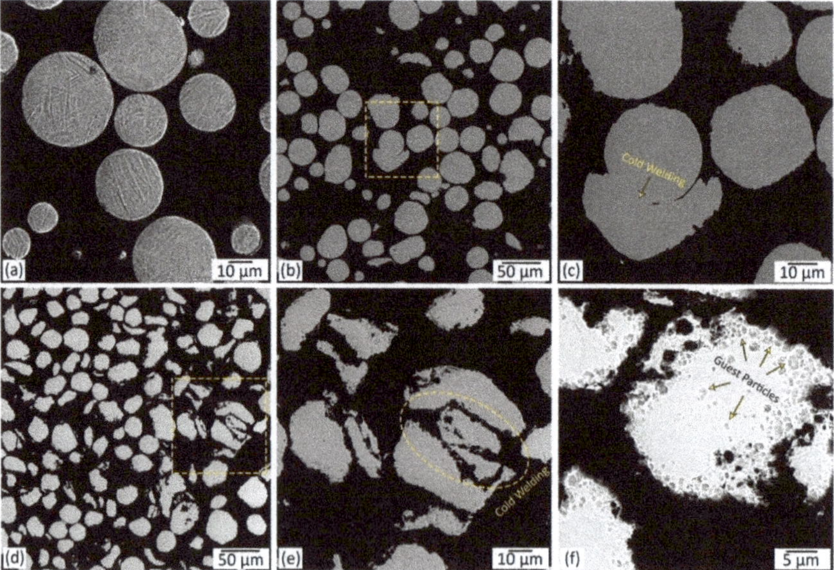

Figure 6. Cross-sectional SEM micrographs of: starting Ti-6Al-4V powder showing a martensitic microstructure after etching (**a**); 5 wt.% B$_4$C/Ti-6Al-4V system subjected to ball milling for: (**b,c**) 2 and (**d–f**) 6 h. (**c,e**) are higher magnification micrographs of the regions shown in (**b,d**), respectively.

As can be observed in Figure 4, the Ti64 host particles maintained their spherical morphology when employing the regular mixing method even with mixing times as long as 6 h (R6). However, even after a long mixing time of 6 h (R6), many of the guest particles were not attached to the host powder particles (Figure 4e), meaning that noticeably longer mixing times may still be required to provide more guest-to-host attachment. The guest particles that are not attached to the host powder particles tend to form agglomerates (as depicted in Figure 4a,c,e). Despite this fact, the regular mixing method has been adopted in numerous studies to produce composite powders due to its relative simplicity [30–32,68,69].

Depending on the employed mixing time, the host powder particles experienced different levels of plastic deformation during the ball-milling process (Figures 5 and 6). At a relatively short milling time of 1 to 2 h, the host power particles preserved their spherical shape (Figures 5a and 6b,c). Higher amounts of plastic deformation imparted to the system by a longer milling time of 3 h (B3) resulted in some spherical to quasi-spherical/irregular shape change (Figure 5c). Also, due to the extended time that the hard guest particles were hitting their surface, the enhanced milling time increased the surface roughness of the host powder particles. In the prolonged milling time of 6 h (B6), the desired spherical shape of the host particles altered to a flattened/irregular shape (Figures 5e,f and 6d–f). Microstructural observations also revealed the cold-welding of the ductile host powder particles during the ball milling process. Longer mixing times were found to cause intensified cold-welding in the applied range of ball milling time (Figures 5 and 6). Even at relatively short mixing time of 1 h, a significant guest-to-host attachment was obtained in the ball milling process, while still a few free B_4C particles could be observed in the composite powder (Figure 5a,b). Increasing the mixing time to 3 h eliminated the non-attached B_4C particles and led to the host particles fully decorated by the guest ones (Figure 5c,d). Further enhancement of the mixing time to 6 h caused the embedment of the guest particles into the host powder particles (Figures 5e,f and 6d–f).

The observed shape change and the agglomeration of the particles both observed at relatively long milling times are known as the main issues limiting the application of such composite powder systems in PBF–AM processes [29,36]. Sieving has been suggested as one of the strategies that could be employed to tailor the particle size in the powder systems subjected to the extended milling times [70]. While successful with particle size, this technique does not control the particle shape. Therefore, depending on the ductility of the host powder particles, appropriate milling times need to be found for each system to control the final shape of the particles.

3.3. Particle Size, Size Distribution and Sphericity

Figure 7 presents the results of particle size distribution for the starting Ti64 powder as well as the developed 5 wt.% B_4C/Ti64 powder feedstocks prepared by the regular mixing and ball-milling methods for 2 and 6 h of mixing. As observed in Figure 7b,c, the regularly mixed powder systems have a bimodal size distribution, while the ball-milled composite powders show a mono-modal size distribution (Figure 7d,e). This can be attributed to the non-attached B_4C particles in the regularly mixed case as opposed to the full attachment of the guest particles to the host powder particles in the ball-milled powder systems. It is worth noting that even a relatively long mixing time of 6 h in the regular mixing method was not successful in full guest-to-host attachment (Figure 4e,f). However, a relatively short mixing time of 1 h in the ball-milling method led to the B_4C particles being well-attached to the host powder particles (Figure 5a,b). The enhanced milling time provided attachment of more guest particles to the surface of the host particles, eliminating the free guest particles in the composite powder (Figure 5c,d). Application of longer milling times (B6) resulted in the embedment of B_4C particles into the host powder particles (Figures 5e,f and 6f). The severe weakening (or even disappearance) of the B_4C diffraction peaks in this system (B6 powder in Figure 2a) may be attributed to their embedment into the ductile host powder particles.

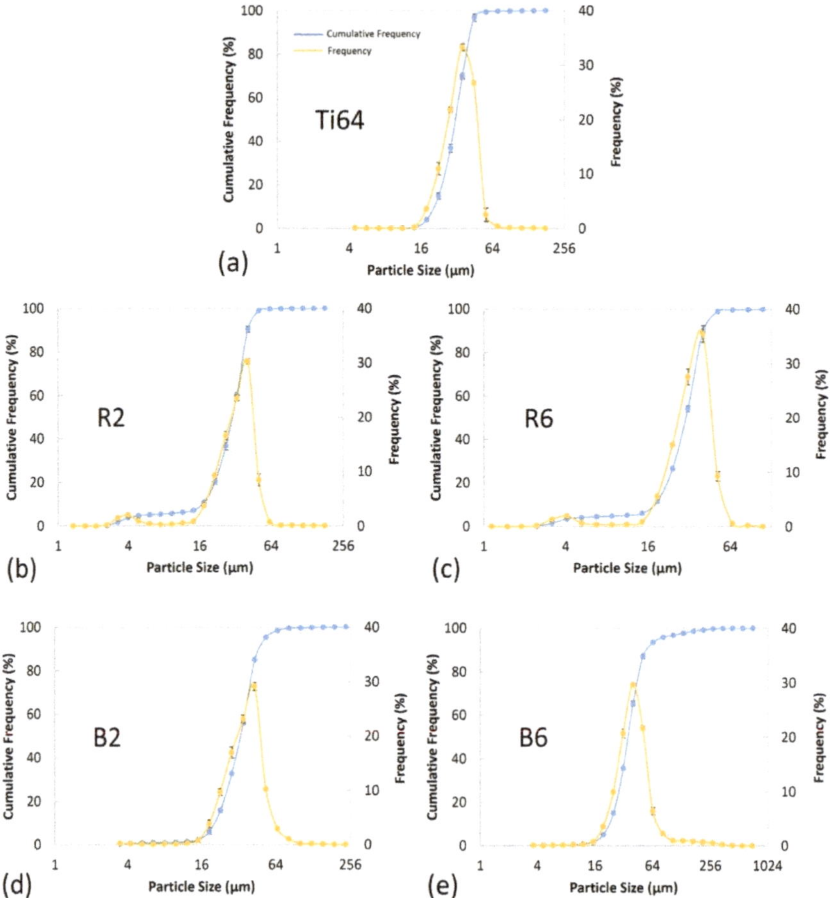

Figure 7. The particle size distribution curves of: (**a**) starting Ti-6Al-4V (Ti64), (**b**) 2 h (R2) and (**c**) 6 h (R6) regularly mixed; (**d**) 2 h (B2) and (**e**) 6 h (B6) ball milled 5wt.% B$_4$C/Ti64 composite powder feedstocks.

The D10, D50, and D90 of the powder systems derived from the cumulative frequency shown in Figure 7 are provided in Table 2. Referring to Figure 7 and Table 2, the regularly mixed composite powders showed almost the same particle size and particle size distribution as that of the starting Ti64 powder. The slight deviation in D10 and D50 of R2 and R6 systems from those for Ti64 is caused by the presence of free guest B$_4$C particles with a significantly smaller particle size compared to the host Ti64 particles (Figure 4). Due to its non-equilibrium nature, the ball-milling process involves persistent deformation, cold-welding, and fracture of powder particles [71,72]. The size of the powder particles subjected to the ball-milling process is determined by the competition between two major mechanisms, namely, cold-welding and fracture [71]. While the cold-welding mechanism facilitates the formation of larger-sized particles by attachment of the host powder particles, the fracture mechanism favors the decrease in particle size. Hence, the refining or coarsening of powder particles during the ball-milling process depends on whether the cold-welding or the fracture mechanism is dominant [73]. For the applied range of milling times, the particle size showed an ascending trend upon increasing the mixing time as a result of the cold-welding mechanism being dominant (Figures 5e and 6b,c,e). For instance, D90 for the B2 and B6 samples is 12% and 35% higher than that of Ti64 powder, respectively. The particle coarsening in the ball-milled composite samples is caused by the guest-to-host attachment as well as the cold-welding of host particles. The decoration of host particles is the dominant factor in increasing the particle size in the B2 sample due to the limited cold-welding caused by the relatively short mixing time (Figures 5 and 6). However, the significant cold-welding induced by the prolonged mixing time in the B6 sample is the main factor governing the particle coarsening, since the guest particles are embedded into the host Ti64 powder (not decorating the host particles). Based on the D90 of the B6 sample (Table 2) and Figure 7, 10% of particles have sizes in the range of 62–275 μm. These particle sizes are noticeably larger than the starting Ti64 powder particles, with 0.3% of particles exceeding 62 μm in size.

Table 2. The particle size distribution results presenting the cumulative distribution of D10, D50, and D90 for the Ti-6Al-4V (Ti64) and the developed composite powders.

Sample	D10	D50	D90
Ti64	23.39 ± 0.515	35.04 ± 0.360	45.71 ± 0.878
R2	18.81 ± 0.564	33.91 ± 0.420	45.24 ± 0.497
R6	20.12 ± 0.070	33.92 ± 0.310	45.49 ± 0.570
B2	22.89 ± 0.061	37.01 ± 0.362	51.28 ± 0.210
B6	25.20 ± 0.207	40.30 ± 0.221	61.94 ± 1.437

Figure 8 shows the sphericity of the powder particles with respect to their size for the starting Ti64 powder, as well as the developed composite powders. In the regular mixing case shown in Figure 8a, the sphericity of the host powder particles is the same as the starting Ti64 particles (18–50 μm). This can be ascribed to the absence of plastic deformation in the regular mixing method as well as the limited attachment of the guest B_4C particles to the host ones (Figure 4). The non-attached B_4C particles existing in the R2 and R6 samples showed particle sizes in the range of 3–18 μm, representing the formation of agglomerated guest particles (Figure 4). The relatively low sphericity of these agglomerates originates from their irregular shape (Figure 3c). According to Figure 8b, three different zones can be defined in the sphericity–particle-size curves for the B2 and B6 samples. Zones (I), (II) and (III) refer to the particle sizes in the range of (5–18), (18–65) and (>65) μm for B2, and (10–25), (25–80) and (>80) μm for B6, respectively. Since the B2 and B6 samples are free from non-attached B_4C particles, Zone (I) signifies the fractured host powder particles showing lower sphericity than the starting Ti64 particles. In Zone (II), the deformation and cold-welding, as well as the decoration of the host particles by the guest ones, lead to the decreased sphericity compared to the Ti64 powder. Since the short milling time of 2 h resulted in the limited cold-welding, the host particles decoration and deformation are believed to be the dominant factors, slightly decreasing the sphericity of the B2 sample. According to Figures 5a,b and 6b,c, most of the host powder particles preserved their spherical shape at short milling times. However, due to the embedment of the guest particles into the host ones in the B6 sample (absence of decoration), deformation and cold-welding of the host particles are responsible for the sphericities lower than the B2 sample. Zone (III) which is mostly visible in the B6 sample represents the cold-welded quasi-spherical/irregular shape powder particles (agglomerates) with sizes much larger than the starting Ti64 particles (Figures 5e,f and 6d–f). The sphericity of these agglomerates follows a decreasing trend by increasing their size.

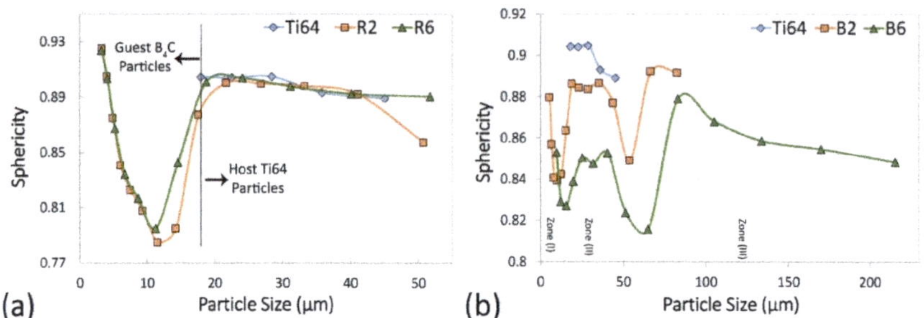

Figure 8. The sphericity of the particles as a function of their size for: (**a**) regularly mixed and (**b**) ball-milled 5 wt.% B$_4$C/Ti64 composite powder feedstocks. In each case, the sphericity of the starting Ti-6Al-4V (Ti64) powder is provided as the reference.

3.4. Flow Behavior and Conditioned Bulk Density (CBD)

3.4.1. Flowability

As can be observed in Figure 9a,b, all of the developed composite powder systems showed lower flowability (higher BFE and SE) as compared to the host powder (Ti64). The flow response of the composite powders was found to be dependent on the blade movement direction. In downward movement (BFE), the ball-milled composite powders exhibited higher flow energy (lower flowability) compared to the regularly mixed composite systems (Figure 9b). However, considering the upward movement of the blade (SE), the ball-milled powders showed better flowability (Figure 9a). The major difference between the BFE and SE is the confined flow of the powder in the former case due to the effect of the bottom of the vessel. Since the recoater interacts with powder in an unconfined state during the powder layer deposition process, the SE is a better representative of the powder flow in PBF–AM processes.

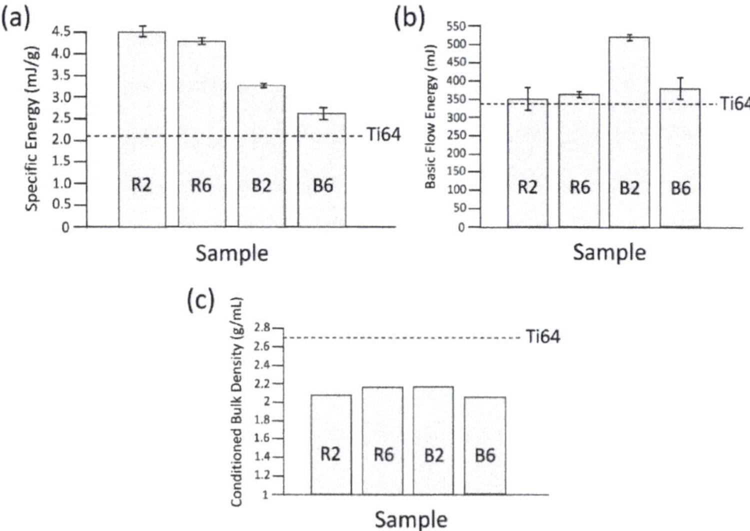

Figure 9. The flow behavior and conditioned bulk density results for 5 wt.% B$_4$C/Ti-6Al-4V systems prepared by regular mixing and ball-milling methods with different mixing times: (**a**) specific energy (SE), (**b**) basic flow energy (BFE), and (**c**) conditioned bulk density (CBD) of the composite powder systems compared to the reference powder (Ti64).

As the microstructural and sphericity characterizations revealed in Figures 5, 6 and 8, the plastic deformation induced with longer milling times led to some degrees of shape change for the host powder particles from spherical to quasi-spherical/moderately flattened. In addition, the surface roughness of the host particles increased. The surface roughening (not the shape change) of Ti64 particles in the ball-milled composite powders subjected to relatively short mixing times (1–3 h) can be attributed mainly to the guest B_4C particles hitting their surface due to two reasons:

i. The guest B_4C powder particles are noticeably harder than the host Ti64 powder particles [22, 74–76]. Accordingly, the noticeably harder B_4C particles have a great potential to scratch, punch and roughen the surface of softer Ti64 particles. Since they have the same hardness, the host–host inter-particle collisions might not affect the surface roughness of the host particles.

ii. As the microstructural observations of the starting powders revealed (Figure 3), the guest B_4C particles have an irregular shape as opposed to the spherical shape of the host Ti64 particles. The collision of irregular-shaped B_4C particles with the spherical-shape Ti64 particles has a higher chance of making the surface of Ti64 particles rough compared to the host–host collisions.

It is worth noting that the guest B_4C particles affect the surface roughness of the host Ti64 particles only if the extreme guest–host collisions are provided by the metallic balls (ball milling). As the microstructural characterizations revealed, the regular mixing did not affect the surface roughness of Ti64 particles even at relatively long mixing times such as 6 h (Figure 4).

The increase in the SE (decreased flowability) in the B2 feedstock compared to the Ti64 reference sample may be attributed to the contribution of two different factors: (i) the change in the particle morphology [77], and (ii) the decoration of the host particles by the guest particles. The slight spherical to quasi-spherical shape change and the surface roughening of the host particles seem not to play major roles on the reduction of powder flowability of the B2 sample. Therefore, the considerable increase in the SE for this system could be related to the presence of the decorating guest powder particles. The corresponding suggested mechanism is schematically illustrated in Figure 10. The guest-decorated host powder particles may experience two different interactions during their flow. While the "mechanical interlocking" mechanism decreases the flowability by resisting the free flow of the powder particles relative to each other (Figure 10b), the "contact surface reduction" mechanism may improve the flowability of the system [78] (Figure 10c). Compared to the non-decorated host powder system, the inter-particle tangling caused by the decorating guest particles leads to the enhanced flow resistivity. On the other hand, if not forming tangles, the presence of the decorating guest particles lowers the contact surface area required for the movement of a guest-decorated host particle (particle 1) from position (I) to position (II) relative to another particle assumed to be fixed (particle 2) (as shown in Figure 10d). By reducing the inter-particle friction and adhesion force, this mechanism can improve the flowability [78–80]. As mentioned earlier, although the slight change in the morphology of the host powder particles in the B2 case could have adverse effects on the flowability, these factors seem not to be predominant in the flowability reduction due to their negligible deviation from the starting host powder particles. Accordingly, the significant decrease in the flowability of the B2 powder system could be due to the dominance of the "mechanical interlocking" mechanism over the "contact surface reduction" mechanism. The contribution of the "mechanical interlocking" and the "contact surface reduction" mechanisms in the overall powder flowability is a major function of the guest particle size. When nano-scale guest powder particles are deposited on the surface of the primarily cohesive host powder particles, the artificially generated nano-scale roughness has been reported to enhance the flowability [79,81–83]. Based on the mechanism provided in Figure 10, this could be attributed to the "contact surface reduction" combined with the lack of active "mechanical interlocking" sites in such composite powder systems, reducing the chance of particle entangling.

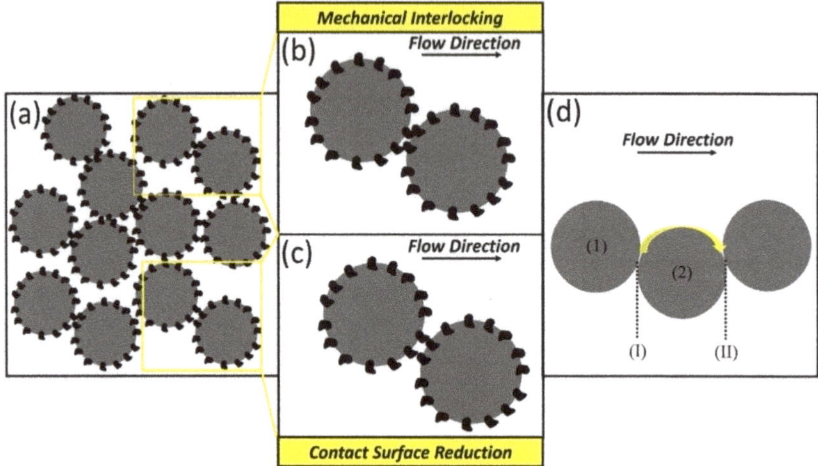

Figure 10. Schematic illustration presenting: (**a**) an overview of composite powder system with guest-decorated host particles; (**b,c**) close view of the selected regions showing "Mechanical Interlocking" and "Contact Surface Reduction" mechanisms, respectively; and (**d**) the contact area of two host powder particles when flowing relative to each other.

As indicated in Figure 9a, the B6 sample showed lower SE (better flowability) than the B2 one. In order to describe this finding, the mechanisms influencing the flowability of the B6 sample need to be explored. Referring to Figure 11, the flowability of the B6 powder feedstock is determined by the contribution of three mechanisms. Although the significant spherical to flattened/irregular shape change as well as the inter-locking of the guest-embedded host particles in such a system act to decrease the flowability [84–86], the increased particle size caused by the dominancy of the cold-welding over the fracture mechanism may reduce the effective surface area of the powder particles and, consequently, favor higher flowability [86] (Figure 7e and Table 2). Therefore, the overall flow behavior of such composite powder systems is governed by the competition among these mechanisms. The slight dominance of the inter-locking and shape change mechanisms over the surface area reduction in the B6 system is believed to be the reason for the small increment in its SE compared to the Ti64 case (Figure 11).

Referring to Figure 9a, the regularly mixed powder systems depicted SEs about twice that of the host particles. The observed phenomenon can be discussed based on the microstructural characterization of the powder systems. As shown in Figures 4 and 8, the regularly mixed powder feedstocks had the guest particles almost non-attached to the completely spherical (non-deformed) host particles. Accordingly, neither the host particle shape change nor the decoration-induced tangling can be the reason behind the lower flowability of regularly mixed powders. However, the enhanced inter-particle friction caused by the presence of fine non-attached guest powder particles with an extremely high surface-to-volume ratio can explain the drastic increase in the SE of these samples.

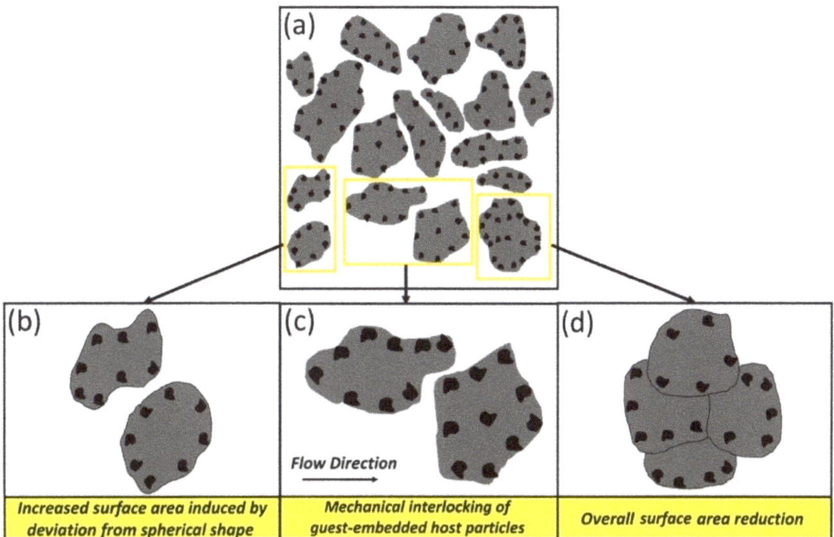

Figure 11. Schematic view presenting: (**a**) an overview of composite powder system with guest-embedded host particles; (**b–d**) close view of the selected regions showing "Increased Surface Area", "Host Particles Mechanical Interlocking" and "Surface Area Reduction" mechanisms, respectively.

3.4.2. Conditioned Bulk Density (CBD)

The packing density of the powder as the starting material in the PBF–AM processes has a significant influence on the quality of the parts produced. The density of the powders in this study was analyzed by their conditioned bulk density (CBD). Referring to Figure 9c, regardless of the mixing method and time, incorporation of the guest powder particles to form composite powder feedstocks decreased the CBD compared to the starting host powder (Ti64). A portion of this decrease is due to the addition of a less-dense material (B_4C) to the Ti64 powder. Moreover, the increased inter-particle friction arising from the irregular shape of the B_4C particles combined with the large particle size distributions acts to reduce the packing density of the composite powder feedstocks compared to Ti64 powder [83,87,88]. In general, the higher random loose packing provided by the lower friction among the powder particles results in higher CBD. In the case of starting host powder particles, there is only the host/host inter-particle friction which determines the powder density. However, the introduction of the guest particles into the host powder leads to the emergence of new friction sources, namely, host/guest, and guest/guest inter-particle frictions which could be responsible for the lower CBD of the composite powders as compared to the Ti64 host powder.

Presence of free (non-attached) guest particles in the composite powder leads to the guest/host and guest/guest inter-particle frictions in addition to the host/host ones. According to Figure 9c, the higher CBD of the R6 sample compared to the R2 one is due to the lower amounts of free B_4C particles existing in the system (Figure 4). Attachment of the guest to the host powder particles eliminates the host/guest inter-particle friction, which leads to the decrease in the host/host inter-particle friction by reducing their contact surface area. The morphology of powder particles also has a significant impact on the packing density of the powder bed and consequently, the density of the final component [89,90]. While the full decoration of the host particles by the guest particles in the B2 system promotes the attainment of a higher CBD, the slight deviation of the host particles from a spherical shape adversely influences the density (Figure 6b,c and Figure 8b). Therefore, the same CBD as that of the R6 system was obtained for the B2 case. The significant morphological change of the host particles in the B6 sample could be the main reason behind its low CBD. The flattened/irregular shape of the powder particles in the B6 system renders a poor packing due to the elevated inter-particle friction [87]. It is also worth noting that the formation of agglomerated guest particles in the regularly mixed composite systems adversely affects their potential in occupying the host particles interstices.

3.5. Material Loss

The mechanical mixing processes involves some material loss due to the cold-welding of the powder particles to the balls and/or jar. Therefore, the amount of the starting and final powder needs to be quantified. Figure 12 presents the variation in the powder mass for both the regularly mixed and ball-milled composite powders as a function of the mixing time. The material loss for both the regularly mixed and ball-milled cases is negligible (<1%). However, the regular mixing resulted in lower material loss than the ball-milling method due to the absence of balls.

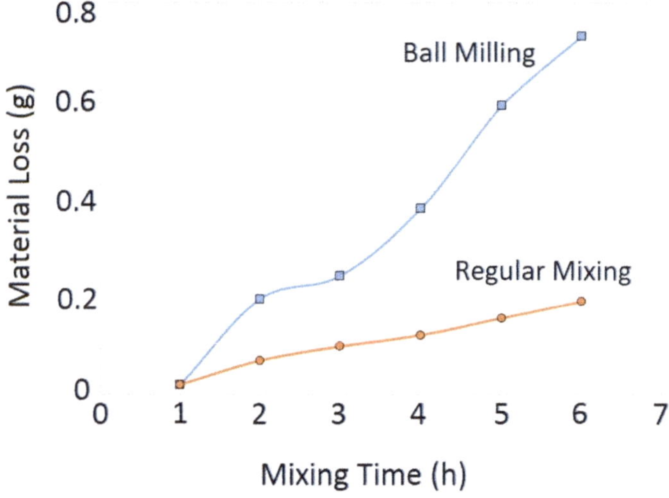

Figure 12. Material loss as a function of the mixing time for the regularly mixed and ball-milled 5 wt.% B_4C/Ti-6Al-4V composite powders. The starting powder mixture was 100 g, and the ball-to-powder ratio was 5:1.

3.6. Selection of the Best Possible Composite Powder

The ideal mixed powder system for the PBF–AM processing of MMCs needs to have: (i) non-free guest powder particles which are uniformly and homogeneously distributed throughout the system, (ii) host powder particles preserving their desired spherical shape, and (iii) the same flow behavior and apparent packing density as the starting host powder constituent. Although preserving the spherical shape of the host particles, the inadequate guest–host attachment associated with the regular mixing results in heterogeneous final MMC parts with improper distribution of the guest particles (or in-situ formed phases) or even their agglomeration. In addition, the regularly mixed composite powders showed noticeably lower flowability compared to the ball-milled composite systems. These issues restrict the implementation of the regularly mixed (R2 and R6) powder feedstocks in the PBF–AM of MMCs. Ball milling of the powders for relatively long milling times (B6) significantly improves the distribution state of the guest particles throughout the system by their embedment into the host powder particles. However, the significant spherical to flattened/irregular shape change and the particle coarsening are the main drawbacks of such powder feedstocks (Figures 7 and 8 and Table 2). While the particle coarsening issue can be solved by sieving, this strategy may not be cost-and-time effective, especially for high-priced materials. When employing shorter milling times (2 h), the composite powder system showed almost spherical host particles fully decorated by the guest powder particles, meeting some of the requirements mentioned for the ideal composite powder feedstock (best possible case). Although such a powder ensures achieving homogenous MMC parts due to the proper host/guest attachment and the lack of non-attached (free) agglomerated guest particles, the decorating guest particles act as obstacles to the free flow of host particles and consequently, sacrifice the flowability. Proper selection of the recoater speed in PBF–AM processes could be a strategy to solve this issue when using such composite powder feedstocks [91]. Based on the above discussions, the composite powder system prepared by a relatively short milling time of 2 h (B2) is suggested as the best possible case for PBF–AM processes.

4. Conclusions

The characteristics of the 5 wt.% B_4C(guest)/Ti-6Al-4V(host) mixed powder systems were studied from a powder bed fusion (PBF)–additive manufacturing (AM) perspective. For this purpose, the regular mixing and the ball-milling methods were employed with a wide range of milling times (1–6 h) to produce composite powder feedstocks. The developed powders were examined using microstructural characterization, phase formation, particle size, size distribution, sphericity, apparent density, and flow behavior analyses. Moreover, the mechanisms that play a role in the flowability of the mixed powder systems were analyzed based on the microstructural observations and the flow measurement results. The main outcomes can be outlined as follows:

1. With the regular mixing, the shape of the host powder particles remained unchanged until 6 h of mixing. The ball-milling method led to the change in the shape of host powder particles from spherical to quasi-spherical and then to a flattened/irregular shape by increasing the milling time, resulting in the decreased particle sphericity compared to the starting host particles.
2. The regular mixing method did not provide acceptable attachment of the guest B_4C particles to the host particles. However, milling times as short as 2 h in the ball-milling case provided the host particles with a full decoration by the guest particles. Longer milling time (6 h) led to the guest particles embedded in the severely deformed host particles.
3. Although the basic flow energy (BFE) results contradict the specific energy (SE) measurements, the SE is believed to be a better representative of the powder layer deposition during PBF–AM process due to the unconfined and low-stress state of the powder.
4. Although being highly dependent on the mixing process variables, the flowability of the developed composite powders was lower than that of the reference Ti-6Al-4V powder. The regularly mixed and ball-milled composite powders exhibited ~110%, and 24–57% increase in SE compared to the Ti-6Al-4V powder, respectively.
5. The ball-milled feedstocks showed lower SE (better flowability) than the regularly mixed powders. The flow behavior of developed composite feedstocks was discussed based on the underlying mechanisms.
6. The produced composite powder systems showed 18–24% decrease in density compared to the reference Ti-6Al-4V powder.
7. The composite powder benefitting from fully decorated spherical-shape host particles is suggested as the best possible mechanically processed feedstock for PBF–AM processes. The relatively low flowability of this powder system should be considered when defining the recoater speed in PBF–AM processes.

Author Contributions: Conceptualization, E.F. and A.G.; methodology, E.F. and A.G.; investigation, E.F. and A.G.; writing—original draft preparation, E.F. and A.G.; writing—review and editing, E.F. and M.E.; supervision, M.E.

Funding: This research received no external funding.

Acknowledgments: The authors would like to thank Henry Ma and Allan Rogalsky from the MSAM (Multi-scale Additive Manufacturing) Laboratory at the University of Waterloo for the flowability and particle size measurement tests.

Conflicts of Interest: The authors declare no conflict of interest.

References

1. Dadkhah, M.; Saboori, A.; Fino, P. An overview of the recent developments in metal matrix nanocomposites reinforced by graphene. *Materials* **2019**, *12*, 2823. [CrossRef] [PubMed]
2. Wu, S.Q.; Wang, H.Z.; Tjong, S.C. Mechanical and wear behavior of an Al/Si alloy metal-matrix composite reinforced with aluminosilicate fiber. *Compos. Sci. Technol.* **1996**, *56*, 1261–1270. [CrossRef]
3. Tjong, S.C. Recent progress in the development and properties of novel metal matrix nanocomposites reinforced with carbon nanotubes and graphene nanosheets. *Mater. Sci. Eng. R Rep.* **2013**, *74*, 281–350. [CrossRef]
4. Zhang, J.; Perez, R.J.; Wong, C.R.; Lavernia, E.J. Effects of secondary phases on the damping behaviour of metals, alloys and metal matrix composites. *Mater. Sci. Eng. R Rep.* **1994**, *13*, 325–389. [CrossRef]
5. Liu, L.; Li, W.; Tang, Y.; Shen, B.; Hu, W. Friction and wear properties of short carbon fiber reinforced aluminum matrix composites. *Wear* **2009**, *266*, 733–738. [CrossRef]
6. Dieringa, H. Properties of magnesium alloys reinforced with nanoparticles and carbon nanotubes: A review. *J. Mater. Sci.* **2011**, *46*, 289–306. [CrossRef]
7. Liu, Y.; Ding, J.; Qu, W.; Su, Y.; Yu, Z. Microstructure evolution of TiC particles In situ, synthesized by laser cladding. *Materials* **2017**, *10*, 281. [CrossRef]
8. Miracle, D. Metal matrix composites–from science to technological significance. *Compos. Sci. Technol.* **2005**, *65*, 2526–2540. [CrossRef]
9. Geandier, G.; Vautrot, L.; Denand, B.; Denis, S. In situ stress tensor determination during phase transformation of a metal matrix composite by high-energy X-ray diffraction. *Materials* **2018**, *11*, 1415. [CrossRef]
10. Mertens, A.I.; Lecomte-Beckers, J. On the role of interfacial reactions, dissolution and secondary precipitation during the laser additive manufacturing of metal matrix composites: A review. In *New Trends in 3D Printing*; Shishkovsky, I.V., Ed.; InTech: Rijeka, Croatia, 2016; Chapter 9. [CrossRef]
11. Kainer, K.U. Basics of metal matrix composites. In *Metal Matrix Composites*; Wiley-VCH Verlag GmbH & Co. KGaA: Weinheim, Germany, 2006; Chapter 1, pp. 1–54. [CrossRef]
12. Suresh, S. *Fundamentals of Metal-Matrix Composites*; Elsevier: Amsterdam, The Netherlands, 2013.
13. Rosso, M. Ceramic and metal matrix composites: Routes and properties. *J. Mater. Process. Technol.* **2006**, *175*, 364–375. [CrossRef]
14. Wang, J.; Guo, X.; Qin, J.; Zhang, D.; Lu, W. Microstructure and mechanical properties of investment casted titanium matrix composites with B4C additions. *Mater. Sci. Eng. A* **2015**, *628*, 366–373. [CrossRef]
15. Fereiduni, E.; Elbestawi, M. Process-structure-property relationships in additively manufactured metal matrix composites. In *Additive Manufacturing of Emerging Materials*; Springer: Berlin, Germany, 2019; pp. 111–177.
16. Fereiduni, E.; Yakout, M.; Elbestawi, M. Laser-based additive manufacturing of lightweight metal matrix composites. In *Additive Manufacturing of Emerging Materials*; Springer: Cham, Switzerland, 2019; pp. 55–109.
17. Sutton, A.T.; Kriewall, C.S.; Leu, M.C.; Newkirk, J.W. Powder characterisation techniques and effects of powder characteristics on part properties in powder-bed fusion processes. *Virtual Phys. Prototyp.* **2017**, *12*, 3–29. [CrossRef]
18. Ali, U.; Mahmoodkhani, Y.; Shahabad, S.I.; Esmaeilizadeh, R.; Liravi, F.; Sheydaeian, E.; Huang, K.Y.; Marzbanrad, E.; Vlasea, M.; Toyserkani, E. On the measurement of relative powder-bed compaction density in powder-bed additive manufacturing processes. *Mater. Des.* **2018**, *155*, 495–501. [CrossRef]
19. Rausch, A.; Küng, V.; Pobel, C.; Markl, M.; Körner, C. Predictive simulation of process windows for powder bed fusion additive manufacturing: Influence of the powder bulk density. *Materials* **2017**, *10*, 1117. [CrossRef]
20. Jacob, G.; Donmez, A.; Slotwinski, J.; Moylan, S. Measurement of powder bed density in powder bed fusion additive manufacturing processes. *Meas. Sci. Technol.* **2016**, *27*, 115601. [CrossRef]
21. Narvan, M.; Al-Rubaie, K.S.; Elbestawi, M. Process-structure-property relationships of AISI H13 tool steel processed with selective laser melting. *Materials* **2019**, *12*, 2284. [CrossRef]
22. Fereiduni, E.; Ghasemi, A.; Elbestawi, M. Selective laser melting of hybrid ex-situ/in-situ reinforced titanium matrix composites: Laser/powder interaction, reinforcement formation mechanism, and non-equilibrium microstructural evolutions. *Mater. Des.* **2019**, *184*, 108185. [CrossRef]

23. Calignano, F.; Minetola, P. Influence of process parameters on the porosity, accuracy, roughness, and support structures of hastelloy X produced by laser powder bed fusion. *Materials* **2019**, *12*, 3178. [CrossRef]

24. Mirkoohi, E.; Seivers, D.E.; Garmestani, H.; Liang, S.Y. Heat source modeling in selective laser melting. *Materials* **2019**, *12*, 2052. [CrossRef]

25. Neubauer, E.; Kitzmantel, M.; Hulman, M.; Angerer, P. Potential and challenges of metal-matrix-composites reinforced with carbon nanofibers and carbon nanotubes. *Compos. Sci. Technol.* **2010**, *70*, 2228–2236. [CrossRef]

26. Bakshi, S.R.; Lahiri, D.; Agarwal, A. Carbon nanotube reinforced metal matrix composites-a review. *Int. Mater. Rev.* **2010**, *55*, 41–64. [CrossRef]

27. Campbell, T.A.; Ivanova, O.S. 3D printing of multifunctional nanocomposites. *Nano Today* **2013**, *8*, 119–120. [CrossRef]

28. Zhao, X.; Song, B.; Fan, W.; Zhang, Y.; Shi, Y. Selective laser melting of carbon/AlSi10Mg composites: Microstructure, mechanical and electronical properties. *J. Alloys Compd.* **2016**, *665*, 271–281. [CrossRef]

29. Yu, W.; Sing, S.; Chua, C.; Kuo, C.; Tian, X. Particle-reinforced metal matrix nanocomposites fabricated by selective laser melting: A state of the art review. *Prog. Mater. Sci.* **2019**. [CrossRef]

30. Song, B.; Dong, S.; Coddet, C. Rapid in situ fabrication of Fe/SiC bulk nanocomposites by selective laser melting directly from a mixed powder of microsized Fe and SiC. *Scr. Mater.* **2014**, *75*, 90–93. [CrossRef]

31. Hao, L.; Dadbakhsh, S.; Seaman, O.; Felstead, M. Selective laser melting of a stainless steel and hydroxyapatite composite for load-bearing implant development. *J. Mater. Process. Technol.* **2009**, *209*, 5793–5801. [CrossRef]

32. Cooper, D.E.; Blundell, N.; Maggs, S.; Gibbons, G.J. Additive layer manufacture of Inconel 625 metal matrix composites, reinforcement material evaluation. *J. Mater. Process. Technol.* **2013**, *213*, 2191–2200. [CrossRef]

33. Gu, D.; Wang, H.; Dai, D.; Chang, F.; Meiners, W.; Hagedorn, Y.-C.; Wissenbach, K.; Kelbassa, I.; Poprawe, R. Densification behavior, microstructure evolution, and wear property of TiC nanoparticle reinforced AlSi10Mg bulk-form nanocomposites prepared by selective laser melting. *J. Laser Appl.* **2015**, *27*, S17003. [CrossRef]

34. Gu, D.; Wang, H.; Dai, D. Laser additive manufacturing of novel aluminum based nanocomposite parts: Tailored forming of multiple materials. *J. Manuf. Sci. Eng.* **2016**, *138*, 021004. [CrossRef]

35. Hu, Y.; Cong, W.; Wang, X.; Li, Y.; Ning, F.; Wang, H. Laser deposition-additive manufacturing of TiB-Ti composites with novel three-dimensional quasi-continuous network microstructure: Effects on strengthening and toughening. *Compos. Part B Eng.* **2018**, *133*, 91–100. [CrossRef]

36. Attar, H.; Bönisch, M.; Calin, M.; Zhang, L.-C.; Scudino, S.; Eckert, J. Selective laser melting of in situ titanium–titanium boride composites: Processing, microstructure and mechanical properties. *Acta Mater.* **2014**, *76*, 13–22. [CrossRef]

37. AlMangour, B.; Grzesiak, D. Selective laser melting of TiC reinforced 316L stainless steel matrix nanocomposites: Influence of starting TiC particle size and volume content. *Mater. Des.* **2016**, *104*, 141–151. [CrossRef]

38. AlMangour, B.; Grzesiak, D.; Yang, J.-M. Selective laser melting of TiB 2/H13 steel nanocomposites: Influence of hot isostatic pressing post-treatment. *J. Mater. Process. Technol.* **2017**, *244*, 344–353. [CrossRef]

39. Zhang, B.; Bi, G.; Wang, P.; Bai, J.; Chew, Y.; Nai, M.S. Microstructure and mechanical properties of Inconel 625/nano-TiB 2 composite fabricated by LAAM. *Mater. Des.* **2016**, *111*, 70–79. [CrossRef]

40. Li, X.; Ji, G.; Chen, Z.; Addad, A.; Wu, Y.; Wang, H.; Vleugels, J.; Van Humbeeck, J.; Kruth, J.-P. Selective laser melting of nano-TiB 2 decorated AlSi10Mg alloy with high fracture strength and ductility. *Acta Mater.* **2017**, *129*, 183–193. [CrossRef]

41. Chen, M.; Li, X.; Ji, G.; Wu, Y.; Chen, Z.; Baekelant, W.; Vanmeensel, K.; Wang, H.; Kruth, J.-P. Novel composite powders with uniform TiB2 nano-particle distribution for 3D printing. *Appl. Sci.* **2017**, *7*, 250. [CrossRef]

42. Zhou, W.; Sun, X.; Kikuchi, K.; Nomura, N.; Yoshimi, K.; Kawasaki, A. Carbon nanotubes as a unique agent to fabricate nanoceramic/metal composite powders for additive manufacturing. *Mater. Des.* **2018**, *137*, 276–285. [CrossRef]

43. Tan, H.; Hao, D.; Al-Hamdani, K.; Zhang, F.; Xu, Z.; Clare, A.T. Direct metal deposition of TiB 2/AlSi10Mg composites using satellited powders. *Mater. Lett.* **2018**, *214*, 123–126. [CrossRef]

44. Ma, C.; Chen, L.; Cao, C.; Li, X. Nanoparticle-induced unusual melting and solidification behaviours of metals. *Nat. Commun.* **2017**, *8*, 14178. [CrossRef]

45. Martin, J.H.; Yahata, B.D.; Hundley, J.M.; Mayer, J.A.; Schaedler, T.A.; Pollock, T.M. 3D printing of high-strength aluminium alloys. *Nature* **2017**, *549*, 365. [CrossRef]

46. Xia, M.; Liu, A.; Hou, Z.; Li, N.; Chen, Z.; Ding, H. Microstructure growth behavior and its evolution mechanism during laser additive manufacture of in-situ reinforced (TiB+ TiC)/Ti composite. *J. Alloys Compd.* **2017**, *728*, 436–444. [CrossRef]

47. Gu, D.; Yuan, P. Thermal evolution behavior and fluid dynamics during laser additive manufacturing of Al-based nanocomposites: Underlying role of reinforcement weight fraction. *J. Appl. Phys.* **2015**, *118*, 233109. [CrossRef]

48. Ning, J.; Sievers, D.E.; Garmestani, H.; Liang, S.Y. Analytical modeling of in-process temperature in powder bed additive manufacturing considering laser power absorption, Latent heat, scanning strategy, and powder packing. *Materials* **2019**, *12*, 808. [CrossRef] [PubMed]

49. Pinkerton, A.J.; Li, L. Direct additive laser manufacturing using gas-and water-atomised H13 tool steel powders. *Int. J. Adv. Manuf. Technol.* **2005**, *25*, 471–479. [CrossRef]

50. DebRoy, T.; Wei, H.; Zuback, J.; Mukherjee, T.; Elmer, J.; Milewski, J.; Beese, A.M.; Wilson-Heid, A.; De, A.; Zhang, W. Additive manufacturing of metallic components–process, structure and properties. *Prog. Mater. Sci.* **2018**, *92*, 112–224. [CrossRef]

51. Muñiz-Lerma, J.; Nommeots-Nomm, A.; Waters, K.; Brochu, M. A Comprehensive approach to powder feedstock characterization for powder bed fusion additive manufacturing: A case study on AlSi7Mg. *Materials* **2018**, *11*, 2386. [CrossRef]

52. Mahmoodkhani, Y.; Ali, U.; Imani Shahabad, S.; Rani Kasinathan, A.; Esmaeilizadeh, R.; Keshavarzkermani, A.; Marzbanrad, E.; Toyserkani, E. On the measurement of effective powder layer thickness in laser powder-bed fusion additive manufacturing of metals. *Prog. Addit. Manuf.* **2019**, *4*, 109–116. [CrossRef]

53. Wang, P.; Gammer, C.; Brenne, F.; Niendorf, T.; Eckert, J.; Scudino, S. A heat trea[TiB2/Al-3.5 Cu-1.5 Mg-1Si composite fabricated by selective laser melting: Microstructure, heat treatment and mechanical properties. *Compos. Part. B Eng.* **2018**, *147*, 162–168. [CrossRef]

54. AlMangour, B.; Grzesiak, D.; Borkar, T.; Yang, J.-M. Densification behavior, microstructural evolution, and mechanical properties of TiC/316L stainless steel nanocomposites fabricated by selective laser melting. *Mater. Des.* **2018**, *138*, 119–128. [CrossRef]

55. Xi, L.; Wang, P.; Prashanth, K.; Li, H.; Prykhodko, H.; Scudino, S.; Kaban, I. Effect of TiB2 particles on microstructure and crystallographic texture of Al-12Si fabricated by selective laser melting. *J. Alloys Compd.* **2019**, *786*, 551–556. [CrossRef]

56. AlMangour, B.; Kim, Y.-K.; Grzesiak, D.; Lee, K.-A. Novel TiB2-reinforced 316L stainless steel nanocomposites with excellent room-and high-temperature yield strength developed by additive manufacturing. *Compos. Part. B Eng.* **2019**, *156*, 51–63. [CrossRef]

57. Takacs, L.; McHenry, J.S. Temperature of the milling balls in shaker and planetary mills. *J. Mater. Sci.* **2006**, *41*, 5246–5249. [CrossRef]

58. Retsch Camsizer X2 (Retsch Technology GmbH, H.; Germany). Available online: https://www.retsch-technology.com/products/dynamic-image-analysis/camsizer-x2/function-features/ (accessed on 1 November 2019).

59. ISO. *9276–6 Representation of Results of Particle Size Analysis—Part 6: Descriptive and Quantitative Representation of Particle Shape and Morphology*; ISO: Geneva, Switzerland, 2008.

60. Spierings, A.B.; Voegtlin, M.; Bauer, T.; Wegener, K. Powder flowability characterisation methodology for powder-bed-based metal additive manufacturing. *Prog. Addit. Manuf.* **2016**, *1*, 9–20. [CrossRef]

61. Krantz, M.; Zhang, H.; Zhu, J. Characterization of powder flow: Static and dynamic testing. *Powder Technol.* **2009**, *194*, 239–245. [CrossRef]

62. Snow, Z.; Martukanitz, R.; Joshi, S. On the development of powder spreadability metrics and feedstock requirements for powder bed fusion additive manufacturing. *Addit. Manuf.* **2019**, *28*, 78–86. [CrossRef]

63. Freeman, R. Measuring the flow properties of consolidated, conditioned and aerated powders—A comparative study using a powder rheometer and a rotational shear cell. *Powder Technol.* **2007**, *174*, 25–33. [CrossRef]

64. Jeyasimman, D.; Sivasankaran, S.; Sivaprasad, K.; Narayanasamy, R.; Kambali, R. An investigation of the synthesis, consolidation and mechanical behaviour of Al 6061 nanocomposites reinforced by TiC via mechanical alloying. *Mater. Des.* **2014**, *57*, 394–404. [CrossRef]

65. Zhao, N.; Nash, P.; Yang, X. The effect of mechanical alloying on SiC distribution and the properties of 6061 aluminum composite. *J. Mater. Process. Technol.* **2005**, *170*, 586–592. [CrossRef]

66. Williamson, G.; Hall, W. X-ray line broadening from filed aluminium and wolfram. *Acta Metall.* **1953**, *1*, 22–31. [CrossRef]

67. Özbilen, S. Satellite formation mechanism in gas atomised powders. *Powder Metall.* **1999**, *42*, 70–78. [CrossRef]

68. Song, B.; Dong, S.; Coddet, P.; Zhou, G.; Ouyang, S.; Liao, H.; Coddet, C. Microstructure and tensile behavior of hybrid nano-micro SiC reinforced iron matrix composites produced by selective laser melting. *J. Alloys Compd.* **2013**, *579*, 415–421. [CrossRef]

69. Wei, Q.; Li, S.; Han, C.; Li, W.; Cheng, L.; Hao, L.; Shi, Y. Selective laser melting of stainless-steel/ nano-hydroxyapatite composites for medical applications: Microstructure, element distribution, crack and mechanical properties. *J. Mater. Process. Technol.* **2015**, *222*, 444–453. [CrossRef]

70. Lorusso, M.; Aversa, A.; Manfredi, D.; Calignano, F.; Ambrosio, E.P.; Ugues, D.; Pavese, M. Tribological behavior of aluminum alloy AlSi10Mg-TiB 2 composites produced by direct metal laser sintering (DMLS). *J. Mater. Eng. Perform.* **2016**, *25*, 3152–3160. [CrossRef]

71. Suryanarayana, C. Mechanical alloying and milling. *Prog. Mater. Sci.* **2001**, *46*, 1–184. [CrossRef]

72. Prabhu, B.; Suryanarayana, C.; An, L.; Vaidyanathan, R. Synthesis and characterization of high volume fraction Al–Al2O3 nanocomposite powders by high-energy milling. *Mater. Sci. Eng. A* **2006**, *425*, 192–200. [CrossRef]

73. Zhang, D. Processing of advanced materials using high-energy mechanical milling. *Prog. Mater. Sci.* **2004**, *49*, 537–560. [CrossRef]

74. Mukhanov, V.; Kurakevych, O.; Solozhenko, V. Thermodynamic model of hardness: Particular case of boron-rich solids. *J. Superhard Mater.* **2010**, *32*, 167–176. [CrossRef]

75. Ji, W.; Rehman, S.S.; Wang, W.; Wang, H.; Wang, Y.; Zhang, J.; Zhang, F.; Fu, Z. Sintering boron carbide ceramics without grain growth by plastic deformation as the dominant densification mechanism. *Sci. Rep.* **2015**, *5*, 15827. [CrossRef]

76. Zeng, X.; Liu, W. Aqueous tape casting of B4C ceramics. *Adv. Appl. Ceram.* **2016**, *115*, 224–228. [CrossRef]

77. Ghadiri, M.; Pasha, M.; Nan, W.; Hare, C.; Vivacqua, V.; Zafar, U.; Nezamabadi, S.; Lopez, A.; Pasha, M.; Nadimi, S. Cohesive powder flow: Trends and challenges in characterisation and analysis. *Kona Powder Part. J.* **2020**, *2020018*. [CrossRef]

78. Jallo, L.J.; Chen, Y.; Bowen, J.; Etzler, F.; Dave, R. Prediction of inter-particle adhesion force from surface energy and surface roughness. *J. Adhes. Sci. Technol.* **2011**, *25*, 367–384. [CrossRef]

79. Yang, J.; Sliva, A.; Banerjee, A.; Dave, R.N.; Pfeffer, R. Dry particle coating for improving the flowability of cohesive powders. *Powder Technol.* **2005**, *158*, 21–33. [CrossRef]

80. Xie, H.-Y. The role of interparticle forces in the fluidization of fine particles. *Powder Technol.* **1997**, *94*, 99–108. [CrossRef]

81. Ramlakhan, M.; Wu, C.Y.; Watano, S.; Dave, R.N.; Pfeffer, R. Dry particle coating using magnetically assisted impaction coating: Modification of surface properties and optimization of system and operating parameters. *Powder Technol.* **2000**, *112*, 137–148. [CrossRef]

82. Mei, R.; Shang, H.; Klausner, J.F.; Kallman, E. A contact model for the effect of particle coating on improving the flowability of cohesive powders. *Kona Powder Part. J.* **1997**, *15*, 132–141. [CrossRef]

83. Sivasankaran, S.; Sivaprasad, K.; Narayanasamy, R.; Iyer, V.K. An investigation on flowability and compressibility of AA 6061100– xx wt.% TiO2 micro and nanocomposite powder prepared by blending and mechanical alloying. *Powder Technol.* **2010**, *201*, 70–82. [CrossRef]

84. Lumay, G.; Boschini, F.; Traina, K.; Bontempi, S.; Remy, J.-C.; Cloots, R.; Vandewalle, N. Measuring the flowing properties of powders and grains. *Powder Technol.* **2012**, *224*, 19–27. [CrossRef]

85. Ziegelmeier, S.; Christou, P.; Wöllecke, F.; Tuck, C.; Goodridge, R.; Hague, R.; Krampe, E.; Wintermantel, E. An experimental study into the effects of bulk and flow behaviour of laser sintering polymer powders on resulting part properties. *J. Mater. Process. Technol.* **2015**, *215*, 239–250. [CrossRef]

86. Fu, X.; Huck, D.; Makein, L.; Armstrong, B.; Willen, U.; Freeman, T. Effect of particle shape and size on flow properties of lactose powders. *Particuology* **2012**, *10*, 203–208. [CrossRef]

87. Din, R.U.; Shafqat, Q.A.; Asghar, Z.; Zahid, G.; Basit, A.; Qureshi, A.; Manzoor, T.; Nasir, M.A.; Mehmood, F.; Hussain, K.I. Microstructural evolution, powder characteristics, compaction behavior and sinterability of Al 6061–B 4 C composites as a function of reinforcement content and milling times. *Russ. J. Non Ferrous Met.* **2018**, *59*, 207–222. [CrossRef]

88. Fogagnolo, J.B.; Velasco, F.; Robert, M.H.; Torralba, J.M. Effect of mechanical alloying on the morphology, microstructure and properties of aluminium matrix composite powders. *Mater. Sci. Eng. A* **2003**, *342*, 131–143. [CrossRef]

89. Gibson, I.; Rosen, D.W.; Stucker, B. *Additive Manufacturing Technologies*; Springer: Amsterdam, The Netherlands, 2014; Volume 17.

90. Manfredi, D.; Calignano, F.; Krishnan, M.; Canali, R.; Ambrosio, E.; Atzeni, E. From powders to dense metal parts: Characterization of a commercial AlSiMg alloy processed through direct metal laser sintering. *Materials* **2013**, *6*, 856–869. [CrossRef] [PubMed]

91. Azizi, H.; Ghiaasiaan, R.; Prager, R.; Ghoncheh, M.H.; Samk, K.A.; Lausic, A.; Byleveld, W.; Phillion, A.B. Metallurgical and mechanical assessment of hybrid additively-manufactured maraging tool steels via selective laser melting. *Addit. Manuf.* **2019**, *27*, 389–397. [CrossRef]

 *materials*

Article

Research on the Thermal Behaviour of a Selectively Laser Melted Aluminium Alloy: Simulation and Experiment

Zhonghua Li [1,†], Bao-Qiang Li [2,†], Peikang Bai [2,*], Bin Liu [2] and Yu Wang [2]

[1] School of Mechanical Engineering, North University of China, Taiyuan 030051, China; lzh2017@nuc.edu.cn
[2] School of Materials Science and Engineering, North University of China, Taiyuan 030051, China; lbqlalala@gmail.com (B.-Q.L.); liubin3y@nuc.edu.cn (B.L.); wangyu@nuc.edu.cn (Y.W.)
* Correspondence: baipeikang@nuc.edu.cn; Tel.: +86-0351-3925-618
† These authors contributed equally to this work.

Received: 10 June 2018; Accepted: 6 July 2018; Published: 9 July 2018

Abstract: A 3D Finite Element (FE) model was developed to investigate the thermal behaviour within the melt pool during point exposure to Selective Laser Melting (SLM) processed AlSi10Mg powder. The powder–solid transition, temperature-dependent thermal properties, melt pool convection, and recoating phase were taken into account. The effects of Exposure Time (ET) and Point Distance (PD) on SLM thermal behaviour were also investigated and showed that the short liquid phase time and high cooling rate of the melt pool reduced the viscosity of the melt pool at a lower ET or a higher PD. This resulted in poor wettability and the occurrence of balling and micropores. At a higher ET or lower PD the melt pool became unstable and allowed for easy formation of the self-balling phenomenon, as well as further partial remelting in the depth direction resulting in the creation of larger pores. The proper melt pool width (119.8 μm) and depth (48.65 μm) were obtained for a successful SLM process using an ET of 140 μs and a PD of 80 μm. The surface morphologies and microstructures were experimentally obtained using the corresponding processing conditions, and the results aligned with those predicted in the simulation.

Keywords: selective laser melting; numerical analysis; thermal behaviour; AlSi10Mg alloy

---

1. Introduction

Al-Si alloys have excellent properties such as low density, specific strength, stiffness, and electrical and thermal conductivity. Due to this, they have become one of the desirable lightweight materials in aerospace, automotive, microelectronics, and other mechanical instrument industries. The hypoeutectic alloy is close to the eutectic composition (12.5% Si) and therefore has excellent castability and weldability in comparison to other Al alloys. Hypoeutectic alloy AlSi10Mg is a fitting example in this regard. Owing to the combination of its light weight and good mechanical properties, this alloy has been widely used for satellites, space stations, and other space vehicles. Precipitation hardening also plays an important role in its performance, as it enables precipitating Mg_2Si during age hardening [1]. However, it is difficult to manufacture complex integrated components through conventional manufacturing technologies [2]. This restricts further development and applications of AlSi10Mg. Therefore, new processing methods are in high demand.

Selective Laser melting (SLM) is an emerging additive manufacturing process that provides substantial flexibility in the production of components with complex shapes [3]. In SLM, there are two different laser scan patterns: point exposure scan and continuous exposure scan. Both processes of the metallic powders are based on a complete melting/solidification mechanism, which indicates suitability in producing fully dense components approaching 99.9% relative density [4]. Moreover, the technology can produce near-net-shaped parts with customised and complicated structures in a direct way; it is extremely suitable for manufacturing the final part with almost complete geometric freedom. Despite its great economic and technological advantages, the technology still has some drawbacks in terms of processing stability, processing accuracy, and surface quality. The quality of the formed part largely depends on the stability of the melting process [5,6]. The temperature and stability of the melt pool play a crucial role in the SLM process, which directly affects the shape of the melt pool and the mechanical properties of the final parts. At present, part quality is determined mainly by optimising parameters (such as laser power and scanning speed). As trial and error requires a long period and is costly, the numerical simulation method is preferred to process optimisation [7]. Combining the predicted results of numerical simulation with a small amount of experimental optimisation will significantly reduce the cycle of production and costs.

There has been some recent valuable research aimed at simulating the SLM process. Li et al. [8] established a Finite Element (FE) model and studied the temperature fields during the SLM process with AlSi10Mg powder under different laser powers and scanning speeds. The results indicated that relatively higher scanning speeds produced a narrower melt pool width, causing large gaps and pores along the scanning direction. Following on, Loh et al. [9] proposed a FE model to achieve volume shrinkage and vaporisation in SLM-processed AA6061 parts. The results showed that laser power had a significant influence on the evaporation of molten materials because the heat in the centre of the laser beam could not spread out quickly. Yu et al. [4] established a 3D mesoscopic model using the Finite Volume Method so as to investigate the thermal behaviour of the melt pool during the SLM of AlSi10Mg powder under different laser powers and scanning speeds. It was found that both low and high laser power led to a poor surface quality. High volume energy density caused the balling phenomenon due to excessive liquid formation. Along similar lines, Han et al. [10] developed a FE model and simulated the transient temperature distribution and configurations of the melt pool during the SLM of Al–Al$_2$O$_3$ composite powder by changing the hatch spacing and scanning speed. The maximum temperature showed a slight increase along the scanning route; this trend was more obvious at low scanning speeds.

The abovementioned investigations into SLM-processed Al alloy powder are all focused on thermal behaviours caused by parameters which are all in respect of the laser scan pattern of the continuous exposure scan. Limited research exists into the point exposure process and defect prediction. During the continuous exposure scan process, a continuous laser travels over the surface of the powder with a constant scanning speed. While in the point exposure scan process, the material is heated intermittently with a succession of short-duration pulses, the laser moves step by step with a specific Exposure Time (ET) and Point Distance (PD). In general, the scanning speed is determined by the PD and ET in some literature [2,11]. In actual fact, the pulsed laser can easily melt powder with higher peak temperature. This kind of SLM process offers the advantage of very low heat input to the melt pool, which helps to deposit temperature-sensitive components with low distortion.

In this paper, a 3D FE model was introduced to simulate the thermal behaviour and melt pool dimensions of AlSi10Mg alloy parts in the process of point exposure scan and to predict the relationship between processing parameters and microstructure defects. Parametric analysis was developed under varying ET and PD. Moreover, corresponding experiments were carried out to study the microstructure of components produced by SLM, in order to verify the numerical results.

2. Model for SLM Process

2.1. Mathematical Modelling of Heat Transfer

Figure 1 depicts the schematic diagram of the SLM process. During the scan, the laser beam is irradiated on the top surface of the powder bed. Here, a series of complicated physical and chemical phenomena occur in the interaction process of the laser and powder materials, which includes two phases: (1) the reflectivity and absorption of the material's surface to the laser and (2) heat transfer between powder particles, the processed part and the surrounding environment [8].

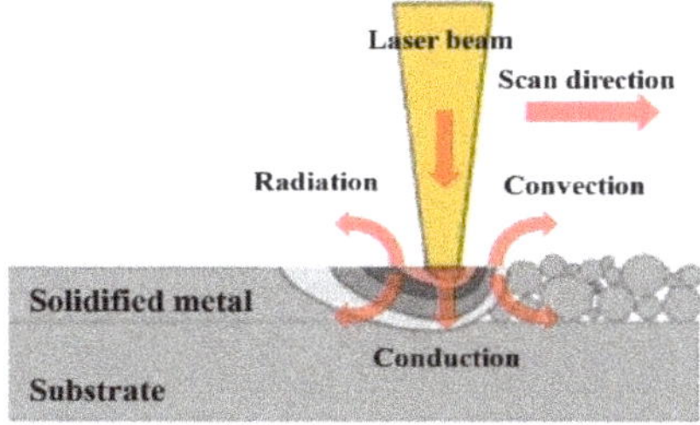

Figure 1. Schematic of thermal behaviour of powder bed under laser irradiation.

The powder is firstly melted by the laser beam, following which the metallurgic bonding is formed. In the process of heat transfer, it is assumed that the powder bed and its surrounding environment constitute a closed and insulated system. The energy balance follows the first law of thermodynamics, which includes thermal conduction, and the heat losses due to convection and radiation. In Cartesian coordinates, the spatial and temporal distribution of the temperature field of D can be represented by a three-dimensional heat transfer differential equation, which can be expressed as [12]:

$$\rho c \frac{\partial T}{\partial t} = k_p \left(\frac{\partial^2 T}{\partial x^2} + \frac{\partial^2 T}{\partial y^2} + \frac{\partial^2 T}{\partial z^2} \right) + Q, \quad (x,y,z \in D) \tag{1}$$

where ρ is the material density (kg/m^3); c is the specific heat capacity (J/(kg·°C)); T is the temperature (°C); t is the interaction time (s); k_p is the effective thermal conductivity of the powder bed; and Q is the heat generated per volume within the component (W/m^3).

Before the SLM process begins, the initial temperature, T_0, in the powder bed and substrate is equal to 25 and 80 °C respectively. The heat relationship between them follows the formula below:

$$k_p \frac{\partial T}{\partial n} = Q + Q_c + Q_r, \quad (x,y,z \in S) \tag{2}$$

where S denotes the surfaces attached to the applied heat flux, convection and radiation; n is the normal vector of the powder surface; Q is the input heat flux from the laser, which is described specifically in the following section; and Q_c is the heat convection and can be expressed as:

$$Q_c = h(T_s - T_m) \tag{3}$$

Q_r is the heat radiation and can be defined by:

$$Q_r = \sigma \varepsilon \left(T_s^4 - T_m^4 \right) \tag{4}$$

Since the heat radiation process is a highly non-linear process, Equations (3) and (4) can be substituted into Equation (2). The following equation is thus derived:

$$k_p \frac{\partial T}{\partial n} = H(T_s - T_m), (x,y,z \in S) \tag{5}$$

In Equations (3)–(5), h is the coefficient for heat convection; σ is the Stefan–Boltzmann constant; ε is the emissivity and equal to 0.3 [9]; T_s is the model surface temperature; T_m is the ambient temperature (25 °C); and H is the effective heat transfer coefficient.

2.2. Numerical Model Establishment

In order to align with the real operating conditions for the SLM process of AlSi10Mg, a 3D transient numerical model was developed using ANSYS 16.0 commercial FEM software. The FE model and the laser scanning pattern in the SLM process are shown in Figure 2. The model in Figure 2a includes an ENAW-5083 Al alloy substrate positioned underneath and an upside powder bed with dimensions of 1.6 mm × 0.9 mm × 0.3 mm and 1.2 mm × 0.5 mm × 0.075 mm, respectively. Figure 2b shows the movement of the laser in a reciprocating raster pattern. In order to improve the calculation accuracy of the model, the SOLID 70 hexahedron elements with a fine mesh of 0.01 mm × 0.01 mm × 0.0125 mm were used in the powder bed with a layer thickness 0.025 mm.

The laser beam had a spot size of 80 μm and was treated with a surface heat flux boundary condition in analysis; its energy density nearly followed a Gaussian function as shown in Figure 2a. The laser moved step by step with the specific ET and PD in the X-direction on the powder bed (named the modulated pulsed laser by Renishaw PLC, Wotton-under-Edge, UK). Parameter ET, represented by T_1, concerns the time for which each point is radiated by the laser before jumping to the next point, while the PD data indicates the distance between the adjacent points. The schematic diagram is shown in Figure 3. At a fixed PD value equal to 80 μm, the varying ET values were applied from 100 μs to 180 μs in increments of 20 μs. At a fixed ET value equal to 140 μs, the PD values varied from 60 μm to 100 μm in increments of 10 μm. The process parameters used in the study are shown in Table 1.

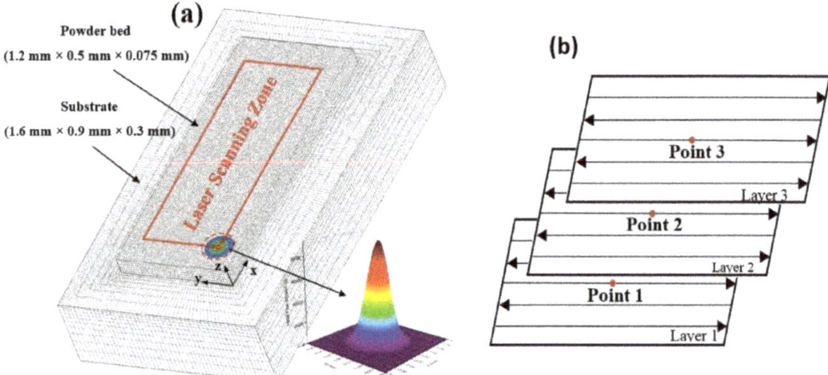

Figure 2. 3D Finite Element (FE) model and Gaussian laser energy density (**a**) and laser scanning pattern in Selective Laser Melting (SLM) process (**b**). (Point 1 at the centre of layer 1, Point 2 at the centre of layer 2 and Point 3 at the centre of layer 3 respectively).

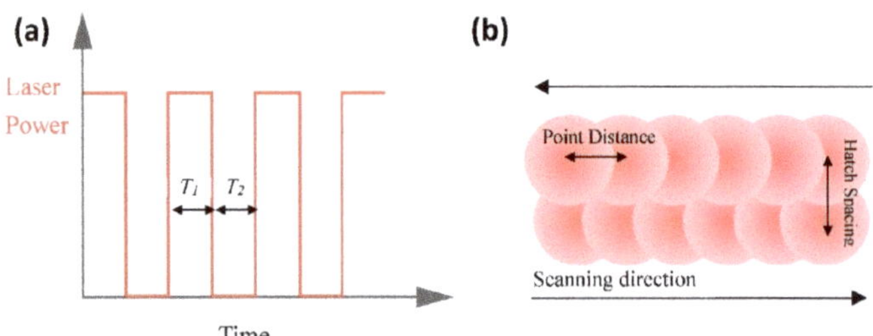

Figure 3. The waveform of the modulated pulsed laser power output (**a**) and the laser working mode (**b**).

Table 1. The process parameters in simulation and experiments.

Parameter	Value
Laser power, P	200 W
Hatching space, S	80 µm
Spot diameter, D	80 µm
Layer thickness	25 µm
Laser absorptivity of the Al powder, A	18% [13]
Spreading time	5 s
Exposure Time, ET	100, 120, 140, 160, 180 µs
Point Distance, PD	60, 70, 80, 90, 100 µm

2.3. Modelling of Laser Energy

During the SLM process, the material was molten through a laser beam. The laser intensity distribution almost followed a Gaussian relationship, which is mathematically expressed as:

$$Q(r) = \frac{2AP}{\pi R^2} exp\left(-\frac{2r^2}{R^2}\right) \tag{6}$$

where P is the laser power; r is the real-time distance to the centre of the laser spot; R is the effective radius of the heat source on the material surface where the energy density is reduced to $1/e^2$ at the centre of the laser spot; and A is the laser energy absorptivity of the material. The absorptivity of fibre laser radiation by a polished AlSi10Mg surface is around 9% at room temperature. However, the laser absorptivity of a powder material is 2–3 times higher than the counterpart block material. An absorptivity level of 12% for material AlSi10Mg is used by Ding et al. [14]. In the simulation, an absorptivity of 18% was used, as per Wei et al. [13]

During the SLM process, the latent heat of the phase change must not be ignored, and is usually treated by enthalpy, H (J/m^3), which can be written as a function of temperature, T, and the specific heat, C:

$$H = \int_{T_{ref}}^{T} \rho C dT \tag{7}$$

2.4. Thermal Physical Parameters

Temperature has a fundamental influence on the microscopic structures and mechanical properties formed during the radiation of the AlSi10Mg alloy powders by the laser in the SLM processing. The density, specific heat capacity, and thermal conductivity are indispensable for the purpose of obtaining the temperature field and the effective thermal conductivity of powders determines the validity of the simulation of SLM among them.

The temperature-dependent thermophysical parameters of AlSi10Mg are shown in Figure 4 [15]. The twofold thermal conductivity is used to account for melt pool convection beyond the melting temperature [16]. The effective thermal conductivity of powder k_p can be estimated as follows:

$$k_p = k_s \frac{pn}{\pi} x \tag{8}$$

where k_s is the thermal conductivity of solid AlSi10Mg; p is the relative density of the powder bed assigned at 0.6; n is the coordination number equal to 6; and x is the contact size ratio.

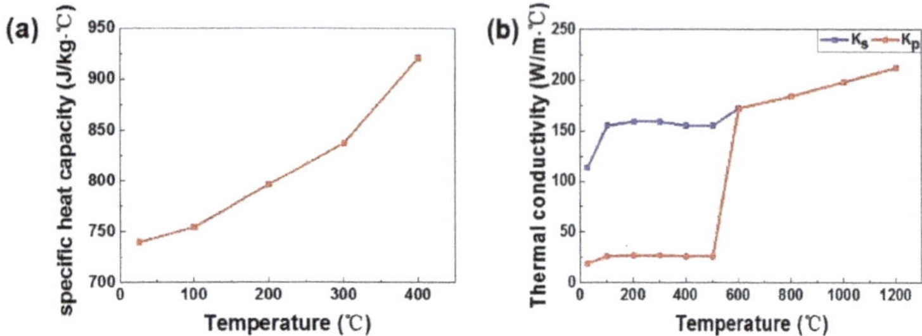

Figure 4. Temperature-dependent thermal properties: specific heat capacity (**a**) and the effective thermal conductivity of AlSi10Mg powder and solid (**b**).

3. Experiments

In this study, an AlSi10Mg powder with an average particle size of 45 μm, supplied by LPW Technology (LPW Technology Ltd., Widnes, Cheshire, UK) was used. A Renishaw AM 400 SLM system (Renishaw PLC, Wotton-under-Edge, UK) that employs a modulated ytterbium fibre laser with a wavelength of 1070 nm was utilised to fabricate cubic samples in order to verify the results of the simulation. The entire SLM process was also conducted in an argon atmosphere. The spot size of the laser beam focused on the powder bed was 80 μm. The process parameters employed were the same as those in the simulations (Table 1). Samples for metallographic examinations were cut, ground, and polished according to standard procedures, and then etched with Keller's reagent for 10 s. Mechanical polishing was adopted by means of a high-speed polishing round and the samples were polished up to 1.5 μm using diamond polishing paste. A ZEISS Scope. A1 OM microscope (Zeiss, Jena, Germany) was used to observe the low-magnification cross-sectional microstructures of specimens. The OM images were captured from the central area of the observation plane and magnified 100 times. The characteristic top surface morphologies of the SLM-fabricated AlSi10Mg parts were achieved using a Hitachi SU-5000 scanning electron microscopy (SU-5000 SEM, Hitachi Ltd., Chiyoda-ku, Tokyo, Japan) device in secondary electron mode at 20 kV.

4. Results and Discussion

4.1. Temperature Distribution

Figure 5 shows the transient temperature distribution of the top surface and the cross section of the melt pool at the centre of the different layers during the SLM process, using ET values of 140 μs and PD of 80 μm. We can see that the melt pool had an elliptical shape, was more intensive at the forepart of the ellipses, and also included a larger temperature gradient. This was primarily due to the thermal conductivity of the powder being less than that of the solid and heat loss which occurred mainly during solidification. The dotted black line represents the isotherm of the melting temperature of AlSi10Mg (600 °C), and the temperature gradient inside of the dotted lines area is larger than that in the surrounding area. This resulted in a fine cellular structure inside of the melt pool and a heat-affected zone around the melt pool was formed during the SLM process [17].

The predicted characteristics of the melt pool varied according to the mobile laser exposure. With the laser exposed at point 1, the maximum temperature reached 1758.53 °C with a width and depth of 114.5 μm and 44.54 μm respectively (Figure 5a,b). When the laser moved to point 2, the maximum temperature of the melt pool decreased to 1729.01 °C, while the width and depth increased to 116.7 μm and 46.46 μm, respectively (Figure 5c,d). However, as the laser reached point 3, the maximum temperature increased slightly to 1740.24 °C, with the width (119.8 μm) and depth (48.65 μm) increasing by 2.67% and 4.71%, respectively (Figure 5e,f). It was clear that, although the recoating phase provided sufficient cooling time, there was still a slight heat accumulation effect during the SLM process, thus leading to an increase in the melt pool dimensions. Meanwhile, the dimensions of the melt pool were larger than those in the continuous exposure scan. Ding et al. [14] explained that there are two vortexes in the velocity field, thus leading to intense convective motion. Moreover, the range in depth of the melt pool was greater than its range in width. This was because the thermal conductivity was higher in the previous layer than that of the adjacent tracks, and more heat was conducted in the depth direction.

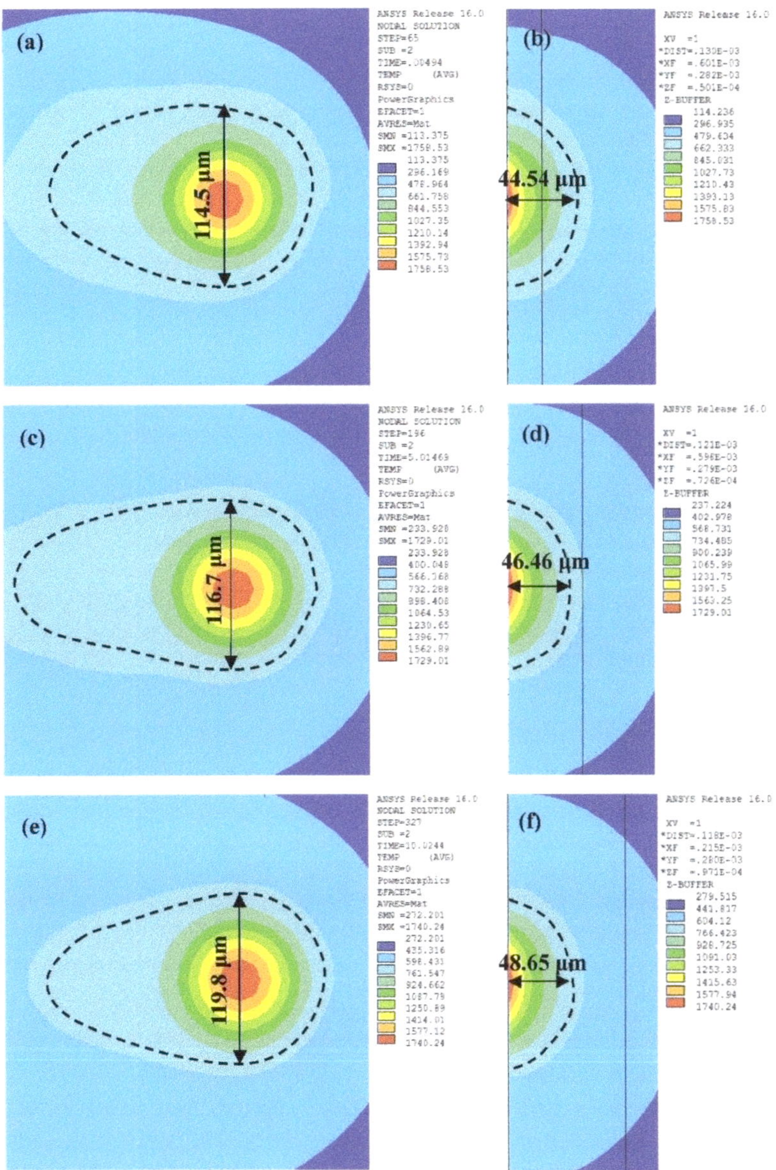

Figure 5. Temperature distributions of melt pool at Point 1 (**a,b**); Point 2 (**c,d**); and Point 3 (**e,f**) during SLM process at Exposure Time (ET) = 140 μs and Point Distance (PD) = 80 μm. The left is the top surface and the right is the cross-section.

4.2. Thermal Behaviours

Figure 6 shows the temperature variation with time at point 2 using different ET and PD values. The slope of the curve represents its cooling rate. As the ET increased from 100 μs to 180 μs, the cooling rate of the melt pool decreased from 7.93×10^6 °C/s to 3.61×10^6 °C/s. When increasing the PD from 60 μm to 100 μm the cooling rate rose significantly from 3.25×10^6 °C /s to 7.48×10^6 °C/s. Based on these results, it is easy to conclude that the cooling rate is more sensitive to the PD value than to that of the ET. By comparing this data with relevant literature on AlSi10Mg [8,18,19], which is built through the continuous exposure SLM, it is found that the higher cooling rate is in the point exposure SLM process. This contributes to obtaining fine grain. The temperature variation showed an apparent fluctuation behaviour and each peak represented the laser beam passing through. Each fluctuation is a scanning cycle of the track. The breaks on the x-axis represent the recoating phase in SLM (the recoating time is equal to 5 s). It is noted that the cooling rate in this phase rose with increasing ET and decreasing PD values. The differences between the cooling rates of the melt pool were due to an increased heat accumulation in the underlying solid material; therefore, more time is required to cool down. When the laser arrived at point 2, the temperature of the third peak is the highest. The powder in the area directly under the laser melted in a shorter time, helping towards fine-grain strengthening [20,21]. With the laser leaving, the region experienced a rapid fall in temperature and underwent solidification. After the third powder was spread, the region experienced a similar variation from the first track to the fifth track as the laser scanned in the same way. When the laser reached the centre of layer 3 (point 3), the temperature of point 2 was over the melting line, thus indicating that the remelting phenomenon occurred in layer 2 and metallurgical bonding took place between the adjacent layers.

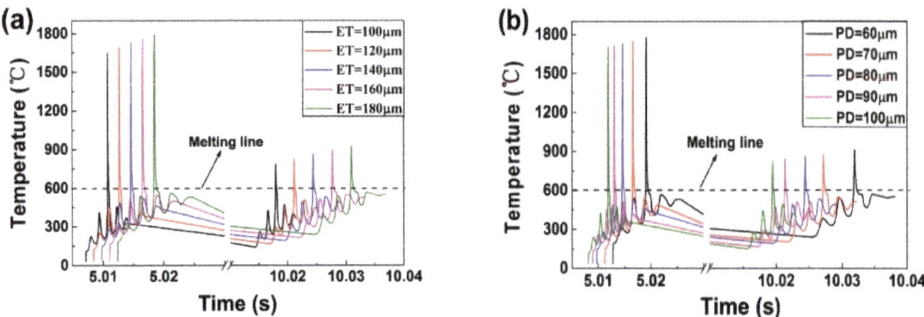

Figure 6. The temperature variation with time at point 2. Using different ET (PD = 80 μm) (a) and different PD (ET = 140 μs) (b).

Figure 7 shows the maximum temperature and the liquid lifetime at point 2 with the varying process. It can clearly be seen that the maximum temperature increased from 1692.06 °C to 1761.64 °C when the ET increased from 120 μs to 160 μs. Meanwhile, the liquid lifetime also increased from 287.56 μs to 424.28 μs, rising by 4.11% and 47.54%, respectively (Figure 7a). At the same time, as the PD increased from 60 μm to 100 μm, the maximum temperature decreased from 1777.75 °C to 1703.02 °C, decreasing by 4.20%; moreover, the liquid lifetime shortened from 502.41 μs to 269.15 μs, decreasing by 46.43% (Figure 7b). As noticed, the PD had a more pronounced impact on the liquid lifetime under the same amplitudes of variation, and this finding has a primary direct meaning for process optimisation.

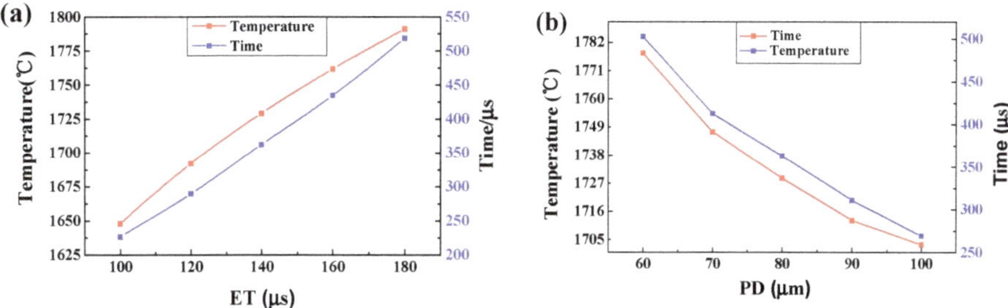

Figure 7. The maximum temperature and the liquid lifetime of the melt pool at point 2. Using different ET (PD = 80 μm) (**a**) and different PD (ET = 140 μs) (**b**).

In the SLM process, the maximum temperature, the cooling rate and the liquid lifetime of the melt pool depend on the applied ET and PD, which determine the formation of a stable melt pool and good metallurgical bonding. Poor melt pool configurations can reduce the density of parts and can even mean that the forming does not finish. It is therefore vital to ascertain the appropriate ET and PD in order to achieve a stable melt pool and smooth overlap.

4.3. Melt Pool Configurations

The melt pool shape in layer 3 has reached a steady state (Figure 5). Figure 8 shows temperature variation with different ET and PD values when the laser moved to point 3. The temperature gradients at the centre of the melt pool were much larger than those in the edge area. To improve the precision of the data, the data points with spacing of 0.5 μm were acquired using linear interpolation via the ANSYS software. As the ET increased from 100 μs to 180 μs, the width and depth increased from 107.2 μm to 134 μm and from 39.8 μm to 57.5 μm, increasing by 25% and 44.47%, respectively. The corresponding width-to-depth ratio decreased from 2.69 to 2.33, dropping by 13.38% (Figure 9a). As the PD increased from 60 μm to 100 μm, the width and depth decreased from 131.9 μm to 114 μm and from 56.5 μm to 44.5 μm, decreasing by 13.57% and 21.24%, respectively. The corresponding width-to-depth ratio increased from 2.33 to 2.56, rising by 9.87% (Figure 9b). It can be seen that both ET and PD had a more notable effect in the depth direction, which was due to the solidified layers with higher heat conductivity. Besides this, the width and depth were more sensitive to the PD than to the ET. These results were also found by Ding et al. [7]. The aforementioned authors postulated that the overall temperature around the portion of laser scanning is elevated when a larger PD is applied, which supports heat transfer in the horizontal direction. Moreover, in their study, it was found that, with regard to the ET, changing the ET from 20 μm to 280 μm had a negligible effect on heat transfer. Li et al. [8] concluded that higher laser power will penetrate deeper in the SLM process of AlSi10Mg and the morphology of melt pool is more sensitive to laser powder compared with scanning speed. The same was concluded by Zhang et al. [22]. Along similar lines, Cherry et al. [11] studied the effect of ET and PD on the quality of 316L stainless steel. They found that ET had little effect on roughness, whilst variation in the PD caused a large change in roughness.

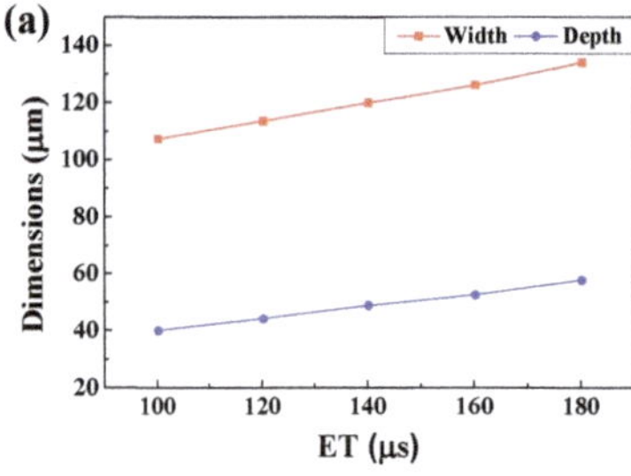
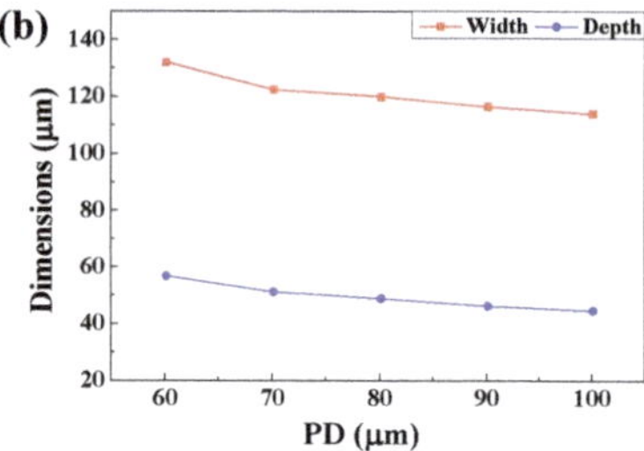

Figure 8. The width and depth of melt pool with different ET (PD = 80 μs) (**a**) and different PD (ET = 140 μs) (**b**).

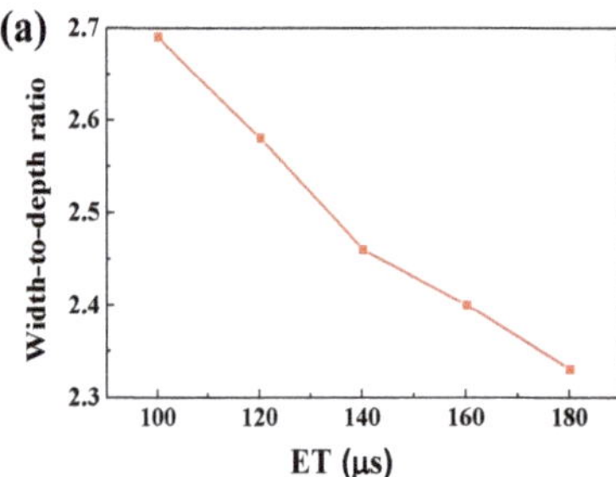
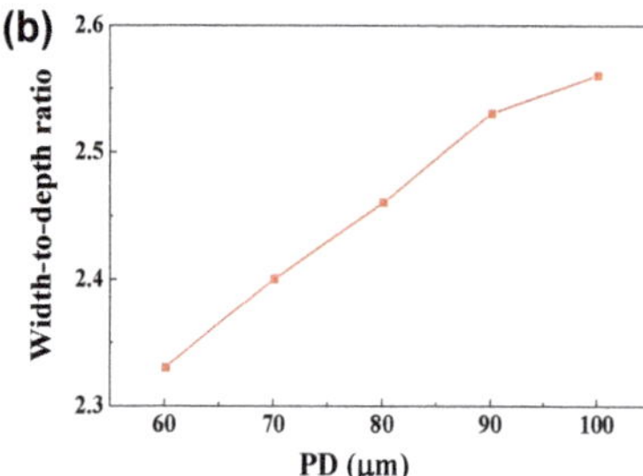

Figure 9. The ratio of width to depth of the melt pool with different ET (PD = 80 μm) (**a**) and different PD (ET = 140 μs) (**b**).

As the ET increased and the PD reduced, except for the enlarged melt pool with the increasing laser energy density, the width-to-depth ratio was also enhanced. The same trend in the length-to-depth ratio was obtained by elevating the laser power and reducing the scanning speed in a continuous exposure scan pattern [4,8]. This is because that the melt pool overlap conditions transformed from the conduct mode to the keyhole mode. Under such circumstances, the greater the amount of the molten powder that evaporated, the deeper the melt pool became, and this resulted in a decreasing width-to-depth ratio. Stable melt pool dimensions are vital for forming an appropriate overlap between tracks. When the overlapping ratio is around 30% it ensures strong overlapping between melt pools and helps to obtain a smooth surface. In this study, the overlapping ratio was approximately 31.25% when the ET and PD were assigned to 140 μs and 80 μm, respectively. The corresponding laser energy density was 175 J/mm^3, which is less than the optimised value (250 J/mm^3) in the continuous exposure scan in Yu et al.'s works [8]. This is attributed to the fact that the pulsed laser has a higher peak temperature and, thus, can more easily melt powder in low energy density. However, the predicted results are in accordance with the optimal process of the point exposure scan SLM conducted by Wang et al. [2] and Aboulkhair et al. [23].

Good metallurgic bonding between melt pools is vital when it comes to decreasing the pores in the workpiece and to improving processing quality [24,25]. Therefore, under good overlapping conditions, a larger melt pool is important in order to obtain highly dense products. However, when the size of the melt pool exceeds the threshold level the melt pool becomes unstable, thus causing micropores, holes, cracks and so on. Due to the excessive energy density, a high temperature gradient is formed and causes the large thermal stress in the melt pool. At the initial position of the scan track, the cooling rate and the solidification rate remain high because of the low overall temperature. A hot crack easily leads to warpage, which has been verified for SLM-processed iron powder by Li et al. [26].

4.4. Experimental Investigation

In order to verify modelling results, corresponding experiments were carried out. The layered structures of the SLMed AlSi10Mg samples can be observed in Figure 10; furthermore, at the same PD value, we can also see that both the depth and the width of the melt pools both increased with an increasing ET. In Figure 10a, plenty of spherical gas pores (<20 μm) and an irregular lack of fusion pores (>100 μm) can be seen. These were induced by low input energy densities. The short liquid phase time and the high cooling rate of the melt pools reduced the viscosity of said melt pools (Figure 7a) and, as such, it was not possible to fully wet the powder particles around the melt pool. Therefore, some micropores and unmelted areas remained after alloy solidification, thus leading to a low relative density. When the ET was equal to 140 μs, the melt pool was in a good overlap due to large dimensions and flowability. As a result of longer liquid phase times and lower cooling rate, some bubbles escaped from the melt pool under Marangoni flow. Therefore, few gas pores (<20 μm) were observed (Figure 10b). However, when the laser ET reached 180 μs, the number of pores (<50 μm) in the melt pool began to increase (Figure 10c). Xiao et al. found that the gas solubility of the melt pool increased with increasing energy density during laser welds of the Al–Li alloy. In SLM, gas including evaporative magnesium [27], hydrogen [28], and argon will dissolve out but be detained in the melt pool under the condition of a low solidification rate; moreover, gas pores diffuse and expand easily, and larger keyhole pores are induced. Weingarten et al. [29] concluded that each time the following layer was scanned, the consolidated material was heat-treated if the penetration depth led to a more partial remelting at high energy density; as such, it is possible to conclude that trapped gas pores can enlarge and reduce the relative density. Besides this, a large of particles adhered to the melt pool or the evaporation of local material due to excessive laser energy input; as a result, a lack of powder and keyhole pores can also be caused in the neighbouring region.

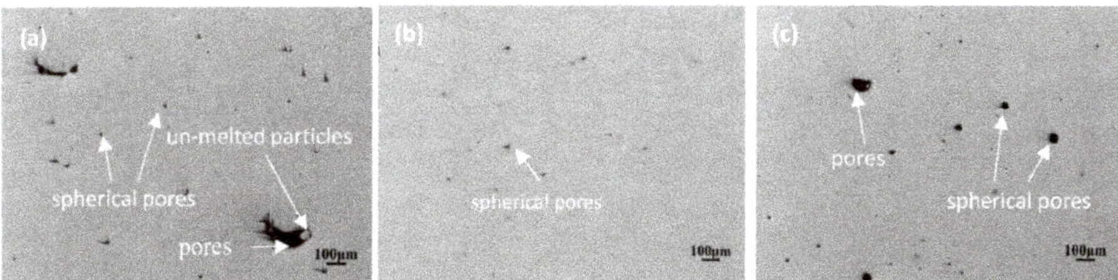

Figure 10. OM images of the cross-section microstructures of AlSi10Mg parts fabricated at different ET: (**a**) 100 μs; (**b**) 140 μs and (**c**) 180 μs. The PD is fixed at 80 μm.

Figure 11 shows the top surface morphologies of AlSi10Mg parts produced by different PD values, with an ET fixed at 140 μs. It can be clearly seen that variation exists in these SEM images. As shown in Figure 11a, obvious "self–balling" can be seen at a PD of 60 μm. This was caused by the amount of energy density, which increased the liquid phase time and the maximum temperature of the melt pool. Here, an excessive liquid phase induced the formation of the balling phenomenon, with balling inevitably leading to pores. Besides, plenty of micrometre-scaled balls can be seen in between melt tracks. This was due to melt splashes from inside of the melt pool inevitably splattering on its edges under the action of the laser. When the PD was equal to 80 μm, a smooth surface was obtained (Figure 11b) and there were fewer pores and better properties in the cross section (Figures 10b and 11b). However, when further increasing the PD to 100 μm, apparent balling occurred (Figure 11c). As previously stated, at a high PD value, the lower maximum temperature gave rise to small amounts of liquid formation, with higher cooling rates and a lower liquid lifetime and viscosity (Figures 6 and 7). It can therefore be concluded that, the liquid metal transforms into near balling in a short period of time and will decrease the relative density and increase surface roughness. While increasing the laser power or decreasing the scanning speed, the same variation trend in respect of the balling phenomenon was found by Li et al. [8] and Han et al. [30].

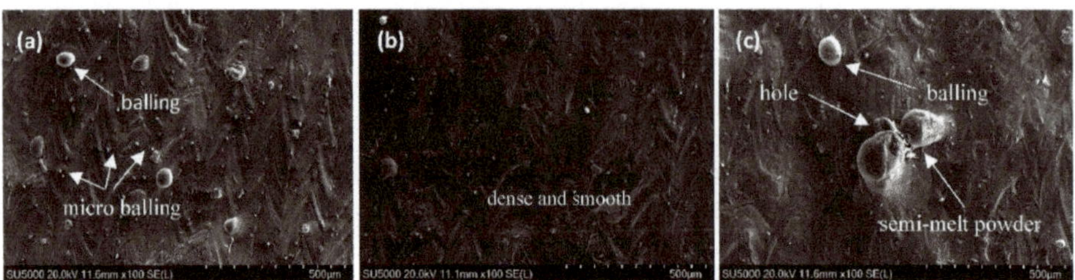

Figure 11. SEM images of the top surface morphologies of AlSi10Mg parts fabricated at different PD: (a) 60 µm; (b) 80 µm; and (c) 100 µm. The ET is fixed at 140 µs.

In a track-by-track, layer-by-layer SLM process, the ability of metallurgical bonding plays an important role in the densification behaviour of the parts produced by SLM. Too much or too little energy input can cause serious pores and balling in parts, which ultimately degrades performance.

5. Conclusions

(1) As the laser moved from the centre of the first layer to the centre of the third layer, the maximum temperature of the melt pool decreased firstly and then increased slightly. The width and the depth of the melt pool increased continuously but in small amounts. This is mainly due to the heat accumulation phenomenon although the recoating phase.

(2) The cooling rate of the melt pool decreased from 7.93×10^6 °C/s to 3.61×10^6 °C/s as the ET increased from 100 µs to 180 µs. However, the rate rose significantly from 3.25×10^6 °C/s to 7.48×10^6 °C/s as the PD increased from 60 µm to 100 µm. The cooling rate during the recoating phase elevated as the ET increased and the PD decreased. This was due to increasing heat accumulation in the underlying solid material.

(3) The maximum temperature and the liquid lifetime rose as the ET increased and the PD decreased. However, the PD had more notable effects on the liquid lifetime.

(4) The dimensions of the melt pool increased (the width from 107.2 µm to 134 µm and the depth from 39.8 µm to 57.5 µm) as the ET elevated from 100 µs to 180 µs, however, said dimensions decreased (the width from 131.9 µm to 114 µm and the depth from 56.5 µm to 44.5 µm) as the PD elevated from 60 µm to 100 µm. It can be seen that the depth and the width were more sensitive to the PD than to the ET. The proper melt pool width (119.8 µm) and depth (48.65 µm) were obtained for a successful SLM process with a combination of ET = 140 µs and PD = 80 µm.

(5) The best forming quality—free of apparent pores, holes, cracks, and the balling phenomenon—was obtained at the optimised combination of ET = 140 µs and PD = 80 µm.

Author Contributions: P.B. conceived the idea of this study. Z.L. and B.-Q.L. built and validated the finite element model. B.-Q.L. and B.L. performed the experiments. B.-Q.L. prepared the manuscript and Z.L. and Y.W. revised it. All authors discussed the results and commented on the manuscript.

Funding: This work was funded by the National Natural Science Foundation of China (61475056 and 51701185).

Acknowledgments: The authors thank the Modern Analytical and Testing Center of NUC for the experimental activities.

Conflicts of Interest: The authors declare no conflicts of interest.

References

1. Fousová, M.; Dvorský, D.; Michalcová, A.; Vojtěch, D. Changes in the microstructure and mechanical properties of additively manufactured AlSi10Mg alloy after exposure to elevated temperatures. *Mater Charact.* **2018**, *137*, 119–126. [CrossRef]
2. Wang, L.; Wang, S.; Wu, J. Experimental investigation on densification behavior and surface roughness of AlSi10Mg powders produced by selective laser melting. *Opt. Laser Technol.* **2017**, *96*, 88–96. [CrossRef]
3. Almangour, B.; Grzesiak, D.; Cheng, J.; Ertas, Y. Thermal behavior of the molten pool, microstructural evolution, and tribological performance during selective laser melting of TiC/316L stainless steel nanocomposites: Experimental and simulation methods. *J. Mater. Process. Technol.* **2018**, *257*, 288–301. [CrossRef]
4. Yu, G.; Gu, D.; Dai, D.; Xia, M.; Ma, C.; Shi, Q. On the role of processing parameters in thermal behavior, surface morphology and accuracy during laser 3D printing of aluminum alloy. *J. Phys. D Appl. Phys.* **2016**, *49*. [CrossRef]
5. Ilin, A.; Logvinov, R.; Kulikov, A.; Prihodovsky, A.; Xu, H.; Ploshikhin, V.; Günther, B.; Bechmann, F. Computer Aided Optimisation of the Thermal Management During Laser Beam Melting Process. *Phys. Procedia* **2014**, *56*, 390–399. [CrossRef]
6. Dai, D.; Gu, D. Tailoring surface quality through mass and momentum transfer modeling using a volume of fluid method in selective laser melting of TiC/AlSi10Mg powder. *Int. J. Mach. Tools Manuf.* **2015**, *88*, 95–107. [CrossRef]
7. Ding, X.; Wang, L. Heat transfer and fluid flow of molten pool during selective laser melting of AlSi10Mg powder: Simulation and experiment. *J. Manuf. Proc.* **2017**, *26*, 280–289. [CrossRef]
8. Li, Y.; Gu, D. Parametric analysis of thermal behavior during selective laser melting additive manufacturing of aluminum alloy powder. *Mater. Des.* **2014**, *63*, 856–867. [CrossRef]
9. Loh, L.E.; Chua, C.K.; Yeong, W.-Y.; Song, J.; Mapar, M.; Sing, S.L.; Liu, Z.H.; Zhang, D.Q. Numerical investigation and an effective modelling on the Selective Laser Melting (SLM) process with aluminium alloy 6061. *Int. J. Heat Mass Transf.* **2015**, *80*, 288–300. [CrossRef]
10. Han, Q.; Setchi, R.; Lacan, F.; Gu, D.; Evans, S.L. Selective laser melting of advanced Al-Al2O3 nanocomposites: Simulation, microstructure and mechanical properties. *Mater. Sci. Eng. A* **2017**, *698*, 162–173. [CrossRef]
11. Cherry, J.A.; Davies, H.M.; Mehmood, S.; Lavery, N.P.; Brown, S.G.R.; Sienz, J. Investigation into the effect of process parameters on microstructural and physical properties of 316L stainless steel parts by selective laser melting. *Int. J. Adv. Manuf. Technol.* **2015**, *76*, 869–879. [CrossRef]
12. Hussein, A.; Hao, L.; Yan, C.; Everson, R. Finite element simulation of the temperature and stress fields in single layers built without-support in selective laser melting. *Mater. Des.* **2013**, *52*, 638–647. [CrossRef]
13. Pei, W.; Wei, Z.; Zhen, C.; Li, J.; Zhang, S.; Du, J. Numerical simulation and parametric analysis of selective laser melting process of AlSi10Mg powder. *Appl. Phys. A* **2017**, *123*, 540. [CrossRef]
14. Ding, X.; Wang, L.; Wang, S. Comparison study of numerical analysis for heat transfer and fluid flow under two different laser scan pattern during selective laser melting. *Optik Int. J. Light Electron Opt.* **2016**, *127*, 10898–10907. [CrossRef]
15. Pan, F.; Zhang, D. *Aluminum Alloys and Their Application*; Chemical Industry Press: Beijing, China, 2006.
16. Safdar, S.; Pinkerton, A.J.; Li, L.; Sheikh, M.A.; Withers, P.J. An anisotropic enhanced thermal conductivity approach for modelling laser melt pools for Ni-base super alloys. *Appl. Math. Model.* **2013**, *37*, 1187–1195. [CrossRef]
17. Thijs, L.; Kempen, K.; Kruth, J.-P.; Van Humbeeck, J. Fine-structured aluminium products with controllable texture by selective laser melting of pre-alloyed AlSi10Mg powder. *Acta Mater.* **2013**, *61*, 1809–1819. [CrossRef]
18. Tang, M.; Pistorius, P.C.; Narra, S.; Beuth, J.L. Rapid Solidification: Selective Laser Melting of AlSi10Mg. *JOM* **2016**, *68*, 960–969. [CrossRef]
19. Liu, Y.J.; Liu, Z.; Jiang, Y.; Wang, G.W.; Yang, Y.; Zhang, L.C. Gradient in microstructure and mechanical property of selective laser melted AlSi10Mg. *J. Alloys Compd.* **2017**. [CrossRef]
20. Wu, J.; Wang, X.Q.; Wang, W.; Attallah, M.M.; Loretto, M.H. Microstructure and strength of selectively laser melted AlSi10Mg. *Acta Mater.* **2016**, *117*, 311–320. [CrossRef]
21. Zhao, Z.; Guan, R.; Zhang, J.; Zhao, Z.; Bai, P. Effects of Process Parameters of Semisolid Stirring on Microstructure of Mg-3Sn-1Mn-3SiC (wt %) Strip Processed by Rheo-rolling. *Acta Metall. Sin. Engl. Lett.* **2017**, *30*, 66–72. [CrossRef]

22. Zhang, B.; Dembinski, L.; Coddet, C. The study of the laser parameters and environment variables effect on mechanical properties of high compact parts elaborated by selective laser melting 316L powder. *Mater. Sci. Eng. A* **2013**, *584*, 21–31. [CrossRef]

23. Aboulkhair, N.T.; Maskery, I.; Tuck, C.; Ashcroft, I.; Everitt, N.M. Improving the fatigue behaviour of a selectively laser melted aluminium alloy: Influence of heat treatment and surface quality. *Mater. Des.* **2016**, *104*, 174–182. [CrossRef]

24. Di, W.; Yang, Y.; Su, X.; Chen, Y. Study on energy input and its influences on single-track, multi-track, and multi-layer in SLM. *Int. J. Adv. Manuf. Technol.* **2012**, *58*, 1189–1199. [CrossRef]

25. Cao, J.; Liu, F.; Lin, X.; Huang, C.; Chen, J.; Huang, W. Effect of overlap rate on recrystallization behaviors of Laser Solid Formed Inconel 718 superalloy. *Opt. Laser Technol.* **2013**, *45*, 228–235. [CrossRef]

26. Li, J.; Shi, Y.; Lu, Z.; Huang, S. Numerical simulation of transient temperature field in selective laser melting. *Chin. Mech. Eng.* **2008**, *19*, 2492–2495. [CrossRef]

27. Haboudou, A.; Peyre, P.; Vannes, A.B.; Peix, G. Reduction of porosity content generated during Nd:YAG laser welding of A356 and AA5083 aluminium alloys. *Mater. Sci. Eng. A* **2003**, *363*, 40–52. [CrossRef]

28. Wu, H.; Ren, J.; Huang, Q.; Zai, X.; Liu, L.; Chen, C.; Liu, S.; Yang, X.; Li, R. Effect of laser parameters on microstructure, metallurgical defects and property of AlSi10Mg printed by selective laser melting. *J. Micromech. Mol. Phys.* **2018**, *02*, 1750017. [CrossRef]

29. Weingarten, C.; Buchbinder, D.; Pirch, N.; Meiners, W.; Wissenbach, K.; Poprawe, R. Formation and reduction of hydrogen porosity during selective laser melting of AlSi10Mg. *J. Mater. Process. Technol.* **2015**, *221*, 112–120. [CrossRef]

30. Han, X.; Zhu, H.; Nie, X.; Wang, G.; Zeng, X. Investigation on Selective Laser Melting AlSi10Mg Cellular Lattice Strut: Molten Pool Morphology, Surface Roughness and Dimensional Accuracy. *Materials* **2018**, *11*, 392. [CrossRef] [PubMed]

 *materials*

Article

Effect of Selective Laser Melting Process Parameters on the Quality of Al Alloy Parts: Powder Characterization, Density, Surface Roughness, and Dimensional Accuracy

Ahmed H. Maamoun *, Yi F. Xue, Mohamed A. Elbestawi * and Stephen C. Veldhuis

Department of Mechanical Engineering, McMaster University, 1280 Main Street,
West Hamilton, ON L8S 4L7, Canada; xueyf4@mcmaster.ca (Y.F.X.); veldhu@mcmaster.ca (S.C.V.)
* Correspondence: maamouna@mcmaster.ca (A.H.M.); elbestaw@mcmaster.ca (M.A.E.)

Received: 26 October 2018; Accepted: 20 November 2018; Published: 22 November 2018

Abstract: Additive manufacturing (AM) of high-strength Al alloys promises to enhance the performance of critical components related to various aerospace and automotive applications. The key advantage of AM is its ability to generate lightweight, robust, and complex shapes. However, the characteristics of the as-built parts may represent an obstacle to the satisfaction of the parts' quality requirements. The current study investigates the influence of selective laser melting (SLM) process parameters on the quality of parts fabricated from different Al alloys. A design of experiment (DOE) was used to analyze relative density, porosity, surface roughness, and dimensional accuracy according to the interaction effect between the SLM process parameters. The results show a range of energy densities and SLM process parameters for AlSi10Mg and Al6061 alloys needed to achieve "optimum" values for each performance characteristic. A process map was developed for each material by combining the optimized range of SLM process parameters for each characteristic to ensure good quality of the as-built parts. This study is also aimed at reducing the amount of post-processing needed according to the optimal processing window detected.

Keywords: additive manufacturing; selective laser melting; AlSi10Mg; Al6061; SLM process parameters; performance characteristics

---

1. Introduction

High-strength aluminum alloys (Al alloys) are typically used for the production of lightweight critical components for a variety of applications in space, aerospace, automotive, military, and biomedical fields [1]. Additive manufacturing (AM) offers additional flexibility in the design and manufacturing of parts, particularly the ability to fabricate complex geometries without the need for custom tools [2]. Selective laser melting (SLM) offers superior dimensional accuracy and material quality of the fabricated parts [3].

SLM is a layer-by-layer process, in which the laser beam selectively melts the powder layer according to slices generated from the three-dimensional designed model. SLM possesses rapid melting and solidification rates and thus is applicable for a narrow selection of materials according to their coefficient of thermal expansion (CTE). In addition, the optimization of the SLM process parameters of Al alloys is hampered by part defects due to energy loss in the laser beam projected to the powder bed surface. The quality of Al alloys produced by SLM could be influenced by the chemical composition and CTE of the material used. Galy et al. [4] showed that porosity, hot cracking, anisotropy, and surface quality are the principal defects of Al alloy parts. They also demonstrated that the selection of SLM process parameters and the laser beam energy loss due to Al reflectivity are the primary causes of porosity and hot crack formation. The pre-mixing of Al alloy powder with some composite materials might improve the material properties of the SLM as-built parts [5,6]. However, the quality and material properties of Al alloy parts can be customized according to the SLM parameters selected without pre-mixing with other elements.

Some of the SLM process parameters can be controlled, such as laser power, scan speed, hatch spacing, and powder layer thickness. The energy density is a function of these parameters. Optimization of the SLM process parameters is an essential step for controlling material characteristics and the quality of the fabricated parts. Sufiiarov et al. [7] showed that a 30 µm powder layer thickness could result in higher strength and lower elongation for Inconel 718 than a 50 µm layer thickness. Nguyen et al. [8] also studied the effect of the powder layer thickness within a range from 20 to 50 µm. Their results showed that, as the thickness of the powder layer diminishes, part density and dimensional accuracy increase. Cheng et al. [9] investigated the effect of the scanning strategy on the stress and deformation of parts. Their results showed that minimum stress and deformation values are obtained using a layer orientation strategy with an angle of 45° or 67°. The powder feedstock quality also represents an essential parameter that might affect part characteristics. Sutton et al. [10] reported that the powder morphology, microstructure, and chemical characteristics could change depending on the production method, i.e., gas, water, or plasma atomization. This could generate a difference in quality between the parts produced using different feedstock powders [11,12].

Various studies [13–17] utilized a design of experiment (DOE) approach to investigate the effect of SLM process parameters on AlSi10Mg part quality, by evaluating their density, surface roughness, and dimensional accuracy. Read et al. [14] used the response surface methodology (RSM) to evaluate the influence of SLM process parameters on part porosity. Their study was limited by the use of a laser power up to 200 W. The results showed that minimum porosity was obtained at a critical energy density of 60 J/mm^3. Abouelkhair et al. [15] used the one-factor-at a-time (OFAT) method to optimize the SLM process parameters for producing dense parts. They achieved an optimum combination of laser power, scan speed, and hatch spacing, which resulted in a 99.77% relative density. Hitzler et al. [17] demonstrated that the surface roughness of the as-built samples varies according to their position on the build plate. They also concluded that the increase of energy density resulted in higher values of roughness on the surface of the side faces, compared to the roughness values measured on the top surface. Calignano et al. [16] used the Taguchi method to investigate the effect of the SLM process parameters on the surface roughness of the parts. They found that the laser scan speed has a significant influence on the surface roughness. Lower surface roughness was obtained using a scan speed of 900 mm/s, 120 W laser power, and 0.1 mm hatch spacing. Han et al. [18] reported that a decrease in surface roughness, combined with an increase of the laser scan speed, results in better dimensional accuracy. It is worthwhile to note that the previous studies used the DOE within a range of laser power up to 200 W. However, the post-processing treatment is also considered to be an essential stage for reducing the defects inside the as-built parts. This, in turn, raises the final production cost of the parts [19–21]. Consequently, the optimization of the SLM process parameters has a significant role in optimizing the steps of the manufacturing process. This might lead to a cost-effective process for specific applications which are compatible with the characteristics of the as-built parts.

In general, Al6061 is seldom used for SLM. Fulcher et al. [22] reported that Al6061 parts have a lower dimensional accuracy compared to AlSi10Mg parts because of a higher CTE. High-strength Al alloys such the Al6061 and Al7075 series have low Mg and Si content, which might result in hot cracking and formation of large columnar grains [23]. Louvis et al. [24] reported that low-relative-density parts of Al6061 might be produced via SLM because of the effect of oxide formation inside these parts. This might result from the relatively low laser power used, (100 W) which may not be enough to achieve complete melting. In general, more research is required to evaluate the effect of SLM process parameters on the as-built Al6061 characteristics such as density, surface roughness, and dimensional accuracy. In addition, the effect of Si content requires further investigation aimed at optimizing the process parameters.

In this study, a comprehensive experimental work using the DOE approach was performed to evaluate the influence of the SLM process parameters on the quality of as-built Al alloys. The current work focuses on investigating the density, surface topology, and dimensional accuracy of AlSi10Mg and Al6061. SLM process parameters were selected over a wide range of laser power, scanning speed, and hatch spacing values. Part characteristics were evaluated for various SLM parameters to develop a process map which displayed the effect of Si content on part quality. Maamoun et al. [25] also covered the impact of the SLM process parameters on the microstructure and mechanical properties of the same Al alloys. This work aims to investigate the limits of SLM in fabricating critical components for the aerospace industry using these alloys. In particular, the current research is focused on producing high-quality metallic optics and optomechanical components to improve the performance of telescopes and laser systems.

2. Experimental Procedure

2.1. Material

Powder characterization was performed according to ASTM F3049-14. The powders' chemical composition was evaluated using Energy X-ray dispersive Spectroscopy (EDS) equipped in TESCAN VP Scanning Electron Microscope (SEM). The powder size distribution (PSD) was measured using laser diffraction by dispersing the powder in water. PSD is attributed by D (α), which represents the diameter of the measured particle, where α is the particles volume percentage that have a smaller diameter than the D value. The powder morphology was investigated using the SEM instrument. A Bruker D8 DISCOVER diffractometer equipped with a cobalt sealed tube source and an area detector was used to obtain the X-ray diffraction (XRD) phase pattern for both powders.

2.2. Design of Experiment

A DOE was developed to evaluate the response of the SLM process parameters and the volumetric energy density with respect to the as-built parts' quality. The volumetric energy density is defined as follows:

$$E_d = \frac{P}{V_s * D_h * T_l} \tag{1}$$

where E_d is the energy density (J/mm^3), P is the laser beam power (W), V_s is the laser scan speed (mm/s), D_h is the hatch spacing between scan passes, and T_l is the deposited layer thickness (μm). The OFAT method was used to analyze the performance of AlSi10Mg samples. Eight different samples were produced with six replications for each. Several SLM parameters were selected to build the AlSi10Mg samples, as listed in Table 1, with a constant layer thickness of 30 μm. The effect of the laser power, scan speed, hatch spacing, and energy density on the as-built part characteristics were evaluated with regression analysis.

Table 1. The selective laser melting (SLM) process parameters used for building the AlSi10Mg samples. P: laser beam power, V_s: laser scan speed, D_h: hatch spacing between scan passes, E_d: energy density.

Sample #	P (W)	V_s (mm/s)	D_h (mm)	E_d (J/mm^3)
AS1	370	1000	0.19	65
AS2	370	1300	0.15	63.2
AS3	370	1300	0.19	50
AS4	350	1300	0.19	47.2
AS5	370	1500	0.19	43.3
AS6	300	1300	0.19	40.5
AS7	370	1300	0.25	38
AS8	200	1300	0.19	27

A full factorial DOE was developed using the response surface over a wide range of SLM parameters. Two sets of three SLM parameters (laser power, scanning speed, and hatch spacing) were selected, as presented in Table 2. Three samples for each SLM parameters group were fabricated for a total of 48 samples. The energy density (E_d) for the Al6061 study was selected within a higher range (40–125 J/mm^3) compared to the E_d used for AlSi10Mg (27–65 J/mm^3). This was due to the higher reflectivity of the laser power for Al6061, which resulted in less energy absorption by the powder particles. The overlap of SLM parameters for some samples of AlSi10Mg and Al6061 enabled the investigation of each material at equal parameters. The correlation coefficient (R^2) was used to indicate how the regression models fit with the measured data, and this factor was added to each performance characteristic map for each material.

Table 2. The SLM process parameters applied for fabricating the Al6061 samples.

Sample #	P (W)	V_s (mm/s)	D_h (mm)	E_d (J/mm^3)	Sample #	P (W)	V_s (mm/s)	D_h (mm)	E_d (J/mm^3)
1A	370	1000	0.1	123.3	11A	370	800	0.15	102.8
2A	300	1000	0.1	100	12A	350	800	0.15	97.2
3A	370	1300	0.1	95	13A	370	800	0.19	81.1
4A	300	1300	0.1	76.9	14A	350	800	0.19	76.8
5A	370	1000	0.19	65	15A	370	1300	0.15	63.2
6A	300	1000	0.19	52.6	16A	350	1300	0.15	59.8
7A	370	1300	0.19	50	17A	370	1300	0.19	50
8A	300	1300	0.19	40.5	18A	350	1300	0.19	47.2

2.3. SLM Process Parameters

The AlSi10Mg and Al6061 parts were fabricated by an EOSINT M290 machine equipped with a 400 W Yb-fiber laser using a 100 μm laser beam diameter. The same layer thickness of 30 μm and layer orientation angle of 67° were selected for all samples undergoing strip scan, using a 0.02 mm laser beam offset. The build chamber was vacuumed with argon to reduce the oxygen content below 0.1%, and thus the possibility of oxide formation in the produced parts. All samples were fabricated as 15 mm cubes according to the SLM parameters listed in Tables 1 and 2. A preheating technique was applied to the build platform at 200 °C before starting the build, to minimize the thermal residual stresses (by reducing the thermal gradient between the deposited layers).

2.4. Sample Characterization Method

In the current study, the as-built part characterization focuses on relative density, internal porosity, surface roughness, and dimensional accuracy. Archimedes method was used to measure the density of the as-built cubes for both AlSi10Mg and Al6061 samples. The relative density was also evaluated after sample surface polishing to investigate the percentage of internal porosity. Density measurement via water displacement, according to ASTM B962-17, was performed with an electronic densimeter (MD-200S).

Surface roughness measurements were performed according to ASTM D7127-17 with a Mitutoyo SJ-410 surface tester. Five measurements at intervals of 4.5 mm were conducted on the cubic specimen's top surface, and their average was taken at each location. A light microscope (Alicona Infinite Focus G5) was used to capture the surface texture of some of the AlSi10Mg and Al6061 samples. The area tested was 10 mm × 10 mm using a 10× magnification lens, and surface roughness was also measured to validate the values obtained by the mechanical stylus.

The measurement of geometric dimensions and tolerances (GD&T) was conducted with a Mitutoyo CRYSTA-Apex S544 Coordinate Measuring Machine (CMM) which includes an SP25M stylus. This machine has a resolution of 0.1 μm within a working zone of 500 mm × 400 mm × 400 mm. The tested surface was probed at 10 measurement points along each sample's face. Flatness, perpendicularity, and parallelism were measured for all sample faces, except for the bottom.

3. Results and Discussion

3.1. Powder Characterization

The characteristics of the gas-atomized AlSi10Mg and Al6061 powders, supplied by the LPW Technology, Imperial, USA, were examined according to ASTM F3049-14. The powders were sieved with a 75 μm mesh before being characterized. The morphology of both powders was detected using SEM, as illustrated in Figure 1. The SEM observations showed a relatively higher percentage of elongated or irregular-shape particles in the AlSi10Mg powder compared to the same alloy provided by a different supplier, which presented spherical particles, as reported by Maamoun et al. [19,20]. Figure 1a,b shows that the Al6061 powder also had a greater percentage of spherical particles compared to the AlSi10Mg powder shown in Figure 1c,d. The existence of irregular or elongated particles might reduce powder flowability and the homogeneity of the powder layer distribution and thus negatively affect the quality and density of the fabricated parts [11]. The combination of a wide range of fine and coarse particles could increase the powder packing density but it reduces the flowability as a result of the effect of powder cohesion and inter-particle forces [11]. The weight percentages of the chemical elements of both powders were detected with EDS, as listed in Table 3. The results revealed higher Si content and relatively lower weight percentages of Mg, Cu, and Fe inside the AlSi10Mg powder compared to the Al6061 powder. The influence of this difference in chemical composition, in addition to the effect of the powders' particle shape, on the characteristics of the as-built parts will be discussed in the following sections.

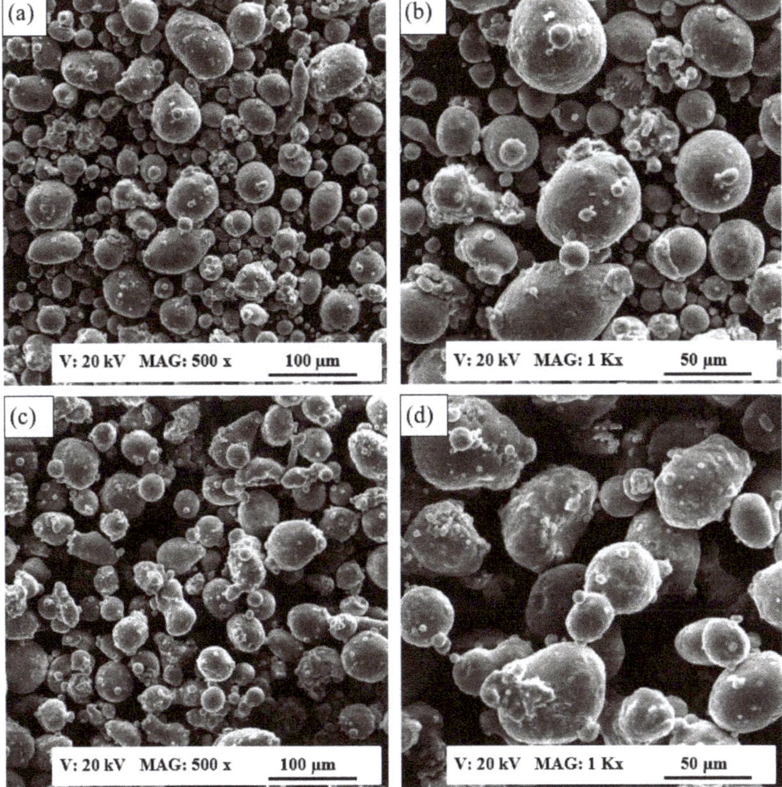

Figure 1. SEM observations of the powder morphology: (**a**,**b**) Al6061 powder, (**c**,**d**) AlSi10Mg powder.

Table 3. Energy X-ray dispersive spectroscopy (EDS) analysis of the Al6061 and AlSi10Mg powders' chemical composition.

Element	Si	Mg	Cu	Fe	Al
Al6061 wt %	1.2	0.77	0.32	0.90	Balance
AlSi10Mg wt %	11.34	0.28	0.08	0.32	Balance

Figure 2 illustrates the particle size distribution (PSD) profile of the AlSi10Mg and Al6061 powders, showing a positively skewed profile. This PSD profile could achieve a better surface quality and higher density compared to negatively skewed and Gaussian distribution grades, by increasing the laser energy absorption [11,26]. The quantitative data of PSD presented in Table 4 show that the particle size ranged from 12 to 110 μm for AlSi10Mg and from 12 to 120 μm for the Al6061 powder. These results indicated the presence of larger sized particles compared to the mesh size used for sieving. This might be related to the detected elongated particles with a smaller cross section which permits filtration through the mesh during sieving. Table 4 data also illustrate that in the AlSi10Mg and Al6061 powders, 90% of the particles were smaller than the sieving mesh (75 μm), with D (0.9) corresponding to 66.55 and 71.92 μm, respectively.

The XRD phase patterns of the AlSi10Mg and Al6061 powders were detected as shown in Figure 3. The Al and Si peaks were identified according to the Joint Committee on Powder Diffraction Standards (JCPDS) patterns of 01-089-2837 and 01-089-5012, respectively. The low-intensity of the Si peaks in the Al6061 phase pattern was due to its small weight percentage inside that alloy. For the AlSi10Mg powder, the higher intensity of the Si peaks and the slight shift of the Al peaks to the left indicated a lower solubility of Si in AlSi10Mg compared to Al6061 [27].

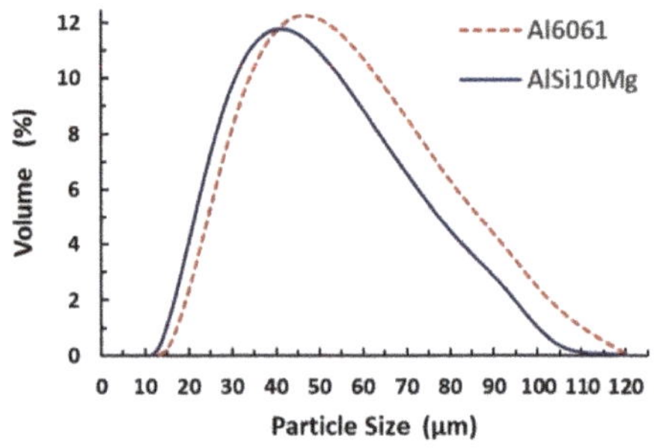

Figure 2. Particle size distribution of the Al6061 and AlSi10Mg powders.

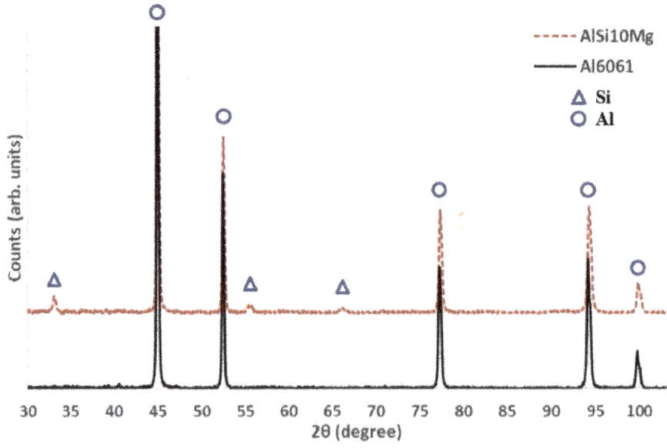

Figure 3. XRD phase patterns of the Al6061 and AlSi10Mg powders.

Table 4. Values measured for the particle size distribution of the Al6061 and AlSi10Mg powders.

Sample Type		D (0.1)	D (0.5)	D (0.9)
Al6061 Powder	Diameter (µm)	22.83	41.27	71.92
AlSi10Mg Powder		23.16	39.62	66.55

3.2. Relative Density

Figure 4 shows the effect of energy density on the formation of pores inside the as-built AlSi10Mg part. The results showed that the low energy density of 27 J/mm^3 in the AS8 sample could cause significant keyhole porosity due to the lack of fusion of the powder particles during the SLM process, as shown in Figure 4a. The keyhole pores were observed along the building direction within the layer boundaries, which had an irregular elongated shape due to the low energy density. The keyhole pores might form either because of insufficient energy delivered to the powder particles or because of the entrapment of gas bubbles between the interlayers during laser scanning [28]. The keyhole pore size reached 200 µm and gradually decreased as the energy density increased, until disappearing when the energy density value exceeded 50 J/mm^3. Inside the keyhole pores, partially melted particles were visible, as shown in Figure 4d. This might have occurred owing to the trapping of the consolidated powder inside the keyhole pores as a result of the low energy incident on the powder particle surface. However, spherical hydrogen pores or metallurgical pores with a size within 10 µm were observed at 50 J/mm^3, as shown in Figure 4b,e. The average size of these spherical pores tended to grow to more than 20 µm at 65 J/mm^3, as illustrated in Figure 4c,f. The mechanism of pore formation at high fusion rates might be related to the pores existing inside the gas-atomized powder particles [4]. They may also result from the balling phenomena where the melted powder fails to wet the previously deposited layer [29]. The melt pool viscosity might also change according to the applied energy density, and this could affect the porosity formation inside the as-built parts [30].

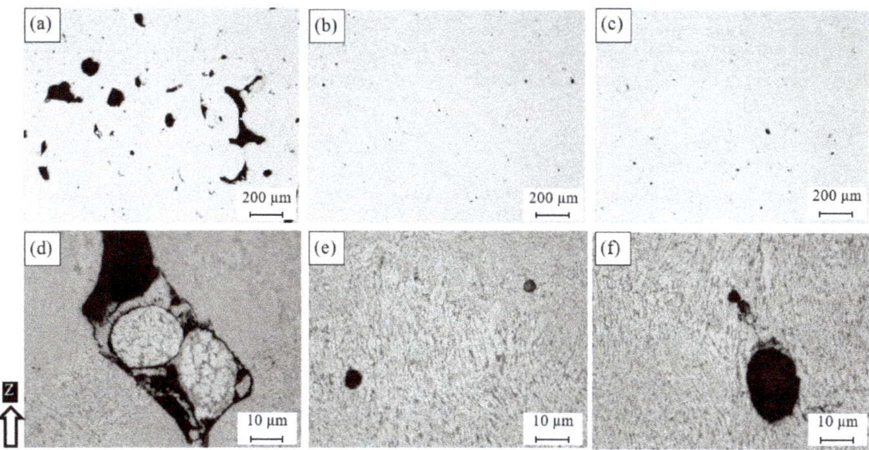

Figure 4. Pores observed inside the as-built AlSi10Mg sample fabricated with different SLM parameters; (**a,d**) AS8, (**b,e**) AS3, and (**c,f**) AS1.

The mechanism of pore formation inside the as-built AlSi10Mg samples, according to the SLM process parameters presented in Figure 4, was validated after evaluating the relative density. Figure 5 illustrates the map developed by the regression model generated from the DOE analysis. This map describes the effect of laser power, scan speed, hatch spacing, and energy density on the relative density of the as-built AlSi10Mg parts. The results showed the optimum range of the process parameters which allowed the least amount of spherical and keyhole pores to reach the highest possible relative density value of the part. An energy density value between 50 and 60 J/mm^3 produced a high relative density reaching 99.7%. Beyond this range, the relative density diminished because of the lack of fusion at the lower energy density, balling formation at the higher energy density, or hydrogen gases trapped inside the powder particles. It is worthwhile to note that higher values of the as-built part density could be obtained according to optimized SLM process parameters compared to the values reported by different literature studies [14,31–33]. In order to evaluate the internal porosity inside the as-built cube samples, their outer sides were polished before the relative density was re-measured. As shown in Figure 5b, the higher relative density obtained from the polished samples reached 99.9% at an energy density of 50 J/mm^3 (sample AS3), with a 0.1–1% reduction in porosity. By comparing the relative density between the as-built and the polished samples, it can be concluded that an increase in hatch spacing or scan speed parameters significantly increased the porosity on the sample surface, as illustrated in Figure 5c,d. This effect might result from the reduction of the material solidification rate at a higher scan speed and hatch spacing due to heat accumulation. The effect of the laser power indicated a significant impact of the increase of the melting rate and energy on the relative density of the as-built part. It is worth noting that the porosity percentage could be reduced after preheating the build platform prior to the sample build, as reported by Siddique et al. [34].

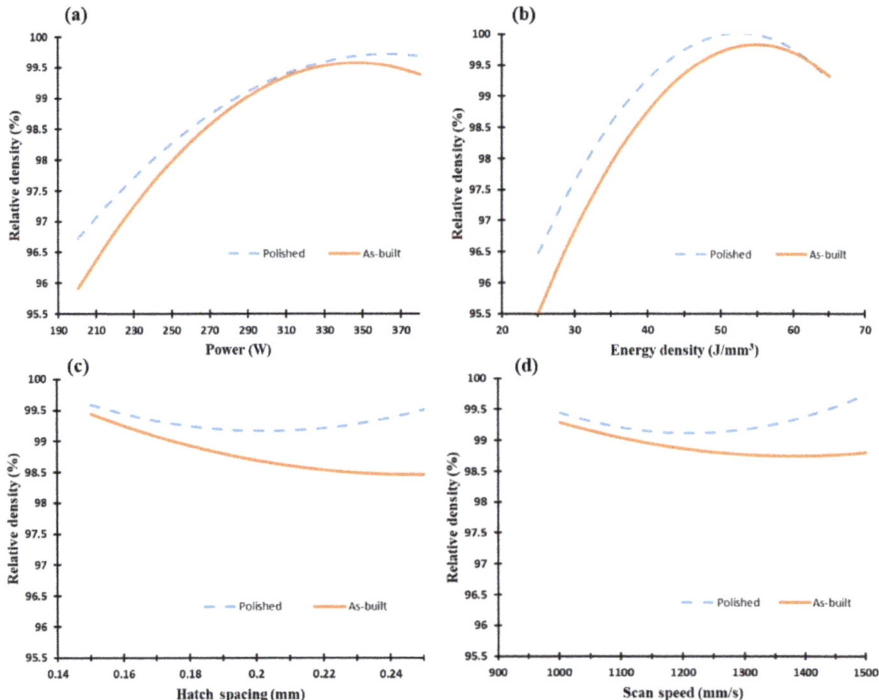

Figure 5. Relative density of the as-built AlSi10Mg samples vs. (**a**) laser power (W), (**b**) energy density (J/mm^3), (**c**) hatch spacing (mm), and (**d**) scan speed (mm/s).

Figure 6 shows microscopic observations of the polished as-built Al6061 samples fabricated using the different SLM process parameters listed in Table 2. Different sizes of micro-cracks were observed between the samples along the Z-direction and the XY-plane. As shown in Figure 6, a lower porosity percentage was observed compared to the as-built AlSi10Mg samples. The keyhole pores were also reduced until they were hardly noticeable, with the exception of some spherical hydrogen pores. However, the relative density was relatively lower than that of the AlSi10Mg samples because of the presence of micro-cracks. The size of these micro-cracks depended on the thermal gradient between the deposited layer in addition to the CTE of the alloy and was also affected by the values of the SLM process parameters applied. Figure 6a–c shows the longitudinal microcracks formed along the Z-direction. Cracks with different sizes were obtained according to the applied SLM parameters. The smallest size and density of the cracks were observed after the applied energy density reached 102.8 J/mm^3 in the 1A sample. However, no specific trend was detected between the energy density and the density of cracks, which is in agreement with Debroy et al. [35]. As illustrated in Figure 6d–f, the micro-cracks along the XY plane were shaped as semi-closed loops, similar in form to an equiaxed grain, but they were not entirely closed or sharp-edged. The results showed that the laser scan speed was the leading parameter affecting crack formation. The crack density, along with the building direction, increased along with the energy density from 40.5 to 76.9 J/mm^3 at the same scan speed (1300 mm/s), as shown in Figure 6a,b respectively. However, the crack density displayed in Figure 6c was significantly reduced at a higher energy density (102.8 J/mm^3) with a lower scan speed of 800 mm/s. Consequently, the scan speed had a more substantial effect on hot crack formation than the applied energy density, since it controlled the rate of solidification. The size of the semi-closed cracks formed in the XY plane tended to grow alongside the scan speed reduction, as noted in Figure 6d–f. The indents presented in Figure 6d–f were formed during the microhardness measurement that was reported by Maamoun et al. [25].

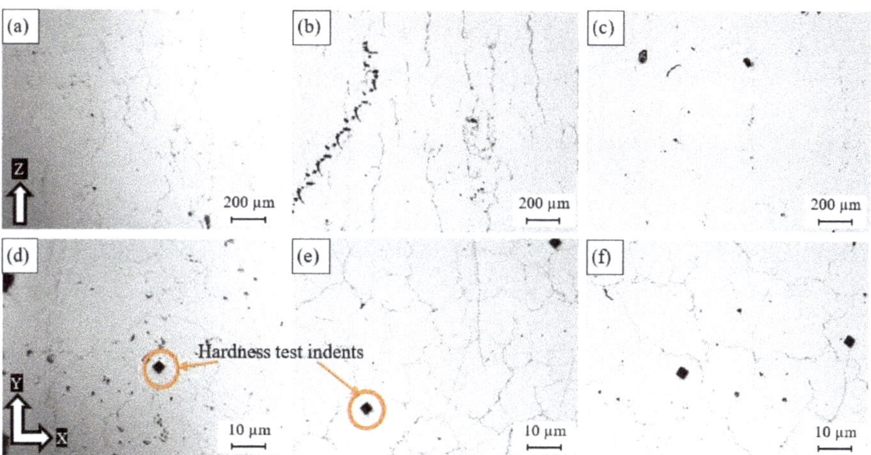

Figure 6. Pores observed inside the as-built Al6061 samples processed through different SLM parameters; (**a,d**) 8A, (**b,e**) 4A, and (**c,f**) 1A.

Figure 7 presents the plots generated via the DOE analysis using the effect of the two combined process parameters on the relative density of the as-built part. It can be concluded that relative density tended to increase along with the laser power and energy density, while a lower rate of the laser scan speed led to denser parts. A significant relationship can be seen between laser power and scan speed, as illustrated in Figure 7c,d. A relative density average of 98.2 ± 0.5% was measured according to the selected process parameters, with the maximum relative density reaching 98.72% at energy density of 102.8 J/mm^3 (sample 11A). These plots validated the trend obtained from the microscopic observations in Figure 6 and confirmed the effect of the laser scan speed on crack formation and relative density. The cracks observed inside the as-built parts could result from the hot crack phenomena which occur during material solidification owing to the combined chemical composition of the material. Kou et al. [36] reported that adding filler materials during welding to Al alloys susceptible to crack formation could eliminate the cracks and enhance the alloys' weldability. This explains the crack-free structure obtained in the as-built AlSi10Mg parts, which had a high Si content compared to Al6061.

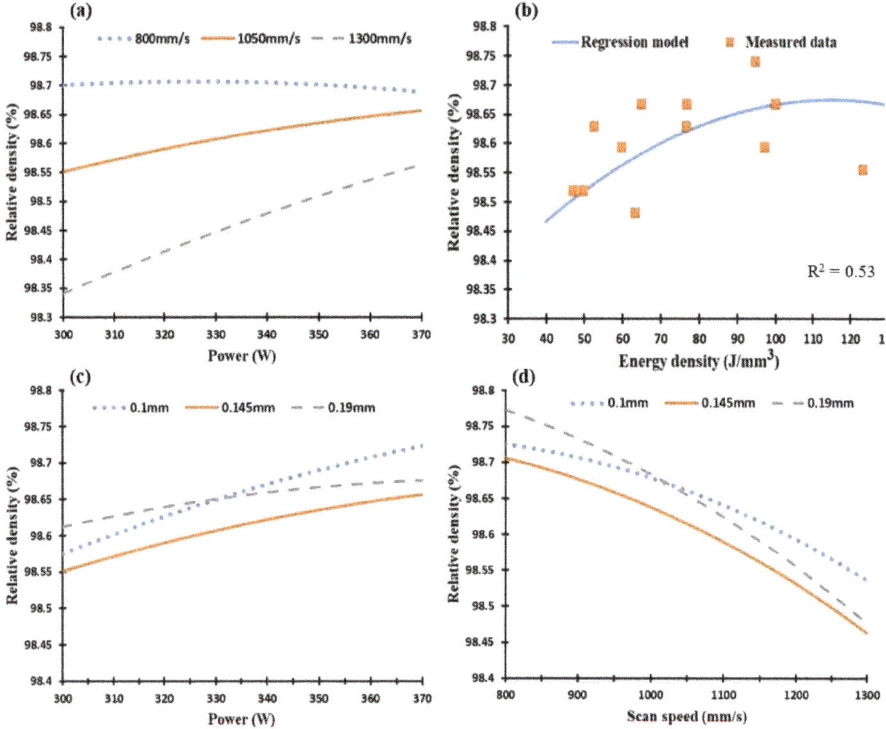

Figure 7. Relative density of the as-built Al6061 samples vs. (**a**) laser power (W), (**b**) energy density (J/mm^3), (**c**) hatch spacing (mm), and (**d**) scan speed (mm/s).

3.3. Surface Topology

The surface topology analysis of the as-built AlSi10Mg and Al6061 parts was conducted with SEM, displaying the 3D surface texture and mapping the relationship of the surface roughness with the SLM process parameters, according to the DOE analysis regression model. The surface defects of the as-built AlSi10Mg parts are exhibited in Figure 8 for different samples alongside the energy density increase. Figure 8a,d shows the rough surface obtained from the AS8 sample fabricated using a low energy density of 27 J/mm^3. According to SEM observations, this high roughness resulted from surface pores forming because of a lack of fusion and of partially melted powder adhering to the surface. As shown in Figure 8b,e, an increase of energy density in the AS3 sample to 49.9 J/mm^3, improved the surface roughness by eliminating noticeable surface pores and by reducing the density of the partially melted powder attached to the surface. However, the tracks of laser scanning were still visible with the commencement of balling phenomena. Figure 8c,f shows a better surface on the AS1 sample after applying a higher energy density of 63 J/mm^3. This eliminated the tracks of laser scanning, but the balling effect was still present. The balling phenomena occur at higher energy density levels as a consequence of the surface tension generated around the melted powder particles. This represents an obstacle to the wetting of the underlying substrate layer by the melted powder [35]. It is also worthwhile to note that the effect of the balling phenomena increased as the energy density exceeded 65 J/mm^3. As a result, the part build failed because of the detachment of the powder layer which had melted on the top of the underlying layer.

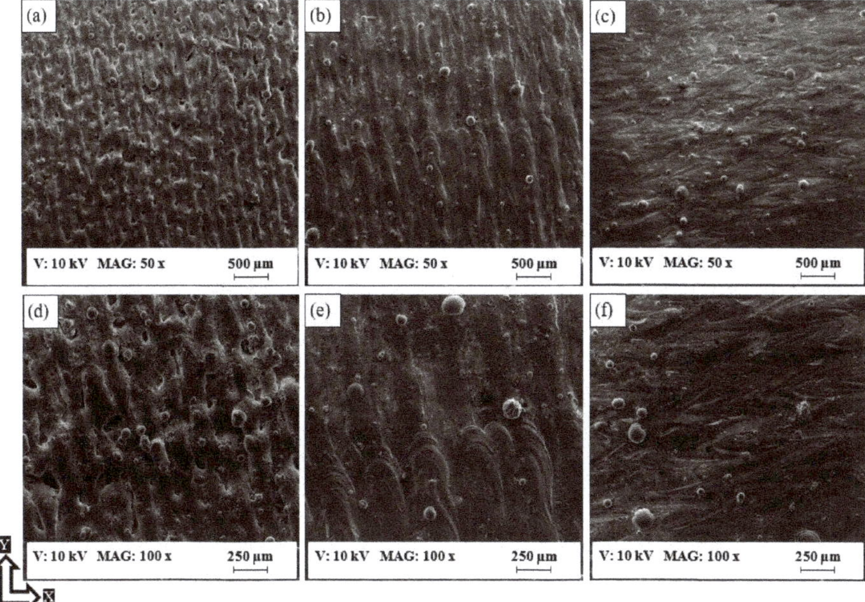

Figure 8. SEM observations of the as-built surface of AlSi10Mg samples; (**a,d**) AS8, (**b,e**) AS3, and (**c,f**) AS1.

Figure 9 exhibits the 3D surface texture of the as-built AlSi10Mg samples; the results showed a significant improvement of the surface roughness alongside the increase of the energy density up to a specific limit. As shown in Figure 9a, applying a low energy density of 27 J/mm^3 resulted in a rough texture with an average of 15 µm surface roughness. As illustrated in Figure 9b, the surface roughness decreased to 10 µm at a relatively high energy density of 40.5 J/mm^3. The surface roughness continued to decrease until reaching the lowest value of 4.5 µm at an energy density of 65 J/mm^3, as presented in Figure 9c,d.

The mapping of the SLM process parameter effect on the surface roughness of the as-built AlSi10Mg parts is illustrated in Figure 10. The regression model generated by the energy density effect on the surface roughness showed a good agreement with the measured values. The laser power effect revealed the same trend as the energy density influence on the samples' surface roughness, as shown in Figure 10a,b. The map displayed in Figure 10c also shows that the increasing of the hatch spacing value resulted in a more rough surface due to the decreasing overlap between the melted tracks, which agrees with the trend presented by Foster et al. [37]. The surface roughness also increased with the laser scan speed because of the reduction of the molten layer solidification rate, as presented in Figure 10d. A superior surface roughness of 4.5 µm was achieved with an E_d of 65 J/mm^3 at 370 W laser power, 1000 mm/s scan speed, and 0.19 mm hatch spacing, which is in good agreement with the regression model.

Figure 11 shows that the surface defects of as-built Al6061 parts were more significant than those of the AlSi10Mg parts. These defects were present in the partially melted powder adhering to the surface at a low energy density, surface porosity, and course shape of solidified tracks of laser scanning, as illustrated in Figure 11a. The surface finish gradually improved as the energy density increased from 50 to 123.3 J/mm^3, as illustrated in Figure 11a–c. In Figure 11d–f, micro-cracks were also observed at a high microscopic magnification within a size of 50–200 µm, concentrated at the end of the laser tracks along the XY plane as a result of high thermal stress. These cracks adversely affected the surface roughness of the as-built Al6061 parts, which is why the SLM process parameters need to be optimized to reduce micro-crack formation.

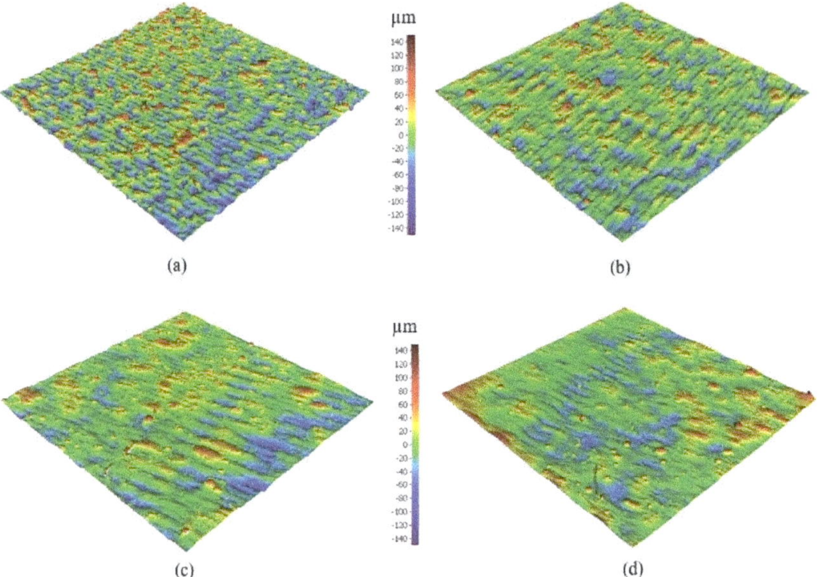

Figure 9. The 3D surface texture of the as-built AlSi10Mg samples; (**a**) AS8, (**b**) AS6, (**c**) AS3, and (**d**) AS1.

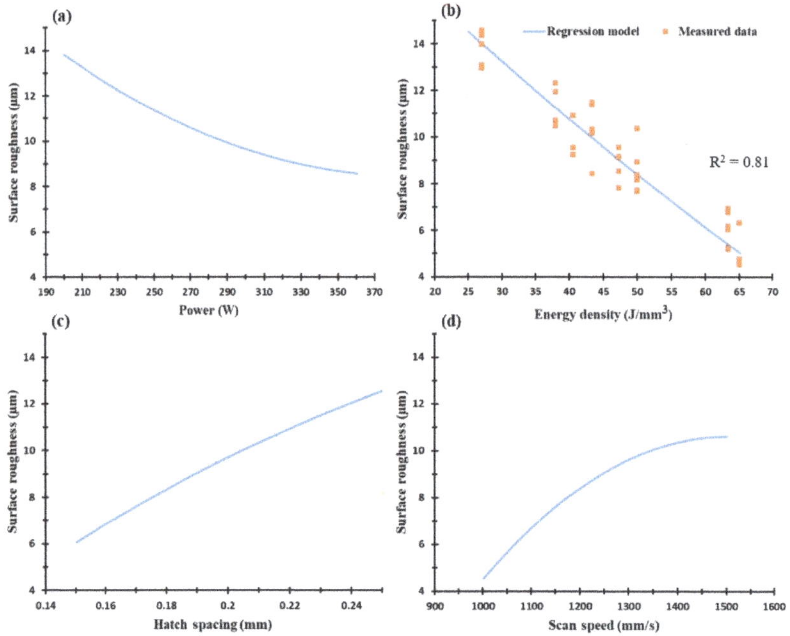

Figure 10. Surface roughness of the as-built AlSi10Mg samples vs. (**a**) laser power (W), (**b**) energy density (J/mm^3), (**c**) hatch spacing (mm), and (**d**) scan speed (mm/s).

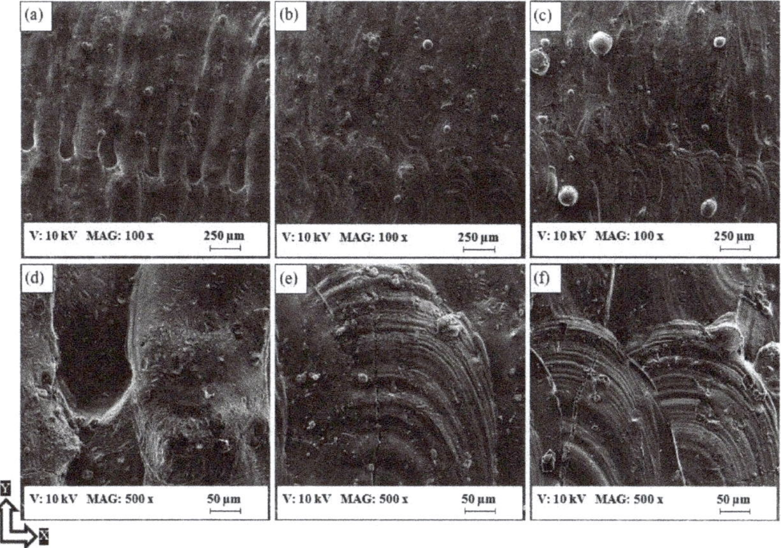

Figure 11. SEM observations of the as-built surface of Al6061 samples; (**a,d**) 7A, (**b,e**) 14A, and (**c,f**) 1A.

The 3D surface texture of the Al6061 samples in Figure 12 confirms the trend of surface finish improvement with the application of a higher energy density. The energy density range of Al6061 (40.5–123.3 J/mm^3) was shifted to a higher value compared to the limited E_d range of the AlSi10Mg alloy (27–65 J/mm^3). This was due to the higher reflectivity and CTE of Al6061 compared to AlSi10Mg, which required more energy to completely melt the powder layer. However, the balling phenomena effect propagated at higher energy densities, limiting the applicable values of E_d.

The regression model derived from the surface roughness values versus the SLM process parameters is presented in Figure 13. The plots illustrate that the higher the laser power, the lower the roughness of the sample surface. The lowest surface roughness of 3 µm was obtained at 370 W laser power, 800 mm/s of scan speed, and 0.15 mm hatch spacing, which is in good agreement with the surface roughness measured for parts fabricated at an energy density of 102.8 J/mm^3. In addition, no relationship was detected between the effect of the laser power on the surface roughness and the change in both scan speed and hatch spacing parameters, as illustrated in Figure 13b,c. However, a substantial relationship was noted between the scan speed and the hatch spacing effect on the surface roughness at a constant laser power value. The parabolic shape of the energy density impact indicated an optimum value of 102.8 J/mm^3, which resulted in a better surface finish, as shown in Figure 13b.

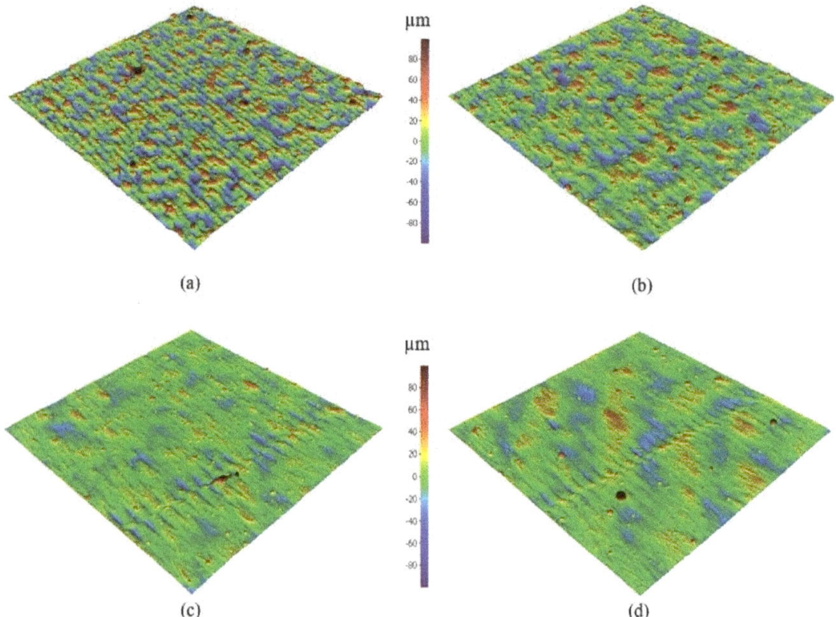

Figure 12. 3D surface texture of the as-built Al6061 samples; (**a**) 8A, (**b**) 6A, (**c**) 14A, and (**d**) 11A.

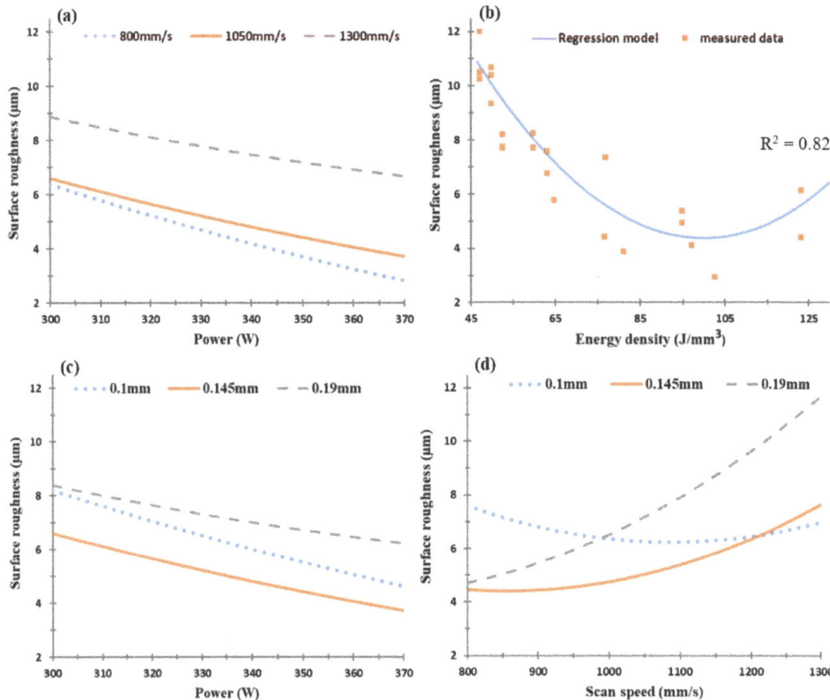

Figure 13. Surface roughness of the as-built Al6061 samples vs. (**a**) laser power (W), (**b**) energy density (J/mm^3), (**c**) hatch spacing (mm), and (**d**) scan speed (mm/s).

3.4. Dimensional Accuracy

The dimensional accuracy analysis was performed according to the CMM measurements for both dimensional length tolerance and top surface flatness of the as-built AlSi10Mg and Al6061 parts. The values measured along the XY plane were used to generate the regression models which represented the effect of the SLM process parameters on each characteristic. Figure 14 shows the dimension tolerance of the average cube length for each sample. According to the recorded results, oversize dimension measurements were compared to the designed values, and there was no contraction observed in the cube sample length. The oversize in the cube length resulted from the balling effect and partial melted particles on the sample surface, which thus affected the outer surface stair-step profile [18]. After excluding the 0.02 mm laser beam offset, the dimension tolerance ranged from 0.15 up to 0.195 mm. Figure 14c shows that hatch spacing was the leading parameter affecting the dimension tolerance accuracy in addition to the laser power. The surface flatness difference between the samples tested showed a smaller range of change from 0.035 to 0.09 mm, due to the application of a small 30 μm layer thickness. Figure 15 shows the surface flatness behavior according to the change of the SLM process parameters. It was observed that the hatch spacing and scan speed were the main parameters affecting the parts' surface flatness. The higher the scan speed and hatch spacing, the lower the surface flatness error obtained. Figure 15 also shows a good agreement between the regression model of the energy density effect on the surface flatness and the measured values. The results indicated that the surface flatness error increased together with an increase of the energy density.

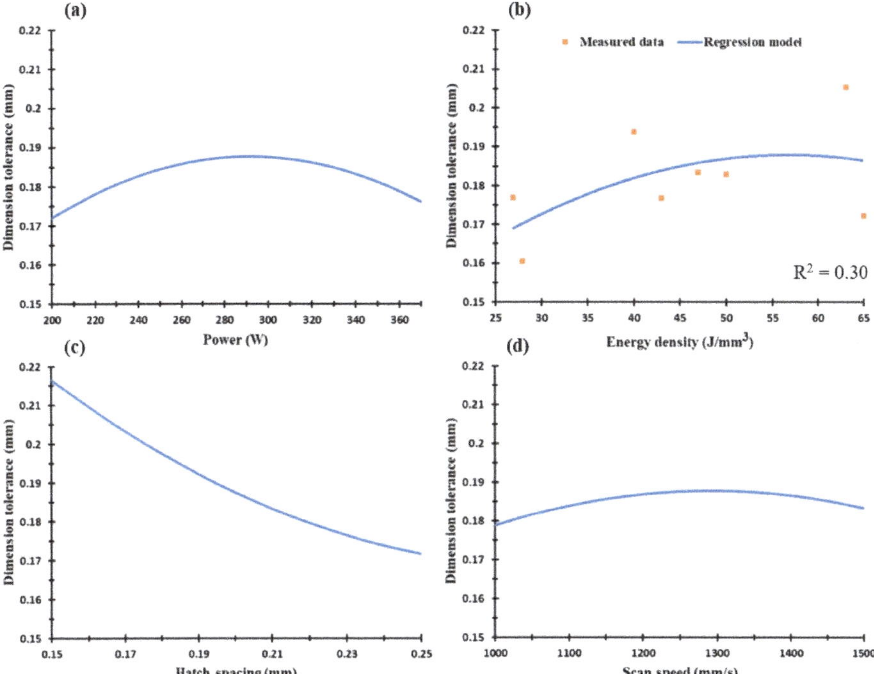

Figure 14. Dimension tolerance of the as-built AlSi10Mg samples vs. (**a**) laser power (W), (**b**) energy density (J/mm³), (**c**) hatch spacing (mm), and (**d**) scan speed (mm/s).

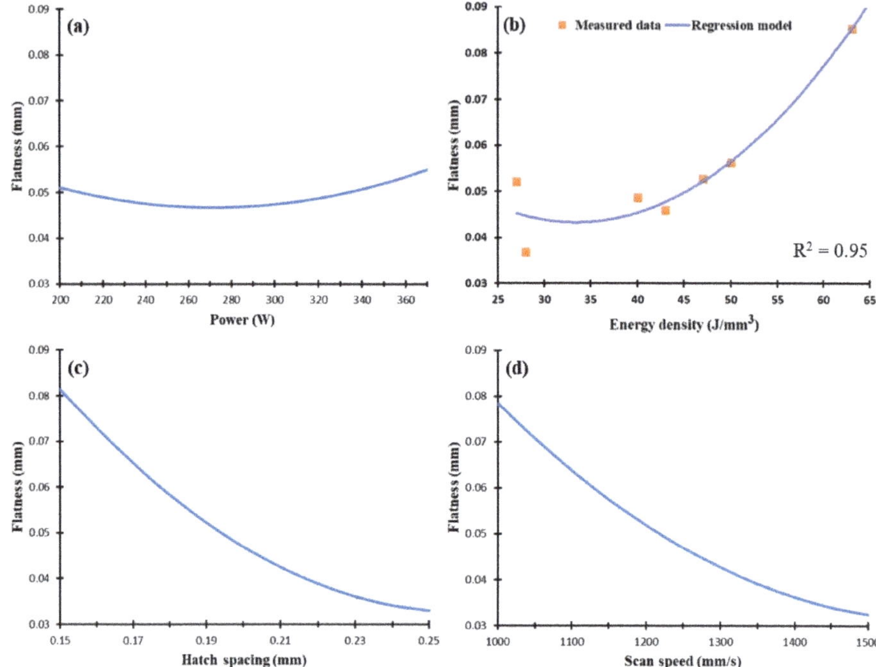

Figure 15. Surface flatness of the as-built AlSi10Mg samples vs. (**a**) laser power (W), (**b**) energy density (J/mm^3), (**c**) hatch spacing (mm), and (**d**) scan speed (mm/s).

The as-built Al6061 parts showed a different behavior for the dimensional tolerance values compared to the AlSi10Mg parts. As illustrated in Figure 16, the SLM parameters can affect the dimension tolerance by either expanding or contracting the dimensions, which could thus depend on the applied energy density. This might be caused by a change in the melt pool dimensions generated by the energy density [18]. The sample dimension tolerance showed a good agreement with the regression model curve. Figure 16b shows that an energy density higher than 76.8 J/mm^3 resulted in higher dimension tolerance than the original energy density. However, an energy density below this level could lead to part dimension contraction due to the high CTE of Al6061, which resulted in an increased rate of heat dissipation and solidification. It is also noticed that part contraction occurred at lower rates of hatch spacing and at higher scan speeds.

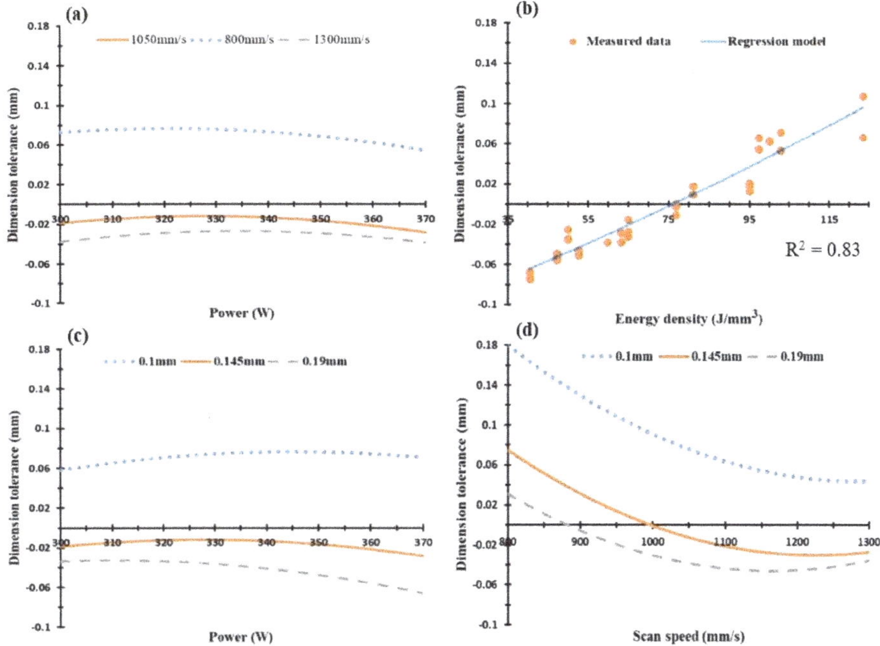

Figure 16. Dimension tolerance of the as-built Al6061 samples vs. (**a**) laser power (W), (**b**) energy density (J/mm^3), (**c**) hatch spacing (mm), and (**d**) scan speed (mm/s).

In Figure 17, the surface flatness of the Al6061 samples was in the range of 0.05–0.24 mm, which was significantly higher than in the AlSi10Mg sample. This elevated surface flatness disparity might be due to the higher CTE of the Al6061 material which reduces heat accumulation inside the part. This difference in the surface flatness might also result from hot cracks forming inside the part after solidification, low Si content in Al6061, and the high reflectivity of Al6061.

A combination of the optimized range for each performance characteristic is presented in the process parameter map of the scan speed and laser power at a constant hatch spacing of 0.19 mm, as illustrated in Figures 18 and 19. The process map for the as-built AlSi10Mg parts is displayed in Figure 18. This map presents an optimized range for the SLM process parameters to satisfy a surface roughness range from 5.5 to 9 μm, a relative density between 99.3% and 99.8%, and a range of dimensional tolerance from +0.18 to +0.2 mm.

Figure 19 illustrates the process map for the as-built Al6061 parts that displays the optimized range for the scan speed and the laser power. The optimized process window shows a surface roughness improvement from 3.2 to 6 μm compared to the values obtained from the AlSi10Mg part process map. The dimensional tolerance is also optimized within a smaller range from −0.03 to +0.03 mm, with minimum reduction of dimensions compared to the severe contraction in Figure 16, which is avoided within the optimized process parameter range. However, the relative density of the optimized range has lower values that vary between 98.6% and 98.7%.

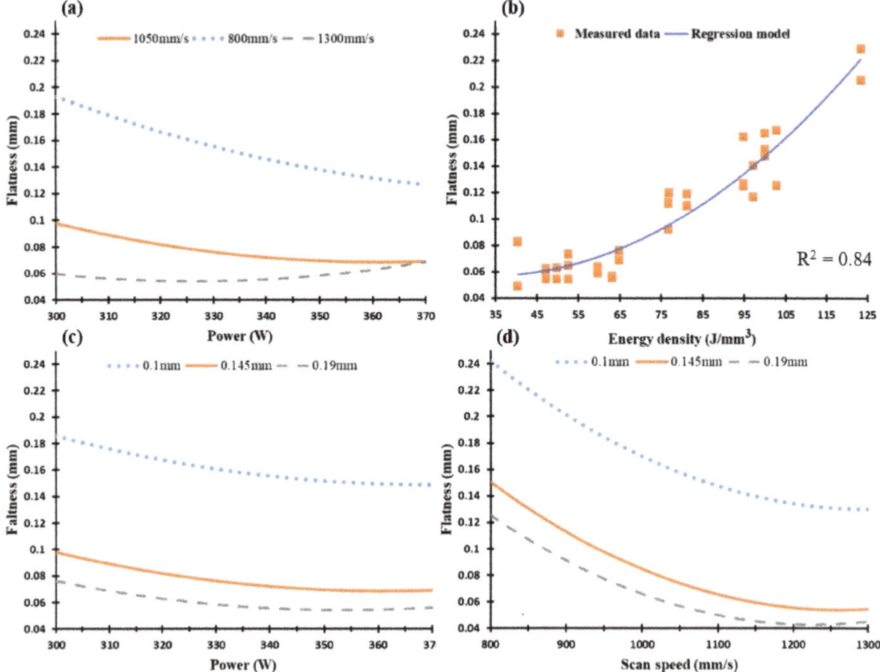

Figure 17. Surface flatness of the as-built Al6061 samples vs. (**a**) laser power (W), (**b**) eenergy density (J/mm^3), (**c**) hatch spacing (mm), and (**d**) scan speed (mm/s).

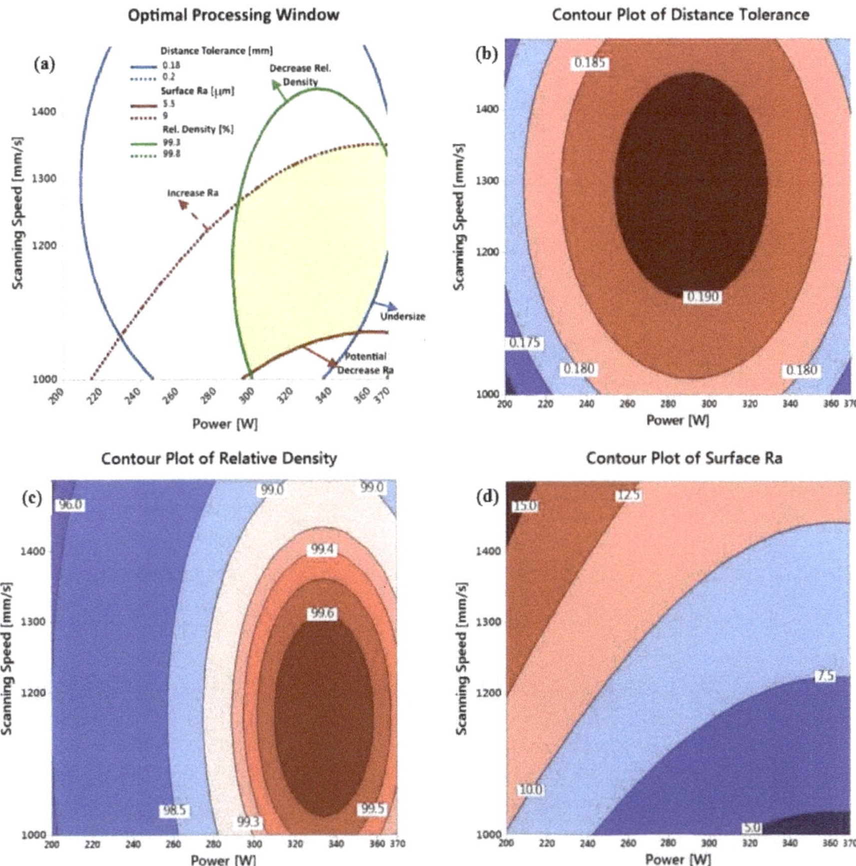

Figure 18. (**a**) Optimal processing window generated for the AlSi10Mg alloy at the hatch spacing value of 0.19 mm and the effect of laser power (W) and scan speed (mm/s) on (**b**) distance tolerance (mm), (**c**) relative density (%), and (**d**) surface roughness Ra (μm).

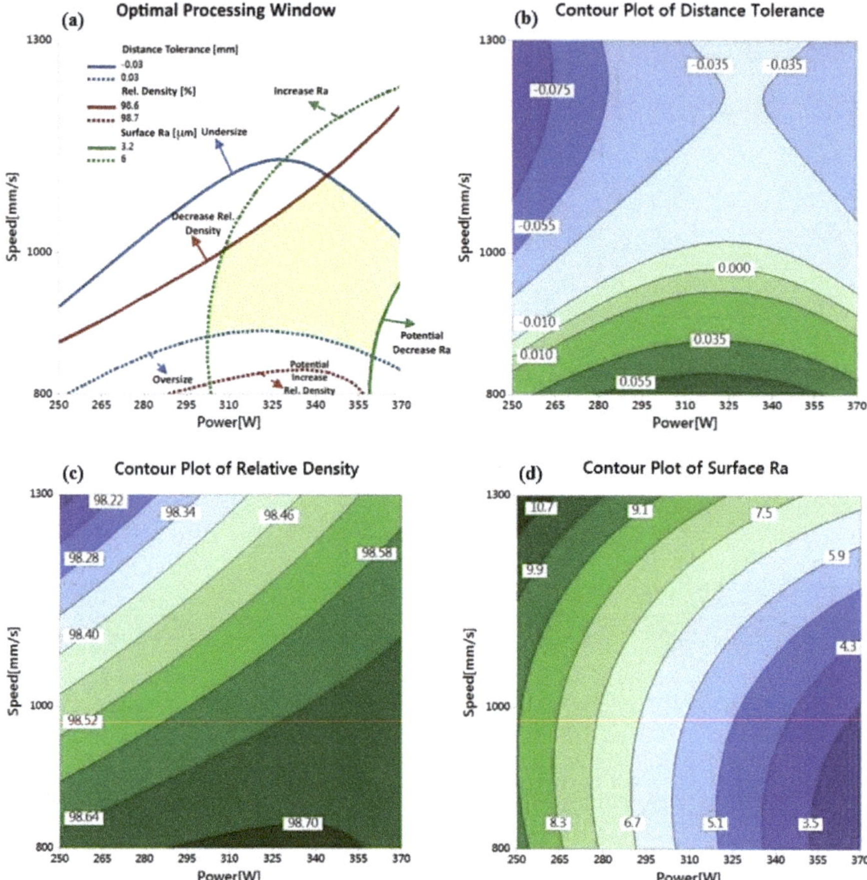

Figure 19. (**a**) Optimal processing window generated for the Al6061 alloy at the hatch spacing value of 0.15 mm and the effect of laser power (W) and scan speed (mm/s) on (**b**) distance tolerance (mm), (**c**) relative density (%) and (**d**) surface roughness Ra (μm).

4. Summary and Conclusions

The current study represents a comprehensive work that investigates the effect of SLM process parameters on the relative density, surface roughness, and dimensional accuracy of as-built AlSi10Mg and Al6061 parts [38]. A full characterization of both materials' powders was presented. DOE was used to investigate relative density, porosity, surface roughness, surface defects, and dimensional accuracy. Regression models and trends were obtained from the measured data. The results showed the following different characteristic behaviors for each material:

1. Powder morphology revealed that AlSi10Mg and Al6061 possess a spherical particle shape interspersed with of elongated particles in a considerable percentage. PSD showed a positively skewed distribution within a range of 12–120 μm.

2. The rate of energy density affected the relative density and porosity formation inside the as-built parts. The optimum range of energy density was 50–60 J/mm^3, which resulted in relative density reaching 99.7%. The relative density of the polished samples reached 99.9%, with a 0.1% internal porosity. The higher rates of energy densities contributed to the formation of large

hydrogen spherical pores, while the lower rates resulted in keyhole pores because of the lack of powder fusion.

3. For the Al6061, the maximum relative density measured was 98.72% using an energy density of 102.8 J/mm^3 and an 800 mm/s scan speed. A relationship between the scan speed and the laser power was noted, whereby the highest relative density was achieved at a low scan speed and high laser power. The relative density of the Al6061 parts showed lower values compared to the relative density of AlSi10Mg, due to lower Si content, which increased the CTE and caused the formation of hot cracks inside the as-built Al6061 parts.

4. The surface topology was significantly affected by the energy density applied to both materials. The surface roughness reduced alongside the increase of energy density. For the AlSi10Mg samples, the minimum surface roughness was 4.5 μm at 65 J/mm^3. For the Al6061 parts, an energy density of 102.8 J/mm^3 resulted in the best surface roughness of 3 μm. The energy density was limited to a maximum of 65 J/mm^3 for AlSi10Mg and to 123.3 J/mm^3 for Al6061, to avoid delamination and failure of part building.

5. For AlSi10Mg, the dimensional tolerance varied between an oversize of 0.15 and 0.195 mm. The best surface flatness could be obtained with higher hatch spacing and scan speeds.

6. For the Al6061 parts, the lowest dimensional tolerance was achieved using an energy density of 76.8 J/mm^3. Contraction of the part dimension was observed at lower energy densities, and oversized part dimension was detected at higher energy densities. The surface flatness of Al6061 was superior to that of the AlSi10Mg parts.

7. An optimal processing window was developed for each material to illustrate the mutual connection between relative density, surface topology, and dimensional accuracy, with the goal of achieving a high-quality end product.

Author Contributions: Formal analysis, A.H.M. and Y.F.X.; Investigation, M.A.E. and Y.F.X.; Methodology, A.H.M.; Supervision, M.A.E. and S.C.V.; Validation, A.H.M.; Writing-original draft, A.H.M. and Y.F.X.; Writing-review & editing, A.H.M., M.A.E., and S.C.V.

Funding: This research received no external funding.

Acknowledgments: The authors would like to acknowledge the Additive Manufacturing Innovation Centre at Mohawk College, Hamilton, Ontario, Canada, and the XRD measurement analysis at McMaster Analytical X-ray (MAX) diffraction facility. The authors would also like to acknowledge Learning Factory, W Booth School at McMaster University for using their metrology facility.

References

1. Manfredi, D.; Calignano, F.; Krishnan, M.; Canali, R.; Paola, E.; Biamino, S.; Ugues, D.; Pavese, M.; Fino, P. Additive Manufacturing of Al Alloys and Aluminium Matrix Composites (AMCs). In *Light Metal Alloys Applications*; InTech: London, UK, 2014; ISBN 978-953-51-1588-5.

2. Srivatsan, T.S.; Sudarshan, T.S. *Additive Manufacturing: Innovations, Advances, and Applications*; CRC Press: Boca Raton, FL, USA, 2015; ISBN 9781498714778.

3. Hassanin, H.; Elshaer, A.; Benhadj-Djilali, R.; Modica, F.; Fassi, I. Surface Finish Improvement of Additive Manufactured Metal Parts. In *Micro and Precision Manufacturing*; Springer: Cham, Switzerland, 2018; pp. 145–164. [CrossRef]

4. Galy, C.; Le Guen, E.; Lacoste, E.; Arvieu, C. Main defects observed in aluminum alloy parts produced by SLM: From causes to consequences. *Addit. Manuf.* **2018**, *22*, 165–175. [CrossRef]

5. Prashanth, K.G.; Scudino, S.; Chaubey, A.K.; Löber, L.; Wang, P.; Attar, H.; Schimansky, F.P.; Pyczak, F.; Eckert, J. Processing of Al-12Si-TNM composites by selective laser melting and evaluation of compressive and wear properties. *J. Mater. Res.* **2016**, *31*, 55–65. [CrossRef]

6. Prashanth, K.G.; Shakur Shahabi, H.; Attar, H.; Srivastava, V.C.; Ellendt, N.; Uhlenwinkel, V.; Eckert, J.; Scudino, S. Production of high strength Al85Nd8Ni5Co2alloy by selective laser melting. *Addit. Manuf.* **2015**, *6*, 1–5. [CrossRef]

7.   Sufiiarov, V.S.; Popovich, A.A.; Borisov, E.V.; Polozov, I.A.; Masaylo, D.V.; Orlov, A.V. The effect of layer thickness at selective laser melting. *Procedia Eng.* **2017**, *174*, 126–134. [CrossRef]

8.   Nguyen, Q.B.; Luu, D.N.; Nai, S.M.L.; Zhu, Z.; Chen, Z.; Wei, J. The role of powder layer thickness on the quality of SLM printed parts. *Arch. Civ. Mech. Eng.* **2018**, *18*, 948–955. [CrossRef]

9.   Cheng, B.; Shrestha, S.; Chou, Y.K. Stress and deformation evaluations of scanning strategy effect in selective laser melting. In Proceedings of the ASME 2016 11th International Manufacturing Science and Engineering Conference, Blacksburg, VA, USA, 27 June–1 July 2016; American Society of Mechanical Engineers: New York, NY, USA; p. V003T08A009.

10.   Sutton, A.T.; Kriewall, C.S.; Leu, M.C.; Newkirk, J.W. Powder characterisation techniques and effects of powder characteristics on part properties in powder-bed fusion processes. *Virtual Phys. Prototyp.* **2017**, *12*, 3–29. [CrossRef]

11.   Tan, J.H.; Wong, W.L.E.; Dalgarno, K.W. An overview of powder granulometry on feedstock and part performance in the selective laser melting process. *Addit. Manuf.* **2017**, *18*, 228–255. [CrossRef]

12.   Scipioni Bertoli, U.; Guss, G.; Wu, S.; Matthews, M.J.; Schoenung, J.M. In-situ characterization of laser-powder interaction and cooling rates through high-speed imaging of powder bed fusion additive manufacturing. *Mater. Des.* **2017**, *135*, 385–396. [CrossRef]

13.   Yang, K.V.; Rometsch, P.; Jarvis, T.; Rao, J.; Cao, S.; Davies, C.; Wu, X. Porosity formation mechanisms and fatigue response in Al-Si-Mg alloys made by selective laser melting. *Mater. Sci. Eng. A* **2018**, *712*, 166–174. [CrossRef]

14.   Read, N.; Wang, W.; Essa, K.; Attallah, M.M. Selective laser melting of AlSi10Mg alloy: Process optimisation and mechanical properties development. *Mater. Des.* **2015**, *65*, 417–424. [CrossRef]

15.   Aboulkhair, N.T.; Everitt, N.M.; Ashcroft, I.; Tuck, C. Reducing porosity in AlSi10Mg parts processed by selective laser melting. *Addit. Manuf.* **2014**, *1*, 77–86. [CrossRef]

16.   Calignano, F.; Manfredi, D.; Ambrosio, E.P.; Iuliano, L.; Fino, P. Influence of process parameters on surface roughness of aluminum parts produced by DMLS. *Int. J. Adv. Manuf. Technol.* **2013**, *67*, 2743–2751. [CrossRef]

17.   Hitzler, L.; Hirsch, J.; Merkel, M.; Hall, W.; Öchsner, A. Position dependent surface quality in selective laser melting: Positionsabhängige Oberflächenqualität im selektiven Laserstrahlschmelzen. *Materwiss. Werksttech.* **2017**, *48*, 327–334. [CrossRef]

18.   Han, X.; Zhu, H.; Nie, X.; Wang, G.; Zeng, X. Investigation on selective laser melting AlSi10Mg cellular lattice strut: Molten pool morphology, surface roughness and dimensional accuracy. *Materials (Basel)* **2018**, *11*, 392. [CrossRef] [PubMed]

19.   Maamoun, A.H.; Elbestawi, M.; Dosbaeva, G.K.; Veldhuis, S.C. Thermal Post-processing of AlSi10Mg parts produced by Selective Laser Melting using recycled powder. *Addit. Manuf.* **2018**, *21*, 234–247. [CrossRef]

20.   Maamoun, A.; Elbestawi, M.; Veldhuis, S. Influence of Shot Peening on AlSi10Mg Parts Fabricated by Additive Manufacturing. *J. Manuf. Mater. Process.* **2018**, *2*, 40. [CrossRef]

21.   Maamoun, A.H.; Veldhuis, S.C.; Elbestawi, M. Friction stir processing of AlSi10Mg parts produced by selective laser melting. *J. Mater. Process. Technol.* **2019**, *263*, 308–320. [CrossRef]

22.   Fulcher, B.A.; Leigh, D.K.; Watt, T.J. Comparison of AlSi10Mg and Al 6061 Processed through DMLS. In Proceedings of the Solid Freeform Fabrication (SFF) Symposium, Austin, TX, USA, 4–6 August 2014; pp. 404–419.

23.   Martin, J.H.; Yahata, B.D.; Hundley, J.M.; Mayer, J.A.; Schaedler, T.A.; Pollock, T.M. 3D printing of high-strength aluminium alloys. *Nature* **2017**, *549*, 365–369. [CrossRef] [PubMed]

24.   Louvis, E.; Fox, P.; Sutcliffe, C.J. Selective laser melting of aluminium components. *J. Mater. Process. Technol.* **2011**, *211*, 275–284. [CrossRef]

25.   Maamoun, A.H.; Xue, Y.F.; Elbestawi, M.A.; Veldhuis, S.C. Effect of SLM Process Parameters on the Quality of Al Alloy Parts; Part II: Microstructure and Mechanical Properties. *Preprints* **2018**. [CrossRef]

26.   Lee, Y.S.; Nandwana, P.; Zhang, W. Dynamic simulation of powder packing structure for powder bed additive manufacturing. *Int. J. Adv. Manuf. Technol.* **2018**, *96*, 1507–1520. [CrossRef]

27.   Ma, P.; Prashanth, K.; Scudino, S.; Jia, Y.; Wang, H.; Zou, C.; Wei, Z.; Eckert, J. Influence of Annealing on Mechanical Properties of Al-20Si Processed by Selective Laser Melting. *Metals (Basel)* **2014**, *4*, 28–36. [CrossRef]

28. Olakanmi, E.O.T.; Cochrane, R.F.; Dalgarno, K.W. A review on selective laser sintering/melting (SLS/SLM) of aluminium alloy powders: Processing, microstructure, and properties. *Prog. Mater. Sci.* **2015**, *74*, 401–477. [CrossRef]

29. Sames, W.J.; List, F.A.; Pannala, S.; Dehoff, R.R.; Babu, S.S. The metallurgy and processing science of metal additive manufacturing. *Int. Mater. Rev.* **2016**, *61*, 315–360. [CrossRef]

30. Attar, H.; Ehtemam-Haghighi, S.; Kent, D.; Dargusch, M.S. Recent developments and opportunities in additive manufacturing of titanium-based matrix composites: A review. *Int. J. Mach. Tools Manuf.* **2018**, *133*, 85–102. [CrossRef]

31. Damon, J.; Dietrich, S.; Vollert, F.; Gibmeier, J.; Schulze, V. Process dependent porosity and the influence of shot peening on porosity morphology regarding selective laser melted AlSi10Mg parts. *Addit. Manuf.* **2018**, *20*, 77–89. [CrossRef]

32. Raus, A.A.; Wahab, M.S.; Ibrahim, M.; Kamarudin, K.; Ahmed, A.; Shamsudin, S. Mechanical and Physical Properties of AlSi10Mg Processed through Selective Laser Melting. *Int. J. Eng. Technol.* **2016**, *8*, 2612–2618. [CrossRef]

33. Thijs, L.; Kempen, K.; Kruth, J.P.; Van Humbeeck, J. Fine-structured aluminium products with controllable texture by selective laser melting of pre-alloyed AlSi10Mg powder. *Acta Mater.* **2013**, *61*, 1809–1819. [CrossRef]

34. Siddique, S.; Imran, M.; Wycisk, E.; Emmelmann, C.; Walther, F. Influence of process-induced microstructure and imperfections on mechanical properties of AlSi12 processed by selective laser melting. *J. Mater. Process. Technol.* **2015**, *221*, 205–213. [CrossRef]

35. DebRoy, T.; Wei, H.L.; Zuback, J.S.; Mukherjee, T.; Elmer, J.W.; Milewski, J.O.; Beese, A.M.; Wilson-Heid, A.; De, A.; Zhang, W. Additive manufacturing of metallic components—Process, structure and properties. *Prog. Mater. Sci.* **2018**, *92*, 112–224. [CrossRef]

36. Kou, S. A Simple Index for Predicting the Susceptibility to Solidification Cracking. *Weld. J.* **2015**, *94*, 374s–388s.

37. Foster, S.J.; Carver, K.; Dinwiddie, R.B.; List, F.; Unocic, K.A.; Chaudhary, A.; Babu, S.S. Process-Defect-Structure-Property Correlations During Laser Powder Bed Fusion of Alloy 718: Role of In Situ and Ex Situ Characterizations. *Metall. Mater. Trans. A* **2018**, *49*, 5775–5798. [CrossRef]

38. Maamoun, A.H.; Xue, Y.F.; Elbestawi, M.A.; Veldhuis, S.C. Effect of SLM Process Parameters on the Quality of Al Alloy Parts; Part I: Powder Characterization, Density, Surface Roughness, and Dimensional Accuracy. *Preprints* **2018**. [CrossRef]

Article

Impacts of Defocusing Amount and Molten Pool Boundaries on Mechanical Properties and Microstructure of Selective Laser Melted AlSi10Mg

Suyuan Zhou [1,2], Yang Su [1], Rui Gu [1], Zhenyu Wang [1], Yinghao Zhou [1], Qian Ma [3] and Ming Yan [1,*]

[1] Department of Materials Science and Engineering Shenzhen Key Laboratory for Additive Manufacturing of High-Performance Materials, South University of Science and Technology of China, Shenzhen 518055, China; zhousy@sustc.edu.cn (S.Z.); 2161160220@email.szu.edu.cn (Y.S.); gurn@mail.sustc.edu.cn (R.G.); wangzy7@mail.sustc.edu.cn (Z.W.); 11649022@mail.sustc.edu.cn (Y.Z.)

[2] Key Laboratory of Artificial Micro- and Nano-Materials of Ministry of Education and School of Physics and Technology, Wuhan University, Wuhan 430072, China

[3] Centre for Additive Manufacture, School of Engineering, RMIT University, Melbourne, VIC 3000, Australia; ma.qian@rmit.edu

* Correspondence: yanm@sustc.edu.cn; Tel.: +86-0755-8801-8967

Received: 23 November 2018; Accepted: 18 December 2018; Published: 26 December 2018

Abstract: The influences of processing parameters such as volumetric energy density (ε) and, particularly, defocusing amount (DA) on densification, microstructure, tensile property, and hardness of the as-printed dense AlSi10Mg alloy by selective laser melting (SLM) were studied systematically. The molten pool boundaries (MPBs) were found overwhelmingly at regular and complex spatial topological structures affected by DA value to exist in two forms, while the "layer–layer" MPB overlay mutually and the "track–track" MPBs intersect to form acute angles with each other. The microstructure of MPBs exhibits a coarse grain zone near the MPBs and the characteristics of segregation of nonmetallic elements (O, Si) where the crack easily happened. The DA value (-2 to 2 mm) affected both the density and the tensile mechanical properties. High tensile strength (456 ± 14 MPa) and good tensile ductility ($9.5 \pm 1.4\%$) were achieved in the as-printed condition corresponding to DA = 0.5 mm. The tensile fracture surface features were analyzed and correlated to the influence of the DA values.

Keywords: aluminum alloys; selective laser melting (SLM); mechanical properties

---

1. Introduction

Additive layer manufacturing (ALM) technology as a layer-additive approach has been widely used and developed for more than 30 years [1,2]. Selective Laser Melting (SLM) is an ALM technique to produce Fe-, Ti-, Ni-, and Al-based alloys by traces of one cross section of the CAD-data [3,4]. The near-net-shaped and near-fully-dense parts can be made by building on the successive layers by using fast-moving laser beam as illustrated in Figure 1a, where a high intensity laser scanned the powder bed selectively layer by layer. The metal powder particles are melted and form molten pool, which rapidly solidifies during the short-term non-equilibrium phase transition. Bandar AlMangour et al. (2018) pointed out that the volumetric laser energy influenced the microstructures and porosity of SLMed TiC/316L nanocomposite parts [5,6]. Wu et al. (2016) revealed the SLM process offers a wide range of advantages including significant microstructural refinement by small grain sizes as compared to other conventional processes, due to the intrinsic high thermal gradients of the SLM process as discussed [7]. Murr et al. (2009) found that the yield and tensile strength of dense SLMed

Ti6Al4V alloy increased by nearly 50% greater than forgings [8]. Michele et al. (2016) revealed the SLMed metallurgy of high-silicon steel can be changed from <001> fibre-texture to a cube-texture at higher energy input [9].

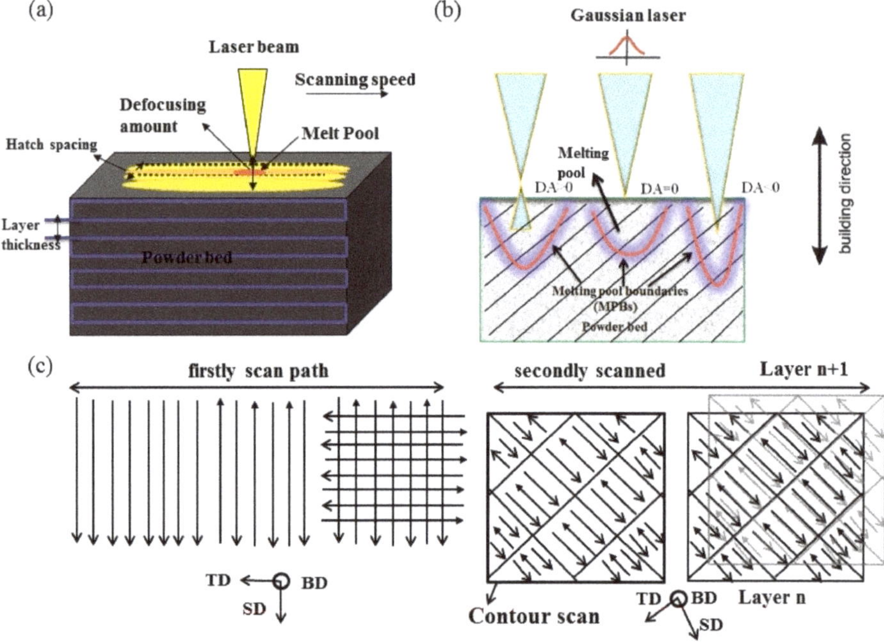

Figure 1. Schematic diagram of Selective Laser Melting (SLM) (**a**); Schematic illustration of the melt pools produced in different defocusing amounts (DAs) with variation of shape and discontinuity (**b**); the scan strategy "the island" (**c**).

Al-Si based traditional light-weight, castable alloys have been widely used in aerospace and automotive industries because of their good strength, low density, and good castability. Li et al. (2016) found SLMed NTD-Al have strong interfacial bonding in the Al matrix with rod-like nano-Si and TiB_2 particles due to solidification of non-equilibrium and rapid cooling [10]. Prashanth et al. (2017) observed the supersaturated Al solid solution of the SLM-processed samples with fibrous eutectic Al/Si has good weldability [11]. Ma et al. (2016) pointed out the SLMed Al-20Si alloy with eutectic Al/Si at high pressure solidification can improve the mechanical properties [12]. Read et al. (2015) acquired near-fully-dense SLMed AlSi10Mg parts by suitable process parameters optimization resulting in good mechanical properties [13]. Aboulkhair et al. (2014) reduced porosity in AlSi10Mg parts by used the different parameters and scan strategies [14]. Brandl et al. (2012) found the fatigue life of AlSi10Mg samples at different directions can be changed at a peak-hardened (T6) state [15], Amato et al. (2012) observed that the yield strength and UTS of SLMed Inconel 718 along the horizontal direction was higher than in the vertical plane mainly due to the difference in the crystallography and microstructure [16,17]. Thijs et al. (2013) showed that the SLMed AlSi10Mg alloy has a different texture, using rotation of 90° of the scanning vectors [17]. Abe et al. (2003) found that the low density and high porosity of the SLMed samples lead to the low yield tensile strength [18]. Li et al. (2012) found that the selective laser melting Al12Si alloy has 25% tensile ductility refined by eutectic microstructure after solution heat treatment [19]. Rosenthal et al. (2017) found that the SLMed AlSi10Mg alloy has a strain rate that is sensitive to changes in strain hardening and flow stress [20].

To maximize the potential of the SLM process for this alloy and other similar metallic materials, most of the relevant studies have focused on processing parameters optimization, e.g., energy density, scanning strategy, hatch space, to explain the appearance of micropores, microstructures, crystal texture, and internal residual stress. Khairallah et al. (2012) showed that the melt flow has three sections such as a topological depression, a transition, and a tail region to generate pore defects [21]. However, limited attention has been addressed to the influence of two other important factors, namely the molten pool boundaries (MPBS) and the defocusing amount (DA). Some studies have proposed the existence of MPBs in the as-built SLMed parts because the rapid high heat input by a moving laser beam repeats the deposition layer after layer results in overlapping of molten pools. Lin et al. (2006) showed the directional heat transfer along the molten pools relative to the bottom has a lower solidification rate [22]. Yadroitsev et al. (2014) observed the MPBs existed clearly and interconnected to form a specific spatial topological structure in the Ti6Al4V SLM parts [23]. Wen et al. (2014) discussed how the performance of SLM parts was greatly affected by the whole MPBs structure in different building directions, because the properties of individual MPBs easily form plane grains which differ from that of other regions [24].

The defocusing amount (DA) is defined as the distance that the laser radiation penetrated into the powder layer and up to a certain distance from the surface. It is recommended as a critical factor for the SLM process because the overlapping of multi-tracks or multi-layers during SLM leads to MPBs with a special shape in the microstructure as illustrated in Figure 1b. To the authors' best knowledge, the influence of either factor on the mechanical properties and density of the SLMed AlSi10Mg has not been investigated in detail yet. This study is therefore carried out to fill this knowledge gap. The research findings highlight the importance of both factors in affecting the mechanical performance and microstructure of the SLMed AlSi10Mg alloy.

2. Experimental

Commercial as-purchased AlSi10Mg powder was used (AMC powders GmbH, Beijing, China). The chemical component of the AlSi10Mg powder is listed in Table 1. The size ranges from 20–60 μm, as measured by a laser diffraction particle analyzer Coulter LS230 (manufacturer, city, country). The SLM process is based on a Concept Laser M2 (Concept Laser GmbH, Lichtenfels, Germany) with a build chamber of 280 mm × 250 mm × 250 mm. The build chamber was filled with high-purity argon and the oxygen content inside of the chamber was below 15 ppm. The operating maximum power of the IPG fiber laser was 400 W. The thickness (30 μm) of powder layer and hatch spacing (100 μm) were kept constant during the SLM processing. The laser processing parameters such as energy density and DA were optimized. Sixty cylindrical samples and tensile samples were produced with scan speed (v) at 1400 mm/s and a varying laser power (P) in order to obtain the highest density as shown in Table 2. The laser beam is raster-scanned individually with small islands in order to reduce the residual stresses as for the laser scanning mode in Figure 1c [17]. The island size was about 5×5 mm^2 being built continuously and randomly and the scanning vectors shift 1 mm in both x and y directions. The defocusing amount can be controlled by manipulating the movement of the sample platform up and down via using the laser displacement sensor (Figure 1b). The displacement accuracy is ±0.01 mm.

Table 1. Chemical composition of the AlSi10Mg powder.

Weight (%)	Si	Fe	Zn	Mn	Mg	Ni	O	Cu	Al
AlSi10Mg	9.0–11.0	≤0.55	≤0.10	≤0.45	0.2–0.45	≤0.05	0.02	0.03	balance

Table 2. Parameters for optimizing the density of the as-printed AlSi10Mg and the mechanical properties ultimate tensile strength (UTS), yield strength (YS), and elongation of AlSi10Mg fabricated by SLM at different DAs.

Batch Number	Power (W)	Scan Speed (mm/s)	Defocusing Amount DA	Ultimate Tensile Strength (MPa)	Yield Strength (MPa)	Microhardness (HV)	Young Modulus (GPa)	Elongation (%)
P1	200	1400	0	390 ± 16	170 ± 3	112 ± 10	62.1	6.5 ± 1.3
P2	260	1400	0	385 ± 18	165 ± 2	114 ± 6	63.7	6.8 ± 1.6
P3	320	1400	0	429 ± 20	175 ± 6	121 ± 11	66.9	7.5 ± 1.5
P4	370	1400	0	415 ± 12	178 ± 8	124 ± 8	70.6	6.9 ± 1.4
P3D1	320	1400	-2	370 ± 10	160 ± 6	117 ± 9	65.4	6.1 ± 1.3
P3D2	320	1400	-1.5	395 ± 12	177 ± 6	119 ± 6	63.7	7.3 ± 1.2
P3D3	320	1400	-1	390 ± 16	150 ± 6	120 ± 10	65.9	5.8 ± 1.3
P3D4	320	1400	-0.5	385 ± 17	182 ± 6	121 ± 10	62.3	7 ± 1.2
P3D5	320	1400	0	429 ± 20	175 ± 6	121 ± 11	66.9	7.5 ± 1.5
P3D6	320	1400	0.5	468 ± 14	191 ± 6	122 ± 12	71.2	9.5 ± 1.4
P3D7	320	1400	1	430 ± 18	182 ± 6	116 ± 12	69.5	7.8 ± 1.2
P3D8	320	1400	1.5	440 ± 15	177 ± 6	114 ± 16	63.5	8.2 ± 1.6
P3D9	320	1400	2	425 ± 16	189 ± 6	119 ± 12	68.2	7.6 ± 1.2

In this work, specimens to characterize the fractography were polished and finally etched in a solution consisting of distilled water (90 mL) with HF (10 mL) for 25 s. The macro- and microstructures of the tensile specimens were analyzed by a Zeiss Axioscop 40 Pol polarising microscope (Zeiss, Oberkochen, Germany) and scanning electron microscope (SEM) at 20 kV (Quanta 200, FEI Company, Eindhoven, the Netherlands). Chemical compositions of the crystalline phases in the as-printed alloy were investigated by an energy dispersive X-ray (EDX) spectroscope (EDAX Inc., Mahwah, NJ, USA). Phase identification was observed by a D8 Advanced X-ray diffractometer (XRD) with Cu Kα radiation (λ = 0.15418 nm) (Bruker AXS GmbH, Karlsruhe, Germany) at 20 kV and 20 mA using a continuous scan mode at 2°/min, The micro-CT experiments were analyzed at the high-resolution comprehensive microfocus CT detection system diondo d2 at 100 kV with the best resolution at ~2 μm. The crystallographic orientations of the samples were investigated via Electron Backscattered Diffraction (EBSD) by an SEM (ZEISS Merlin, Oberkochen, Germany) with a EDAX/Digiview 4 system. The relative density was determined by the Archimedes method as a percentage of the material's bulk density of 2.68 g/cm^3.

The crosshead speed of 1 mm/min was used in the tensile tests with an Instron 25 mm dynamic extensometer (Norwood, MA, USA) for controlled displacement. Microhardness tests were conducted with five repetitions per sample using a Vickers hardness tester FV-700 (FUTURE-TECH, Kawasaki, Japan) with 100 gf load for 15 s.

3. Results and Discussion

3.1. Crystalline Phases in the as-Printed AlSi10Mg

Figure 2a shows the SEM results of the spherical powder of an average size of 42 μm. Figure 2b shows two dominant XRD peaks belong to face-centered cubic (fcc) α-Al and eutectic Si-particles. The presence of the Mg$_2$Si precipitates is also detected, although the corresponding XRD peak is rather weak. The phases revealed by the XRD are typical of the as-printed AlSi10Mg alloy.

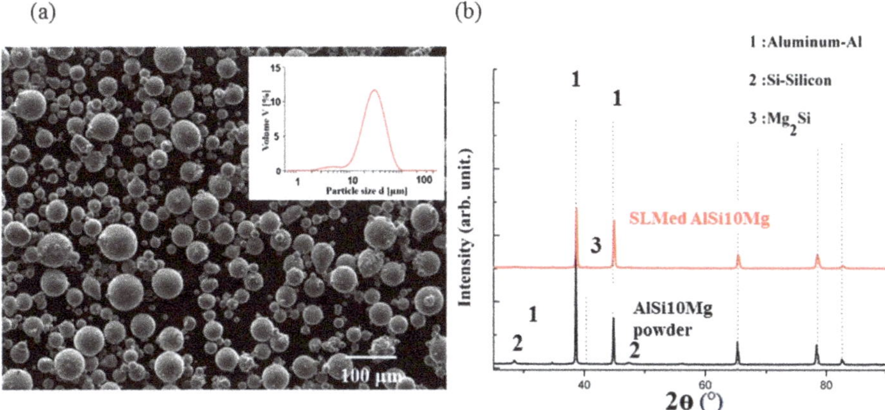

Figure 2. SEM images of the AlSi10Mg spherical powder and size distribution (**a**); Diffraction patterns of powdered samples in as-built conditions (**b**).

3.2. Effect of Laser Energy Density and Defocusing on Densification

In Figure 3a,b, the relative density of the as-printed AlSi10Mg vs. the "volumetric energy density" (ε) and DA, is shown, respectively. The ε is defined as follows:

$$\varepsilon = \frac{P}{vhd}$$

where P is the laser powder, v is the scanning rate, h is the layer thickness, and d is the layer width used.

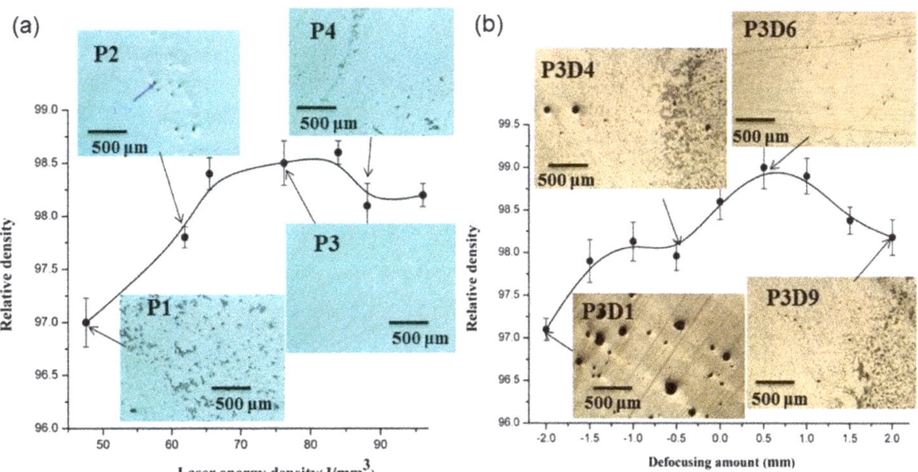

Figure 3. Relative density curves plotted against laser energy density and DA values and optical microscopy (OM) images of polished SLM-processed specimens (**a**,**b**).

In total, 60 samples parallel to the substrate were built and the SLM processing parameters are shown in Table 2. Optical micrographs of the samples are shown in Figure 3a for demonstrative purposes of the samples with low and high density. The general trend revealed by Figure 3a is that increase of ε (i.e., using a higher laser power at lower scan speeds) can lead to better densification. Figure 4a–c shows the porosity of as-printed AlSi10Mg samples from the micro-CT reconstructed profile at ε = 61.9 J/mm^3, 76.2 J/mm^3, 88.1 J/mm^3 respectively. The main reason for the interior porosity formation was that insufficient energy gave unmelted spots, forming voids between the melt pool line and hatch pattern. On the other hand, there are large cavities, a "balling" effect, and irregular long channels caused by highly dynamic and unstable melt pool tracks at an excess of laser energy [25].

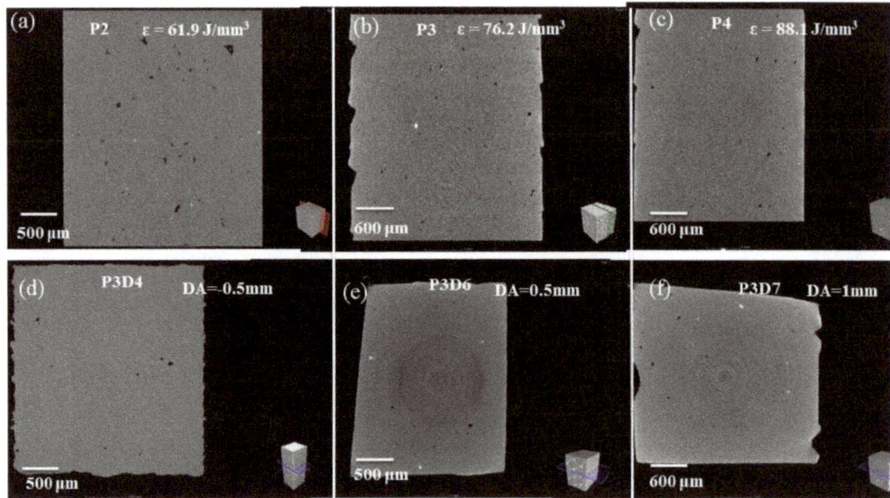

Figure 4. Tomographic images of SLM-processed AlSi10Mg samples at volumetric energy density (ε) about, 61.9 J/mm^3 (**a**), 76.2 J/mm^3 (**b**), 88.1 J/mm^3 (**c**), and DA values about, (**d**): −0.5 mm; (**e**): 0.5 mm; (**f**): 1 mm.

Figure 3b indicates that the DA has a prominent influence on density and increasing the DA to about 0.5 mm has resulted in the maximal density (via the sample of P3D6). In contrast, there is a large amount of pores existing in the microstructure if a negative DA is used, while the pulsed laser recoil force influenced by DA value has changed greatly from the micro-CT images, Figure 4d–f. If using a CW laser at negative DA, under certain conditions the so-called "keyhole-mode" laser welding is motivated [26,27], which forces melt pool oscillations and flow instability by disturbances from thermocapillary convection and the pulsed laser recoil force, and its depth undergoes severe oscillations, some bubbles from the metallic evaporation are trapped at the solidifying front of the tracks, leading to more near-spherical defects [28–30].

The surface roughness is reduced as shown in Figure 5a–d. Andrew Townsend et al. (2010) checked the roughness of additive manufactured (AM) Ti6Al4V samples used X-ray computed tomography (CT) [31]. The melt pool is forced to oscillations and flow instability by disturbances from thermocapillary convection and the pulsed laser recoil force, which results in a rough surface as shown in Figure 5a,b. The temperature gradient are intensified and in turn created surface tension gradient caused by the resultant turbulence of Marangoni flow within the molten pool, with a further increase in the laser energy, the melting pool size becomes bigger and the flow inside would be stabilized to eliminate this type of pores, followed by smoother surface roughness as shown in Figure 5c,d. For different DA values, there is no remarkable influence on the surface roughness because the volumetric energy density was a major influence factor for stability of the molten pool [32].

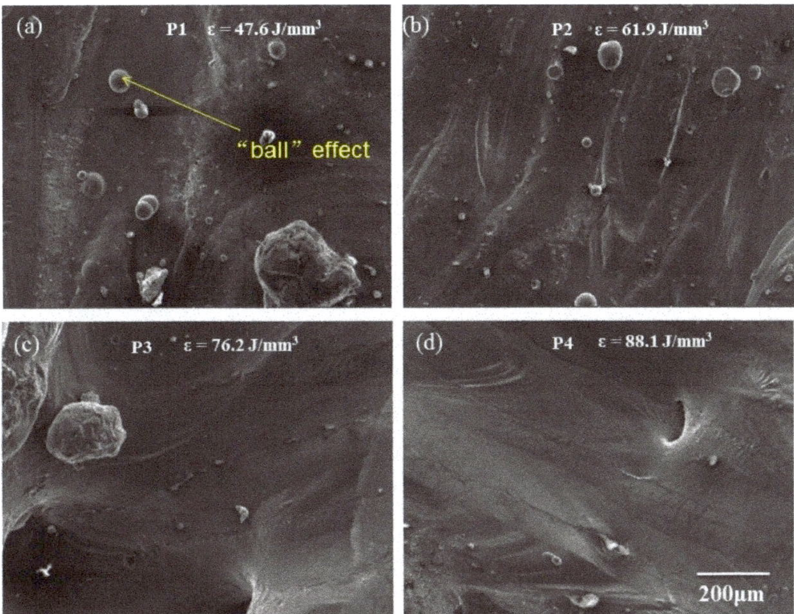

Figure 5. SEM images of the top surface of SLM-processed AlSi$_{10}$Mg samples P1–P4 (**a–d**), all images are at the same magnification.

3.3. MPBs in the as-Printed AlSi10Mg

The cross section of a melt pool formed during SLM can be described approximately as an arched structure schematically shown in Figure 6a. By means of the metallography, two types of MPBs can be found in the SLMed samples. One is the arched "layer–layer" parallel type of MPBs (Figure 6b) and the other is the "track–track" overlapping type of MPBs (Figure 6c). It is found that the parallel "layer–layer" MPBs connect with short "track–track" MPBs and form acute angles. Aside from sample density, the DA value also significantly affects the shapes of the MPBs (Figure 6d).

When DA value is negative (i.e., −2 mm < DA < 0 mm), as illustrated in Figure 1b, the laser melting seems to lead to the formation of a deep penetration in the powder bed, and the corresponding MPBs are closer to a cone-shaped "layer–layer" MPBs heaped up and "track–track" MPBs adjacently intersected with each other to form acute angles ($0° < \theta_T < 45°$). In the case of DA = 0, the MPBs cross-section of a melt pool has small "fish scale" morphology. When the DA increases from 0 mm to 2 mm, the plasma is more likely to be generated and the laser melting mode transits from the so called 'keyhole mode' to the 'conduction mode' as known in laser welding. In this case, the MPBs are prone to be the arched type (Figure 5d). It is found that the "track–track" MPBs usually intersect and form big angles ($45° < \theta_T < 90°$) while arcuate "layer–layer" MPBs are heaped up almost in parallel.

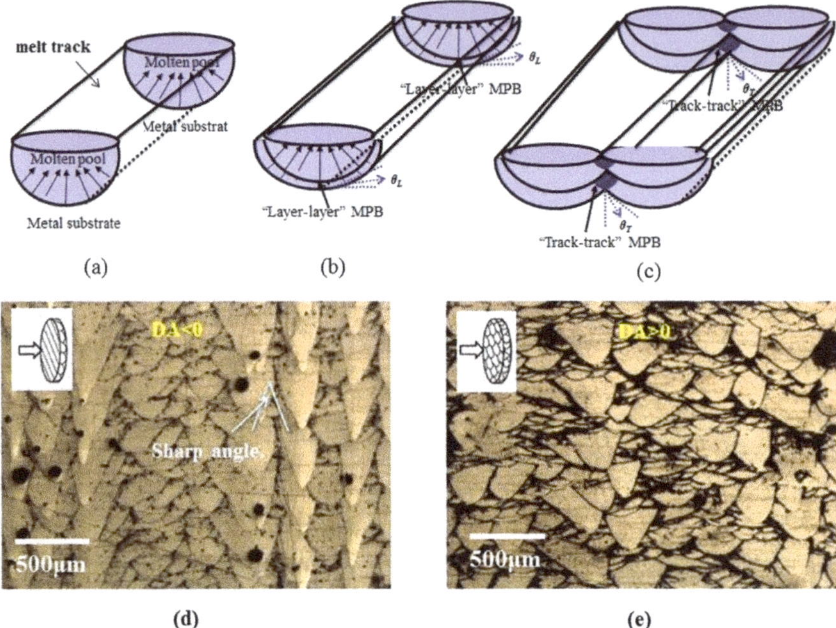

Figure 6. Schematic overview of the solidification of molten pools during the SLM process: (**a**) single half-cylindrical molten pool; (**b**) "layer–layer" molten pool; (**c**) "track–track" molten pool. The overlaying of molten pools with differently shaped molten pool boundaries (MPBs) of "fish scale" morphology produced in different DAs; (**d**) DA < 0; (**e**) DA > 0, respectively.

The slipping shear stress ψ of the horizontal tensile specimens along the slipping direction can be expressed as follows [24] (1):

$$\psi = \frac{F}{S}\cos\theta\cos\lambda \tag{1}$$

In Equation (1), S is the cross section, which the tensile load F is applied to, θ is the angle between the tensile load and the slip direction, and λ is the angle between the tensile load and the normal direction of the slip plane. The critical shear stresses ψ_k up to the yield point σs when the tensile stress (F/S) increases to a critical value. The yield limit σs and the critical shear stress ψ_k are expressed as follows:

$$\psi_k = \sigma s\cos\theta\cos\lambda \tag{2}$$

$$\sigma s = \psi_k\frac{1}{\cos\theta\cos\lambda} \tag{3}$$

When the loading force F of the horizontal samples is parallel with the X–Y plan, the angle λ between the loading force F and the slipping direction is near 0°. The critical shear stress ψ_k is only connected with the interfacial binding force of the slipping surface regardless of the change of the applied loads F, therefore the yield limit σs depends only on the variation of the angle θ.

Figure 7a shows the SEM images of two kinds of molten pool boundaries (MPBs): "layer–layer" and "track–track" MPBs on the cross section. The MPBs shown a honeycomb appearance and the original dendrite arm, which intersect each other to form acute angles shown in Figure 7b. The growth rate (R, m/s) and thermal gradient (G, K/m) during the SLM processing determine the microstructure of samples, when the thermal gradient is perpendicular to the laser scan movement direction, solidification mode varies at the melt pool boundaries (MPBs) edge leading to the coarse grain zone with roughly the width of 2 µm, which was similar with the weld heat-affected zone in

Figure 7c. The energy spectrum along a line (AB) shown the nonmetallic elements (O, Si) cross the coarse grain zone perpendicular to the MPBs as shown in Figure 7d, where cracks are easily initiated and have a deteriorative effect on mechanical properties of the SLM samples.

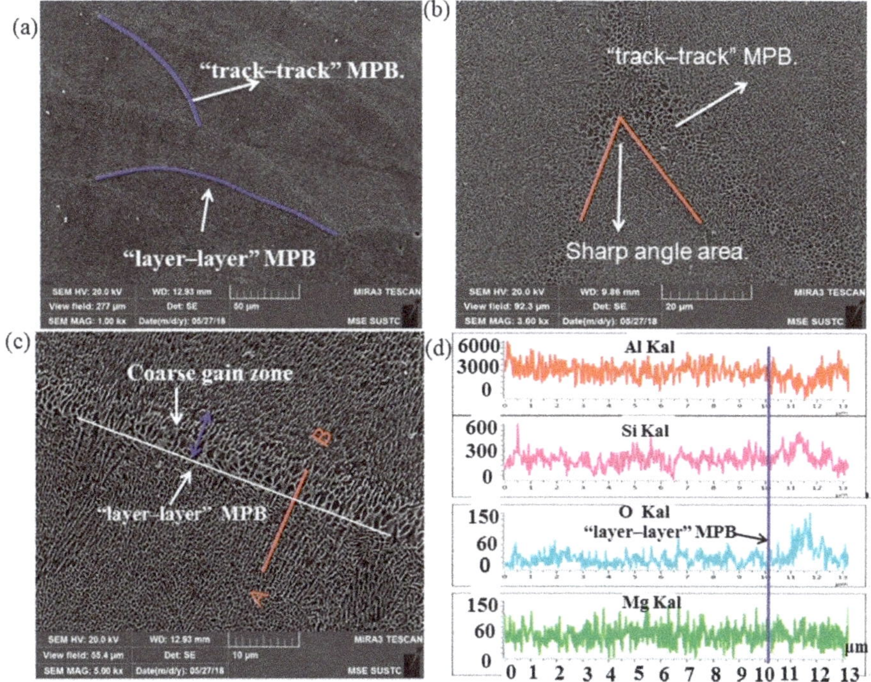

Figure 7. (**a**) SEM images of the MPBs in SLM samples in a section plane: (**a**) the cross section of low magnification morphology of the MPBs at the build direction; (**b**) SEM image of the morphology of "track–track" MPBs; (**c**) high magnification of morphologies of "layer–layer" MPBs; (**d**) the distribution of elements near MPB on the A–B line.

The cross-section microstructure of the as-printed specimen along the building direction is shown by the EBSD image (Figure 8a), where columnar grains with the size of 10–20 μm and intergranular spacing are generated. The orientation difference distribution of adjacent grains shown in Figure 8b, the misorientation angle about 35°–60°, accounted for most of the large angle grain boundary, which makes growth and propagation of cracks difficult, and consequently increases the toughness of the material.

3.4. Tensile Mechanical Properties

Table 2 summarizes the detailed tensile properties obtained from the thirteen batches of the SLMed AlSi10Mg alloy samples printed under different DA values (see Table 2) along with their micro-hardness results, provided by ASTM Standard B26/B26M. The best tensile properties achieved are tensile strengths of ~456 ± 14 MPa and elongation of 9.5 ± 1.4%, corresponding to the DA of 0.5 mm. Figure 9a provides representative tensile stress–strain curves of the as-printed AlSi10Mg alloy. It is noted that the variations of the DA have resulted in markedly different tensile properties.

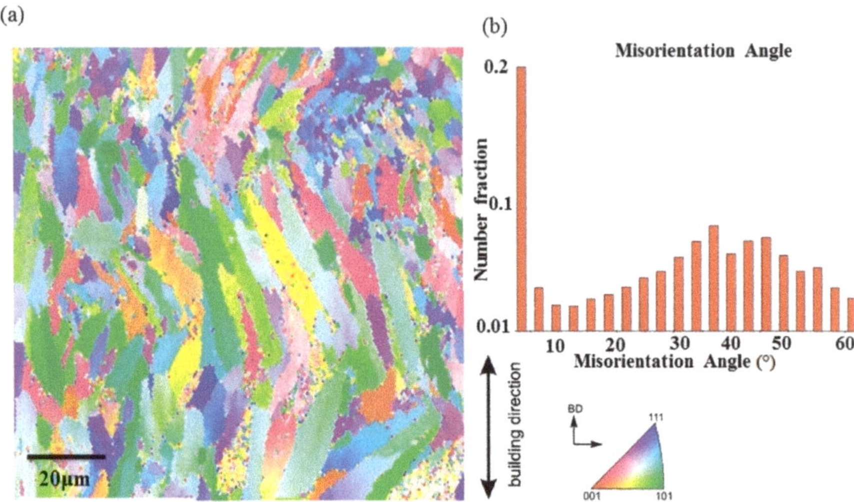

Figure 8. (**a**) The Electron Backscattered Diffraction (EBSD) orientation map of as-built AlSi10Mg sample at the building direction (BD); (**b**) Misorientation angle distribution.

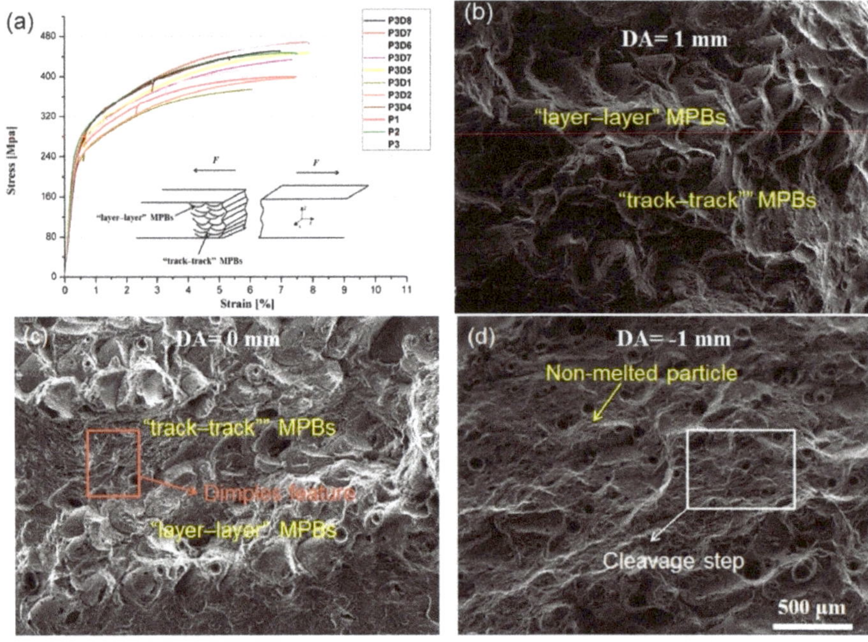

Figure 9. Engineering stress–strain curves of the as-built samples of horizontal direction at different laser power and DA, the Illustrations is the stress cross sections of the horizontal SLMed samples (**a**); the SEM image shows the fractured surface of SLM AlSi10Mg samples with the characteristic microstructure at laser power of P = 320 W and fixed scanning speed of 1400 mm/s but various DA of (**b**): 1 mm; (**c**): 0 mm; (**d**): −1 mm.

Besides grain slipping, The SLMed parts slip along the MPBs form ductile deformation preferentially because of poor adhesion and low stability in the junction between the MPBs. The tensile fracture surfaces display distinctly different features due to different DA values. When the DA increases from 0 mm to 2 mm, the cross section of horizontal processing specimens shows a particular wavy and serrated morphology, related to the surface of MPBs comprising of a mixture of arched parallel "layer–layer" and "track–track" overlapping MPBs, as shown in Figure 9b,c, which is supposed to be a contributing factor to the combination of high strength and good ductility. When DA value is negative (i.e., −2 mm < DA < 0 mm), the crack surfaces appear smoother than those of the other tensile samples. The fracture surfaces contain a mixture of smooth areas with porosity or shrinkage cavities and small dimples as shown in Figure 9d. Cleavage step and non-melted particles are also observed in Figure 9b–d; furthermore, the equiaxed submicrometer dimples with the same hot crack are observed at the whole fracture surfaces shown in Figure 10a,b.

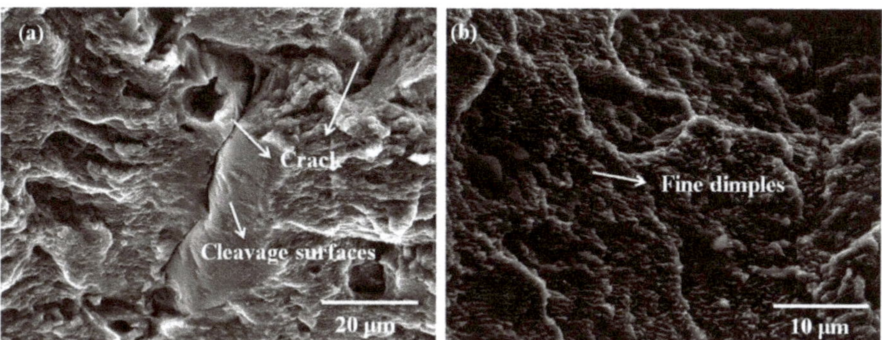

Figure 10. The high magnification SEM images shown fracture surface with crack (**a**) and mixture of small dimples (**b**).

Regardless of how the DA values changes when increase from −2 mm to 0 mm, the sliding surfaces containing "layer–layer" MPBs are almost planes and always vertical to the tensile loading ($\theta_L = 0°$) for the horizontal specimens (Figure 6c). It is difficult along the arched parallel "layer–layer" MPBs slipping surfaces and mainly to slip along overlapping "track-track" MPBs surfaces with the continuous increase of the tensile load ($0° < \theta_T < 45°$) until the occurrence of fracture. Such cracks extend along the "layer–layer" MPBs and the formation of cleavage surfaces occurs, indicating a brittle fracture feature.

When the DA increases from 0 mm to 2 mm, the sliding surfaces consisting of the actual "layer–layer" MPBs surfaces are not regular planes and arched shape "layer–layer" MPBs are also not vertical to the tensile loading ($45° < \theta_L < 90°$). On the other hand, the ductile deformation is easily formed by the slipping along "track–track" MPBs surfaces. Therefore, the tensile fracture SLM samples achieve the best ductility and highest elongation as Equation (3) due to the slipping alone both "track–track" ($45° < \theta_T < 90°$) and "layer–layer" MPBs ($45° < \theta_L < 90°$). With a combination of appropriate DA values, the as-printed AlSi10Mg can show good facture strength as well as good ductility, revealing a ductile fracture mechanism. The P3D6 sample is such a typical example. In summary, the DA has significant impacts on the fracture mode, macroscopic plastic behavior, and microscopic slipping of the as-printed alloy by influencing the shape of MPBs in SLM process.

4. Conclusions

In this study, the crack-free and near fully dense AlSi10Mg samples were fabricated using optimized SLM parameters. The processing-microstructure-mechanical properties correlation was established, allowing us to draw the following conclusions:

(1) It has been shown that the densification behavior of SLMed AlSi10Mg parts was significantly affected by the volumetric energy density (ε) and, particularly, defocusing amount (DA) during SLM.

the minimum porosity and the relatively high relative densities of over 98.5% are possible at the critical energy density of ~76.2 J/mm^3 at high laser power of about 320 w and scan speed of 1400 mm/s. With further increase of the laser energy, the melting pool size became bigger and the flow inside would to be stabilized to eliminate this type of pores and then smoother surface roughness. Simultaneously, the combination of ε and appropriate DA values of about 0.5 mm ensures the highest relative densities of over 99% for this alloy.

(2) The multi-layer and multi-track molten pools track overlapped to generate "layer–layer" and "track–track" MPBs, which have a significant impact on the plastic deformation behavior of the as-printed alloy by the slip theory analysis, the coarse grain zone can be found below the MPBs containing the segregation of nonmetallic elements where cracks are initiated easily.

(3) The defocusing amount (DA) is one of the main factors except for energy density that affects both MPBs' spatial topological structure as well as the density and the mechanical properties of the as-printed AlSi10Mg. The tensile fracture surfaces display distinctly different features due to different DA values. The tensile fracture SLM samples achieve the best ductility and highest elongation due to the slipping along both "track–track" ($45° < \theta_T < 90°$) and "layer–layer" MPBs ($45° < \theta_L < 90°$) when the DA increases from 0 mm to 2 mm. A ductile, dimpled failure mode along the two interlaced distribution of types of MPBs is observed in these as-printed specimens. High tensile strength (456 ± 14 MPa) and good tensile ductility ($9.5 \pm 1.4\%$) and microhardness of 122 ± 2 HV0.1 were achieved in the as-printed condition corresponding to DA = 0.5 mm.

Author Contributions: Conceptualization, S.Z.; Methodology, S.Z.; Software, S.Z.; Validation, M.Y.; Formal Analysis, M.Y.; Investigation, S.Z., Y.S.; Resources, S.Z., Y.S.; Data Curation, S.Z., Y.S.; Writing—Original Draft Preparation, S.Z.; Writing—Review & Editing, S.Z., Y.S., R.G., Z.W., Y.Z., Q.M. and M.Y.; Visualization, S.Z., Y.S.; Supervision, Q.M., M.Y.; Project Administration, Q.M., M.Y.; Funding Acquisition, M.Y.

Funding: This work was funded by Shenzhen Science and Technology Innovation Commission grant number [ZDSYS201703031748354], and the Development and Reform Commission of Shenzhen Municipality.

Conflicts of Interest: The authors declare no conflict of interest.

References

1. Berman, B. 3-D printing: the new industrial revolution. *Bus. Horiz.* **2012**, *55*, 155–162. [CrossRef]
2. Stucker, B.; Rosen, D.W.; Gibson, I. *Additive Manufacturing Technologies: 3D Printing, Rapid Prototyping, and Direct Digital Manufacturing*; Springer: Berlin, Germany, 2014.
3. Yap, C.; Chua, C.; Dong, Z.; Liu, Z.; Zhang, D.; Loh, L.; Sing, S. Review of selective laser melting: Materials and applications. *Appl. Phys. Rev.* **2015**, *2*, 041101. [CrossRef]
4. Gu, D. Laser Additive Manufacturing (AM): Classification, Processing Philosophy, and Metallurgical Mechanisms. In *Laser Additive Manufacturing of High-Performance Materials*; Springer: Berlin, Germany, 2015; pp. 15–71.
5. AlMangour, B.; Grzesiak, D.; Borkar, T.; Yang, J.-M. Densification behavior, microstructural evolution, and mechanical properties of TiC/316L stainless steel nanocomposites fabricated by selective laser melting. *Mater. Des.* **2018**, *138*, 119–128. [CrossRef]
6. AlMangour, B.; Grzesiak, D.; Cheng, J.; Ertas, Y. Thermal behavior of the molten pool, microstructural evolution, and tribological performance during selective laser melting of TiC/316L stainless steel nanocomposites: Experimental and simulation methods. *J. Mater. Process. Technol.* **2018**, *257*, 288–301. [CrossRef]
7. Wu, J.; Wang, X.; Wang, W.; Attallah, M.; Loretto, M. Microstructure and strength of selectively laser melted AlSi10Mg. *Acta Mater.* **2016**, *117*, 311–320. [CrossRef]
8. Murr, L.E.; Quinones, S.A.; Gaytan, S.M.; Lopez, M.I.; Rodela, A.; Martinez, E.Y.; Her-nandez, D.H.; Martinez, E.; Medina, F.; Wicker, R.B. Microstructure and mechanical behavior of Ti-6Al-4V produced by rapid-layer manufacturing, for biomedical applications. *J. Mech. Behav. Biomed. Mater.* **2009**, *2*, 20–32. [CrossRef]
9. Garibaldi, M.; Ashcroft, I.; Simonelli, M.; Hague, R. Metallurgy of high-silicon steel parts produced using Selective Laser Melting. *Acta Mater.* **2016**, *110*, 207–216. [CrossRef]

10. Li, X.P.; Ji, G.; Chen, Z.; Addad, A.; Wu, Y.; Wang, H.W.; Vleugels, J.; Van Humbeeck, J.; Kruth, J.P. Selective laser melting of nano-TiB$_2$ decorated AlSi10Mg alloy with high fracture strength and ductility. *Acta Mater.* **2017**, *129*, 183–193. [CrossRef]

11. Prashanth, K.G.; Scudino, S.; Eckert, J. Defining the tensile properties of Al-12Si parts produced by selective laser melting. *Acta Mater.* **2017**, *126*, 25–35. [CrossRef]

12. Ma, P.; Jia, Y.; Prashanth, K.G.; Scudino, S.; Yu, Z.; Eckert, J. Microstructure and phase formation in Al-20Si-5Fe-3Cu-1Mg synthesized by selective laser melting. *J. Alloys Compd.* **2016**, *657*, 430–435. [CrossRef]

13. Read, N.; Wang, W.; Essa, K.; Attallah, M.M. Selective laser melting of AlSi10Mg alloy: Process optimization and mechanical properties development. *Mater. Des.* **2015**, *65*, 417–424. [CrossRef]

14. Aboulkhair, N.T.; Everitt, N.M.; Ashcroft, I.; Tuck, C. Reducing porosity in AlSi10Mg parts processed by selective laser melting. *Addit. Manuf.* **2014**, *1–4*, 77–86. [CrossRef]

15. Brandl, E.; Heckenberger, U.; Holzinger, V.; Buchbinder, D. Additive manufactured AlSi10Mg samples using Selective Laser Melting (SLM): Microstructure, high cycle fatigue, and fracture behavior. *Mater. Des.* **2012**, *34*, 159–169. [CrossRef]

16. Amato, K.N.; Gaytan, S.M.; Murr, L.E.; Martinez, E.; Shindo, P.W.; Hernandez, J.; Collins, S.; Medina, F. Microstructures and mechanical behavior of Inconel 718 fabricated by selective laser melting. *Acta Mater.* **2012**, *60*, 2229–2239. [CrossRef]

17. Thijs, L.; Kempen, K.; Kruth, J.-P.; Humbeeck, J.V. Fine-structured aluminium products with controllable texture by selective laser melting of pre-alloyed AlSi10Mg powder. *Acta Mater.* **2013**, *61*, 1809–1819. [CrossRef]

18. Abe, F.; Costa Santos, E.; Kitamura, Y.; Osakada, K.; Shiomi, M. Influence of forming conditions on the titanium model in rapid prototyping with the selective laser melting process. *Proc. Inst. Mech. Eng. Part C-J. Mech. Eng. Sci.* **2003**, *217*, 119–126. [CrossRef]

19. Li, X.; Wang, X.; Saunders, M.; Suvorova, A.; Zhang, L.; Liu, Y.; Fang, M.; Huang, Z.; Sercombe, T.B. A selective laser melting and solution heat treatment refined Al-12Si alloy with a controllable ultrafine eutectic microstructure and 25% tensile ductility. *Acta Mater.* **2015**, *95*, 74–82. [CrossRef]

20. Rosenthal, I.; Stern, A.; Frage, N. Strain rate sensitivity and fracture mechanism of AlSi10Mg parts produced by selective laser melting. *Mater. Sci. Eng. A* **2017**, *682*, 509–517. [CrossRef]

21. Khairallah, S.A.; Anderson, A.T.; Rubenchik, A.; King, W.E. Laser powder-bed fusion additive manufacturing: Physics of complex melt flow and formation mechanisms of pores, spatter, and denudation zones. *Acta Mater.* **2016**, *108*, 36–45. [CrossRef]

22. Lin, X.; Yang, H.O.; Chen, J.; Huang, W.D. Microstructure evolution of 316L stainless steel during laser rapid forming. *Acta Metall. Sin.* **2006**, *42*, 361–368.

23. Yadroitsev, I.; Krakhmalev, P.; Yadroitsava, I. Selective laser melting of Ti6Al4V alloy for biomedical applications: Temperature monitoring and microstructural evolution. *J. Alloys Compds.* **2014**, *583*, 404–409. [CrossRef]

24. Wen, S.; Li, S.; Wei, Q.; Yan, C.; Sheng, Z.; Shi, Y. Effect of molten pool boundaries on the mechanical properties of selective laser melting parts. *J. Mater. Process. Technol.* **2014**, *214*, 2660–2667.

25. Zhou, X.; Liu, X.; Zhang, D.; Shen, Z.; Liu, W. Balling phenomena in selective laser melted tungsten. *J. Mater. Process. Technol.* **2015**, *222*, 33–42. [CrossRef]

26. Zhu, X.; Villeneuve, D.M.; Naumov, A. Experimental study of drilling sub-10 μm holes in thin metal foils with femtosecond laser pulses. *Appl. Surf. Sci.* **1999**, *152*, 138–148. [CrossRef]

27. Ng, G.K.L.; Li, L. The effect of laser peak power and pulse width on the hole geometry repeatability in laser percussion drilling. *Opt. Laser Technol.* **2001**, *33*, 393–402. [CrossRef]

28. Panwisawas, C.; Qiu, C.L.; Sovani, Y.; Brooks, J.W.; Attallah, M.M.; Basoalto, H.C. On the role of thermal fluid dynamics into the evolution of porosity during selective laser melting. *Scr. Mater.* **2015**, *105*, 14–17. [CrossRef]

29. Khairallah, S.A.; Anderson, A. Mesoscopic simulation model of selective laser melting of stainless steel powder. *J. Mater. Process. Technol.* **2014**, *214*, 2627–2636. [CrossRef]

30. Bauereiss, A.; Scharowsky, T.; Korner, C. Defect generation and propagation mechanism during additive manufacturing by selective beam melting. *J. Mater. Process. Technol.* **2014**, *214*, 2522–2528. [CrossRef]

31. Townsend, A.; Racasan, R.; Leach, R.; Senin, N.; Thompson, A.; Ramsey, A.; Bate, D.; Woolliams, P.; Brown, S.; Blunt, L. An interlaboratory comparison of X-ray computed tomography measurement for texture and dimensional characterisation of additively manufactured parts. *Addit. Manuf.* **2018**, *23*, 422–432. [CrossRef]
32. Alrbaey, K.; Wimpenny, D.; Tosi, R.; Manning, W.; Moroz, A. On optimization of surface roughness of selective laser melted stainless steel parts: A statistical study. *J. Mater. Eng. Perform.* **2014**, *23*, 2139–2148. [CrossRef]

Article

Influence of Selective Laser Melting Machine Source on the Dynamic Properties of AlSi10Mg Alloy

Ben Amir [1], Shmuel Samuha [2] and Oren Sadot [1,*]

[1] Department of Mechanical Engineering, Ben-Gurion University of the Negev, Beer-Sheva 84105, Israel; ben333amir@gmail.com
[2] Department of Materials Engineering, NRCN, P.O. Box 9001, Beer-Sheva 84190, Israel; samuha@post.bgu.ac.il
* Correspondence: sorens@bgu.ac.il; Tel.: +972-8-6477110

Received: 12 March 2019; Accepted: 3 April 2019; Published: 8 April 2019

Abstract: Selective laser melting (SLM) AlSi10Mg alloy has been thoroughly investigated in terms of its microstructure and quasi-static properties, owing to its broad industrial applications. However, the effects of the SLM process on the dynamic behavior under impact conditions remain to be established. This research deals with the influences of manufacturing process parameters on the dynamic response of the SLM on AlSi10Mg at a high strain rate of 700 to 6700 s^{-1} by using a split Hopkinson pressure bar apparatus. Examinations were performed on vertically and horizontally built samples, processed individually by two manufacturers using a different laser scanning technique on the same powder composition. It was concluded that the fabrication technique does not influence the true stress–true strain dependency at strain rates of 700 to 2800 s^{-1}. However, at higher strain rates (4000 to 6700 s^{-1}), this study revealed different plastic behavior, which was associated only with the horizontally built samples. Moreover, this study found different failure demeanors at true strains exceeding 0.8. The dynamic response was correlated with the as-built microstructure and crystallographic texture, characterized using the electron backscattered diffraction technique.

Keywords: selective laser melting; AlSi10Mg alloy; dynamic properties; impact; crystallographic texture

---

1. Introduction

Additive manufacturing (AM) is an advanced fabrication technique in which a component is built up in successive layers of material to create a three-dimensional (3D) structure under a computer control system. The process can be conducted from a wire feedstock, metallic sheets, or a powder that is selectively melted or sintered using an energy source [1]. In the selective laser melting (SLM) process, a thin powder bed layer is scattered on top of the previous melted layer. A laser beam selectively melts the powder at the locations that need to be solidified. The thickness of each powder layer is approximately 20 to 60 μm [2]. This step is followed by the depositing of fresh powder. The laser melts the new powder and a partial volume of the previous layer to consolidate the two layers together. This process continues until the part is completed [3].

Aluminum alloys are of significant interest for lightweight applications in the aerospace and automotive sectors. In particular, AlSi10Mg alloy is compatible with SLM as it is close to a eutectic Al–Si composition, promoting effective castability and strong weldability, as well as exhibiting very high density (99.7% pure aluminum). Moreover, age hardening is enabled by Mg and Fe, which increases the strength by means of the Mg_2Si and Al5FeSi precipitation sequence [3–7]. For SLM–ALSi10Mg, the enhanced or at least similar mechanical properties compared to conventional high-pressure die-cast AlSi10Mg material can be attributed to the extremely fine microstructure resulting from the manufacturing process [3,8–12]. Numerous process parameters in SLM directly or indirectly alter the final product, and therefore, various studies have attempted to determine the parameters with the greatest impact on the manufacturing process. Some of the parameters that have been identified are the laser power, scanning velocity, hatching distance, and scanning pattern. Substantial efforts have been made to optimize the processing parameters to obtain similar or even superior mechanical properties with respect to cast AlSi10Mg alloy [9,13–15]. It has been suggested that the hatching distance has a significant effect on the mechanical properties and porosity, by increasing the volumetric energy, the porosity can be reduced at the scanning center and increased at scan edges [12,13,16]. The volumetric energy porosity together with the scanning strategy affects the surface topology, dimensional accuracy, residual stresses, and crystallographic texture [14,15,17]. In order to optimize the fatigue resistance, reduction of the porosity must occur mainly at the surface or the subsurface of the samples, along with post-processing heat treatment [3,16]. In addition to optimization research, extensive inquiries into the quasi-static mechanical properties of SLM aluminum alloy have been conducted [3,4,6,18–21]. These studies have suggested that the SLM aluminum yield stress, elongation, and hardness are approximately the same or even higher than those of conventional cast aluminum.

A study performed by Rosenthal et al. revealed sensitivity to the strain rate in the range of 2.77×10^{-6} to 2.77×10^{-1} s^{-1} [6]. Moreover, they and others identified a noticeable dependency of the mechanical properties on the building orientation under quasi-static stress–strain conditions [17,18].

To date, the majority of studies have focused on the quasi-static mechanical properties of SLM–AlSi10Mg alloy. However, to the best of the authors' knowledge, a serious gap exists in the available literature regarding the mechanical behavior of SLM–AlSi10Mg alloy under high strain rates. It should be emphasized that it is imperative to investigate the dynamic behavior of SLM–AlSi10Mg alloy, as unpredicted impact situations in service conditions (such as accidental events including vehicle crashes) may expose this alloy to dynamic loading.

Recent studies by Asgari et al. [22] and Hadadzadeh et al. [23] investigated the behavior of SLM–AlSi10Mg alloy under dynamic loading in a strain rates range of approximately 150 to 4300 s^{-1}, and noted that increasing the strain rate results in a higher dynamic yield, as well as modification of the alloy substructure and crystallographic texture. Moreover, the compressive deformation at a strain rate of 1400 s^{-1} exhibits an increase in the dislocation density and development of accumulated dislocation networks. However, no dependency in the building orientation was identified [24].

In an early study by our group [8], we focused on the effect of heat treatment on the dynamic mechanical properties. In [8], the investigated SLM–AlSi10Mg alloy was fabricated by EOS (described in Section 2) using a similar set of SLM parameters as used in the present study, under strain rates ranging from 700 to 7900 s^{-1}. It was found that the dynamic behavior is sensitive to heat treatment and varies drastically as a function of the building orientation. Nurel et al. [8] noted that the heat treatment reduced the dynamic behavior differences but did not eliminate it. Moreover, it was shown that the strain rate had a minor effect on the mechanical properties. Moreover, Zaretsky et al. characterized the dynamic properties of SLM–AlSi10Mg, under strain rates ranging from 5000 to 100,000 s^{-1}, using planar impact tests and found that the dynamic response of SLM-processed alloys is virtually independent of the processing orientation [25].

To the best of the authors' knowledge, the dynamic response of SLM–AlSi10Mg alloy has been studied on samples that differed only in their built-up direction, regardless of the pattern strategy and other process parameters. Thus, the novelty of the present work is to provide an improved understanding of the roles of the process parameters and building direction on the dynamic properties of AlSi10Mg alloy. For this purpose, the samples were fabricated by the EOS M280 and the Concept Laser (CL) X line 1000R. Both of these implement similar SLM-powder bed technology using AlSi10Mg alloy and known for their unique fabrication strategy. The dynamic properties were experimentally characterized using the split Hopkinson pressure bar (SHPB) on samples in the "as-built" state (without heat treatment), at two different building orientations, XY- and Z-oriented, under strain rates ranging from 700 to 6700 s^{-1}—which were achieved by impact velocities in the range of 10–30 m/s, similar to those encountered in car accidents. It should be clarified upfront that the aim of this research is not to determine the "best printer", but rather to understand the manner in which the printing parameters affect the dynamic mechanical properties. Therefore, we chose to test samples manufactured by each machine in its optimal setup. This study combines a systematic investigation of the microstructure and crystallographic texture in the initial state, employing the electron back-scattered diffraction (EBSD) technique coupled with a high-resolution scanning electron microscope (SEM). The remainder of this paper is arranged as follows—the experimental and building methods are described, followed by characterization of the sample microstructures and the results from the SHPB systems. Finally, concluding remarks are provided.

2. Experimental Method

2.1. Machines and Material

Two machines, distinguished by their unique processing strategies, were selected—the EOS M280 and CL X line 1000R. Each machine fabricated samples of AlSi10Mg alloy using fresh powder, with the chemical compositions presented in Table 1. A comparison between the manufacturer powder particle size and sphericity is presented in Table 2 as a cumulative distribution of the undersized 10%, 50%, and 90% of the tested powder as reported by the manufacturers. In the SLM process, the particle size and the sphericity reflect on the powder quality and have a major influence on the product's mechanical properties [26]. The slight variation in the fresh powder was considered negligible in the present study.

Table 1. Chemical composition, in wt.%, of AlSi10Mg alloy [27,28].

Al	Si	Mg	Fe	Mu	Ti	Cu	Zu
Balance	9–11	0.2–0.45	0–0.55	0–0.45	0–0.15	0–0.1	0–0.1

Table 2. Cumulative distribution of the AlSi10Mg alloy powder particle size and sphericity.

Manufacturer	Property	10%	50%	90%
EOS	particle size (m)	39.5	63.8	87.1
	sphericity	0.79	0.91	0.95
CL	particle size (m)	47.0	64.7	80.2
	sphericity	0.73	0.91	0.93

The EOS M280 and the CL X line 1000R machines are both fiber laser based SLM systems, distinguished by the set of process parameters, the main parameters are presented in Table 3. Other process parameters can be found on the manufacturer's web pages [27,28]. The exact set of parameters used to fabricate the samples for the present study was the optimal default setting recommended by the manufacturers. The scanning strategy is different between the two machines. EOS uses the bi-directional scanning strategy of rotating each layer by 67°, as illustrated in Figure 1a; while the "chessboard strategy" built-up process of the CL machine included a 90° rotation between each layer, as illustrated in Figure 1b. The quasi-static tensile mechanical properties of the alloy are presented in Table 4.

Table 3. Selective laser melting (SLM) process parameters as reported by manufacturers [27,28].

SLM Machine	Chamber Atmosphere	Build Platform Size (mm)	Focus Diameter (μm)	Laser Power (W)	Maximum Scanning Speed (m/s)	Layer Thickness (μm)
EOS M280	argon	250 × 250 × 300	80	400	1	~60
CL X line 1000R	nitrogen	630 × 400 × 500	100 to 500	1000	7	30 to 200

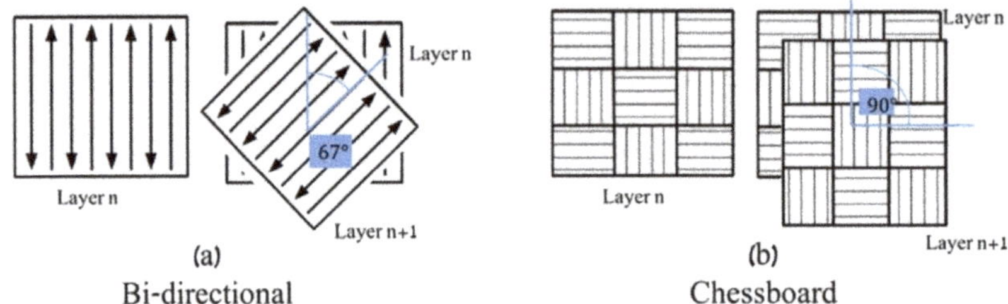

Figure 1. Schematic of scanning strategies by (**a**) bi-directional, EOS M280; and (**b**) chessboard, CL X line 1000R.

Table 4. Quasi-static tensile mechanical properties, as reported by manufacturers [27,28].

Manufacturer	Build Direction	Yield Stress (MPa)	UTS (MPa)	Elongation to Fracture (%)	Modulus of Elasticity (GPa)
EOS	Horizontal (XY)	270 ± 10	460 ± 20	9 ± 2	75 ± 10
	Vertical (Z)	240 ± 10	460 ± 20	6 ± 2	75 ± 10
CL	Horizontal (XY)	218 ± 7	345 ± 11	3 ± 1	75
	Vertical (Z)	214 ± 19	345 ± 8	3 ± 1	75

2.2. Description of Samples

The samples used for the high strain rate tests were machined in two orientations. The vertically built samples, designated by 'Z', had their long axes oriented parallel to the building direction; while the horizontally built samples, designated by 'XY', were oriented normal to the building direction. The vertical (Z) and horizontal (XY) rod-shaped samples are illustrated in Figure 2. To cancel shell effects and make the standard sample size suitable for our SHPB system, the samples were machined from circular rod-shaped cylinders of 13 mm in diameter and 100 mm in height into circular discs of 7 mm in diameter and 3.5 or 7 mm in height.

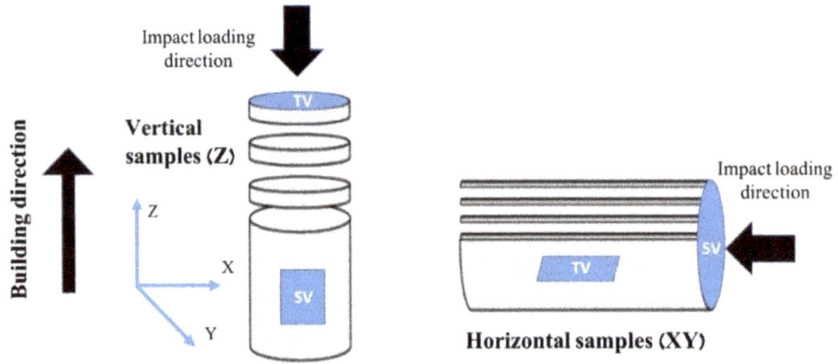

Figure 2. Schematic of vertical and horizontal samples, describing building direction, impact loading direction, and sectioning for microstructural studies. Top view (TV) and side view (SV) denote top view and side view, respectively.

2.3. SHPB System

Each sample was dynamically loaded using the SHPB system, located at the Shock Wave Laboratory at the Ben Gurion University of the Negev. Figure 3 presents a schematic of the system. The experiment was conducted by placing a sample between the incident and transmission bars. A third, shorter striker bar was accelerated using a gas gun. The striker bar impacted the incident bar, which generated a stress pulse in the incident bar. The striker bar velocity dictated the pressure pulse in the bars. A micro-controller filled the gas gun pressure chamber according to the desired striker velocity. The striker bar velocity was measured in close proximity to the incident bar. The impact between the two bars generated an elastic stress wave that propagated along the incident bar towards the sample. Owing to the acoustic impedance mismatch between the bar and sample, some of the incident pressure waves were reflected back into the incident bar, while the remainder were transmitted into the transmitted bar. The stress wave caused a strain along the bars, which was measured by strain gauges located on the incident and transmission bars. Figure 4 depicts the typical strain signals measured on the two bars. The curves representing the strain signal measured on the incident bar and on the transmitted bar are marked by blue and green lines, respectively. The sample stress, strain, and strain rate were analytically calculated by using the standard SHPB, as expressed in Equations (1)–(3) [29], under the assumptions of a one-dimensional (1D) single-wave analysis and force equilibrium at the specimen surfaces.

$$\dot{\varepsilon}_s(t) = \frac{-2C_B \varepsilon_R(t)}{H_s} \tag{1}$$

$$\varepsilon_s(t) = -2\frac{C_B}{H_s}\int_0^t \varepsilon_R(t)\,dt \tag{2}$$

$$\sigma_s(t) = \frac{A_B}{A_s}E_B\varepsilon_T(t) \tag{3}$$

The subscripts S and B refer to the specimen and bar, respectively, while the subscripts R and T denote the reflected and transmitted bars, respectively. Moreover, ε is the strain, $\dot{\varepsilon}$ is the strain rate, σ is the stress, and A is the area cross-section. H is the specimen height, C_B is the elastic wave velocity in the bar, and E_B is the material elastic modulus of the bars.

Hence, in every experiment, the equilibrium assumption was validated by testing that the strain signal created by the incident stress wave (ε_T) was equal to the combined signals created by the transmitted wave (ε_I) and reflected wave (ε_R).

The strain rates ranged from 700 to 6700 s^{-1}. The experiments were conducted under two strain rates ranges of 700 to 2800 s^{-1} and 4000 to 6700 s^{-1} for each manufacturer (EOS and CL) for each orientation (XY and Z). Each unique experiment was repeated at least three times to ensure reliability and repeatability, 27 experiments were conducted in total. In the present study, the results were presented as the true stress, true strain, and true strain rate, and conversions were performed as detailed in [30]. To confirm that the mechanical properties extracted from the SHPB experiments were reflected solely in the properties of a sample, regardless of its geometry, two sample types distinguished by their geometry were investigated under the same strain rate conditions. The sample heights were 3.5 and 7 mm, with a similar diameter. The relative difference between the two results was less than 1%, and therefore negligible [8]. In order to assure the repetition of the SLM process and its effect on the dynamic properties, two individual batches of samples were fabricated by EOS at Z orientation and tested at the SHPB system. The experiments were performed at strain rates of 4000 to 4300 s^{-1}. The true stress–true strain curves are presented in Figure 5a. The difference between the curves is less than 3% and could be considered to be minor. Figure 5b presents Z-oriented samples fabricated by CL in center and on the east side of the building platform. The difference between the results is less than 4%. Therefore, the effect of the sample's location on the building platform was not taken into consideration. It is well known that 2D/3D effects may exist in SHPB systems that need to be accounted for in certain cases [29]. Such effects cause a dispersion phenomenon, which was investigated and found to be negligible in our system. For further details, see [8].

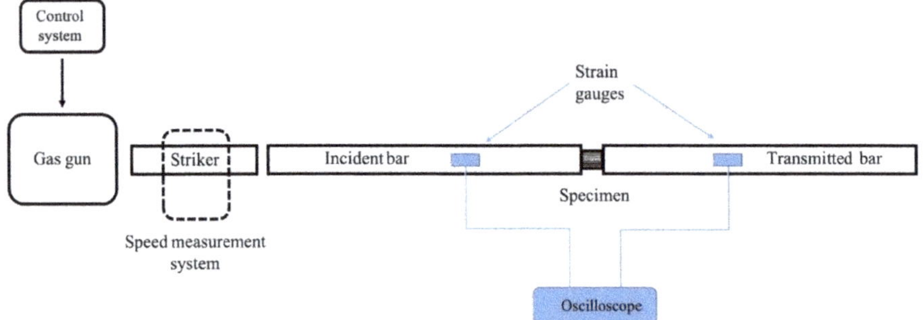

Figure 3. Schematic of the split Hopkinson pressure bar (SHPB) experimental system.

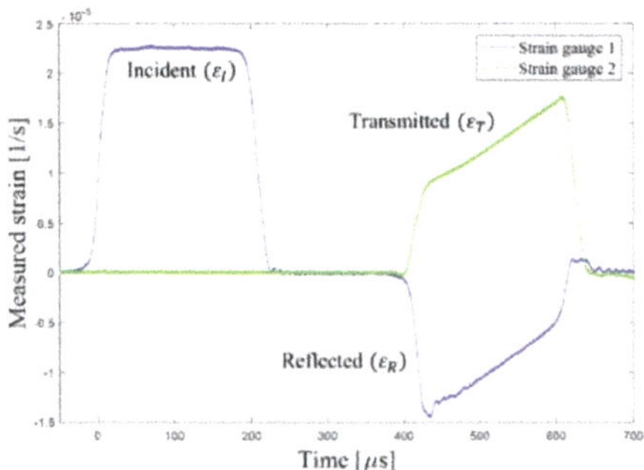

Figure 4. Typical strain signals measured by strain gauges on incident bars (blue) and transmitted bars (green).

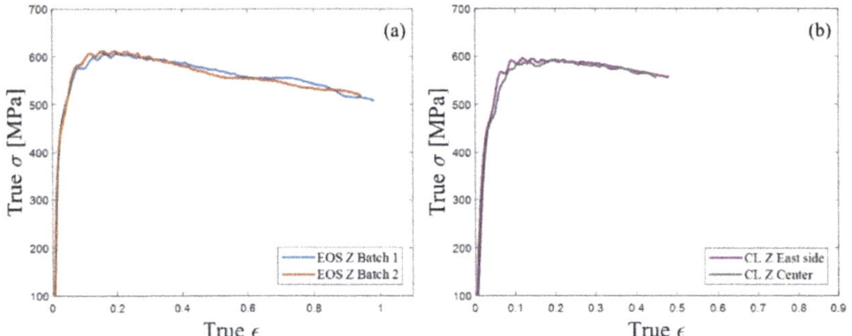

Figure 5. Presentation of the repeatability by comparing, **(a)** 2 EOS's Z-orientation samples from deferent batches and, **(b)** CL's Z orientation samples located at the center and in the east side of the building platform.

2.4. Microstructural Investigation

The microstructures and crystallographic textures of the as-built samples were studied using SEM (MIRA-3, Tescan) with a field-emission gun, equipped with an EBSD detector. The sample surface preparation included mechanical grounding with SiC papers of grit sizes down to 4000, followed by mechanical-chemical polishing up to 0.05 μm using colloidal silica. The principal directions of the SEM-EBSD samples are denoted by the building direction (BD), transverse direction (TD), and normal direction (ND). Successful interpretation of the Kikuchi patterns was obtained by scanning using a step size ranging from 0.5 to 1 μm, in accordance with the severity of the local deformations and grain sizes. The MTEX package, which is a freely available MATLAB toolbox (2017b), was employed for post-processing and visualization of the EBSD data [31]. The characterization also included studying the specimen fracture surfaces following dynamic testing using a standard stereoscope.

3. Results and Discussion

3.1. Microstructural Investigation of As-Built Samples

Figure 6a,b presents top-view images of the fabricated SLM–AlSi10Mg parts, when employing the CL and EOS manufacturing techniques, respectively. For the sake of clarity, enlarged images are presented in Figure 6c,d for the CL and EOS parts, respectively. The unique scan pattern attributed to each manufacturer can be clearly observed. The building surface of the CL part was constructed using an island strategy with 90° rotation (chessboard orientation), as indicated in Figure 6a. A laser-scanning pattern with 67° rotation can be observed in the EOS part in Figure 6b. Typical characteristics of scan vectors and melt pool morphology are evident. It is worthwhile noting the existence of the relatively large pores (in the order of 5 to 50 µm), for the EOS sample. The formation of these pores might be attributed to the process parameters, namely, the rates of scan speed and the hatch spacing, in particular [17,32]. Following a close examination, unmelted powder at the pores' circumference was not observed, suggesting that these microstructure defects can be designated as metallurgical pores rather than keyhole type. These metallurgical pores nucleate at slow scanning speeds from gases trapped within the melt pool or evolved from the powder during consolidation.

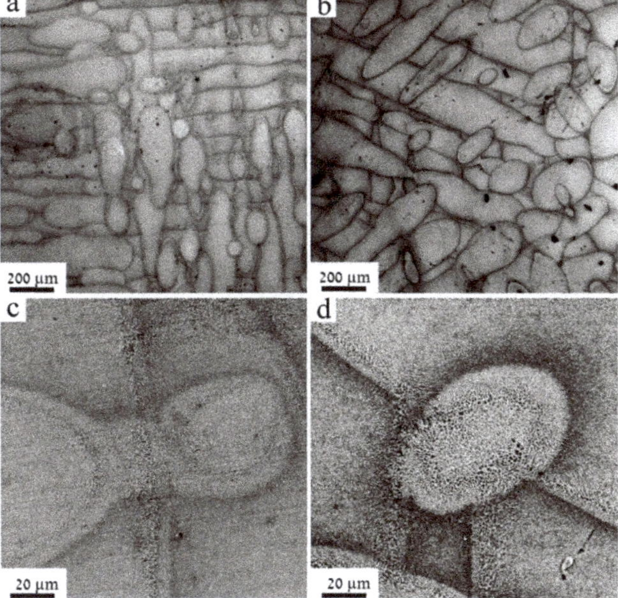

Figure 6. Building surface of SLM–AlSi10Mg part manufactured along the Z orientation, using (**a**) CL and (**b**) EOS techniques, respectively. Images (**c**) and (**d**) are enlarged micrographs of (**a**) and (**b**), respectively.

Figure 7a,c presents top views of the SLM–AlSi10Mg alloy manufactured by the CL and EOS techniques, respectively. The figures are described using band contrast (BC) maps. In this type of map, the contrast is based on the quality of an EBSD pattern obtained from each pixel. As an example, low BC values (blackish color) are assigned to grain boundary or highly deformed regions, while high BC values (bright color) are assigned to dislocation-free zones. Following a close examination of these maps, it could be observed that both structures consisted of equiaxed grains with bimodal grain sizes. Fine grains were formed at the melt pool boundaries, that is, at the borders of scan edges. Their sizes are attributed to rapid solidification, which in turn hinders the grain growth mechanisms. However, a coarser grain size was present at the melt pool, owing to the localized heat flow. Figure 7b,d presents color-coded inverse pole figure (IPF) maps. These types of maps describe the crystal orientation of a grain relative to the sample coordination system. Similar colors indicate grains with close crystallographic orientations. High-angle grain boundaries (misorientation angles higher than 15°) are highlighted in IPFs using black color marks. Upon investigation of these maps, it was found that coarse grains, present at the melt pool centerline, are oriented preferably with their <001> parallel to the BD (cube texture). This dominant texture component—designated using reddish grains in the IPFs—was expected, as the preferred growth direction of cubic crystals during solidification is <100> [14]. However, the fine grains, present at the melt pool boundaries, deviated far from this texture component, as evidenced by the large variety of colors assigned to these grains.

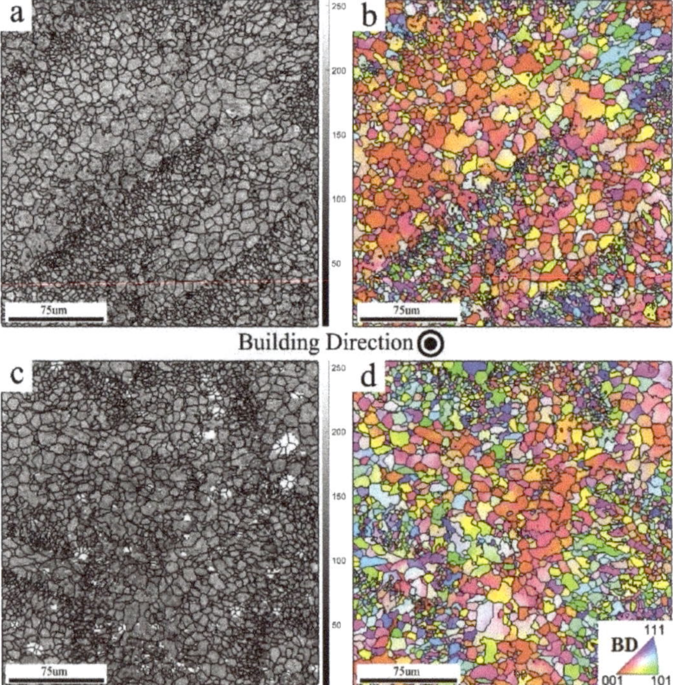

Figure 7. Building surfaces of (**a,b**) CL and (**c,d**) EOS manufactured AlSi10Mg SLM vertical samples, indicating (**a,c**) band contrast (BC) and (**b,d**) inverse pole figure (IPF) maps, respectively. In (**d**), an inset color-coded key triangle indicates the crystallographic orientations.

It is interesting to note that a weaker cube texture was observed for the EOS sample compared to that of the CL. It is plausible to assume that the scanning strategy was the cause of these textural changes. According to [18], a significantly weaker textural component is expected for an SLM product fabricated using a scanning pattern with an axial rotation other than 90°, as in the case of the EOS with a 67° rotation.

A side view of the manufactured SLM–AlSi10Mg alloy is presented in Figure 8a–c and 8d–f for CL and EOS, respectively. This plane describes the microstructure and crystallographic texture perpendicular to the building layers. As such, the cross-section of the melt pools is visible, with its typical half cylindrical morphology (see Figure 8a,d). In contrast to the equiaxed grains characterized for the top view (Figure 7), the majority of grains in the side view solidified in a columnar morphology, suggesting epitaxial growth, enabled by the lack of a nucleation barrier to solidification. These grains were elongated with their long vectors in parallel to (or slightly diverted from) the building direction. Others solidified with a fine equiaxed morphology and could easily be traced, as they were laced along the boundaries of the columnar grains. It is also worth mentioning the crystallographic texture distribution of the grains with respect to their relative sizes. The columnar grains were characterized by a cube texture component, while the fine grains were randomly oriented, as evidenced by various fine grain colors on the corresponding EBSD images (see Figure 8b,d). Therefore, this bi-model structure exhibited a bi-model crystallographic texture, which in turn contributed to the anisotropic properties of the SLM material, at least hypothetically.

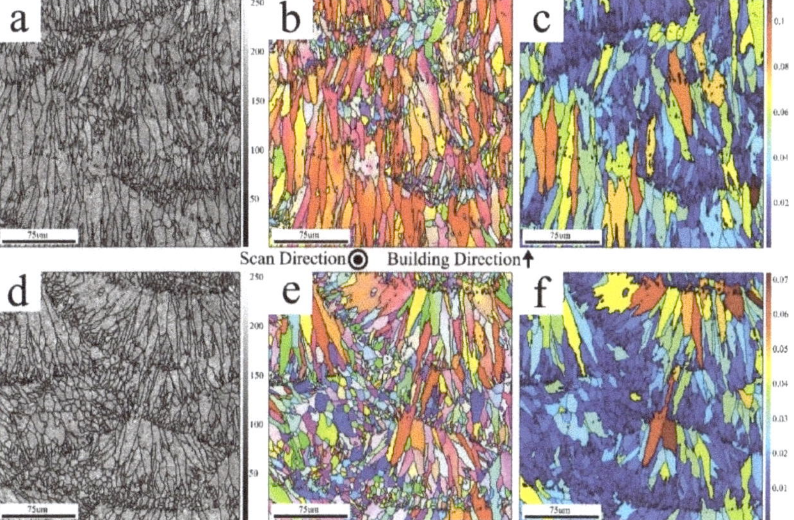

Figure 8. Side view of the SLM–AlSi10Mg samples manufactured by (**a–c**) CL and (**d–f**) EOS, indicating (**a,d**) BC, (**b,e**) IPF (∥ BD), and (**c,f**) grain orientation spread (GOS) maps, respectively. GOS values are located on the right side of each GOS map.

As in the top view, a weaker cube textural component could be observed in the side view for the EOS sample, resulting from the scanning pattern. Figure 8c,f presents qualitative intra-grain strain analyses using the grain orientation spread (GOS) plots on the EBSD data recorded from the region of interest (ROI) indicated in Figure 8a,d. This type of map is color-coded using blue to red colors, which are correlated with the deformation severity for each grain. The assigned colors for an annealed grain, which lacks large deviation from its nominal crystallographic orientation; and for a deformed grain with a high density of the dislocation network, are blue and red, respectively. As can be observed in Figure 8c,f, both structures consisted of fine grains with a low GOS value, which point to their annealed state resulting from repeated local solidification. However, the elongated grains, with a columnar morphology, exhibited substantially higher GOS values and were therefore interpreted with a higher internal distortion. From the comparison of both plots, a further noticeable difference is the higher areal density of grains with a lower GOS in EOS compared with that of CL. That is, the EOS structure was significantly more stress relaxed. This can be associated with the higher solidification rate owing to the faster-moving laser source in the case of the CL.

3.2. Dynamic Mechanical Behavior

Figure 9 represents the mean values of the true stress–true strain curves obtained from the SHPB system. The results are from samples that were manufactured by the EOS machine. Figure 9a depicts the experiment conducted under strain rates ranging from 700 to 2800 s^{-1}; while Figure 9b depicts the results under strain rates from 4000 to 6700 s^{-1}. The strain is a function of the striker velocity, striker length, material, and geometrical properties of the bars and sample [33]. For the same set of geometric parameters, the experiment conducted with an increasing striker velocity resulted in an increasing strain rate, therefore increasing strain and stress in the samples. As reported in previous studies [8], the Z-oriented samples of EOS exhibit an approximately 10% higher yield stress level than the XY-oriented samples. This phenomenon refers to the point at which the sample begins its plastic failure (true $\varepsilon \approx 0.1$). At true strain and higher, $\varepsilon \geq 0.4$, it can be observed that the stress in XY-oriented samples remained constant and could endure the plastic deformation without reducing the stress. However, in the Z-oriented samples, higher stress levels were reached before plastic failure under the same conditions, but as the plastic deformation continued, the stress in the Z-oriented samples exhibited a decline, dropping below the stress level measured in the XY-oriented samples. In the studies of Asgari et al. [22] and Hadadzadeh et al. [23], the XY- and Z-oriented samples had the same dynamic behavior, and no significant differences were found in the stress–strain curves.

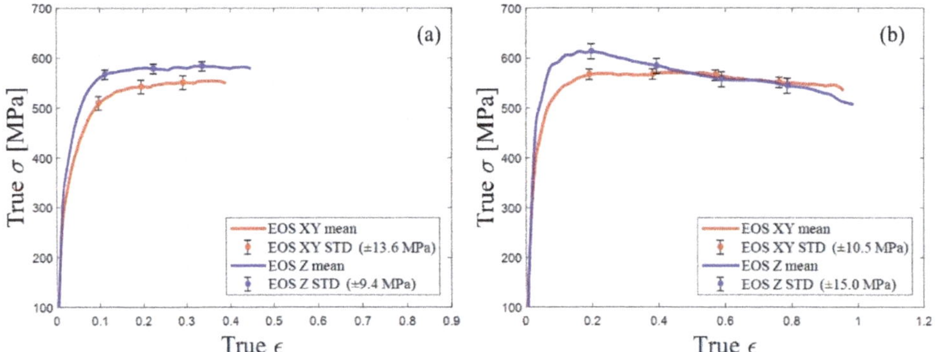

Figure 9. EOS samples mean value and standard deviation (STD) of true stress–true strain curves: (**a**) Comparison between XY- and Z-oriented samples at true strain rate ranged of 700 to 2800 s^{-1}; (**b**) comparison between XY- and Z-oriented samples at true strain rate ranged of 4000 to 6700 s^{-1}.

Similar SHPB experiments were performed with samples manufactured by CL, and the mean values of the true stress–true strain results are presented in Figure 10a,b for strain rates of 700 to 2800 s^{-1} and 4000 to 6700 s^{-1}, respectively. Focusing on the plastic failure point (true $\varepsilon \approx 0.1$), similar phenomena for the building orientation to those found in EOS were also identified in the samples manufactured by CL, and the stress value measured for the Z-oriented samples was approximately 10% higher than that of the XY-oriented samples. In the range of true strain $\varepsilon = 0.35$ to 0.8, the true stress characterization in the XY-oriented samples remained constant, while a decline was observed in the Z-oriented samples. As the true strain reached values of approximately 0.8 and higher (true $\varepsilon \geq 0.8$), a sharp reduction in the true stress appeared in the XY- and Z-oriented samples manufactured by CL. This sharp decline is normally associated with cracks developing within the alloy (Section 3.3).

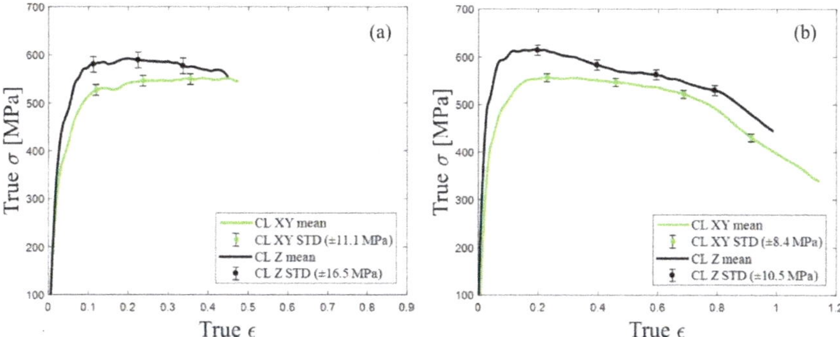

Figure 10. CL samples' mean value and standard deviation (STD) of true stress–true strain curves: (**a**) comparison between XY- and Z-oriented samples at true strain rate range of 700 to 2800 s^{-1}; (**b**) comparison between XY- and Z-oriented samples at true strain rate range of 4000 to 6700 s^{-1}.

To characterize the effect of the manufacturing on the dynamic behavior, this study implemented a comparison between the manufactures attributed to the building orientation and strain rate. Figure 11a,c depicts comparisons of the strain rates in the range of 700 to 2800 s^{-1} for the Z and XY orientations, respectively. In the presented strain rate range, a comparison between the two sample types (EOS and CL) revealed no significant difference in the strain–stress curves regarding the building orientation. This is valid up to a true strain of 0.5, as observed in Figure 11a for the Z orientation and in Figure 11c for the XY orientation (both under strain rates of 700 to 2800 s^{-1}).

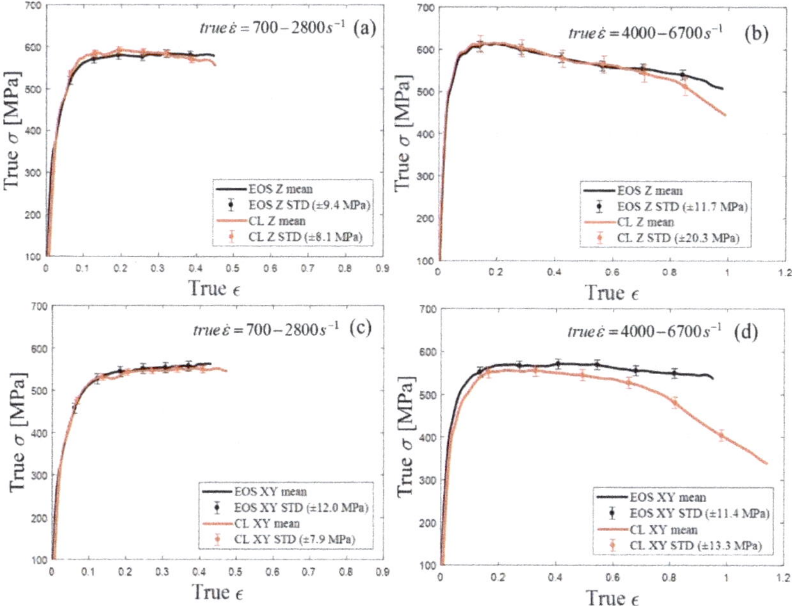

Figure 11. CL vs. EOS samples mean value and standard deviation (STD) of true stress–true strain curves: (**a**) Comparison between Z-oriented samples at true strain rate range of 700 to 2800 s^{-1}; (**b**) comparison between Z-oriented samples at true strain rate range of 4000 to 6700 s^{-1}; (**c**) comparison between Z-oriented samples at true strain rate range of 700 to 2800 s^{-1}; and (**d**) comparison between XY-oriented samples at true strain rate range of 4000 to 6700 s^{-1}.

Figure 11b,d depicts the strain–stress curves in a strain rate range of 4000 to 6700 s−1. In Figure 11b, it is clearly observed that, up to a true strain of approximately 0.8, the two curves lay within each other's standard deviation intervals. This suggests that a strong similarity existed between the EOS Z-oriented samples and CL Z-oriented samples up to a true strain of 0.8. The same applies to the EOS XY-oriented samples and CL XY-oriented samples up to a true strain of 0.6. It should be emphasized that higher internal distortion occurred with the CL, as presented in the GOS plots (see Figure 8c,f for CL and EOS, respectively). As GOS plots describe local characteristics on a micro-scale, the SHBP experiment—which reflected the bulk dynamic mechanical properties—lacked the sensitivity to emphasize those differences. Thus, minor differences that were rather insignificant were obtained in the maximum measured true stress and plastic behavior. In the XY orientation plot (Figure 11b) for a true strain higher than 0.8 up to the end of the measurement, a significant difference can be observed. A more significant difference appeared in the comparison of the XY orientation samples for a true strain of 0.6 and above (see Figure 11d).

The samples manufactured by the CL machine exhibited a decline in the stress–strain curve, which suggests that they experienced the same type of failure, while the samples manufactured using the EOS machine, where no declination occurred, experienced simple plastic deformation. The difference between the curves is beyond the standard deviation and measurement error. An analysis of the maximum true stress measured at the plastic failure point ($\varepsilon \approx 0.1$) suggests that SLM–AlSi10Mg alloy has a very low strain rate sensitivity in contrast to previously reported results [22,24], and compared to precipitation-hardened aluminum alloys, where the increasing strain rate causes an increase in the yield point [34].

3.3. Post-Dynamic Loading Analysis

The strain rate at a striker velocity of nearly 25 m/s was within the range of 5800 to 6700 s^{-1}. For these samples, characterization of the damage was conducted using an optical microscope. Figure 12a,b depicts the EOS and CL XY-oriented samples viewed parallel and perpendicular to the loading direction. In the top view, it can be observed that both samples exhibited significant cracking, caused by severe plastic deformation. In Figure 12c,d, the circumferences of both samples are presented. These exhibited major brittle cracks inclined at 45° with respect to the loading direction. However, for the CL samples, the cracks along the circumference were notched throughout the sample and caused massive separation, as can be observed in Figure 12c. Moreover, the direction in which the cracks were formed was mostly one sided, namely at 45° with respect to the loading direction. In contrast, for the EOS sample in Figure 12d, the cracks along the circumference were inclined at 45° and −45°. Figure 13 illustrates the same phenomena of damage endurance between the EOS and CL samples in the Z-oriented samples. The EOS samples could endure plastic deformation by deflecting the cracks at the circumference in a twisted manner—see Figure 13b,d. The CL samples experienced massive failure when cracks propagated from the circumference into the center of the specimen—see Figure 13a,c. The unique crack nucleation and propagation could point to a different magnitude of anisotropicity inherited through each unique manufacturing strategy. As mentioned previously, the EOS product exhibited a weaker strength of the preferential textural component. It is plausible to assume that lower crack stability was expected for the EOS, as weak planes that pave crack paths are less likely to be continuous owing to the higher distribution of the crystallographic texture. Thus, when subjected to dynamic testing, the EOS will in turn endure damage more homogeneously, as indicated by the structure uniformity of the deformed samples (XY- and Z-oriented, Figures 12b,d and 13b,d, respectively). However, for the CL, the dominant texture or, in other words, lack of local crystallographic alternation, resulted in a lower fracture resistance and higher crack growth, as can be observed in the CL deformed samples (XY- and Z-oriented, Figures 12a,c and 13a,c, respectively), characterized by the spread of catastrophic macro-cracks.

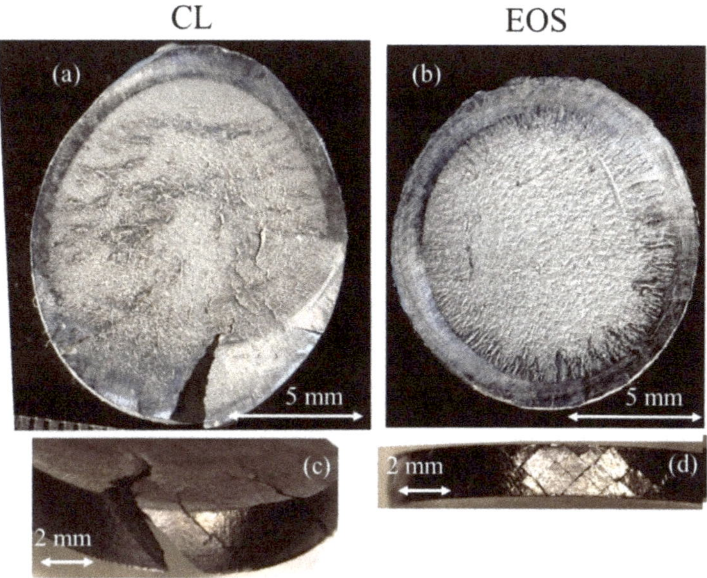

Figure 12. XY-oriented samples after SHPB tests at strain rate of 5800 to 6700 s^{-1}, with plane view of samples for (**a**) CL and (**b**) EOS, and sample circumferences for (**c**) CL and (**d**) EOS.

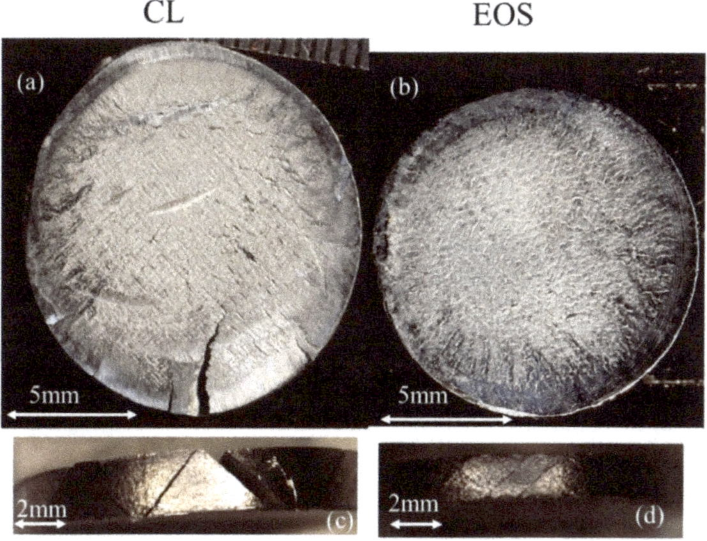

Figure 13. Z-oriented samples after SHPB tests at strain rate of 5800 to 6700 s^{-1}, with plane view of samples for (**a**) CL and (**b**) EOS, and sample circumferences for (**c**) CL and (**d**) EOS.

4. Conclusions

SLM–AlSi10Mg alloy was tested for its dynamic behavior using the SHPB system, at strain rates ranging from 700 to 6700 s^{-1}. The tested SLM samples were fabricated using the CL and EOS 3D printers, which are known for their unique manufacturing processes, mainly distinguished by their scanning pattern, scan speed, and beam energy. Hence, this fundamental study was established to relate dynamic properties to certain manufacturing processes.

The as-built state was characterized by its initial microstructure and crystallographic texture using the SEM–EBSD technique. While similar morphological features were observed in both SLM products, the magnitudes of their crystallographic texture and localized strain were somehow different. A preferential texture was more clearly observed for the CL product.

Under dynamic loading, the vertically built SLM–AlSi10Mg samples exhibited a higher yield stress compared to the horizontally built samples, regardless of the fabrication technique. During early compression strain, the samples individually manufactured by the two machines with the same building orientation exhibited the same stress behavior. However, at higher strains, a significant decline in stress was observed in the CL samples for both building orientations. The difference was much more pronounced in the XY orientation, and appeared at a smaller strain. This deference can be explained by the altered crystallographic texture caused by the manufacturing technique.

The compliance to deformation in the CL product when subjected to dynamic testing was obtained through crack nucleation and propagation, regardless of the building orientation. The damage endurance was traced to a higher magnitude of the preferential texture component for the CL product. The lower probability of local crystallographic alternation was plausibly manifested by continuous weak planes, which paved the crack growth path, and therefore resulted in a lower fracture resistance and higher crack growth in the CL product. However, the EOS product retained its structure without notable micro-crack formation, suggesting higher isotropic properties, as indicated by its lower preferential texture component.

Author Contributions: O.S. conceived & designed the SHPB experiments and B.A. performed & processed SHPB experiments. S.S. was responsible for all metallurgical-crystallographic analysis. Writing-reviewing & editing of this manuscript was accomplished by all authors. S.S. and O.S. supervised the findings of this work.

Funding: This research received no external funding.

Acknowledgments: The authors thank Bar Nurel for his useful advice and remarks. The author (S.S.) acknowledges S. Levi for his technical assistance.

Conflicts of Interest: The authors declare no conflict of interest.

References

1. DebRoy, T.; Wei, H.L.; Zuback, J.S.; Mukherjee, T.; Elmer, J.W.; Milewski, J.O.; Beese, A.M.; Wilson-Heid, A.; De, A.; Zhang, W. Additive manufacturing of metallic components—Process, structure and properties. *Prog. Mater. Sci.* **2018**, *92*, 112–224. [CrossRef]

2. Trevisan, F.; Calignano, F.; Lorusso, M.; Pakkanen, J.; Aversa, A.; Ambrosio, E.P.; Lombardi, M.; Fino, P.; Manfredi, D. On the selective laser melting (SLM) of the AlSi10Mg alloy: Process, microstructure, and mechanical properties. *Materials* **2017**, *10*, 76. [CrossRef]

3. Brandl, E.; Heckenberger, U.; Holzinger, V.; Buchbinder, D. Additive manufactured AlSi10Mg samples using Selective Laser Melting (SLM): Microstructure, high cycle fatigue, and fracture behavior. *Mater. Des.* **2012**, *34*, 159–169. [CrossRef]

4. Kempen, K.; Thijs, L.; van Humbeeck, J.; Kruth, J.P. Mechanical Properties of AlSi10Mg Produced by Selective Laser Melting. *Phys. Procedia* **2012**, *39*, 439–446. [CrossRef]

5. Aboulkhair, N.T.; Tuck, C.; Ashcroft, I.; Maskery, I.; Everitt, N.M. On the Precipitation Hardening of Selective Laser Melted AlSi10Mg. *Metall. Mater. Trans. A Phys. Metall. Mater. Sci.* **2015**, *46*, 3337–3341. [CrossRef]

6. Rosenthal, I.; Stern, A.; Frage, N. Strain rate sensitivity and fracture mechanism of AlSi10Mg parts produced by Selective Laser Melting. *Mater. Sci. Eng. A* **2017**, *682*, 509–517. [CrossRef]

7. Maamoun, A.H.; Elbestawi, M.; Dosbaeva, G.K.; Veldhuis, S.C. Thermal post-processing of AlSi10Mg parts produced by Selective Laser Melting using recycled powder. *Addit. Manuf.* **2018**, *21*, 234–247. [CrossRef]

8. Nurel, B.; Nahmany, M.; Frage, N.; Stern, A.; Sadot, O. Split Hopkinson pressure bar tests for investigating dynamic properties of additively manufactured AlSi10Mg alloy by selective laser melting. *Addit. Manuf.* **2018**, *22*, 823–833. [CrossRef]

9. Buchbinder, D.; Schleifenbaum, H.; Heidrich, S.; Meiners, W.; Bültmann, J. High power Selective Laser Melting (HP SLM) of aluminum parts. *Phys. Procedia* **2011**, *12*, 271–278. [CrossRef]

10. Wu, J.; Wang, X.Q.; Wang, W.; Attallah, M.M.; Loretto, M.H. Microstructure and strength of selectively laser melted AlSi10Mg. *Acta Mater.* **2016**, *117*, 311–320. [CrossRef]

11. Liu, Y.J.; Liu, Z.; Jiang, Y.; Wang, G.W.; Yang, Y.; Zhang, L.C. Gradient in microstructure and mechanical property of selective laser melted AlSi10Mg. *J. Alloy. Compd.* **2018**, *735*, 1414–1421. [CrossRef]

12. Kempen, K.; Thijs, L.; van Humbeeck, J.; Kruth, J.-P. Processing AlSi10Mg by selective laser melting: parameter optimisation and material characterisation. *Mater. Sci. Technol.* **2015**, *31*, 917–923. [CrossRef]

13. Krishnan, M.; Atzeni, E.; Canali, R.; Calignano, F.; Manfredi, D.; Ambrosio, E.P.; Iuliano, L. On the effect of process parameters on properties of AlSi10Mg parts produced by DMLS. *Rapid Prototyp. J.* **2014**, *20*, 449–458. [CrossRef]

14. Buchbinder, D.; Meiners, W.; Wissenbach, K.; Poprawe, R. Selective laser melting of aluminum die-cast alloy—Correlations between process parameters, solidification conditions, and resulting mechanical properties. *J. Laser Appl.* **2015**, *27*, S29205. [CrossRef]

15. Maamoun, A.; Xue, Y.; Elbestawi, M.; Veldhuis, S. The Effect of Selective Laser Melting Process Parameters on the Microstructure and Mechanical Properties of Al6061 and AlSi10Mg Alloys. *Materials* **2018**, *12*, 12. [CrossRef]

16. Yang, K.V.; Rometsch, P.; Jarvis, T.; Rao, J.; Cao, S.; Davies, C.; Wu, X. Porosity formation mechanisms and fatigue response in Al-Si-Mg alloys made by selective laser melting. *Mater. Sci. Eng. A* **2018**, *712*, 166–174. [CrossRef]

17. Maamoun, A.H.; Xue, Y.F.; Elbestawi, M.A.; Veldhuis, S.C. Effect of selective laser melting process parameters on the quality of al alloy parts: Powder characterization, density, surface roughness, and dimensional accuracy. *Materials* **2018**, *11*, 2343. [CrossRef]

18. Thijs, L.; Kempen, K.; Kruth, J.; van Humbeeck, J. Fine-structured aluminium products with controllable texture by selective laser melting of pre-alloyed AlSi10Mg powder. *Acta Mater.* **2013**, *61*, 1809–1819. [CrossRef]

19. Kruth, J.P.; Deckers, J.; Yasa, E.; Wauthlé, R. Assessing and comparing influencing factors of residual stresses in selective laser melting using a novel analysis method. *Proc. Inst. Mech. Eng. Part B J. Eng. Manuf.* **2012**, *226*, 980–991. [CrossRef]

20. Kimura, T.; Nakamoto, T. Microstructures and mechanical properties of A356 (AlSi7Mg0.3) aluminum alloy fabricated by selective laser melting. *Mater. Des.* **2016**, *89*, 1294–1301. [CrossRef]

21. Read, N.; Wang, W.; Essa, K.; Attallah, M.M. Selective laser melting of AlSi10Mg alloy: Process optimisation and mechanical properties development. *Mater. Des.* **2015**, *65*, 417–424. [CrossRef]

22. Asgari, H.; Odeshi, A.; Hosseinkhani, K.; Mohammadi, M. On dynamic mechanical behavior of additively manufactured AlSi10Mg_200C. *Mater. Lett.* **2018**, *211*, 187–190. [CrossRef]

23. Hadadzadeh, A.; Amirkhiz, B.S.; Odeshi, A.; Mohammadi, M. Dynamic loading of direct metal laser sintered AlSi10Mg alloy: Strengthening behavior in different building directions. *Mater. Des.* **2018**, *159*, 201–211. [CrossRef]

24. Li, J.; Hadadzadeh, A.; Amirkhiz, B.S.; Mohammadi, M.; Odeshi, A. Deformation mechanism during dynamic loading of an additively manufactured AlSi10Mg_200C. *Mater. Sci. Eng. A* **2018**, *722*, 263–268.

25. Zaretsky, E.; Stern, A.; Frage, N. Dynamic response of AlSi10Mg alloy fabricated by selective laser melting. *Mater. Sci. Eng. A* **2017**, *688*, 364–370. [CrossRef]

26. Asgari, H.; Baxter, C.; Hosseinkhani, K.; Mohammadi, M. On microstructure and mechanical properties of additively manufactured AlSi10Mg_200C using recycled powder. *Mater. Sci. Eng. A* **2017**, *707*, 148–158. [CrossRef]

27. EOS GmbH—Electro Optical Systems. Material Data Sheet: EOS Aluminium AlSi10Mg. 2014. Available online: https://www.eos.info/material-m (accessed on 11 December 2018).

28. Concept Laser. Cl 30Al/Cl 31Al. 2015. Available online: http://www.conceptlaserinc.com/wp-content/uploads/2014/10/CL-AL30_31AL_Englisch.pdf (accessed on 11 December 2018).

29. Chen, W.; Song, B. *Split Hopkinson (Kolsky) Bar*; Springer: Berlin/Heidelberg, Germany, 2011.

30. Ramesh, K.T. High Rates and Impact Experiments. In *Springer Handbook of Experimental Solid Mechanics*; Springer Handbooks; Springer: Boston, MA, USA, 2008.

31. Bachmann, F.; Hielscher, R.; Schaeben, H. Texture Analysis with MTEX—Free and Open Source Software Toolbox. *Solid State Phenom.* **2010**, *160*, 63–68. [CrossRef]
32. Aboulkhair, N.T.; Everitt, N.M.; Ashcroft, I.; Tuck, C. Reducing porosity in AlSi10Mg parts processed by selective laser melting. *Addit. Manuf.* **2014**, *1*, 77–86. [CrossRef]
33. Gama, B.A.; Lopatnikov, S.L.; Gillespie, J.W. Hopkinson bar experimental technique: A critical review. *Appl. Mech. Rev.* **2004**, *57*, 223. [CrossRef]
34. Oosterkamp, L.D.; Ivankovic, A.; Venizelos, G. High strain rate properties of selected aluminium alloys. *Mater. Sci. Eng. A* **2000**, *278*, 225–235. [CrossRef]

Article

Performance Consistency of AlSi10Mg Alloy Manufactured by Simulating Multi Laser Beam Selective Laser Melting (SLM): Microstructures and Mechanical Properties

Bin Liu [1], Zezhou Kuai [1], Zhonghua Li [2,*], Jianbin Tong [2], Peikang Bai [1], Baoqiang Li [1] and Yunfei Nie [1]

[1] School of Materials Science and Engineering, North University of China, Taiyuan 030051, China; liubin3y@nuc.edu.cn (B.L.); kuaizezhou@163.com (Z.K.); baipeikang@nuc.edu.cn (P.B.); lbqlalala@gmail.com (B.L.); m18834046038@163.com (Y.N.)

[2] School of Mechanical Engineering, North University of China, Taiyuan 030051, China; Evantjb@163.com

*   Correspondence: lizhonghua6868@163.com; Tel.: +86-0351-3925-618

Received: 12 September 2018; Accepted: 16 November 2018; Published: 22 November 2018

Abstract: Multi-laser beam selective laser melting (SLM) technology based on a powder bed has been used to manufacture AlSi10Mg samples. The AlSi10Mg alloy was used as research material to systematically study the performance consistency of both the laser overlap areas and the isolated areas of the multi-laser beam SLM manufactured parts. The microstructures and mechanical properties of all isolated and overlap processing areas were compared under optimized process parameters. It was discovered that there is a raised platform at the junction of the overlap areas and the isolated areas of the multi-laser SLM samples. The roughness is significantly reduced after two scans. However, the surface roughness of the samples is highest after four scans. As the number of laser scans increases, the relative density of the overlap areas of the samples improves, and there is no significant change in hardness. The tensile properties of the tensile samples are poor when the overlap area width is 0, 0.1, or 0.2 mm. When the widths of the overlap areas are equal to or greater than 0.3 mm, there is no significant difference in the tensile strength between the overlap and the isolated areas.

Keywords: AlSi10Mg; multi-laser manufacturing; selective laser melting; microstructure; mechanical property

---

1. Introduction

Selective laser melting (SLM) is one of the most commonly used additive manufacturing (AM) technologies. SLM is a powder-based AM process which utilizes higher laser energy densities to facilitate complete melting and consolidation of successive layers of powder to fabricate 3D components [1,2]. Based on the idea of additive manufacturing, a three-dimensional CAD model is sliced into layers to obtain a 2D contour. Within the 2D contour, the laser selectively scans the metal powder layer and builds high-quality parts. It provides an almost unchallenged freedom of design, without the need for part-specific tooling [3–5]. Customized designs with complex internal and external structures using SLM technology are widely used in a variety of applications, including in automotive, aerospace, and biomedical industries [6].

AlSi10Mg is a typically-cast aluminium alloy that belongs to the Al-Si alloy group. AlSi10Mg alloy is a popular material in the aerospace and automotive industries due to its high specific strength and thermal conductivity [7,8]. In recent years, with the rapid development of additive manufacturing technology, rapid solidification technology has become the focus of research on grain refinement.

3D printing technology has become the research focus of Al-Si alloy manufacturing with SLM. It has unique advantages. Al-Si alloys fabricated by SLM could reach near full density under suitable processing conditions. Due to its rapid solidification with a high cooling rate, SLM Al-Si can exhibit ultrafine unstable and patterned microstructures, leading to some interesting properties, such as very high toughness and high strain-hardening ability. These results imply that SLM alloys might have quite different microstructures and performances compared with materials fabricated by conventional ingot metallurgy or powder metallurgy routes. Some research institutes, both foreign and domestic, have studied the laser-forming technologies of aluminium alloys [9,10]. At present, the SLM manufacturing of Al-Si-based AlSi10Mg has been studied abroad, including research on single-channel, single-layer manufacturing quality [11], the relationship between process parameters and moulding quality, process interval optimization [12,13], characterisation, control of microstructures [14–17], and microstructure and mechanical properties [18–20].

With low building efficiency, most SLM devices currently have the disadvantage of being unable to fabricate large-sized parts. Currently, the most effective way to improve the building efficiency is to increase the scanning speed and the thickness of the powder layer under high laser power. However, improving the laser power causes the laser spot diameter to increase, which may affect the accuracy and therefore the physical and mechanical properties of the part. Multi-laser beam machining is an effective way to increase the size and efficiency of SLM-shaped parts. Multi-laser SLM technology can effectively increase the build rate, reduce manufacturing time, and produce larger parts. In laser manufacturing, each independent but synchronizable laser is responsible for manufacturing different areas. The manufacture time can be greatly reduced and can further expand the forming space. In this way, multiple lasers will form an important development regarding the direction of SLM. Wang Zemin et al. [21] studied the microstructure and mechanical properties of the four-laser fabricated Ti6Al4V overlap areas and isolated areas using self-developed multi-laser equipment. The results illustrated that the stair-step phenomenon of the overlap areas can be eliminated by the compartment rotation scan, and also there was no significant change in the microstructure or mechanical properties of the overlap areas. Multi-laser SLM is feasible for fabrication of large size Ti6Al4V parts. Andreas Wiesner et al. [22] used the SLM 500HL multi-laser device to produce large cylinder heads, each with dimensions 494 × 210 × 143 mm. The manufacturing time was reduced from 170 h for dual lasers to 85 h for four lasers, and the production efficiency was greatly improved. This both highlights an important breakthrough in multi-laser SLM additive manufacturing and proves that multi-laser SLM manufacturing has significant advantages. Buchbinder D et al. [23] used a new prototype machine tool including a 1 kW laser and a multi-beam system to improve the efficiency of aluminum. The results of the investigation demonstrated that the build rate for the production of AlSi10Mg parts can be increased by using a 1 kW laser. There are still many other researchers who have studied multi-laser manufacturing technology. Thorsten Heeling et al. [24,25] studied the effect of the laser compensation process on the stability of SLM formation. This study discovered that the two lasers work together with the melt pool and the proper power compensation, and focus positioning of the second laser can more accurately control the laser melting process. This helps to stabilize the melt pool, obtain dense parts, reduce temperature gradients, and reduce residual stress. Weingarten et al. [26] used just one beam, but they scanned the part two times to dry the aluminium powder bed in the process, prior to melting, to reduce the amount of hydrogen porosity. Abe et al. [27] used a second moving beam in a simple test to selectively reduce the cooling rates and thus influence the microstructure with promising results.

In summary, leading SLM equipment manufacturers are competing to develop their own multi-laser SLM equipment. However, there are few reports on multi-laser SLM manufactured parts. There are few public reports on the performance consistency of multi-laser SLM manufacturing Al alloys. Here we will study the changes of macroscopic morphology, phase distribution, and mechanical properties of the double-laser overlap area, the four-laser overlap area, and the isolated area. This will include analyzing the change mechanism to promote the laser SLM technology and theoretical and technical support in the extensive and in-depth application of shaped aluminium alloy components.

2. Experimental Methods

Multi-laser manufacturing is achieved by simulation of a single laser device (Renishaw AM-400, Renishaw, Gloucestershire, England). The machine is equipped with a modulated ytterbium fiber laser with a wavelength of 1070 nm. This project uses Quant AM software (2016, Renishaw, Gloucestershire, England) to splice two or four parts, and the overlapping area is repeatedly melted under the set process to achieve multi-laser bonding, as shown in Figure 1. Also, the optimized processing parameters employed are listed in Table 1, and the scanning direction is shown in Figure 2. The double-laser overlap area is equivalent to the laser repeating the scanning of one area two times, and the same four-laser overlap area is equivalent to repeating the scanning four times. A single-laser device (Renishaw AM-400) scans one, two, and four times per layer to achieve single laser, double-laser, and four-laser formed parts, respectively. The sizes of the samples are 10 (L) × 10 (W) × 10 (H) mm, and the other two formed samples have sizes of 15 (L) × 10 (W) × 10 (H) mm. The area with an intermediate size of 5 (L) × 10 (W) mm is a double-laser and a four-laser overlapping area. The experimental design of multi-laser manufacturing is shown in Figure 3.

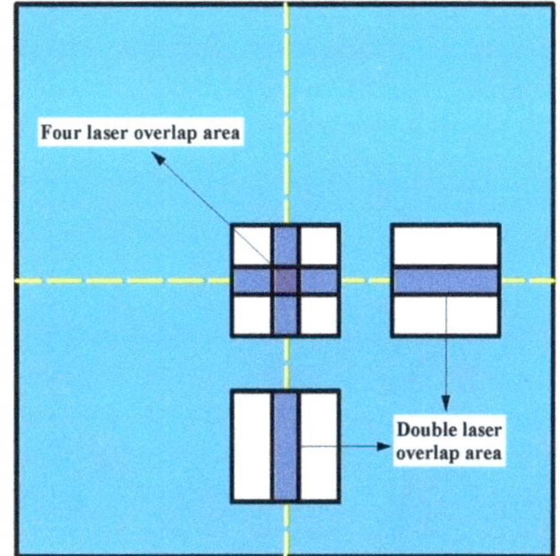

Figure 1. Laser bonding area schematic.

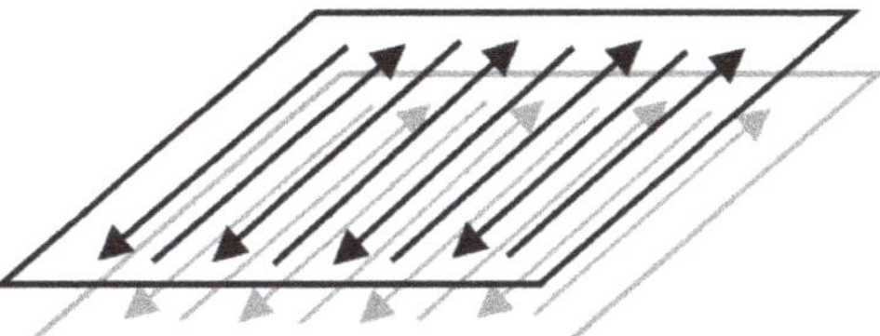

Figure 2. Selective laser melting (SLM) scan direction.

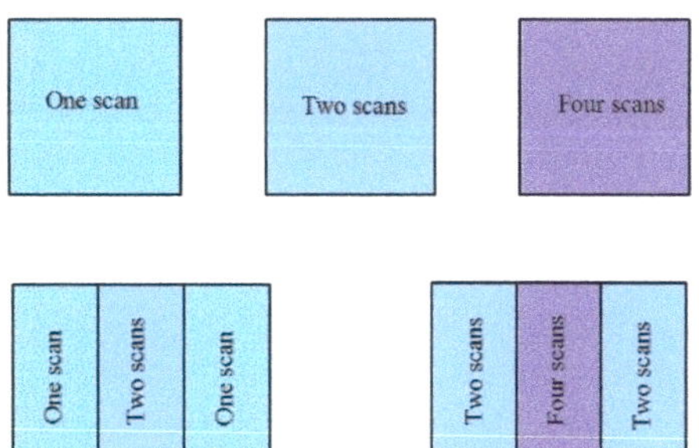

Figure 3. Experimental design of multi-laser manufacturing.

Table 1. SLM process parameters used for manufacturing AlSi10Mg samples.

Manufacturing Parameters	Value
Laser power, P	200 W
Scan space, S	80 μm
Powder layer thickness, h	25 μm
Spot diameter, D	80 μm
Exposure Time, ET	140 μs
Point Distance, PD	80 μm

The AlSi10Mg powder used in this study was provided by LPW Technology with the chemical element composition of the AlSi10Mg powder, as shown in Table 2. In the AlSi10Mg powder alloy, the Al element is a matrix material and the main alloying element is Si. The powder is prepared by gas atomization and has a spherical shape. The average particle size is 45 μm, obeying normal distribution and having good flowability, and its powder electron microscope scanning (SEM, SU5000, HITACHI, Tokyo, Japan) topography is shown in Figure 4. Tensile samples in isolated areas and overlap areas were built according to Figure 5 in this experiment. The overlap width of the tensile samples is 0–1.5 mm. The overlap areas of the tensile samples were scanned twice.

Table 2. The chemical component of AlSi10Mg powders (wt. %).

Elements	Si	Mg	Mn	Cu	Fe	Ni	Zn	Sn	Ti	Pb	Al
wt. %	10.0	0.40	0.40	0.05	0.50	0.05	0.10	0.05	0.15	0.05	Bal

Figure 4. SEM of AlSi10Mg powders.

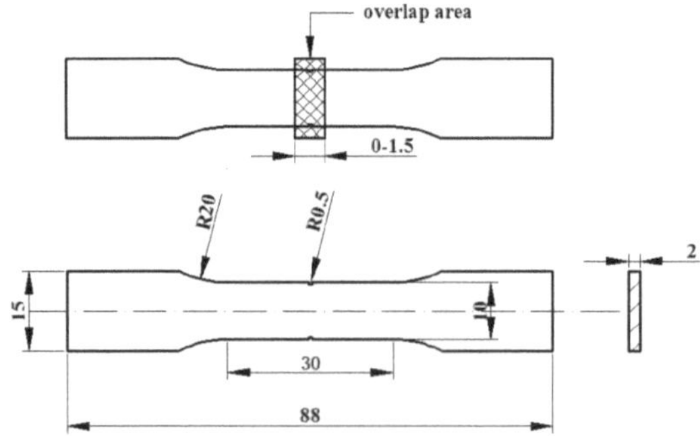

Figure 5. Standard dimension drawing of a tensile specimen (mm).

The surface roughness was measured using the surface roughness step measuring instrument (JB-4C, Taiming Optical Instrument Co., Ltd, Shanghai, China). Metallographic observations were made using a polarising microscope (Axio Scope A1, Carl Zeiss, Heidenheim, Germany) and the phase identification was identified through an X'Pert-Pro X-ray diffractometer (D/max-rB, Rigaku Corporation, Tokyo, Japan). The hardness values were tested (11 times on each type of sample) using a digital micro Vickers hardness tester (TMHVS-1000, Yuming Optical Instrument Co., Ltd, Shanghai, China). Tensile properties of this experiment were tested using a microcomputer-controlled electronic universal testing machine (CMT5105, MTS industrial systems (China) co. LTD, Shenzhen, China). The number of the tensile tests carried out on each type of sample was one. The tensile rate has been used for carrying out the tensile tests of 2 mm/min. Following testing, the fracture surface of the tensile sample was characterized using a scanning electron microscope (SU5000, HITACHI, Tokyo, Japan).

3. Results and Discussion

3.1. Surface topography and Microstructural Analysis

3.1.1. Overlap Boundary Effect

The five images shown in Figure 6 are the super-depth maps of the upper surface of the sample using different remelting times and lap joints under the same process parameters. Figure 6a shows the laser was scanned only once throughout the area (i.e., metal powder only melts–solidifies once). Figure 6b shows the laser was scanned twice in the area, and Figure 6c shows the laser was scanned four times in the area. Figure 6d shows the lap area is laser scanned twice (often called remelting), while the isolated area is scanned only once. Figure 6e shows the lap area was scanned four times, while the isolated area was scanned twice. The areas indicated by the black line frames in Figure 6d,e are the overlap areas.

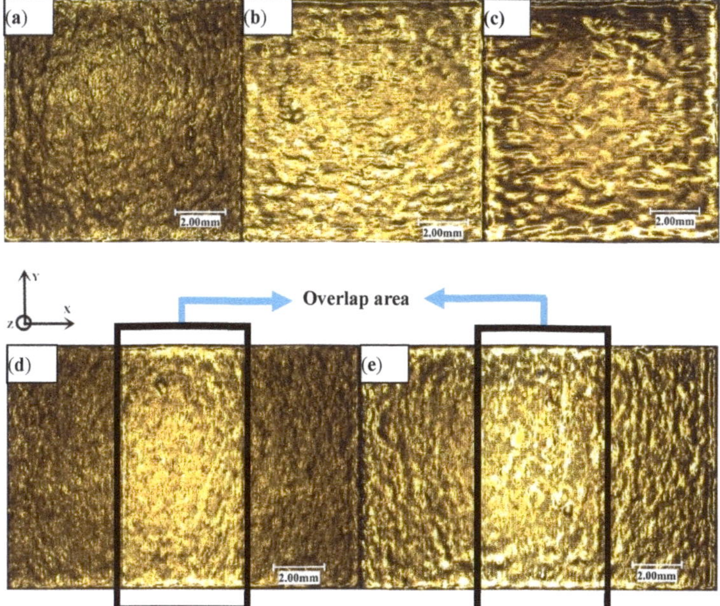

Figure 6. 3D super-depth analysis of field microscopic surface topography at (**a**) one scan, (**b**) two scans, (**c**) four scans, (**d**) overlap—two scans, and (**e**) overlap—four scans.

Figure 6a,b shows the surface topography of a single-layer scan of AlSi10Mg powder after one and two scans. It can be seen that the surface topography after two scans is better than the un-remelting surface topography; not only is the oxide small, but also the surface is relatively clean. The overlap between the molten pools is also improved. Therefore, the single layer after two scans is more conducive to the lower layer connection than the single layer scanned one time so that the layers are more densely bonded together. By observing the three-dimensional microscopic surface topography of the upper surface of the sample, we can see that there is a unique phenomenon in the overlap and the isolated areas, which we call the overlap boundary effect. The overlap area has a distinct convex phenomenon compared to the isolated area (the blue line frame in Figure 7). Figure 7c,d show the three-dimensional contours of samples marked in Figure 7a,b for a more intuitive observation. This phenomenon is mainly due to the fact that the laser melts the metal powder and produces a solid metal layer following the first scan. When the laser scans the overlap area again, the metal powder that is not completely melted absorbs energy, subsequently generating a new melt pool. The molten metal pool solidifies again and generates a new metal layer. The two metal layers are then superimposed on the overlap areas so that the overlap area is increased in the Z-direction dimension when compared with the isolated area, resulting in a convex phenomenon. The bath temperature at the boundary is lower than the central temperature. The viscosity of the molten metal is high and the fluidity is poor, so the phenomenon is more obvious.

Through visual observation we found that with Figure 7b (four scans in the overlap area and two scans in the isolated area), when compared to Figure 7a (two scans in the overlap area and one scan in the isolated area), the overlap boundary effect has a weakening tendency due to the fact that the laser scans the solid metal layer into a liquid state. The liquid metal has improved fluidity, and therefore the molten metal flows to the isolated area and fills the pits in the isolated area, thereby relieving the generation of the overlap boundary effect. Therefore, increasing the number of scans can effectively reduce the overlap boundary effect.

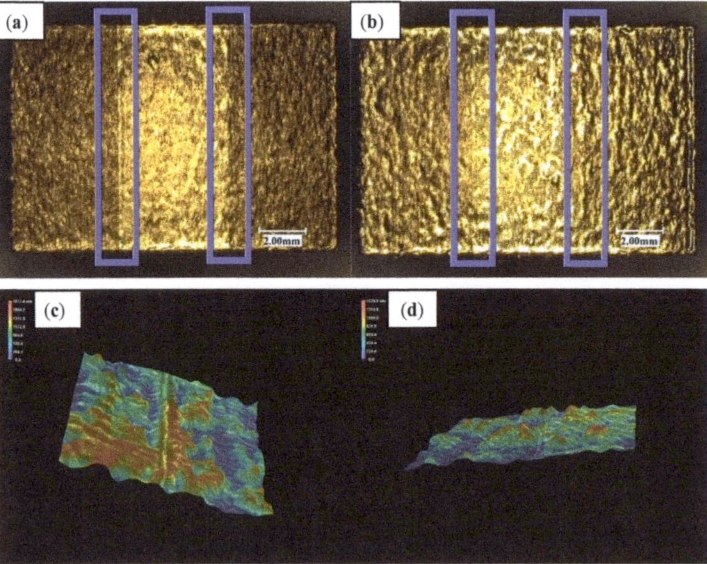

Figure 7. The stair-step effect of overlap area (**a**), overlap—two scans (**b**) overlap—four scans (**c**), and (**d**) three-dimensional contours of (**a**,**b**), which are marked blue.

3.1.2. Surface Roughness

The roughness of the samples mainly includes the upper surface roughness and the side roughness. This paper only studies the surface roughness of multi-laser SLM manufacturing parts. The surface roughness of the part is mainly related to the quality of the manufacturing surface of a single melting channel (the cross-sectional view as shown in Figure 8a). We analyze the splicing process of the SLM melting channel (Figure 8b). Since the melting channel is elliptical in the manufacturing process, the melting channels cannot overlap completely, thereby causing a wave-like distribution between the two melting channels. Finally, the surface roughness is increased. A better way to improve the surface roughness caused by this factor is to properly control the scanning spacing (i.e., the reasonable overlap ratio). Reasonable scanning spacing can press the molten metal between the two melting channels to the trough area, which can effectively reduce the height difference of the trough area and reduce the surface roughness of the parts (the principle is shown in Figure 8c). The optimal situation is when the trough area formed by the tangent and arc of the two melting arcs is equal to the area of the overlapping area. Under such theoretical conditions, the surface roughness of the formed part is the smallest.

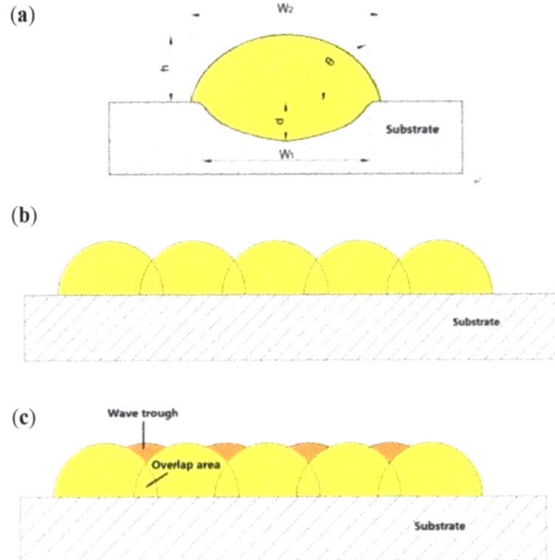

Figure 8. (a) Cross-section of the melting channel (W_1 is the penetration area width, W_2 is the width of the melting track, h is the height of the melting channel, d is the penetration depth, and θ is the wetting angle); (b) overlap schematic melting channels; (c) schematic diagram of the melting channel overlap area.

There are many factors affecting the surface roughness of the parts, including the parameters of the manufacturing process, the movement accuracy of the manufacturing equipment, the particle size distribution of the metal powder material, the sealing of the forming cavity, and the purity of the inert gas. Different parameters in the forming process often have different effects on the roughness of the parts. In view of the above, we only change the number of laser scanning metal powders under the condition that other process parameters are consistent.

According to the data shown in Figure 9, it can be concluded that the average Ra of the upper surface of the sample after one scan is 13.276 μm and after two scans is 12.639 μm. As shown, there is a significant drop, and the surface quality of each area is greatly improved. However, the average Ra of the upper surface of the sample after four scans is 14.339 μm.

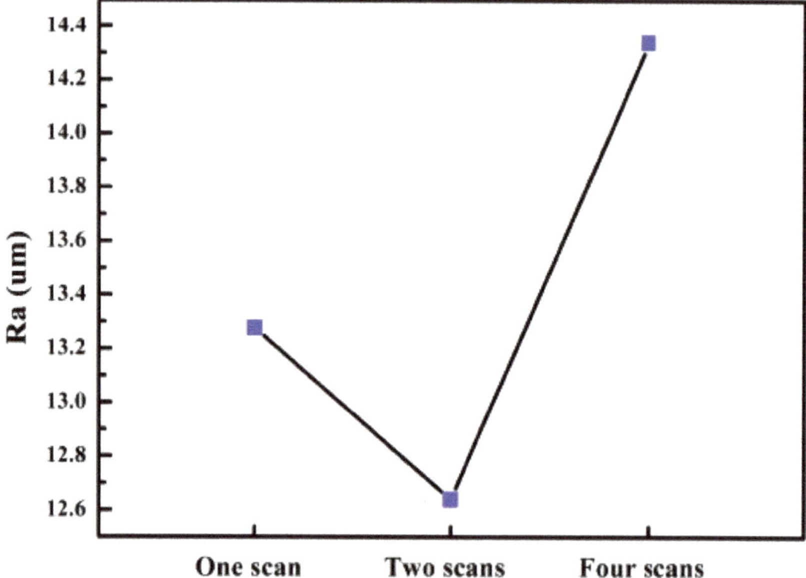

Figure 9. The effect of the number of scans on surface roughness.

The sample with a single laser scan has different particle sizes of metal powder. When the laser functions, the large-particle metal powder, i.e., the surface of the powder particles, melts. However, the inside of the powder cannot be melted due to insufficient energy. The metal powder does not undergo a complete melting-solidification process, although a small particle ball is formed inside. In the metal powder, there are often some powders with smaller particles that cannot be completely melted, and a phenomenon of adhesion occurs between these completely melted metal powders and unmelted metal powders, which results in small particle spheroidisation. SLM is a process in which a layer of metal is superimposed. Since the upper metal surface is not ideal, there are "convex peaks" and "pits" where the thickness of the powder layer is low at the "convex peak" and the thickness of the powder layer at the "pit" is high after the powdering roller is laid. After the laser scan, the metal powder at the thicker powder layer cannot be completely melted, and the molten metal adsorbs the unmelted metal particles outside the molten metal during the solidification process, forming a discontinuous melting channel and increasing the surface roughness. After understanding the melting channel and the formation process of the overlap area (Figure 8), it was determined that one of the factors causing the surface roughness of the upper surface of the laser scanning sample was the scanning direction. Figure 8 shows that one of the factors causing the surface roughness of the upper surface of the laser scanning sample is the scanning direction. The scanning direction of the experimental sample is one-way scanning, that is, the laser scans each layer of metal powder according to a fixed scanning path. The scanning path is the highest position of the melting channel, and the fixed scanning path superimposes the melting channel so that the scanning path becomes increasingly higher, and the wave troughs formed in the middle of the two scanning lines get increasingly lower.

The remelting process (which occurs during double laser scanning) produces smoother, more uniform layers, and efficiently reduces the number of pores formed between the neighboring melt pools at the scan track edges [28,29]. Previous data analysis shows that the surface roughness of the sample after two laser scans can be effectively improved. In the sample scanned by one laser, some of the metal powder particles could not be completely melted, i.e., some of the particles remained in the powder state, which made the sample spheroidised. When the laser was scanned again, these powders could be completely melted to produce a new molten metal which was finally cooled and solidified to produce a metallurgical bond. The spheroidisation of the particles improved, effectively reducing the surface roughness. Due to the scanning direction, the highest point of the melting channel becomes increasingly higher. When the laser is scanned again, the fluidity of the molten metal in the melt pool is increased so that the molten metal originally at the highest point of the melting channel flows towards the "pit", which flattens and reduces the peak between the two melting channels. At the height difference of the valley, the unevenness is alleviated, the surface roughness value is lowered after solidification, and the surface quality is improved.

When the sample is subjected to four scans, the surface roughness value increases and the molten metal is excessively melted, resulting in over-burning. This excessive melting reduces the viscosity of the molten metal, increases the wetting angle of the melt pool, and easily forms a spherical metal liquid, which causes spheroidisation defects as well as discontinuity of the melting channel and increased surface roughness.

3.1.3. Microstructural Analysis

The AlSi10Mg sample was removed from the substrate using a wire cutter, and both the upper surface of the sample and the left side were polished and etched. The processed samples were observed under a metallographic microscope where the microstructure was observed. The left side is shown in Figure 10, and the upper surface is shown in Figure 11.

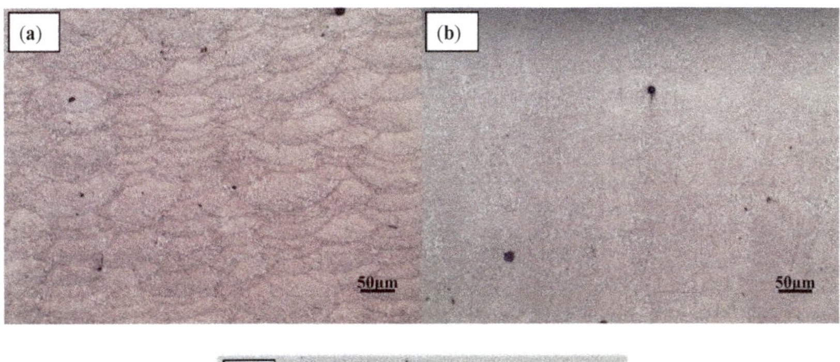

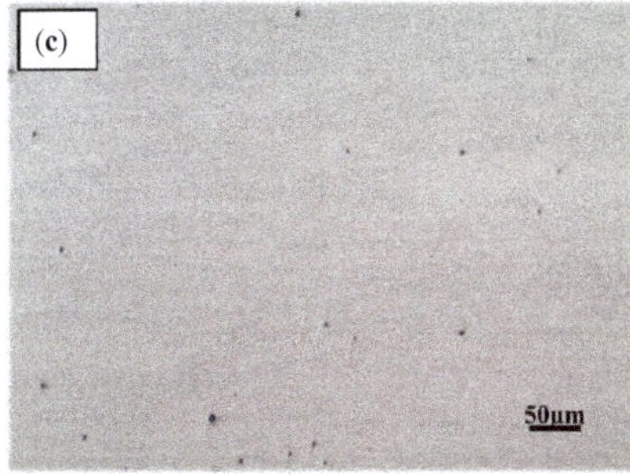

Figure 10. Optical micrograph of the left side of a multi-laser SLM sample (200 times) after (**a**) one scan, (**b**) two scans, and (**c**) four scans.

The microstructure illustrated in Figure 10 shows a dendritic solidification similar to those observed in other studies [14,30]. The brighter areas are aluminium solid solutions, and the darker are silicon or Al-Si-eutectics. Generally, SLM samples show a finer microstructure than, for example, die-cast parts, due to the faster solidification. By observing the shape of the melt pool in Figures 10 and 11, there is no splitting between the neatly arranged melting channels. The overlapping effect between the melting channels is clear, and the pores are relatively small. By comparing the pores in the melt pool under different laser scanning times, it can be found that there are many internal pores—some large—during a single laser scan. When the number of laser scans is increased once, the pores inside the part are significantly reduced. When the number of scans is increased to four, it is found that the internal pores are almost the same as with two scans, and most of the pores are small and rounded. The internal pores are mainly concentrated on the boundary and are mostly rounded. By observing the lap joint of the melted channel, a relatively obvious weld pool boundary can be seen after one laser scan. However, as the number of scans increases, the boundary of the melt pool becomes less obvious, and the higher the number, the less obvious it becomes.

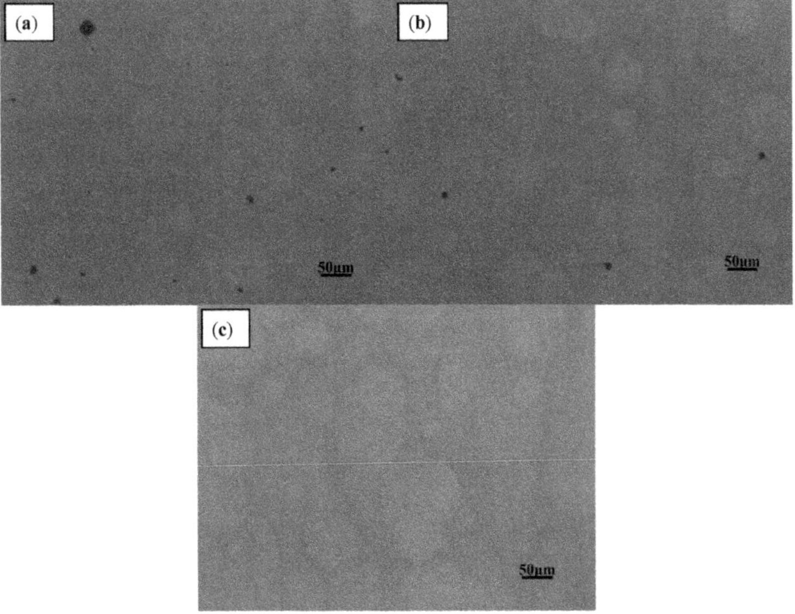

Figure 11. Optical micrograph of the upper surface of multi-laser SLM samples after (**a**) one scan, (**b**) two scans, and (**c**) four scans.

There are two main reasons for the laser to produce a single scan: (1) The internal gas cannot escape completely, and (2) because of the spheroidise effect. When observing the SEM image of the metal powder, the metal powder is mostly spherical. The distance between the valleys of the two melting channels is too small, and the spherical metal powder cannot completely fill the area. When the laser scans these areas, the existing gaps in that area become the internal pores. The reason for the pores may also be due to the inability of the inert protective gas to escape during the solidification of the metal. Although some gaps can be filled with metal powder, others are relatively deep and, therefore, the laser powder cannot penetrate completely and the metal powder is difficult to completely melt. The molten metal liquid encloses the unmelted powder particles and may form large spheroidised particles after solidification, which leads to the generation of pores. Some of the unmelted metal powder is adsorbed around the melt pool to form spheroidised particles, which causes the melting channels to not be completely connected, which is why most of the pores exist in the boundary region with the melting channels.

As the laser scans again, the spheroidisation caused by the previous laser scan can be attenuated. The laser scanning, again, can improve the fluidity of the molten metal which fills the area where the metal powder is difficult to fill and can reduce the generation of voids. According to the analysis of 3.1.1, laser scanning again can reduce the surface roughness of the upper metal layer so that the metal powder existing between the melting channels is reduced, which further reduces the presence of deeper gaps. In this way, the unmelted powder of the deep gap can also be reduced, and the number of pores in the parts can be reduced. This also reduces the porosity inside the part. Laser remelting melts the previously unmelted powder, reducing spheroidisation caused by unmelted metal powder and reducing voids inside the part.

The metal powder is completely melted after four rounds of laser scanning, resulting in a fluid metal liquid. However, during the scanning process, the surface of the melt pool will splash small metal droplets and form small spheroidized particles after rapid cooling and solidification. This explains the more minor pores, as shown in Figure 10c. Part of the reason is that multiple laser scanning can make the molten metal absorb more energy, and the molten metal is vaporized. As the energy elapses, the molten metal vaporizes and then solidifies to form a small spheroidisation phenomenon and subsequently generates pores.

3.2. Phase Distribution

XRD patterns (Figure 12) show that the phase composition of a-Al and Si are found both in AiSi10Mg powders and AlSi10Mg samples (i.e., the microstructure was mainly composed of an Al/Si phase). It can be seen from the spectrum in the XRD that the peak of Al in the SLM sample and the peak corresponding to the powder are shifted to the right. As the number of laser scans increases, the rate of solidification increases. As a result, Si formed a solid solution in the Al matrix, and failed to precipitate in the form of primary silicon and eutectic silicon. This results in a decrease in the peak of Si and an increase in the peak of a-Al. The preferred orientation shown is the (200) plane. This is related to the microstructure and texture characteristics of AlSi10Mg as produced by the SLM process.

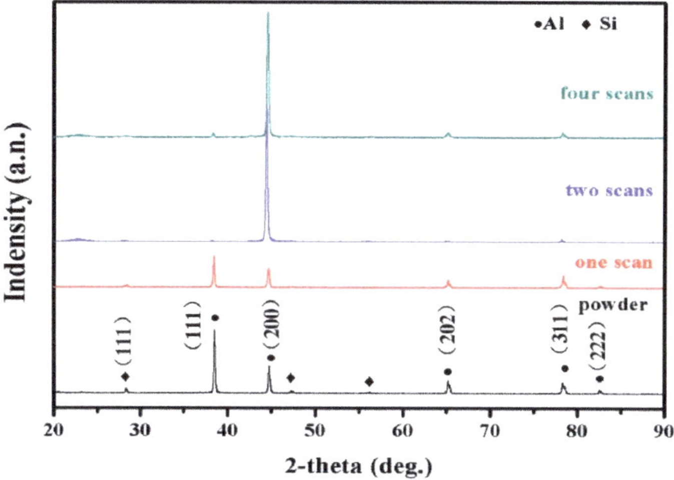

Figure 12. XRD patterns of AiSi10Mg powders and samples.

3.3. Mechanical Properties

3.3.1. Density Measurements

The experimental scheme adopts the drainage method, and the standard density of the material is 2.68. The experimental data is shown in Figure 13. Through experimental data analysis, although the number of scans is different, the relative density of the sample is maintained above 97%. As shown in Figure 13, there are some trends in the number of scans, the number of scan increases, and the relative density increases.

Variations in the thermo-kinetics and thermo-capillary characteristics within the melt pool, such as the viscosity, wettability, and liquid-solid rheological properties, all influenced the densification behavior [31]. The relative density of the part has an important relationship to the presence of internal pores. When there are more internal pores, the relative density of the parts will decrease, and vice versa. When the laser is scanned once, laser scanning fails to completely melt the metal powder, which is prone to spheroidisation, resulting in some very large holes around the melting channel; i.e., the relative density of the parts is decreased. After the laser scans twice, the internal unmelted powder is scanned again by the laser, which can reduce the porosity generated by the unmelted powder and increase the relative density of the parts. By increasing the number of laser scans, the metal powder is fully melted, and the melt pool is fully spread and infiltrated. Therefore, the pores generated in the upper layer are filled, which contributes to the improvement of the internal structure to increase the relative density of the parts. The increase of laser scanning times will increase the relative density of parts. However, there is no significant change in the relative density.

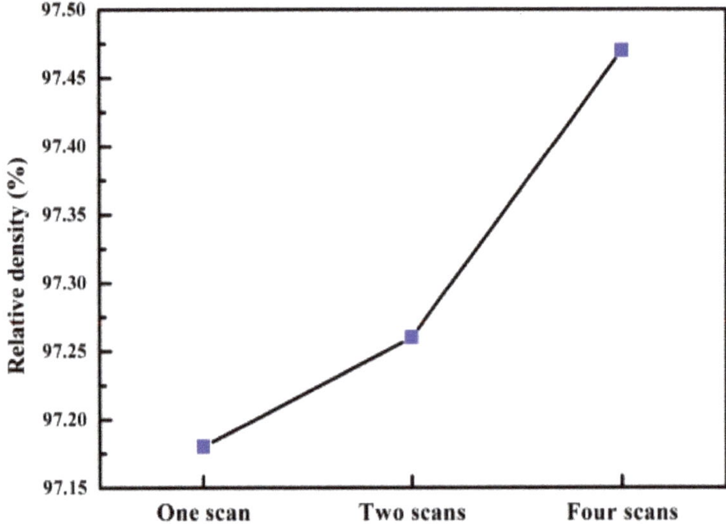

Figure 13. The relative density of different scan times.

3.3.2. Microhardness Tests

The analytical hardness test data were processed with the analysis results shown in Figure 14. We have obtained the average Vickers hardness of each sample, as shown in the bar graph. According to the analysis, the Vickers hardness of the sample scanned by one laser is greater than that of the sample scanned twice and four times, and the difference between the highest point and the lowest point is more than ten. The difference between the other two samples is controlled within ten. By comparing the average Vickers hardness, it can be found that the hardness of the sample after laser remelting decreases, but the change is not obvious and the difference is below five.

Samples that are scanned only once by the laser may feature defects such as voids, spheroidisation, and inclusions due to partial melting of the metal powder, which reduces the microhardness of the material. Some of the defects contained in the area are less rigid and some areas may have larger pores and other defects, causing a decrease in hardness which causes the hardness data to have a larger amplitude. When the laser scans the sample multiple times, the grains and the microstructure become larger. According to the Hall-Petch theory [32,33], grain refinement increases the number of crystal grains contained in the same volume, and internal dislocations reduce this number. When the part is subjected to an external force, the deformation occurs between a large number of crystal grains, and the slip of the dislocation is blocked by the grain boundary so that the expansion cannot be continued, thereby achieving fine grain strengthening. Such grain refinement increases resistance against external forces and increases the hardness.

Although the Vickers hardness of the sample scanned by the laser is decreased, the hardness variation is not particularly remarkable, so the microhardness of the multi-laser overlap area is the same as the isolated area.

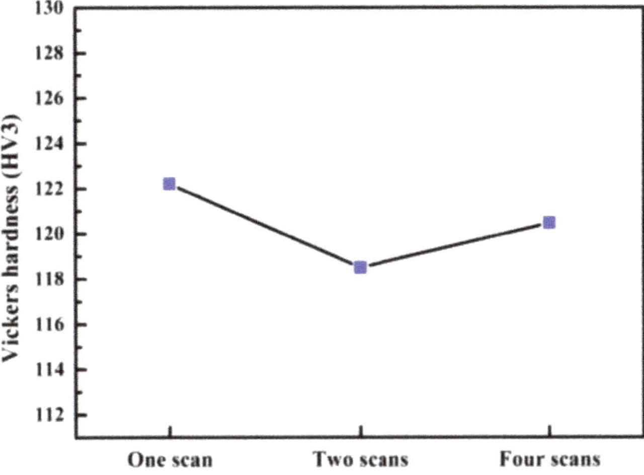

Figure 14. Vickers hardness of different scan times.

3.3.3. Tensile Behavior

In several studies, the mechanical properties of SLM manufacturing parts are affected by the internal melting channel structure, pores, cracks, residual stresses, inclusion defects, etc. The smaller the internal porosity, the smaller the residual stress, and the higher the density, the better the mechanical properties. The influence of these defects is mainly related to the SLM manufacturing process parameters, and the different scanning directions will also result in a structurally different internal melting channel. Kempen K et al. [11] conducted an experimental study on the mechanical properties of SLM-manufactured AlSi10Mg alloy. The results showed that the AlSi10Mg parts manufactured by SLM have improved or equivalent mechanical properties as compared to the cast AlSi10Mg materials. The Vickers hardness of the prepared SLM part is much higher than that of the high-pressure casting (HPDC) AlSi10Mg alloy, and the ultimate tensile strength is always higher than that of the AlSi10Mg alloy under HPDC. In this experiment, we mainly studied the tensile strengths of the overlap areas of two laser scans and the influence of overlap areas of different sizes on tensile properties.

Currently, the SLM-manufactured parts cannot achieve the dimensional accuracy of the tensile samples. There are some small powder bonding particles on the forming surface, and the surface roughness is significant. There was also an error in the wire cutting machine when the samples were removed from the substrate. Taking into account the above factors, we increased the thickness of the samples to 2.5 mm. Test one measures the tensile strength of the standard piece of one scan. The other samples were twice-scanned with different overlap widths. The experimental data was then analyzed and processed, and the experimental results are shown in Table 3 and Figure 15. It can be seen from the table that the tensile strength of the tensile samples is much lower than that of the standard samples when the width of the overlap area is 0, 0.1, or 0.2 mm, and becomes larger as the overlapping width increases. The width of the overlap areas is small so that the powder of the overlap portion is not sufficiently melt-bonded to cause a decrease in tensile strength. When the widths of the overlap areas are equal to or greater than 0.3 mm, the tensile strengths of the samples are remarkably improved.

However, the tensile strength decreased when compared with the standard samples, and the change was not clear. The mechanical properties of the parts are related to the hardness. The tensile strength and the microhardness are analyzed simultaneously. It was found that the number of laser scans has the same trend against the tensile strength and microhardness, whereas when the microhardness of the part increases the tensile strength also increases. There are many pores inside the sample that the laser scanned only once, and the increase of porosity will reduce the tensile strength and mechanical properties. Multiple laser scanning will also make the tissue coarse. According to the Hall-Petch theory, both the part's performance and tensile strength will decrease.

Table 3. The tensile strength of tensile samples with overlapping areas of different sizes.

Test	Overlap Area Width (mm)	Tensile Strength (MPa)
1	0	370
2	0	255
3	0.1	250
4	0.2	325
5	0.3	360
6	0.5	365
7	0.6	340
8	0.7	350
9	0.9	365
10	1.0	355
11	1.1	355
12	1.2	370
13	1.3	340
14	1.4	350
15	1.5	365

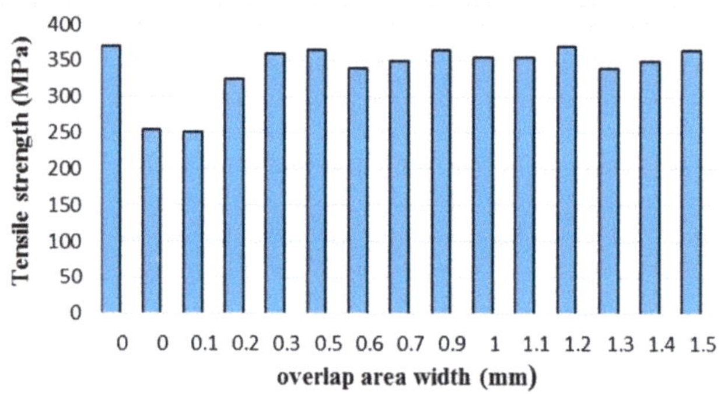

Figure 15. The tensile strength of tensile samples with overlapping areas of different sizes.

In order to further understand the influence of the multi-laser process on the properties of SLM AlSi10Mg alloy, the fracture morphology of SLM AlSi10Mg alloy samples under different scanning times was analyzed. Figure 16a,b show the fracture morphology of the standard samples at one scan. Figure 16c,d show the fracture morphology of the samples at two scans. The overlap area of the sample in Figure 16c,d has a width of 1 mm. It can be seen from the figure that the SLM AlSi10Mg alloy has apparent dimples at the fracture at one scan, which is a typical ductile fracture.

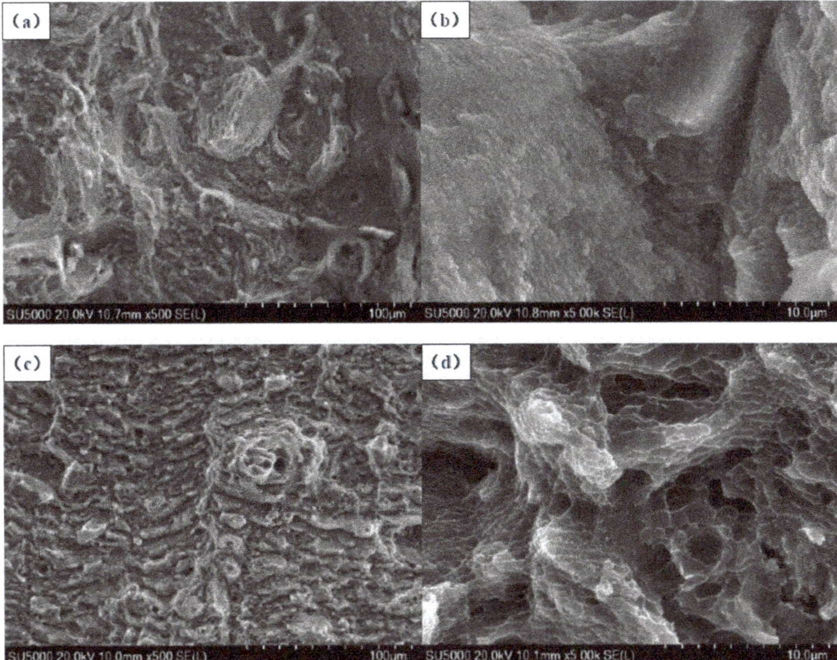

Figure 16. Fractographs of AlSi10Mg tensile samples at different scan times: (**a**) Low and (**b**) high magnification of one-scan SLMed samples; (**c**) low and (**d**) high magnification of two-scan SLMed samples at a high magnification.

As can be seen in Figure 16c, the SLM AlSi10Mg alloy demonstrates smaller dimples. Additionally, the dimple zone on its fracture surface is smaller than that of the SLM AlSi10Mg alloy at two scans in Figure 16a. The formation of dimples implies a ductile rupture, whereas the smooth areas show cleavage facets in the fractured surface indicating a brittle fracture [34]. During the two scans, the number of dimples at the fracture was reduced, and the smooth brittle fracture surface increased. It can be seen from Figure 16b,d that the microstructures become large after two scans, and the internal pores increase, thus the performance of the part will decrease.

In summary, an increase in the number of laser scans will slightly reduce the mechanical properties of the part, but this change is less apparent. Therefore, the overlap areas of the multi-laser SLM samples have no significant change in tensile properties compared to the isolated areas.

4. Conclusions

In this study, the multi-laser beam SLM of AlSi10Mg has been carried out to reveal its microstructure and mechanical properties. The overlap area and isolated area of the multi-laser SLM-manufactured part are systematically studied. The main results can be summarised as follows:

1. The morphology and microstructure of the melt pool of the parts with different scanning times were analyzed. The analysis established that the melting channels were arranged neatly, the overlap boundary effect between the melting channels was clear, and the pores were relatively small. Comparing the internal pores of the part, the inside of the laser scan contains some unmelted metal powder, which causes more pores. Following laser remelting, the pores are reduced, and only some tiny pores exist. After four laser scans, micro-spheroidised particles are produced, and the internal pores are formed as small round hole shapes.

2. The number of laser scans was used as the research object to investigate the influence of the number of scans on the surface quality of the parts. It was found that there is a bulge at the junction of the overlap area and the isolated area of the multi-laser SLM manufactured part, which is called the overlap boundary effect. The average Ra of the upper surface of the sample after one scan is 13.276 μm, and the average Ra of the surface of the sample after two scans is 12.639 μm. However, the average Ra of the upper surface of the sample after four scans is 14.339 μm. A laser scan will result in spheroidisation of the unmelted powder and will increase the surface roughness of the part. Two laser scans can improve this situation and reduce surface roughness. However, increasing the number of scans may lead to spheroidisation and increased surface roughness.

3. It can be seen from the spectrum in XRD that the peak of Al in the SLM sample and the peak corresponding to the powder are shifted to the right. As the number of laser scans increases, the preferential crystallite growth orientation reaches the (200) plane.

4. The relative densities of the multi-laser SLM samples are above 97%, and the increase in the number of laser scans increases the relative density. The microhardness of the samples with different scanning times did not change significantly. The tensile properties of the multi-laser manufactured parts were examined by fabricating tensile specimens of overlap areas with different widths. When the width of the overlap areas is 0, 0.1, or 0.2 mm, the tensile strength of the tensile samples is much lower than that of the standard samples. When the width of the overlap areas is equal to or greater than 0.3 mm, the tensile strengths of the tensile samples are slightly lower than the standard samples.

Author Contributions: For research articles with several authors, a short paragraph specifying their individual contributions must be provided. The following statements should be used "Conceptualization, P.B.; Methodology, B.L. (Baoqiang Li) and Y.N.; Formal Analysis, J.T.; Writing-Original Draft Preparation, B.L. (Bin Lin); Writing-Review & Editing, Z.L. and Z.K."

Funding: This research was funded by [the National Natural Science Foundation of China] grant number [61475056].

Conflicts of Interest: The authors declare no conflict of interest.

References

1. Hooyar, A.; Shima, E.-H.; Damon, K.; Xinhua, W.; Matthew, S.D. Comparative study of commercially pure titanium produced by laser engineered net shaping, selective laser melting and casting processes. *Mater. Sci. Eng. A* **2017**, *705*, 385–393.

2. Attar, H.; Ehtemam-Haghighi, S.; Kent, D.; Matthew, S.D. Recent developments and opportunities in additive manufacturing of titanium-based matrix composites: A review. *Int. J. Mach. Tools Manuf.* **2018**, *133*, 85–102. [CrossRef]

3. Wong, M.; Owen, I.; Sutcliffe, C.J.; Puri, A. Convective heat transfer and pressure losses across novel heat sinks fabricated by Selective Laser Melting. *Int. J. Heat Mass Trans.* **2009**, *52*, 281–288. [CrossRef]

4. Osakada, K.; Shiomi, M. Flexible manufacturing of metallic products by selective laser melting of powder. *Int. J. Mach. Tools Manuf.* **2006**, *46*, 1188–1193. [CrossRef]

5. Mumtaz, K.A.; Hopkinson, N. Selective laser melting of thin wall parts using pulse shaping. *J. Mater. Process Technol.* **2010**, *210*, 279–287. [CrossRef]

6. Li, Z.H.; Xu, R.J.; Zhang, Z.W.; Kucukkoc, I. The influence of scan length on fabricating thin-walled components in selective laser melting. *Int. J. Tool Manuf.* **2017**, *126*, 1–12. [CrossRef]

7. Dadbakhsh, S.; Hao, L. Effect of Al alloys on selective laser melting behaviour and microstructure of in situ formed particle reinforced composites. *J. Alloys Compd.* **2012**, *541*, 328–334. [CrossRef]

8. Joseph, S.; Kumar, S. A systematic investigation of fracture mechanisms in Al–Si based eutectic alloy—Effect of Si modification. *Mater. Sci. Eng. A* **2013**, *588*, 111–124. [CrossRef]

9. Sameljuk, A.V.; Neikov, O.D.; Krajnikov, A.V.; Milman, Y.V.; Thompson, G.E.; Zhou, X. Effect of rapid solidification on the microstructure and corrosion behaviour of Al–Zn–Mg based material. *Corros. Sci.* **2007**, *49*, 276–286. [CrossRef]

10. Karaköse, E.; Keskin, M. Effect of solidification rate on the microstructure and microhardness of a melt-spun Al–8Si–1Sb alloy. *J. Alloys Compd.* **2009**, *479*, 230–236. [CrossRef]

11. Aboulkhair, N.T.; Maskery, I.; Tuck, C.; Ashcroft, I.; Everitt, N.M. On the formation of AlSi10Mg single tracks and layers in selective laser melting: Microstructure and nano-mechanical properties. *J. Mater. Process. Technol.* **2016**, *230*, 88–98. [CrossRef]

12. Li, Y.; Gu, D. Parametric analysis of thermal behavior during selective laser melting additive manufacturing of aluminum alloy powder. *Mater. Des.* **2014**, *63*, 856–867. [CrossRef]

13. Kempen, K.; Thijs, L.; Yasa, E.; Badrossamay, M.; Verheecke, W.; Kruth, J.P. Process optimization and microstructural analysis for selective laser melting of AlSi10Mg. *Solid Freeform Fabr. Symp.* **2011**, *22*, 484–495.

14. Thijs, L.; Kempen, K.; Kruth, J.P.; Van Humbeeck, J. Fine-structured aluminium products with controllable texture by selective laser melting of pre-alloyed AlSi10Mg powder. *Acta Mater.* **2013**, *61*, 1809–1819. [CrossRef]

15. Lam, L.P.; Zhang, D.Q.; Liu, Z.H.; Chua, C.K. Phase analysis and microstructure characterisation of AlSi10Mg parts produced by Selective Laser Melting. *Virtual Phys. Prototyp.* **2015**, *10*, 207–215. [CrossRef]

16. Zhao, Z.Y.; Guan, R.G.; Zhang, J.H. Effects of Process Parameters of Semisolid Stirring on Microstructure of Mg–3Sn–1Mn–3SiC (wt %) Strip Processed by Rheo-rolling. *J. Acta Metall. Sin.* **2017**, *30*, 1–7. [CrossRef]

17. Zhao, Z.Y.; Bai, P.K.; Guan, R.G.; Murugadoss, V. Microstructural evolution and mechanical strengthening mechanism of Mg-3Sn-1Mn-1La alloy after heat treatments. *Mater. Sci. Eng. A* **2018**, *734*, 200–209. [CrossRef]

18. Brandl, E.; Heckenberger, U.; Holzinger, V.; Buchbinder, D. Additive manufactured AlSi10Mg samples using selective laser melting (SLM): Microstructure, high cycle fatigue, and fracture behavior. *Mater. Des.* **2011**, *34*, 159–169. [CrossRef]

19. Kempen, K.; Thijs, L.; Humbeeck, J.V.; Kruth, J.P. Mechanical properties of AlSi10Mg produced by selective laser melting. *Phys. Procedia* **2012**, *39*, 439–446. [CrossRef]

20. Aboulkhair, N.T.; Tuck, C.; Ashcroft, I.; Maskery, I.; Everitt, N.M. On the precipitation hardening of selective laser melted AlSi10Mg. *Metal. Mater. Trans. A* **2015**, *46*, 3337–3341. [CrossRef]

21. Li, F.; Wang, Z.; Zeng, X. Microstructures and mechanical properties of Ti6Al4V alloy fabricated by multi-laser beam selective laser melting. *Mater. Lett.* **2017**, *199*, 79–83. [CrossRef]

22. Wiesnera, D.I.A.; Schwarzea, D. Multi-laser selective laser melting. In Proceedings of the 8th International Conference on Photonic Technologies LANE, Fürth, Germany, 8–11 September 2014.

23. Buchbinder, D.; Schleifenbaum, H.; Heidrich, S.; Meiners, W.; Bültmann, J. High power selective laser melting (HP SLM) of aluminum parts. *Phys. Procedia* **2011**, *12*, 271–278. [CrossRef]

24. Heeling, T.; Wegener, K. Computational investigation of synchronized multibeam strategies for the selective laser melting process. *Phys. Procedia* **2016**, *83*, 899–908. [CrossRef]

25. Heeling, T.; Zimmermann, L.; Wegener, K. Multi-beam strategies for the optimization of the selective laser melting process. In Proceedings of the 27th Annual International Solid Freeform Symposium (Solid Freeform Fabrication 2016), Austin, TX, USA, 8–10 August 2016.

26. Weingarten, C.; Buchbinder, D.; Pirch, N.; Meiners, W.; Wissenbach, K.; Poprawe, R. Formation and reduction of hydrogen porosity during selective laser melting of AlSi10Mg. *J. Mater. Process. Technol.* **2015**, *221*, 112–120. [CrossRef]

27. Abe, F.; Osakada, K.; Shiomi, M. The manufacturing of hard tools from metallic powders by selective laser melting. *J. Mater. Process. Technol.* **2001**, *111*, 210–213. [CrossRef]

28. AlMangour, B.; Grzesiak, D.; Yang, J.M. Scanning strategies for texture and anisotropy tailoring during selective laser melting of TiC/316L stainless steel nanocomposites. *J. Alloys Compd.* **2017**, *728*, 424–435. [CrossRef]

29. Louvis, E.; Fox, P.; Sutcliffe, C.J. Selective laser melting of aluminium components. *J. Mater. Process. Technol.* **2011**, *211*, 275–284. [CrossRef]

30. Read, N.; Wang, W.; Essa, K.; Attallah, M.M. Selective laser melting of AlSi10Mg alloy: Process optimisation and mechanical properties development. *Mater. Des.* **2015**, *65*, 417–424. [CrossRef]

31. AlMangour, B.; Grzesiak, D.; Borkar, T.; Yang, J.M. Densification behavior, microstructural evolution, and mechanical properties of TiC/316L stainless steel nanocomposites fabricated by selective laser melting. *Mater. Des.* **2018**, *138*, 119–128. [CrossRef]

32. Sylwestrowicz, W.; Hall, E.O. The Deformation and Ageing of Mild Steel. *Proc. Phys. Soc.* **2002**, *64*, 495. [CrossRef]

33. Petch, N.L. The cleavage strength of polycrystals. *J. Iron Steel Inst.* **1953**, *174*, 25.

34. Ehtemam-Haghighi, S.; Liu, Y.; Cao, G.; Zhang, L.C. Phase transition, microstructural evolution and mechanical properties of Ti-Nb-Fe alloys induced by Fe addition. *Mater. Des.* **2016**, *97*, 279–286. [CrossRef]

Article

The Effect of Selective Laser Melting Process Parameters on the Microstructure and Mechanical Properties of Al6061 and AlSi10Mg Alloys

Ahmed H. Maamoun *, Yi F. Xue, Mohamed A. Elbestawi * and Stephen C. Veldhuis

Department of Mechanical Engineering, McMaster University, 1280 Main Street West, Hamilton, ON L8S 4L7, Canada; xueyf4@mcmaster.ca (Y.F.X.); veldhu@mcmaster.ca (S.C.V.)
* Correspondence: maamouna@mcmaster.ca (A.H.M.); elbestaw@mcmaster.ca (M.A.E.)

Received: 6 December 2018; Accepted: 17 December 2018; Published: 20 December 2018

Abstract: Additive manufacturing (AM) offers customization of the microstructures and mechanical properties of fabricated components according to the material selected and process parameters applied. Selective laser melting (SLM) is a commonly-used technique for processing high strength aluminum alloys. The selection of SLM process parameters could control the microstructure of parts and their mechanical properties. However, the process parameters limit and defects obtained inside the as-built parts present obstacles to customized part production. This study investigates the influence of SLM process parameters on the quality of as-built Al6061 and AlSi10Mg parts according to the mutual connection between the microstructure characteristics and mechanical properties. The microstructure of both materials was characterized for different parts processed over a wide range of SLM process parameters. The optimized SLM parameters were investigated to eliminate internal microstructure defects. The behavior of the mechanical properties of parts was presented through regression models generated from the design of experiment (DOE) analysis for the results of hardness, ultimate tensile strength, and yield strength. A comparison between the results obtained and those reported in the literature is presented to illustrate the influence of process parameters, build environment, and powder characteristics on the quality of parts produced. The results obtained from this study could help to customize the part's quality by satisfying their design requirements in addition to reducing as-built defects which, in turn, would reduce the amount of the post-processing needed.

Keywords: additive manufacturing; selective laser melting; AlSi10Mg; Al6061; SLM process parameters; quality of the as-built parts

---

1. Introduction

Additive manufacturing (AM) is considered one of the leading sectors of the upcoming industrial revolution "Industry 4.0" [1]. The AM of metals using selective laser melting (SLM) promises significant development of a variety of critical applications in different industrial fields [2]. The AM of Al alloys could produce high-performance lightweight components with relatively high material quality, mechanical properties, and design flexibility. The selection of the SLM process parameters plays an essential role in controlling the material and mechanical properties of products customized according to their function and design requirements. The effect of the SLM process parameters on the quality of Al alloys was previously presented in some studies [3–9]. Prashanth et al. [10] reported that the selected strategy of hatch spacing and contour parameters could significantly affect the microstructure and mechanical properties of the AlSi12 parts fabricated using SLM. Their results showed that applying the contour parameters negatively affects the ductility of parts due to residual stresses along the surface. However, the effect of the SLM parameters on the microstructure of the AlSi12 parts was not reported in that study. Biffi et al. [8] studied the influence of the SLM energy density on the mechanical properties of the AlSi10Mg samples. However, the effect of each process parameter was not studied to detect how the leading parameter affects the microstructure and mechanical properties. Li et al. [11] reported that the high cooling rate during the SLM process of AlSi10Mg results in ultrafine-grained microstructures, and thus leads to superior mechanical properties as compared to casted material of the same alloy. The microstructure characteristic could be affected by the applied cooling rate according to the selected SLM parameters. Akram et al. simulated a model of grain structure evolution in the multi-layer deposition during the SLM process [12]. Their results illustrated the change in grain size and orientation according to the process parameters applied.

In SLM of Al alloys, the chemical composition of the Al alloys could result in variation between their microstructure and mechanical properties, due to the difference in some elements such as Si and Mg. However, SLM of some Al alloys, such as Al6061, results in solidification and liquation cracking due to the material's relatively higher coefficient of thermal expansion (CTE) [3]. This is why AlSi10Mg is the most commonly-used Al alloy for the SLM process due to its lower CTE compared to the Al6061 alloy [13]. The Si content may also play a significant role in microstructure evolution and the elimination of hot cracks. Uddin et al. [14] reported that preheating the build plate at 500 °C using specific process parameters resulted in crack-free parts. However, the mechanical properties of the fabricated components are adversely affected as compared to as-built parts produced without preheating the build platform. Martin et al. [15] studied the effect of adding nanoparticles of some additives to Al6061 and Al7075 alloy powder. Their results showed that the additives added could control the solidification process and reduce the crack formation inside the as-built parts. However, lower mechanical properties were obtained as compared to the as-built components of the original alloy due to the formation of spherical pores. Moreover, the effect of the SLM process parameters for Al6061 alloy was not reported in the literature to investigate the ability to control the hot-crack formation.

Fulcher et al. [13] reported that the SLM process map should be regularly updated for each material as technical capabilities develop. This could help to optimize the SLM process parameters and customize the characteristics of the as-built parts. Consequently, the microstructure and mechanical properties of the additively-manufactured parts can be tailored according to their design requirements. In general, the literature studies show the importance to update the SLM process map according to the capabilities of the upgraded machines. This might help to detect the leading parameter affecting each characteristic to achieve the desired quality of the fabricated parts. However, the laser power of SLM was limited to 200 W, which is a relatively low figure compared to currently-used laser power which can reach 400 W for the majority of existing machines. Therefore, due to the widespread use of Al6061 in the aerospace and automotive fields, study is recommended of the influence of SLM process parameters on this material. This also might reduce the amount of post-processing applied to heal the as-built part defects.

The current study focuses on the effect of SLM process parameters on microstructure and mechanical properties of both AlSi10Mg and Al6061 as-built parts. This work completes the comprehensive study presented by Maamoun et al. [16] to develop a full process map for different Al alloys fabricated with SLM. Microstructure characterization is performed to investigate the evolution of as-built microstructure along with changing the SLM process parameters. An optimum process parameter range is investigated to reduce the microstructure defects and hot-cracks formation. The behavior of mechanical properties is studied for both materials. A regression model is created based on the design of experiment (DOE) analysis for each mechanical property along the applied range of SLM process parameters. The regression model trend for each property of the as-built parts is validated according to experimental results, and is additionally verified with microstructure analysis. A comparison between the obtained results from the current study as compared to literature is conducted to illustrate the effect of powder characteristics, build environment, and process parameters on the properties of parts fabricated. This study presents a process map of the influence of SLM process parameters of AlSi10Mg and Al6061 as-built parts, and this could offer the following:

1. Customization of parts fabricated according to their function and design requirements.
2. Reduction or elimination of the microstructure defects by investigating the optimum range of process parameters, which could reduce the amount of post-processing required.

2. Experimental Procedure

In the current study, the samples were produced using the SLM process parameters listed in Tables 1 and 2. The methodology of the DOE is the same as reported by Maamoun et al. [16]. The technique of one factor at a time (OFAT) is applied for AlSi10Mg parts, and the response surface method is used for Al6061 parts. The correlation coefficient (R^2) is used to indicate how the regression models fit with the measured data; this factor was added to each mechanical property characteristic map for each material. The build plate was preheated to 200 °C before building started under an argon medium. So, AlSi10Mg_200C and Al6061_200C also referred to the as-built AlSi10Mg and Al6061 samples respectively. Powder characteristics, microstructure analysis and the measurement of mechanical properties were performed with the following methods.

Table 1. The SLM process parameters applied for producing the AlSi10Mg_200C samples.

Sample	P (W)	V_s (mm/s)	D_h (mm)	E_d (J/mm^3)
AS1	370	1000	0.19	65
AS2	370	1300	0.15	63.2
AS3	370	1300	0.19	50
AS4	350	1300	0.19	47.2
AS5	370	1500	0.19	43.3
AS6	300	1300	0.19	40.5
AS7	370	1300	0.25	38
AS8	200	1300	0.19	27

Table 2. The SLM process parameters used for building the Al6061_200C samples.

Sample	P (W)	V_s (mm/s)	D_h (mm)	E_d (J/mm^3)	Sample	P (W)	V_s (mm/s)	D_h (mm)	E_d (J/mm^3)
1A	370	1000	0.1	123.3	11A	370	800	0.15	102.8
2A	300	1000	0.1	100	12A	350	800	0.15	97.2
3A	370	1300	0.1	95	13A	370	800	0.19	81.1
4A	300	1300	0.1	76.9	14A	350	800	0.19	76.8
5A	370	1000	0.19	65	15A	370	1300	0.15	63.2
6A	300	1000	0.19	52.6	16A	350	1300	0.15	59.8
7A	370	1300	0.19	50	17A	370	1300	0.19	50
8A	300	1300	0.19	40.5	18A	350	1300	0.19	47.2

2.1. Powder Characteristics

ASTM F3049-14 was used to examine the fresh powder of AlSi10Mg and Al6061 after sieving using a 75 μm mesh. The full powder characterization of the same powders used in this study was reported by Maamoun et al. [16]. A particle size distribution test showed a particle size ranges from 12 to 120 μm for Al6061 and from 12 μm to 110 μm for AlSi10Mg. Powder morphology of a spherical particle shape was detected with the existence of some elongated particles that might affect the flowability. Both powders have a positively skewed profile which could result in a higher density and better surface roughness as compared to Gaussian and negatively-skewed powder distribution [17].

2.2. Microstructure Characterization

The microstructure of both AlSi10Mg and Al6061 as-built samples were characterized with optical microscopy (OM), scanning electron microscope (SEM), and X-ray diffraction (XRD) measurements. A Nikon optical microscope LV100 was used to evaluate the microstructure of the etched parts. The polishing and etching procedures were performed according to the recommendations of Maamoun et al. [18]. A TESCAN VP SEM equipped with an energy dispersive X-ray spectroscopy (EDS) detector, was used to investigate the grain size and structure observations. A Bruker D8 DISCOVER XRD instrument provided with a cobalt sealed tube source was used for the samples' phase analysis. The XRD phase pattern was obtained for each sample along different orientations of the AlSi10Mg and Al6061 samples.

2.3. Mechanical Properties Measurements

The microhardness measurement was performed according to ASTM E384-17 using an automatic Clemex CMT tester. The average values of the samples' microhardness were obtained along the building direction (Z-direction) and along the plane parallel to the deposited layers (XY-plane). Each recorded value was an average of 5–10 indentations along the tested area of a 200 gf load applied over a 10 s dwell time. The residual stress was measured by an XRD instrument using a Vantec500 area detector, and the results were analyzed using LEPTOS software. The tensile rod samples were designed and fabricated according to the geometry and dimension included in ASTM E8/E8M-16a. The tensile test was performed according to ASTM E8 standard procedures using an MTS Criterion 43 universal test system which applies a load capacity up to 50 kN.

3. Results and Discussion

3.1. Microstructure

The optical microscope analysis was performed using the as-built etched samples of AlSi10Mg and Al6061. Figure 1 shows the microstructure defects and observations along the building direction (Z-direction) of AlSi10Mg samples fabricated at different SLM process parameters. Figure 1a illustrates that process-induced porosity or keyhole pores of 100–250 μm size and irregular shapes are formed inside the AS8 sample fabricated at a low energy density of 27 J/mm^3. This results from a lack of fusion due to insufficient powder delivery to the melted layer. Unmelted powder may be visible around these keyhole pores [19]. Figure 1a also shows that the melt pool solidified with an elliptically shaped profile, and that these melt pool shapes overlap in a specific arrangement according to the value of hatch spacing used. This shape is related to the Gaussian distribution of laser beam power [18]. Figure 1b shows a magnified view of the melt pool shape; a fine grain structure is observed inside, while a coarse grain is formed along its borders due to the gradient change of the solidification rate. Figure 1c shows the microstructure of the AS7 sample fabricated at an energy density of 38 J/mm^3. The keyhole pore density and size are decreased due to a higher energy density. The melt pool shape geometry of the AS7 sample is enlarged compared to the AS8 sample due to a diminishing solidification rate together with an energy density increase. In the AS3 sample produced at a 50 J/mm^3 energy density, the keyhole pores almost disappeared as shown in Figure 1d. A coarser grain structure is also present inside and along the borders of the melt pool shape as illustrated in Figure 1f. At a higher rate of energy density of 65 J/mm^3 applied to the AS1 sample, the melt pool borders disappear along some layers, and spherical hydrogen pores can be seen in Figure 1e. The areas where the melt pool borders disappear show a more homogeneous structure with elongated columnar grains oriented along the building direction, Figure 1g. While the areas displaying melt pool borders show the same inhomogeneity of microstructure as in the other samples, they have a larger grain structure, as illustrated in Figure 1h. It is worthwhile to note that the energy density level significantly affects the solidification rate, and thus, creates specific microstructure characteristics corresponding to the applied values [20]. Also, according to the SLM process parameters listed in Table 1 for each sample, the low laser power of

200 W applied to the AS8 sample resulted in low energy density, and thus a lack of fusion according to the definition of the energy density in the following equation:

$$E_d = \frac{P}{V_s * D_h * T_l} \tag{1}$$

where E_d represents energy density in J/mm^3, P is the laser beam power (W), V_s is the laser scan speed (mm/s), D_h is the hatch spacing between scan passes, T_l is the deposited layer thickness, which remains a constant value in this study with a 30 μm height. The disappearance of the melt pool profile borders observed inside the AS1 sample might be related to the reduction of the scan speed and hatch spacing parameters.

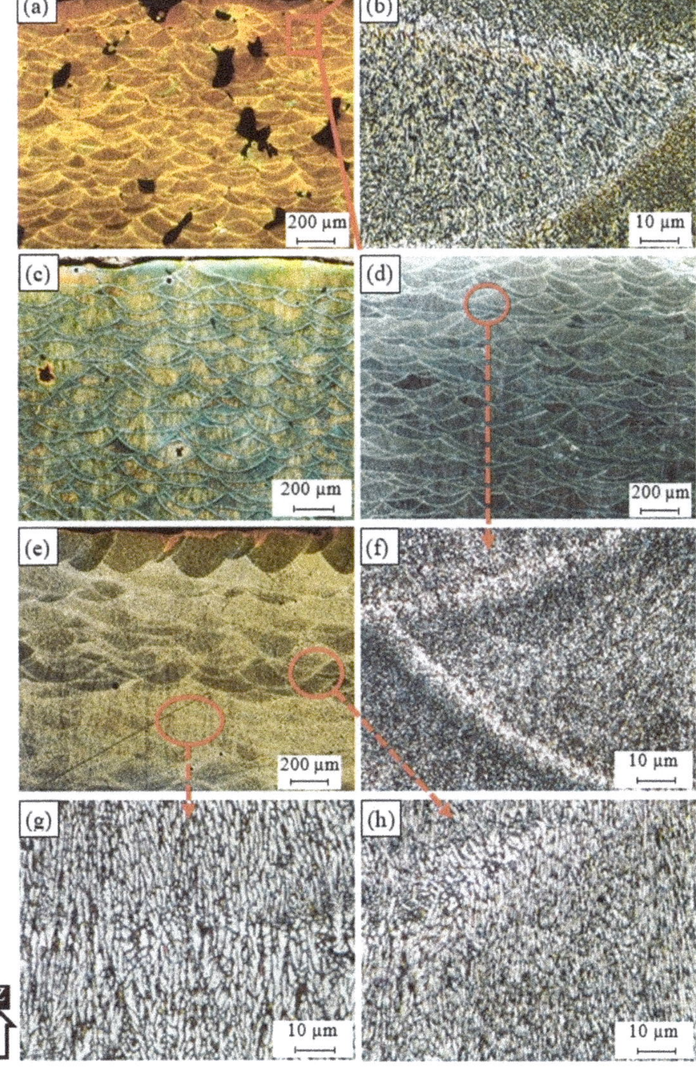

Figure 1. Microstructure of the as-built AlSi10Mg_200C samples processed under different SLM process parameters; (**a,c**) AS8; (**b**) AS7; (**d,f**) AS3, and (**e,g,h**) AS1.

SEM observations in Figures 2 and 3 show the change in the developed microstructure and the evolution of the Al matrix grain size of the as-built AlSi10Mg samples produced at different energy densities and SLM process parameters. Figure 2 displays the microstructure along the Z-direction of the AlSi10Mg samples. In general, the development mechanism of the as-built AlSi10Mg microstructure depends on the mechanism of particle accumulated structure (PAS) formation [21]. The PAS mechanism shows that during the high cooling rate of 106–108 °C/s, Si is ejected out of the solidifying Al matrix to form a fibrous Si network around the Al matrix grain borders. At a lower energy density of 27 J/mm^3, the microstructure shows an ultra-fine elongated grain structure with an inhomogeneous size distribution of Al matrix grains surrounded by a fibrous Si network. The Al matrix grain size ranges from 0.2 μm to 2 μm, where the finer grain structure formed inside the melt pool shape and larger grains located around its borders as displayed in Figure 2a. The increase of energy density to 50 J/mm^3 results in the same microstructure formation with a coarser inhomogeneous microstructure and grain size ranging from 500 nm to 3 μm as illustrated in Figure 2b. Figure 2c,d shows that when the AS1 sample is produced at a higher energy density of 65 J/mm^3, an equiaxed larger grain structure is present with Al matrix grain size varying between 3–4 μm. A more homogeneous microstructure is also obtained compared to the samples produced at a lower energy density. The final top layers in Figure 2c have a finer microstructure compared to the vicinity of the middle of the part in Figure 2d. This is attributed to the thermal gradient difference between these areas during the building of the layers, which affects the solidification rate.

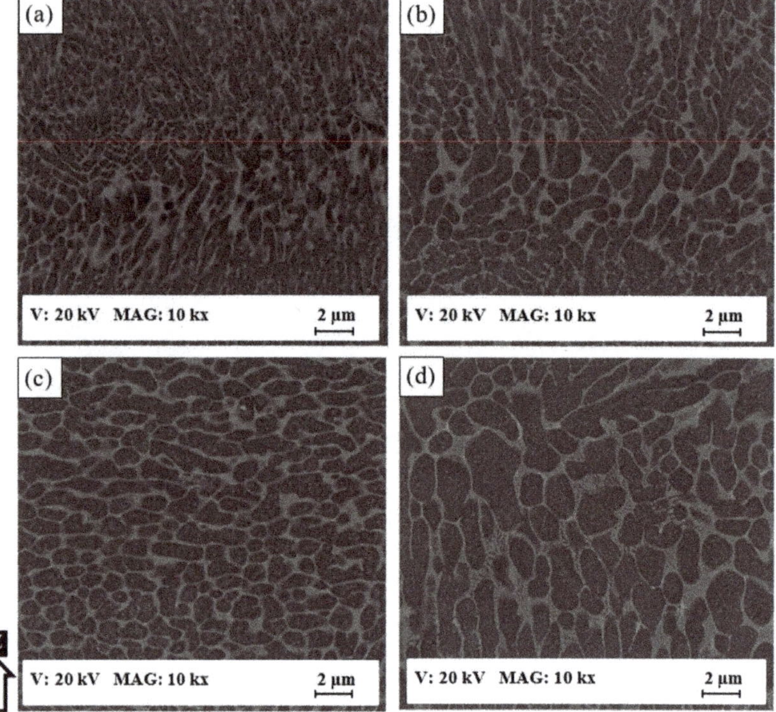

Figure 2. The SEM observations of the as-built AlSi10Mg microstructure along Z-direction; (**a**) AS8; (**b**) AS3; (**c**) AS1 near top surface, and (**d**) AS1 near the center.

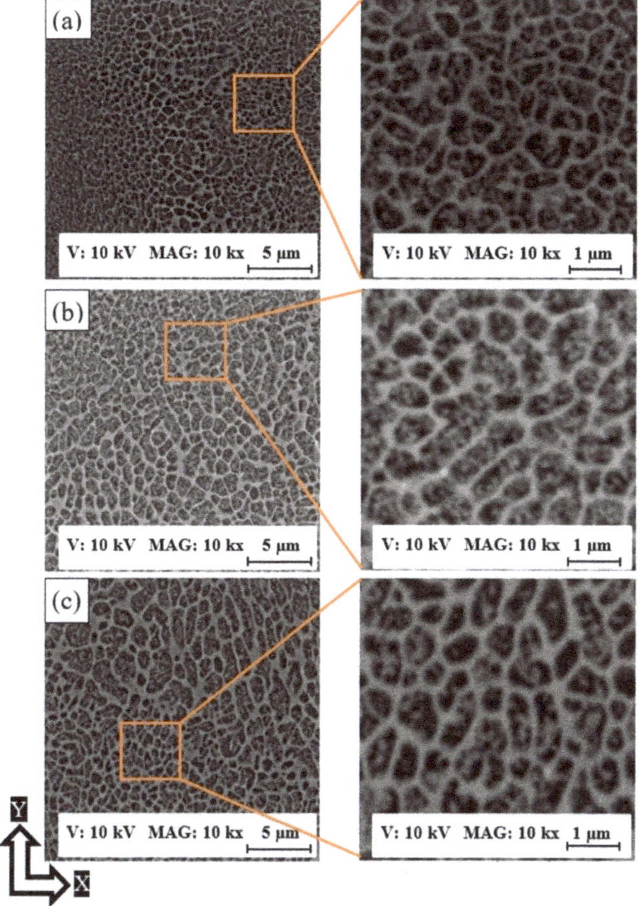

Figure 3. The SEM observations of the as-built AlSi10Mg microstructure along the XY plane; (**a**) AS8; (**b**) AS3, and (**c**) AS1.

The as-built AlSi10Mg samples along the XY plane had an equiaxed grain microstructure, as can be seen in Figure 3. The microstructure is inhomogeneous due to the existence of coarser grains along the border of the melt pool profile compared to the microstructure inside. This confirms the PAS formation mechanism of the microstructure development along the XY plane as well as the Z-direction [3,21]. Figure 3a shows the microstructure of the AS8 sample, where an inhomogeneous grain distribution of 0.15–1 μm size can be seen within the fine and coarser Al matrix grain zone. The grain size slightly increased along with energy density. Figure 3b presents the microstructure of the AS3 sample with a grain size ranging from 0.3 μm to 2 μm. The microstructure evolution of the higher energy density of 65 J/mm^3 applied to the AS1 samples has almost the same Al matrix grain structure value, as illustrated in Figure 3c. The application of energy densities higher than 50 J/mm^3 caused no significant difference in the microstructure. However, the XRD measurements were performed for a more accurate analysis of crystal size change and solubility percentage of the Si inside the Al matrix [22,23]. The zoom-in views of Figure 3a–c show more details about the evolution of the Al matrix grain inside the melt pool profile, which increases along with the energy density applied. The observed evolution of the Al Matrix grain size might be related to the reduction of solidification rate along with increasing the energy density [24].

The XRD phase pattern presented in Figures 4 and 5 shows a comparison of the Al and Si peak characteristics of different AlSi10Mg samples. The Al and Si peak is identified using the Joint Committee on Powder Diffraction Standards (JCPDS) patterns of 01-089-2837, 01-089-5012, respectively. A small peak of Mg_2Si is detected according to the JCPDS pattern of 00-001-1192, and the low intensity of this peak is related to the existence of nano-size Mg_2Si precipitates of 20–40 nm that are hardly detectable with XRD [18,23]. The difference in Al and Si peak width between the samples indicates crystal size change under different SLM process parameters. This can be inferred from Scherrer's equation, where peak broadening varies inversely with crystallite size [25]. According to the phase pattern obtained in Figure 4, the grain size significantly increased along the Z-direction as energy density increase to 50 J/mm^3, before becoming stable at a specific value, which agrees with microstructure observations in Figure 2. The XRD phase pattern in Figure 5 illustrates the peak broadening comparison along the XY plane, where a slight difference of the crystal size is observed between the samples fabricated at different SLM parameters. This corresponds to the SEM observations in Figure 3. By comparing the peak broadening of the same sample along the Z-direction and the XY plane, a significant difference can be seen in peak broadening and intensity. This difference might result from the change in the crystal shape and size, and microstrain [25,26]. The microstructure is inhomogeneous along different orientations. For more accurate values, an full width half maximum (FWHM) analysis was performed according to the phase pattern in Table 3. The results showed a broadened peak of Al and Si in the AS8 sample at the lower energy density, with FWHM values of 0.2111 and 0.5935 degrees respectively. This confirms the finer microstructure observed at the lower rates of energy densities in Figure 2. The significant difference of Al and Si peak broadening in the AS8 sample along the XY plane and Z-direction also confirms the microstructure inhomogeneity at the low energy of 27 J/mm^3. There is no significant difference between the FWHM values detected along the top and side orientations of the AS1 sample produced at a higher energy density of 65J/mm^3. A homogeneous equiaxed grain structure is present along the XY plane and Z-direction of the AS1 sample, which indicates an improvement in the microstructure homogeneity at the higher energy densities. A Rietveld analysis was performed to detect the relative weight percentage of Al and Si according to the XRD phase pattern measured along the top and side surfaces of the AlSi10Mg samples. The results listed in Table 4 indicate that Si becomes more soluble inside the Al matrix along the XY plane as energy density gets higher. The percentage of Si solubility inside the Al matrix is higher along the Z-direction compared to that in the XY plane for AS1 and AS3 samples after an energy density of 27 J/mm^3 and 50 J/mm^3 is respectively applied. In addition, the highest percentage of Si precipitates is obtained at the AS8 sample produced at the higher energy density of 65 J/mm^3. These results validate the thickness increase of the Si network at higher energy densities in Figures 2 and 3. It is worthwhile to note that Rietveld analysis was used to get the indication of the Si solubility change inside the Al matrix along the SLM process parameters applied, where the small amount of Mg_2Si existed can be neglected in this case [18].

Table 3. The average FWHM of Al and Si peaks according to the XRD phase pattern of the as-built AlSi10Mg samples.

Material	Peak	Position	AS1	AS3	AS8
AlSi10Mg	Al (200) FWHM (°)	Top (XY)	0.2111	0.2332	0.2294
		Side (Z)	0.2105	0.2304	0.2269
	Si (220) FWHM (°)	Top (XY)	0.5935	0.7281	0.7137
		Side (Z)	0.5217	0.5531	0.5420

Table 4. Rietveld analysis throughout the top and side surfaces of the as-built AlSi10Mg samples.

Material	Element	Top Surface (XY plane)			Side Surface (Z-direction)		
		AS1	AS3	AS8	AS1	AS3	AS8
AlSi10Mg	Al wt.%	91.11	91.98	90.81	93.49	93.57	90.75
	Si wt.%	8.89	8.02	9.19	6.51	6.43	9.25

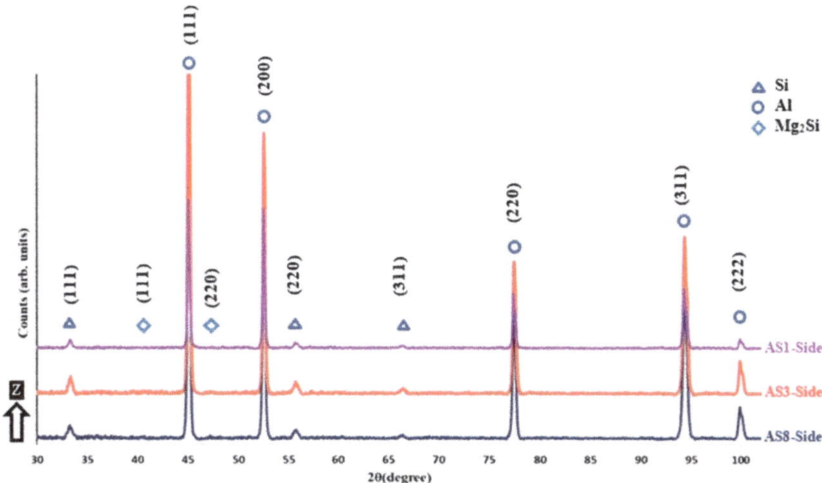

Figure 4. XRD phase pattern measured on the side surface (along the Z-direction) of different as-built AlSi10Mg samples.

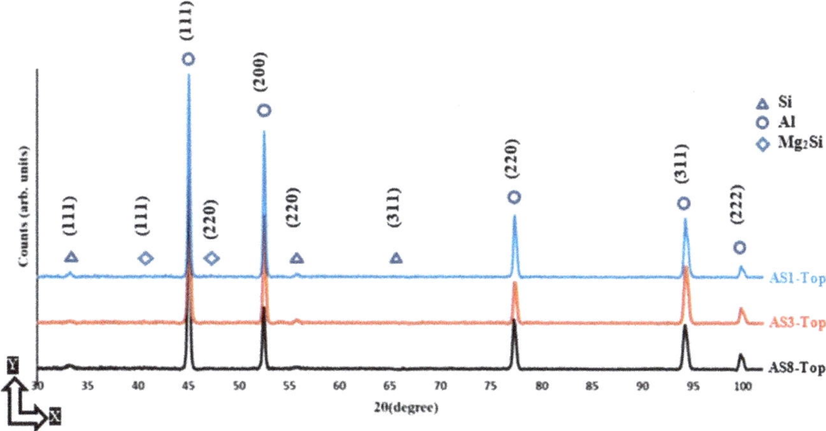

Figure 5. XRD phase pattern measured on the top surface (along the XY plane) of different as-built AlSi10Mg samples.

The microstructure of Al6061 samples shows hot crack formation in both the XY plane and Z-direction, as displayed in Figure 6. These cracks form as a result of solidification shrinkage and thermal contraction, or liquation cracking inside the partially melted zone [3,27]. For the 6A sample, hot cracks are observed along the XY plane within a size of 200–300 μm, and these cracks are connected in a closed loop, as illustrated in Figure 6a,c. The micro-cracks form into an elongated shape within an average size of 200 μm along the Z-direction and propagate through the middle zone of some solidified melt pool shape as shown in Figure 6b,d. A pore of 10–20 μm is also noticed amongst these cracks. The micro-crack formation is caused by high CTE of the Al6061, which, in turn, resulted in significant shrinkage due to the rapid melting and solidification rates of the SLM process [13]. A fine grain structure persists along both XY-plane and Z-direction, as shown in Figure 6e,f. Coarse grains are present around the melt pool profile, which substantiates the thermal gradient inside each melt pool during the solidification process. It is worthwhile to note that no large keyhole pores are observed inside the 6A sample microstructure fabricated with an energy density of 52.6 J/mm^3. The evolution of crack formation behaves differently along the Z-direction, corresponding to the applied energy density and SLM process parameters, as shown in Figure 6b,g,h. Observations indicate an increase of the crack size and distribution density under higher levels of energy densities as illustrated in Figure 6g. Large hydrogen spherical pores were seen forming along the longitudinal micro-cracks as energy density increased. By comparing the microstructure in Figure 6b,h, it can be concluded that a higher laser power and lower scan speed significantly increases the length of the cracks and their distribution density due to the imbalance between the higher melting and lower solidification rates.

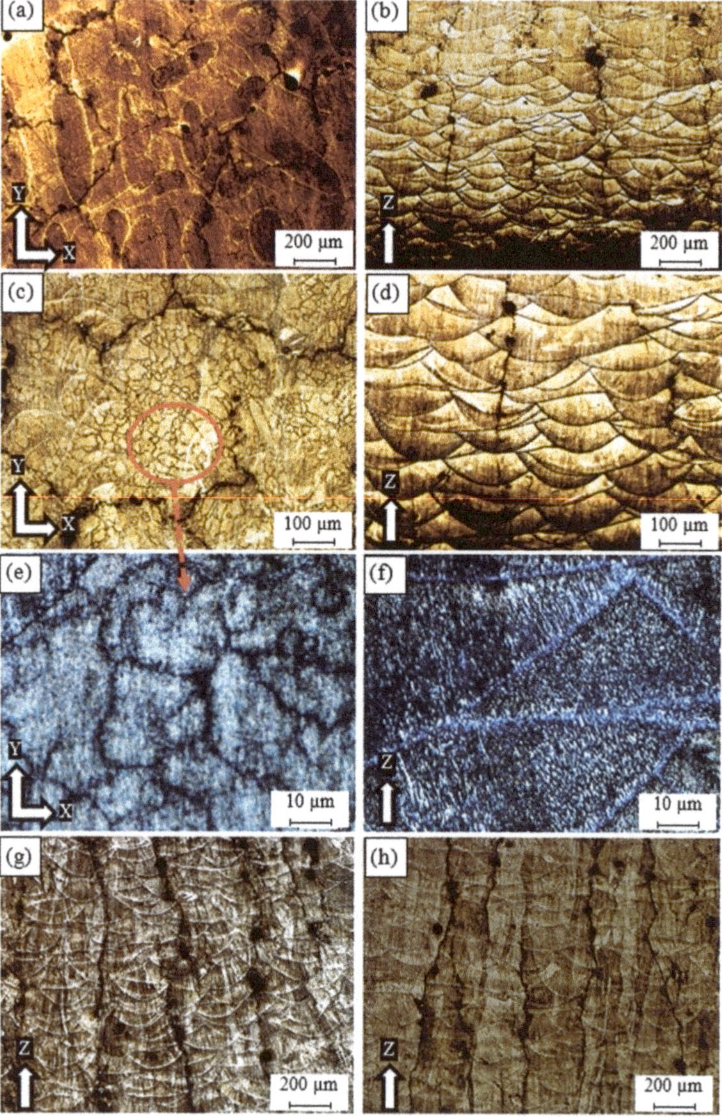

Figure 6. Microstructure of the as-built Al6061 samples processed under different SLM process parameters; (**a,c,e**) 6A along the Z-direction; (**b,d,f**) 6A along the XY plane; (**g**) 14A, and (**h**) 15A.

The as-built Al6061 microstructure in Figure 7 shows the precipitation of nano-size Si particles around the Al matrix grains, which confirms the PAS formation mechanism where the Si particles solidified around the Al matrix [21]. However, the same fibrous Si network is not present in the AlSi10Mg due to Si content in the Al6061 alloy being insufficient to develop this fibrous network. A fine microstructure with an elongated grain form is observed along the Z-direction with a size of 3–5 μm as shown in Figure 7a. Along the XY plane, an equiaxed grain structure is present, with an average grain size of (2–4 μm), Figure 7b. The difference in the grain structure between these orientations reveals microstructure inhomogeneity, which could result in anisotropic structure properties.

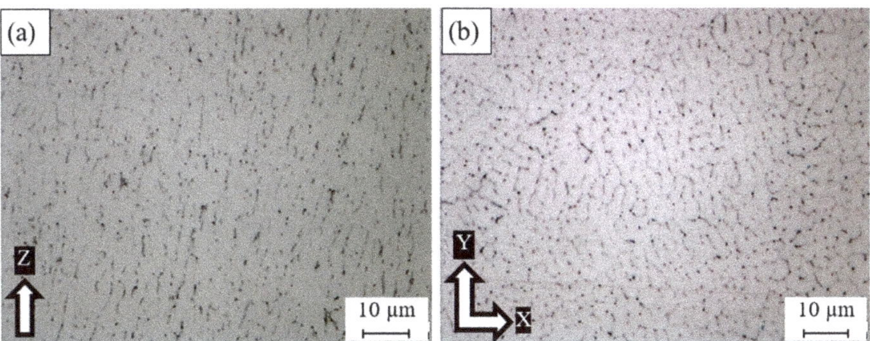

Figure 7. Microstructure grains of the as-built Al6061 sample at a higher magnification: (**a**) along the Z-direction; (**b**) along the XY plane.

The XRD phase pattern in Figure 8 shows Al and Si peak up on the top surface of the as-built Al6061 samples in the XY plane. Figure 9 illustrates the phase pattern up on the side surface along the Z-direction. The Al peak is identified according to the JCPDS pattern of 01-089-2837. According to the JCPDS patterns of 01-089-5012, a Si peak was hardly distinguishable due to the precipitation of the nano-size Si particles inside the as-built microstructure, as displayed in Figure 7. A low-intensity peak of Mg_2Si is also detected according to the JCPDS pattern of 00-001-1192, as indicated in Figures 8 and 9. The change of Al peak broadening along the XY plane and Z-direction indicates Al crystal size change according to the specified SLM process parameters. This change is closely investigated, according to FWHM analysis listed in Table 5. A wider Al peak is obtained at a low energy density of 50 J/mm^3, which confirms the growth of the grain size as energy density increases. According to Scherrer's equation, the sharper peak in the XRD phase pattern indicates a larger crystal size [25]. The FWHM shows a lower value of 0.1874 degrees in the 1A sample produced at an energy density of 123.3 J/mm^3, revealing a coarser grain structure at higher energy densities. There was no significant difference between the FWHM values of the top and side surfaces. Al6061 microstructure is more homogeneous along the applied range of the selected parameters as compared to the considerable microstructure inhomogeneity inside AlSi10Mg samples. It is worthwhile to note that the Al6061 alloy could be processed at higher energy density values than the AlSi10Mg alloy due to the higher reflectivity of Al6061, which decreases the percentage of absorbed energy. However, SLM process parameters need to be optimized to reduce the formation of micro-cracks and the spherical hydrogen pores.

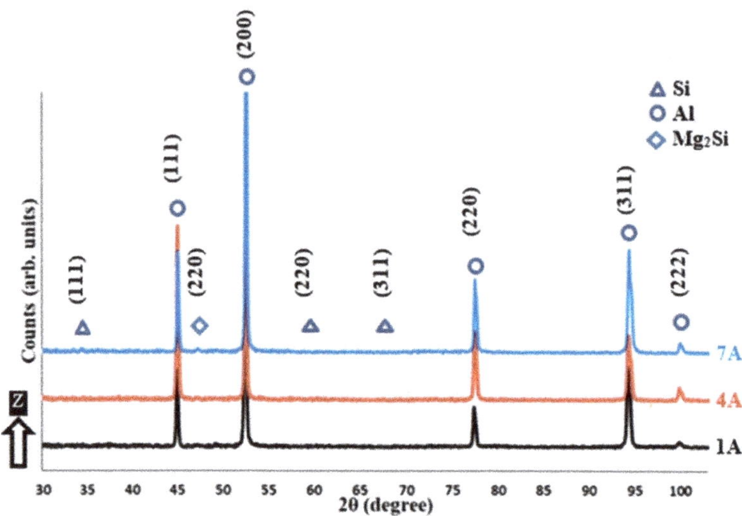

Figure 8. XRD phase pattern measured on the side surface (along the Z-direction) of different as-built Al6061 samples.

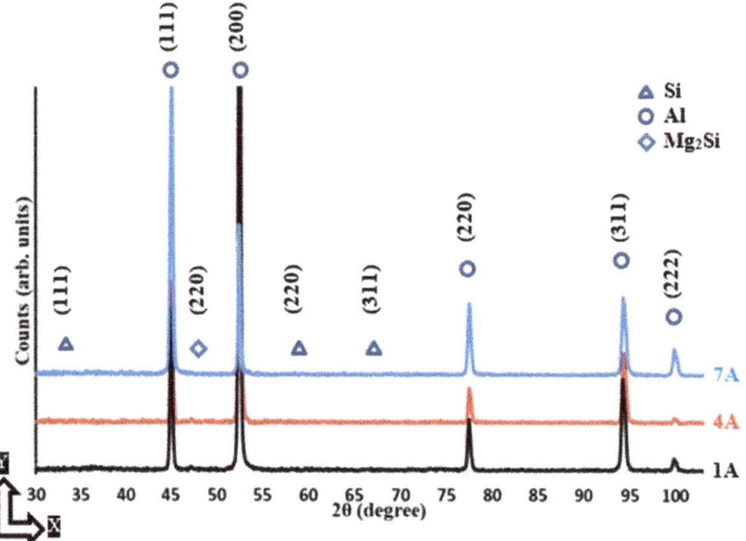

Figure 9. XRD phase pattern measured on the top surface (along the XY plane) of different as-built AlSi10Mg samples.

Table 5. The average FWHM of Al (200) peak of the as-built Al6061 samples.

Material	Peak	Position	1A	4A	7A
Al6061	Al (200) FWHM (°)	Top (XY)	0.1874	0.2086	0.2045
		Side (Z)	0.1838	0.2042	0.2029

3.2. Mechanical Properties

The effect of SLM process parameters on the mechanical properties of the as-built AlSi10Mg and Al6061 parts is investigated according to the regression models developed from experimental results. In the following section, DOE analysis will illustrate microhardness and tensile behavior according to the selected SLM process parameters.

3.2.1. Microhardness

Figure 10 displays the microhardness of the as-built AlSi10Mg parts along the Z-direction within the range of SLM process parameters. Microhardness ranges between 86 and 103 HV; the maximum value is obtained at 27 J/mm^3, due to smaller grain size, as presented in Figure 2b. However, a significant number of keyhole pores are observed at this energy density of the AS8 sample, which underscores the need for SLM process optimization. The results show that microhardness values linearly decrease as laser power and energy density grow, as illustrated in Figure 2a,b. A higher hatch spacing and scan speed improve sample microhardness in Figure 10c,d. Low values of sample microhardness at low scan speeds result from high solidification rates and low hatch spacing due to decreasing overlap between the scanned passes [3,24]. The microhardness profile of AlSi10Mg samples shows a good agreement with microstructure observations and the crystal size change of SLM process parameters.

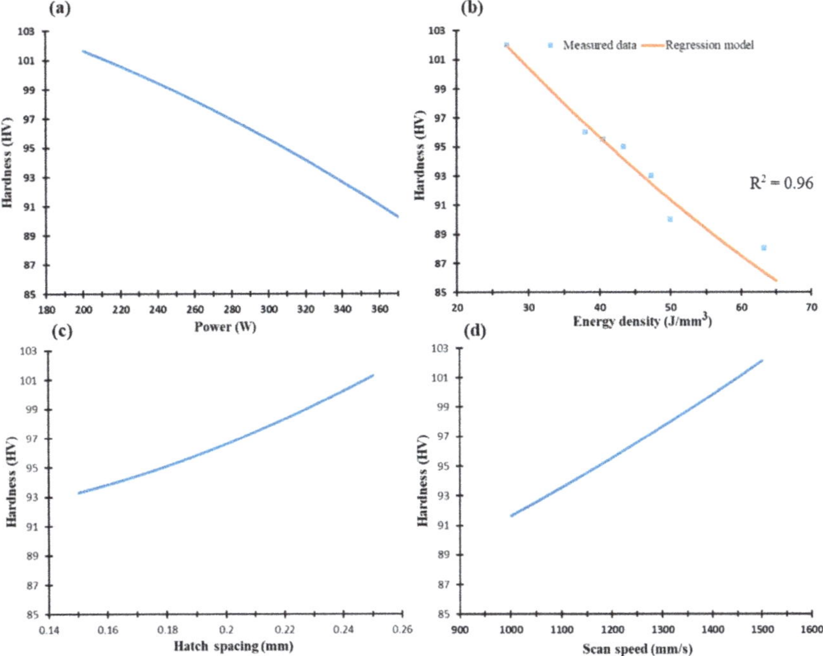

Figure 10. Microhardness of the as-built AlSi10Mg samples along the Z-direction vs. (**a**) Laser power (W); (**b**) Energy density (J/mm^3); (**c**) Hatch spacing (mm), and (**d**) Scan speed (mm/s).

As illustrated in Figure 11, microhardness along the XY plane is relatively higher than in the Z-direction, demonstrating the inhomogeneity of the as-built microstructure. The microhardness is 115 to 118 HV along the range of the SLM parameters, which confirms better homogeneity along the XY direction, Figure 3. This trend agrees with studies in the literature [18,20,22]. The reduction in laser power and greater hatch spacing improves microhardness along the XY plane, as shown in Figure 11a,c. Although the low laser power rates show higher microhardness values, control of SLM process parameters should aim to produce denser parts by reducing porosity. According to Figures 10 and 11, microhardness values correspond to the DOE analysis regression model along both the XY plane and Z-direction.

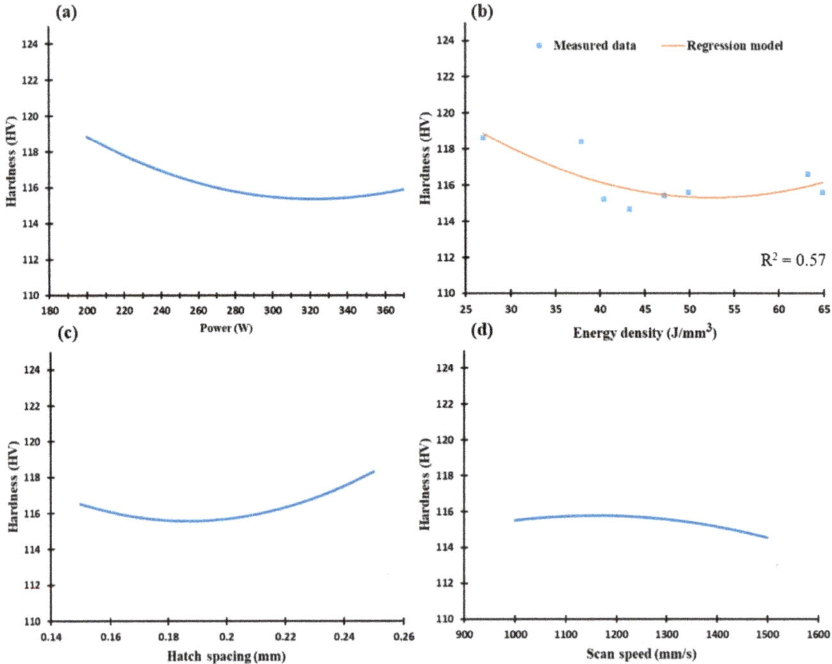

Figure 11. Microhardness of the as-built AlSi10Mg samples along the XY plane vs. (**a**) Laser power (W); (**b**) Energy density (J/mm^3); (**c**) Hatch spacing (mm), and (**d**) Scan speed (mm/s).

Figures 12 and 13 display the microhardness profile of selected SLM process parameters of the Al6061 parts along the Z-direction and XY plane respectively. The map in Figure 12 shows a gradual decrease of microhardness values along the Z-direction from 85 HV to 72 HV at an energy density range of 40.5 J/mm^3 to 97.2 J/mm^3. A slight increase was observed at higher energy densities, e.g., up to 123 J/mm^3, as illustrated in Figure 12b. At a microhardness of 78 HV, a relation is observed between the low laser power of 300 W and scan speeds of 1050 mm/s and 1300 mm/s, Figure 12a. Scan speeds higher than 800 mm/s show a significant increase in microhardness due to the associated higher rate of solidification as illustrated in Figure 12a. Results indicate that a finer microstructure can be obtained at these higher scan speeds. Another interaction between scan speed and hatch spacing occurs at a scan speed of 1050 mm/s and hatch spacing values of 0.145 mm and 0.19 mm at a microhardness value of 77 HV, as shown in Figure 12d. The average microhardness measure has a high scattering pattern around the regression model due to the effect of the micro-cracks formed inside the parts, and thus, results from the microstructure inhomogeneity along the Z-direction [18].

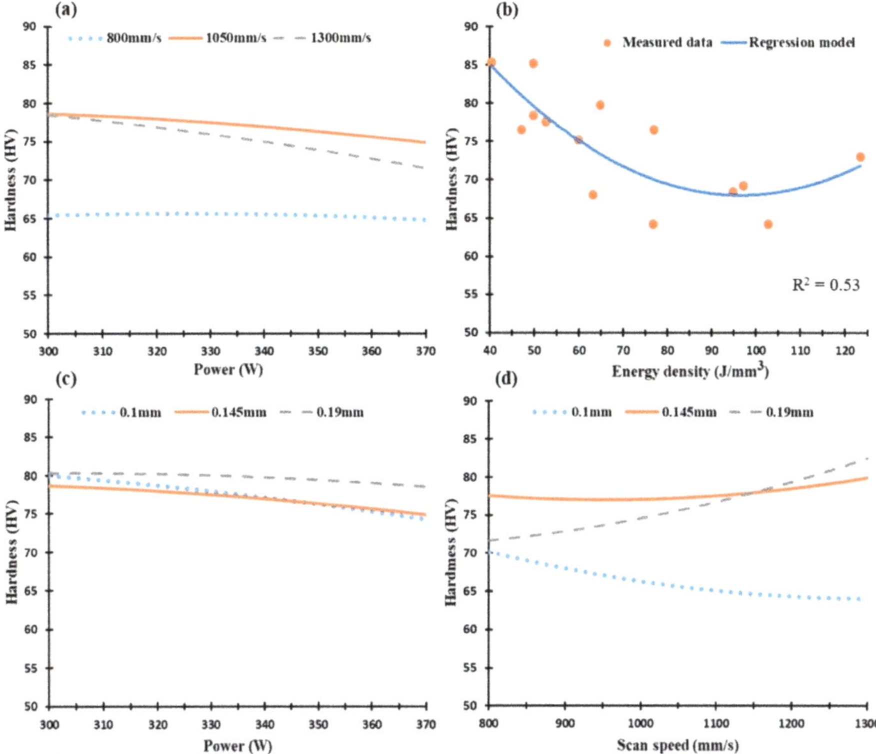

Figure 12. Microhardness of the as-built Al6061 samples along the Z-direction vs. (**a**) Laser power (W); (**b**) Energy density (J/mm^3), (**c**) Hatch Spacing (mm), and (**d**) Scan speed (mm/s).

Figure 13 shows microhardness of the Al6061 samples along the XY plane that varies significantly between 62 HV to 77 HV according to the SLM process parameters. This could be related to the change in micro-crack size, as illustrated in Figure 6. Figure 13a,b shows microhardness decrease along with laser power and energy density increase due to increasing the solidification rate, and thus results in a more coarser grain structure [20]. Figure 13c,d illustrates microhardness increase along with increasing scan speed and reducing laser power, and thus indicates that energy density is the leading parameter affects this property. In contrast with AlSi10Mg samples, hatch spacing significantly affects the microhardness of Al6061. Microhardness gradually decreases with the increase of energy density due to the gradient in microstructure characteristics. This is in agreement with the trend reported in studies in the literature [12,20].

Due to greater Si content, the microhardness of AlSi10Mg samples was significantly higher than that of Al6061 samples. The as-built AlSi10Mg samples have a higher microhardness than the same alloy cast material, which is limited to 75 HV [28]. The particle size distribution of the powder and its shape also might affect the microhardness of the as-built parts. This was demonstrated by comparing the microhardness values in this study with those reported by Maamoun et al. at different powder characteristics [18].

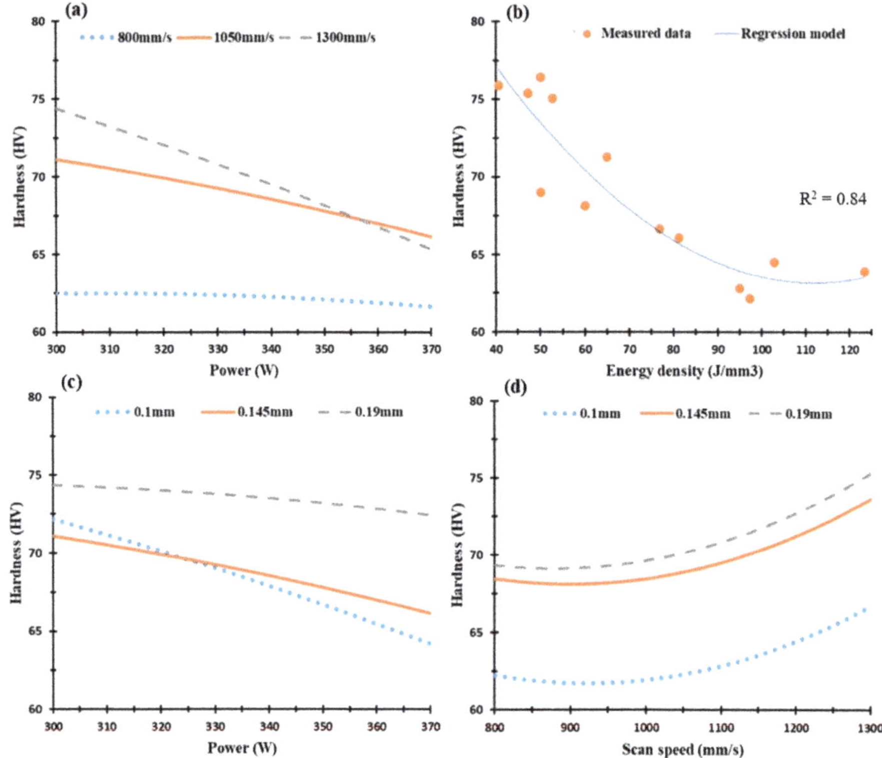

Figure 13. Microhardness of the as-built Al6061 samples along the XY plane vs. (**a**) Laser power (W); (**b**) Energy density (J/mm^3); (**c**) Hatch Spacing (mm), and (**d**) Scan speed (mm/s).

3.2.2. Tensile Properties

The ultimate tensile strength (UTS) of the AlSi10Mg was measured to generate the regression model plots for both as-built and machined tensile samples, as presented in Figure 14. The as-built and machined samples possessed the same tensile profile as the samples produced under SLM process parameters. However, the machined samples had higher UTS values of up to 450 MPa compared to those of the as-built samples (400 MPa). This 20 MPa to 50 MPa difference in UTS values indicates the effect of surface roughness on mechanical properties. However, UTS values of the as-built parts could demonstrate the impact of SLM parameters on tensile properties, taking into consideration the surface roughness of each sample. Figure 14 also shows a good agreement between the experimental measurements and the regression model generated from the DOE analysis, as illustrated in Figure 14b. Also, laser power has a more significant effect on UTS sample properties than changes in hatch spacing and scan speed, Figure 14a,c,d. The optimum UTS value is obtained in the AS3 sample at an energy density of 50 J/mm^3. This agrees with the microstructure observation, which showed minimum defects of the as-built AlSi10Mg sample at these parameters. It is worthwhile to note that better surface roughness of the as-built AlSi10Mg was obtained at higher energy density that 50 J/mm^3 [16]. However, the current results indicate that applying an energy density of 50 J/mm^3 is the optimum condition for processing AlSi10Mg alloy using a scan speed of 1300 mm/s, 370 W laser power, and 0.19 mm hatch spacing. This could result in better quality of the as-built AlSi10Mg parts according to the mutual connection between surface roughness, microstructure, and mechanical properties.

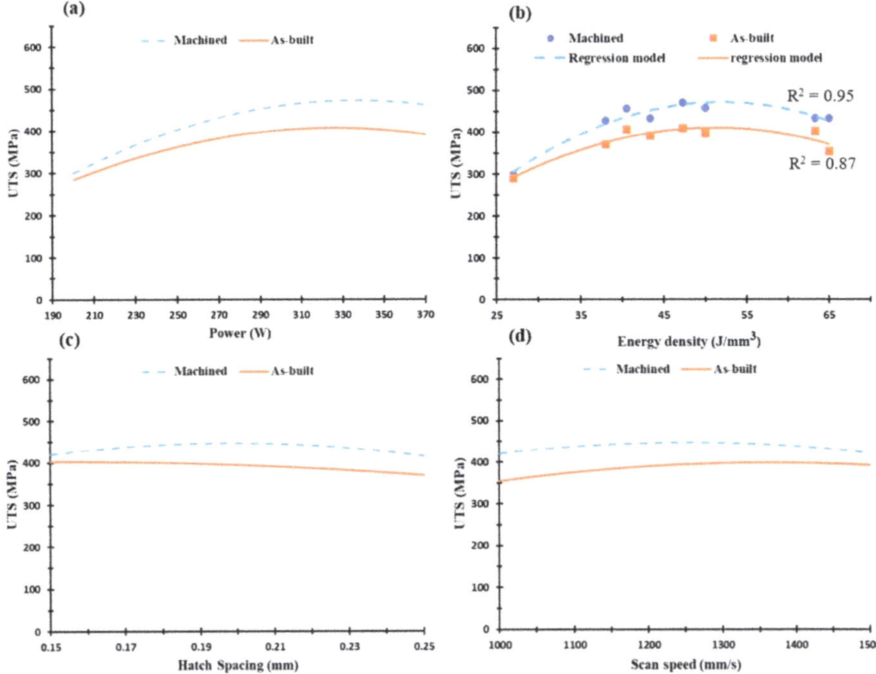

Figure 14. Ultimate tensile strength of the as-built AlSi10Mg samples along the building direction vs. (**a**) Laser power (W); (**b**) Energy density (J/mm^3), (**c**) Hatch Spacing (mm), and (**d**) Scan speed (mm/s).

Figure 15 illustrates yield strength versus the scan speed, laser power, hatch spacing, and the energy density for the as-built AlSi10Mg samples. Results indicate a decrease of yield strength within a range of 240 MPa to190 MPa at increasing energy densities as presented in Figure 15b. A slight difference of 30–50 MPa in yield strength was observed at the range of SLM process parameters, Figure 15a–d. This indicates that a change in SLM process parameters has a greater impact on UTS values than the yield strength. From the results illustrated in Figures 14 and 15, UTS and yield strength trends of the s-built AlSi10Mg parts significantly reflect the microstructure observations in Section 3.1. An increase of energy density creates a coarser microstructure with lower hardness and tensile values. This trend is in agreement with that reported by Ding et al. [29]. Moreover, some results obtained in the current study showed superior values of mechanical properties than those reported in the literature [29–31].

As illustrated in Figure 16a–d, the UTS values of the as-built Al6061 samples were investigated at a range of 150 MPa to 184 MPa. The results indicate a significant reduction in UTS of the Al6061 samples compared to that of AlSi10Mg. This could result from the lower percentage of Si content inside the Al6061 alloy and micro-cracks inside its as-built samples. As the energy density increases, UTS values gradually decrease. Figure 16b shows that a maximum UTS of 184 MPa was obtained in the 18A sample using the higher scan speed (1300 mm/s), hatch spacing (0.19 mm), and energy density of 47.2 J/mm^3. A significant decrease in the UTS values was observed at the lower scan speed of 800 mm/s and smaller hatch spacing of 0.1 mm, Figure 16a,c,d. This decrease in the UTS values might result from the microstructure defects at low rates of scan speed and hatch spacing, such as keyhole pores or areas of unmelted powder.

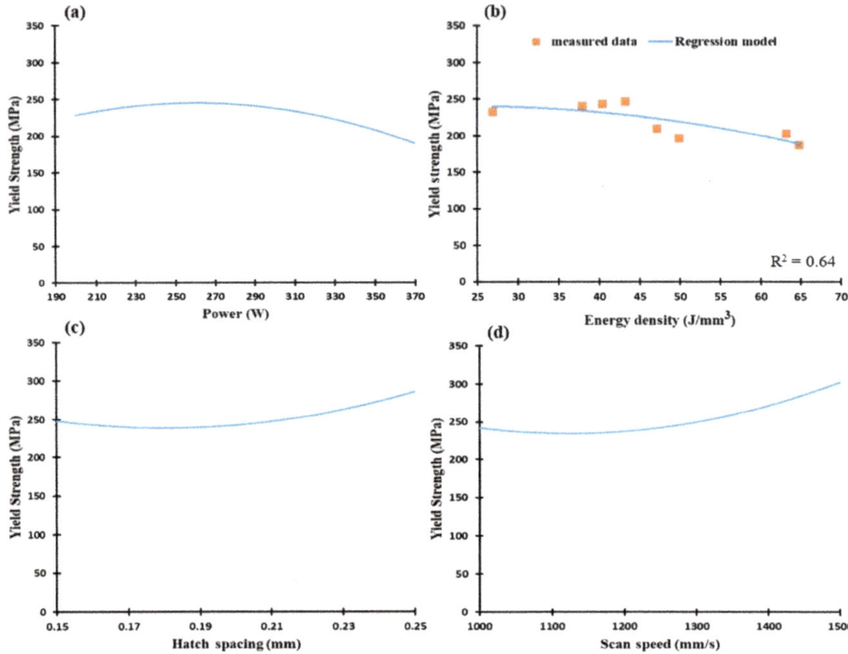

Figure 15. Yield strength of the as-built AlSi10Mg samples vs. (**a**) Laser power (W); (**b**) Energy density (J/mm^3); (**c**) Hatch Spacing (mm), and (**d**) Scan speed (mm/s).

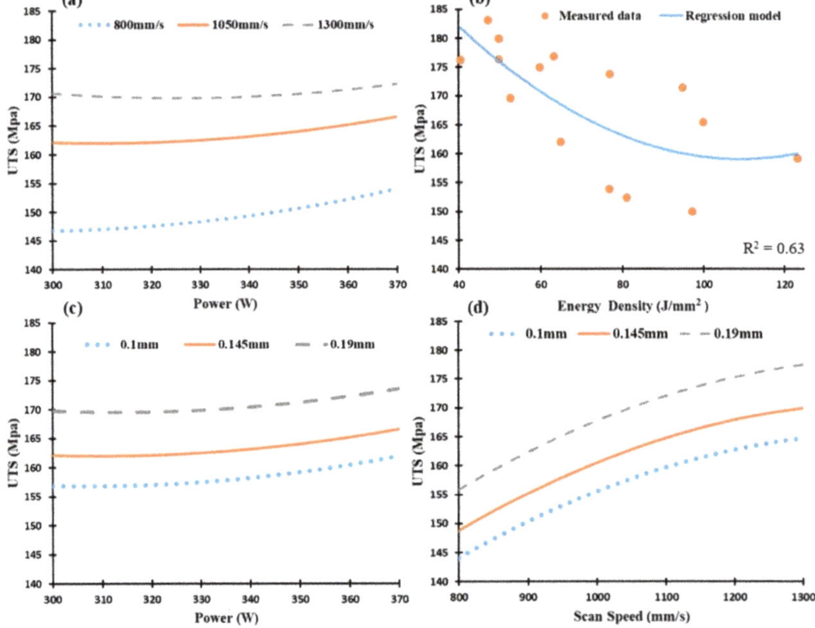

Figure 16. Ultimate tensile strength of the as-built Al6061 samples along the building direction vs. (**a**) Laser power (W); (**b**) Energy density (J/mm^3), (**c**) Hatch Spacing (mm), and (**d**) Scan speed (mm/s).

Yield strength of the Al6061 samples is presented in Figure 17, where a similar trend as in UTS is present. The yield strength values along the SLM parameters vary from 125 MPa to 172 MPa, Figure 17a–d). The maximum yield strength of 172 MPa was detected in the 8A and 18A samples also produced at the higher scan speeds, hatch spacing and energy density range of 40.5–47.2 J/mm^3. The trend obtained for the mechanical properties behavior shows a good agreement with that reported in some previous studies [13,14,32], in addition to detecting the values of SLM parameters that results in a higher values than reported in these studies. It is worthwhile to note that the UTS and yield strength values of the as-built Al6061 samples hardly differ, which indicates the lower ductility of these parts compared to the as-built AlSi10Mg samples.

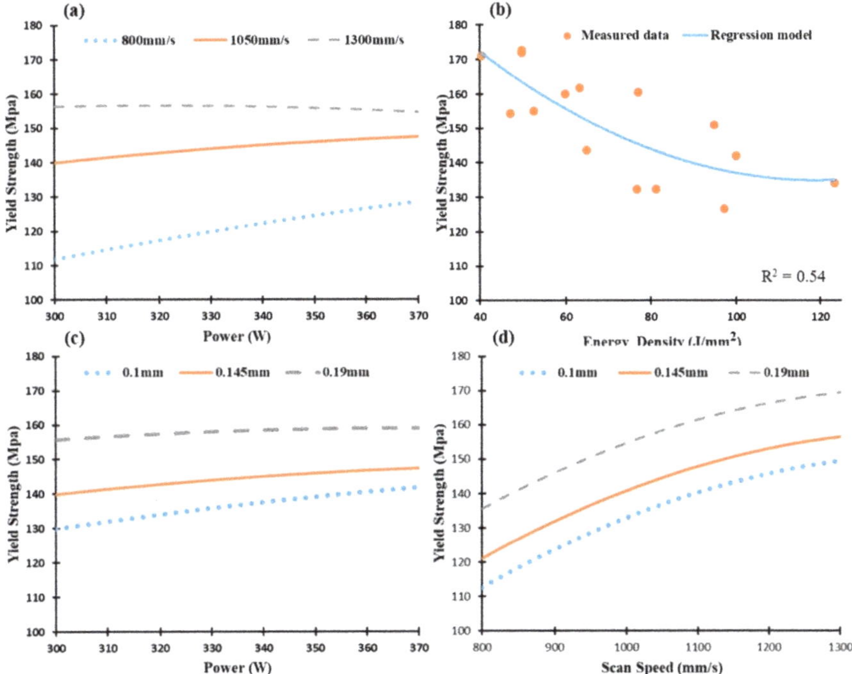

Figure 17. Yield strength of the as-built Al6061 samples vs. (**a**) Laser power (W); (**b**) Energy density (J/mm^3); (**c**) Hatch Spacing (mm), and (**d**) Scan speed (mm/s).

Figure 18 shows the stress-strain curve of the as-built samples for both AlSi10Mg and Al6061 alloys. Figure 18a illustrates the stress-strain behavior of the AS1, AS3, and AS8 AlSi10Mg samples. The maximum UTS and highest ductility was observed in the AS3 sample produced at an energy density of 50 J/mm^3. Microstructure observations confirm that the optimum SLM process parameters of the AlSi10Mg alloy are present in the AS3 sample. The AS1 sample was affected by hydrogen pores and a coarse microstructure that forms at a higher 65 J/mm^3 energy density, resulting in lower stress value. Keyhole pores and lack of fusion negatively affected the quality of the AS8 sample produced at a low energy density of 27 J/mm^3, which resulted in the lowest material strength along with higher brittleness. The strain curves of the 1A, 4A, and 7A Al6061 samples are presented in Figure 18b. Energy density change had no significant effect on the UTS value, whereas laser power proved to be the most influential. The 4A sample produced at a low laser power level of 300 W, exhibited minimum UTS values.

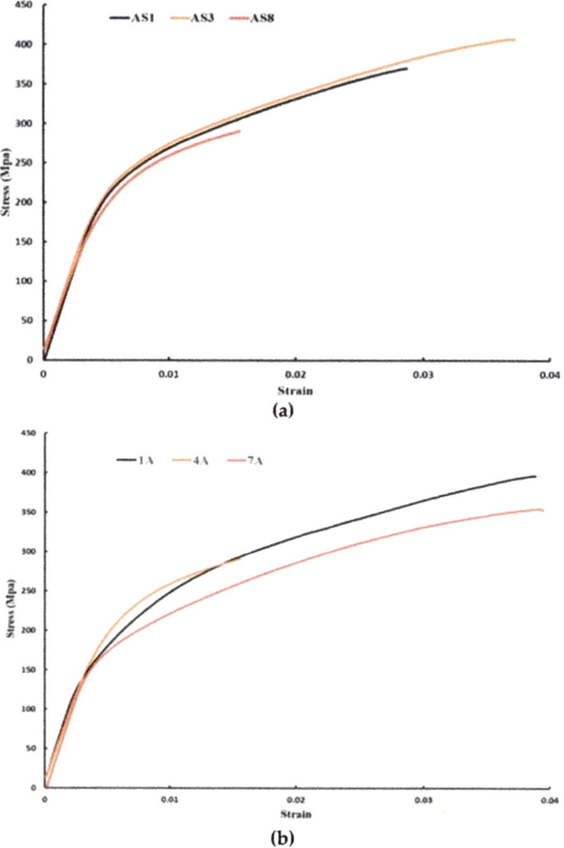

Figure 18. The stress strain diagram for the as-built samples: (**a**) AlSi10Mg; and (**b**) Al6061 samples.

Table 6 summarizes the mechanical property values of the AlSi10Mg and Al6061 samples in the current study, compared to the literature. According to values listed in Table 6, the following insights can be drawn:

1. Mechanical properties and Al matrix grain size are illustrated for the as-built AlSi10Mg_200C samples in the current study. Although the lower rate of energy density created a fine microstructure, mechanical properties were inferior due to the internal defects inside the areas caused by lack of fusion.

2. The microhardness reported in a previous study by the authors [18], using the same preheating technique, shows higher values than those reported in this study. This indicates the effect of powder morphology and its particle size distribution. It can be concluded that a wide range of particle size distribution with a spherical shape resulted in high microhardness values.

3. The mechanical properties of the AlSi10Mg_200C samples have relatively lower values than those of samples produced by build plate preheating [30,33–36]. However, residual stresses are significantly lower due to the preheating technique [4,18].

4. Superior mechanical properties of the AlSi10Mg_200C samples are detected compared to parts produced with a conventional or the high-pressure die cast (HPDC) material of the same alloy [28,37].

5. As-built Al6061_200C parts had better mechanical properties than Al6061_500C. However, no cracks were observed inside the Al6061_500C as reported by Uddin et al. [14], but the mechanical properties of the part were significantly decreased.

6. The mechanical properties of the Al6061_200C samples show comparable values to the T6, and T4 treated Al6061 wrought material [38].

Table 6. A summary of mechanical properties microstructure grain size of the AlSi10Mg and Al6061 parts processed though SLM and the conventional techniques under different conditions.

Material	SLM Process Parameters			Energy Density (J/mm^3)	Treatment	UTS (MPa)	Yield Strength (MPa)	Average Hardness (HV)	Al Matrix Grain Size (µm)
	P (W)	Vs (mm/s)	Dh (mm)						
AlSi10Mg_200C (Current Study)	370	1000	0.19	65	As-built	354.6	186.4	Z 102 XY 118	3–4 0.3–2
	370	1300	0.19	50	As-built	396.5	196	Z 90 XY 115	0.5–3 0.3–2
	200	1300	0.19	27	As-built	290.6	232.3	Z 84.5 XY 116	0.2–2 0.15–1
AlSi10Mg_200C [18]	370	1300	0.19	50	As-built			Z 120 XY 130	0.5–1
					T6			Z 115 XY 116	1–5
AlSi10Mg [36]	250	500	0.5		As-built	350	250	145	
					T6	285	340	116	
AlSi10Mg [33]					As-built	460 ± 20	270 ± 10	119 ± 5	
AlSi10Mg_200C [34]					As-built	390	210		
AlSi10Mg [30]	200	1400	0.105		As-built	391 ± 6		127	
AlSi10Mg [35]	350	1650	0.13	54.4	As-built	412 ± 2	242 ± 5	139	
AlSi10Mg [37]					HPDC	300–350	160–185	95–105	
					HPDC-T6	330–365	285–330	130–133	
A360 [28]					Casting	317	165	75	
Al6061_200C (Current Study)	370	1000	0.1	123.3	As-built	396.5	196	Z 67 XY 71	4–6 4–5
	300	1300	0.1	76.9	As-built	290	232.3	Z 81 XY 77	4–5 3–4
	370	1300	0.19	50	As-built	392	246.7	Z 67 XY 84	3–5 2–4
Al6061 [14,32]					As-built			90 ± 6	
Al6061_500C [14]	400	1400	0.14	20.41	As-built	133	66	54 ± 2.5	
					T6	308	282	119 ± 6	
AA6061-wrought [38]					O	125	55	30	
					T4	240	145	65	
					T6	310	276	95	

4. Summary and Conclusions

The current study focused on the influence of SLM process parameters on the microstructure and mechanical properties of the as-built AlSi10Mg and Al6061 parts. The mechanical behavior of these parts along the range of selected SLM parameters was investigated using DOE regression models. The main results are summarized as follows:

1. The microstructure of the AlSi10Mg parts changes significantly according to the applied energy density. After solidification, the size of the melt pool profile increases together with energy density. An energy density range of 50–60 J/mm^3 was found to be the optimal range of the energy density due to it minimizing keyholes and larger hydrogen spherical pores.

2. The grain size of the Al matrix inside the as-built AlSi10Mg samples grows along with energy density. The microstructure homogeneity is also improved by the development of an equiaxed grain structure at 65 J/mm^3 along the Z-direction and XY plane. However, this can adversely affect the relative density due to the formation of large hydrogen pores.

3. Micro-cracks form inside the microstructure of the as-built Al6061 samples. Size and distribution of these cracks vary according to SLM process parameters. The smallest micro-cracks are obtained at an energy density of 52.6 J/mm^3 and a scan speed of 1000 mm/s.

4. The microstructure of Al6061 parts did not show the same fibrous Si network that formed inside the AlSi10Mg microstructure due to lower Si content in the Al6061 alloy. The microstructure of Al6061 parts followed the PAS mechanism, and nano-size Si particles precipitated along the grain boundary of the AL matrix.

5. Microhardness of AlSi10Mg and Al6061 parts corresponds with microstructure observations along the Z-direction and in the XY plane. However, Al6061 microhardness is affected by already present micro-cracks.

6. UTS and yield strength of the as-built AlSi10Mg and the Al6061 samples are investigated through regression models.

7. The effect of surface texture on UTS of the AlSi10Mg parts was investigated by comparing the results from the as-built and machined tensile samples.

8. The mechanical properties of the studied Al alloys showed different values according to the SLM process parameters, build plate temperature, powder characteristics, and the technique used in Table 6.

The current work, together with presented by Maamoun et al. [16], forms a comprehensive study of the SLM process parameters effect on the quality of Al alloy parts. The results of this study could help customize the properties of the parts according to design and function requirements. This work may also offer a means to reduce the post-processing treatment required for part characteristics in some applications.

Author Contributions: Formal analysis, A.H.M. and Y.F.X.; Investigation, A.H.M. and Y.F.X.; Methodology, A.H.M.; Supervision, M.A.E. and S.C.V.; Validation, A.H.M.; Writing—original draft, A.H.M. and Y.F.X.; Writing—review & editing, A.H.M., M.A.E. and S.C.V.

Acknowledgments: The authors would like to acknowledge the Additive Manufacturing Innovation Centre at Mohawk College, Hamilton, Ontario, L9C 0E5, Canada and the XRD measurement analysis at McMaster Analytical X-ray (MAX) diffraction facility.

Conflicts of Interest: The authors declare no conflict of interest.

References

1. Schwab, K. *The Fourth Industrial Revolution*; Crown Business: New York, NY, USA, 2017.
2. Gibson, I.; Rosen, D.; Stucker, B. Development of Additive Manufacturing Technology. In *Additive Manufacturing Technologies*; Springer: New York, NY, USA, 2015; pp. 19–42, ISBN 978-1-4939-2112-6.
3. DebRoy, T.; Wei, H.L.; Zuback, J.S.; Mukherjee, T.; Elmer, J.W.; Milewski, J.O.; Beese, A.M.; Wilson-Heid, A.; De, A.; Zhang, W. Additive manufacturing of metallic components—Process, structure and properties. *Prog. Mater. Sci.* **2018**, *92*, 112–224. [CrossRef]
4. Buchbinder, D.; Meiners, W.; Pirch, N.; Wissenbach, K.; Schrage, J. Investigation on reducing distortion by preheating during manufacture of aluminum components using selective laser melting. *J. Laser Appl.* **2014**, *26*, 012004. [CrossRef]
5. Tradowsky, U.; White, J.; Ward, R.M.; Read, N.; Reimers, W.; Attallah, M.M. Selective laser melting of AlSi10Mg: Influence of post-processing on the microstructural and tensile properties development. *Mater. Des.* **2016**, *105*, 212–222. [CrossRef]
6. Olakanmi, E.O.; Cochrane, R.F.; Dalgarno, K.W. A review on selective laser sintering/melting (SLS/SLM) of aluminium alloy powders: Processing, microstructure, and properties. *Prog. Mater. Sci.* **2015**, *74*, 401–477. [CrossRef]
7. Siddique, S.; Imran, M.; Wycisk, E.; Emmelmann, C.; Walther, F. Influence of process-induced microstructure and imperfections on mechanical properties of AlSi12 processed by selective laser melting. *J. Mater. Process. Technol.* **2015**, *221*, 205–213. [CrossRef]
8. Biffi, C.A.; Fiocchi, J.; Tuissi, A. Selective laser melting of AlSi10Mg: Influence of process parameters on Mg_2Si precipitation and Si spheroidization. *J. Alloys Compd.* **2018**, *755*, 100–107. [CrossRef]
9. Krishnan, M.; Atzeni, E.; Canali, R.; Calignano, F.; Manfredi, D.; Ambrosio, E.P.; Iuliano, L. On the effect of process parameters on properties of AlSi10Mg parts produced by DMLS. *Rapid Prototyp. J.* **2014**, *20*, 449–458. [CrossRef]

10. Prashanth, K.G.; Scudino, S.; Eckert, J. Defining the tensile properties of Al-12Si parts produced by selective laser melting. *Acta Mater.* **2017**, *126*, 25–35. [CrossRef]

11. Li, W.; Li, S.; Liu, J.; Zhang, A.; Zhou, Y.; Wei, Q.; Yan, C.; Shi, Y. Effect of heat treatment on AlSi10Mg alloy fabricated by selective laser melting: Microstructure evolution, mechanical properties and fracture mechanism. *Mater. Sci. Eng. A* **2016**, *663*, 116–125. [CrossRef]

12. Akram, J.; Chalavadi, P.; Pal, D.; Stucker, B. Understanding grain evolution in additive manufacturing through modeling. *Addit. Manuf.* **2018**, *21*, 255–268. [CrossRef]

13. Fulcher, B.A.; Leigh, D.K.; Watt, T.J. Comparison of AlSi10Mg and Al 6061 Processed Through DMLS. In Proceedings of the Solid Freeform Fabrication (SFF) Symposium, Austin, TX, USA, 4–6 August 2014; pp. 404–419.

14. Uddin, S.Z.; Murr, L.E.; Terrazas, C.A.; Morton, P.; Roberson, D.A.; Wicker, R.B. Processing and characterization of crack-free aluminum 6061 using high-temperature heating in laser powder bed fusion additive manufacturing. *Addit. Manuf.* **2018**, *22*, 405–415. [CrossRef]

15. Martin, J.H.; Yahata, B.D.; Hundley, J.M.; Mayer, J.A.; Schaedler, T.A.; Pollock, T.M. 3D printing of high-strength aluminium alloys. *Nature* **2017**, *549*, 365–369. [CrossRef] [PubMed]

16. Maamoun, A.; Xue, Y.; Elbestawi, M.; Veldhuis, S. Effect of Selective Laser Melting Process Parameters on the Quality of Al Alloy Parts: Powder Characterization, Density, Surface Roughness, and Dimensional Accuracy. *Materials* **2018**, *11*, 2343. [CrossRef] [PubMed]

17. Tan, J.H.; Wong, W.L.E.; Dalgarno, K.W. An overview of powder granulometry on feedstock and part performance in the selective laser melting process. *Addit. Manuf.* **2017**, *18*, 228–255. [CrossRef]

18. Maamoun, A.H.; Elbestawi, M.; Dosbaeva, G.K.; Veldhuis, S.C. Thermal Post-processing of AlSi10Mg parts produced by Selective Laser Melting using recycled powder. *Addit. Manuf.* **2018**, *21*, 234–247. [CrossRef]

19. Sames, W.J.; List, F.A.; Pannala, S.; Dehoff, R.R.; Babu, S.S. The metallurgy and processing science of metal additive manufacturing. *Int. Mater. Rev.* **2016**, *61*, 315–360. [CrossRef]

20. Liu, Y.J.; Liu, Z.; Jiang, Y.; Wang, G.W.; Yang, Y.; Zhang, L.C. Gradient in microstructure and mechanical property of selective laser melted AlSi10Mg. *J. Alloys Compd.* **2018**, *735*, 1414–1421. [CrossRef]

21. Prashanth, K.G.; Eckert, J. Formation of metastable cellular microstructures in selective laser melted alloys. *J. Alloys Compd.* **2017**, *707*, 27–34. [CrossRef]

22. Maamoun, A.H.; Veldhuis, S.C.; Elbestawi, M. Friction stir processing of AlSi10Mg parts produced by selective laser melting. *J. Mater. Process. Technol.* **2019**, *263*, 308–320. [CrossRef]

23. Maamoun, A.; Elbestawi, M.; Veldhuis, S. Influence of Shot Peening on AlSi10Mg Parts Fabricated by Additive Manufacturing. *J. Manuf. Mater. Process.* **2018**, *2*, 40. [CrossRef]

24. Hooper, P.A. Melt pool temperature and cooling rates in laser powder bed fusion. *Addit. Manuf.* **2018**, *22*, 548–559. [CrossRef]

25. Langford, J.I.; Wilson, A.J.C. Scherrer after sixty years: A survey and some new results in the determination of crystallite size. *J. Appl. Crystallogr.* **1978**, *11*, 102–113. [CrossRef]

26. Prevey, P.S. X-ray Diffraction Characterisation of Residual Stresses Produced by Shot Peening. In *Shot Peening Theory and Application*; Niku-Lari, A., Ed.; IITT-International: Paris, France, 1990; pp. 81–93.

27. Carter, L.N.; Attallah, M.M.; Reed, R.C. Laser Powder Bed Fabrication of Nickel-Base Superalloys: Influence of Parameters; Characterisation, Quantification and Mitigation of Cracking. In *Superalloys*; Huron, E.S., Reed, R.C., Hardy, M.C., Mills, M.J., Montero, R.E., Portella, P.D., Telesman, J., Eds.; TMS: Pittsburgh, PA, USA, 2012; pp. 577–586, ISBN 9780470943205.

28. Kaufman, J.G.; Rooy, E.L. *Aluminum Alloy Castings: Properties, Processes, and Applications*; ASM International: Novelty, OH, USA, 2004; ISBN 0871708035.

29. Ding, Y.; Muñiz-Lerma, J.A.; Trask, M.; Chou, S.; Walker, A.; Brochu, M. Microstructure and mechanical property considerations in additive manufacturing of aluminum alloys. *MRS Bull.* **2016**, *41*, 745–751. [CrossRef]

30. Kempen, K.; Thijs, L.; Van Humbeeck, J.; Kruth, J.P. Mechanical Properties of AlSi10Mg Produced by Selective Laser Melting. *Phys. Procedia* **2012**, *39*, 439–446. [CrossRef]

31. Read, N.; Wang, W.; Essa, K.; Attallah, M.M. Selective laser melting of AlSi10Mg alloy: Process optimisation and mechanical properties development. *Mater. Des.* **2015**, *65*, 417–424. [CrossRef]

32. Zia Uddin, S.; Espalin, D.; Mireles, J.; Morton, P.; Terrazas, C.; Collins, S.; Murr, L.E.; Wicker, R. Laser powder bed fusion fabrication and characterization of crack-free aluminum alloy 6061 using in-process powder bed induction heating. In Proceedings of the 28th Annual International Solid Freeform Fabrication Symposium, Austin TX, USA, 7–9 August 2017; pp. 214–227.

33. *Material Data Sheet: EOS Aluminium AlSi10Mg*; EOS GmbH-Electro Optical Systems: Krailling, Germany, 2014.

34. *Material Data Sheet: EOS Aluminium AlSi10Mg_200C*; EOS GmbH-Electro Optical Systems: Krailling, Germany, 2013.

35. Raus, A.A.; Wahab, M.S.; Ibrahim, M.; Kamarudin, K.; Ahmed, A.; Shamsudin, S. Mechanical and Physical Properties of AlSi10Mg Processed through Selective Laser Melting. *Int. J. Eng. Technol.* **2016**, *8*, 2612–2618. [CrossRef]

36. Buchbinder, D.; Meiners, W. *Generative Fertigung von Aluminiumbauteilen für die Serienproduktion*; Fraunhofer Institute: Aachen, Germany, 2010.

37. Lumley, R.N. Technical Data Sheets for Heat-Treated Aluminum High-Pressure Die Castings. *Die Cast. Eng.* **2008**, *52*, 32–36.

38. Metals, A.S. *Properties and Selection: Nonferrous Alloys and Special-Purpose Materials*; American Society for Metals: Geauga, OH, USA, 1990; ISBN 0871703785.

 *materials*

Article

Microstructure and Mechanical Properties of Al–(12-20)Si Bi-Material Fabricated by Selective Laser Melting

Shikai Zhang [1], Pan Ma [1,*], Yandong Jia [2,*], Zhishui Yu [1], Rathinavelu Sokkalingam [3], Xuerong Shi [1], Pengcheng Ji [1], Juergen Eckert [4,5] and Konda Gokuldoss Prashanth [4,6,*]

1   School of Materials Engineering, Shanghai University of Engineering Science, Shanghai 201620, China
2   Laboratory for Microstructures, Institute of Materials, Shanghai University, Shanghai 200444, China
3   Advanced Materials Processing Laboratory, Department of Metallurgical and Materials Engineering, National Institute of Technology, Tiruchirappalli, Tamil Nadu 620015, India
4   Erich Schmid Institute of Materials Science, Austrian Academy of Sciences, Jahnstraße 12, A-8700 Leoben, Austria
5   Department of Materials Science, Montanuniversität Leoben, Jahnstraße 12, A-8700 Leoben, Austria
6   Department of Mechanical and Industrial Engineering, Tallinn University of Technology, Ehitajate tee 5, 19086 Tallinn, Estonia
*   Correspondence: mapan@sues.edu.cn (P.M.); jiayandong2008@163.com (Y.J.); kgprashanth@gmail.com (K.G.P.)

Received: 5 June 2019; Accepted: 29 June 2019; Published: 2 July 2019

Abstract: In this study, a combination of Al–12Si and Al–20Si (Al–(12-20)Si) alloys was fabricated by selective laser melting (SLM) as a result of increased component requirements such as geometrical complexity and high dimensional accuracy. The microstructure and mechanical properties of the SLM Al–(12-20)Si in as-produced as well as in heat-treated conditions were investigated. The Al–(12-20)Si interface was in the as-built condition and it gradually became blurry until it disappeared after heat treatment at 673 K for 6 h. This Al–(12-20)Si bi-material displayed excellent mechanical properties. The hardness of the Al–20Si alloy side was significantly higher than that of the Al–12Si alloy side and the disparity between both sides gradually decreased and tended to be consistent after heat treatment at 673 K for 6 h. The tensile strength and elongation of the Al–(12-20Si) bi-material lies in between the Al–12Si and Al–20Si alloys and fracture occurs in the Al–20Si side. The present results provide new insights into the fabrication of bi-materials using SLM.

Keywords: Al–Si; selective laser melting (SLM); microstructure; mechanical properties

---

1. Introduction

The characteristic properties of aluminum, such as high strength and stiffness-to-weight ratio, good formability, good corrosion resistance, and recycling potential make it an ideal candidate for automobile and aerospace applications [1,2]. The use of aluminum reduces the dead weight of the components, leading to more fuel-efficient vehicles, lower energy consumption and less air pollution [1–4]. Among the families of Al-based alloys, Al–Si is one of the most commonly used cast alloys because of its advantages concerning fluidity [5,6]. With increasing silicon content, especially from hypoeutectic to hypereutectic composition, the strength, tribological properties and corrosion resistance of the alloy increases significantly. However, the ductility, thermal conductivity and machinability show the opposite trend [7–10]. Moreover, the performance requirements of different parts of any particular component are diverse and can be demanding based on the stringent requirements for reduced waste and/or minimal use of resources. Such stringent usage and heavy demands often lead to the use of new alloys replacing traditional alloys produced by casting [11,12]. Hence, parts with different Si contents may be preferred depending on the application, where the outer surface of the components may require good tribological properties (higher Si content) and the core of the part should have good ductility (lower Si content). Hence, application of new methods to produce such components is of primary interest.

In comparison with traditional processes, additive manufacturing (AM) technologies offer significant benefits, such as near-net-shape production capabilities, superior design and geometrical flexibility, reduced tooling, shorter cycle time for design and manufacturing, as well as material, energy and cost efficiency. In particular, selective laser melting (SLM) is an AM process in which functional, complex parts are produced by selectively melting consecutive layers of powder with a laser beam [13–16]. In SLM, layers of atomized powder are spread sequentially on a building platform. The powder bed is then melted selectively by a laser beam. The melt pool is cooled rapidly by the underlying substrate or the previously built metal layer. The parameter optimization is rather simple based on an empirical rule of thumb considering the energy density, which strongly depends on laser power, laser scan speed, layer thickness, hatch distance and laser spot size [16,17]. In general, the parameters of the process are decided by the defect content in the material rather than using the melting point or solidification range as reference [18]. The flexibility in parameter selection offered by SLM enables the exploration of a wide spectrum of possibilities for creating novel alloys or even metal–metal composites with unique microstructures [18–22]. Heat treatments at high temperatures can promote the coalescence of second phases and change their distribution [23,24]. In this research, Al–(12-20)Si bi-materials were fabricated by SLM and heat treated at different temperatures. The microstructure and the mechanical properties of the Al–(12-20)Si bi-material fabricated by SLM were systematically studied.

2. Experimental Methods

Spherical gas atomized Al–12Si and Al–20Si powders with an average particle size of 40 μm were used for the experiments, as illustrated in Figure 1. SLM solutions SLM280^{HL} device equipped with a Nd-YAG laser was used for preparing the Al–12Si and Al–20Si bi-materials. Both Al–12Si and Al–20Si are manufactured using the same SLM processing conditions. The processing parameters have been optimized as follows: the laser power was 320 W, laser scan speed was 1455 mm/s, the powder layer thickness was 50 μm, the hatch spacing was 110 μm and the hatch style rotation was 73°. High purity argon gas was used to ventilate the chamber before and during the building process to maintain a low concentration of oxygen. For comparison, the Al–12Si and Al–20Si alloys were also fabricated separately by SLM using the same experimental parameters. The SLM samples were subsequently heat-treated at 473 ± 1 K, 573 ± 1 K and 673 ± 1 K for 6 h and cooled to room temperature by furnace cooling.

The microstructures and fracture morphologies of the SLM and heat-treated samples were characterized by optical microscopy (OM) using an Olympus optical microscope (Olympus, Hamburg, Germany) and by scanning electron microscopy (SEM) using a Gemini 1530 microscope (Jeol, München, Germany) operated at 20 KV after etching with 0.5% HF solution (0.5% HF, 99.5% H_2O, in vol %). Vickers microhardness measurements were carried out using a HV-1000 Z-type digital microhardness tester (Stuers, Willich, Germany) with 100 g load and 15 s dwell time. Tensile tests were performed with an INSTRON 5569 testing machine (INSTRON, Darmstadt, Germany) along the building direction at a strain rate of 1×10^{-4} s^{-1}, as shown in Figure 2. In order to ensure reproducibility of the tensile results, the mechanical properties were determined by mathematically averaging the measurement results of at least six samples.

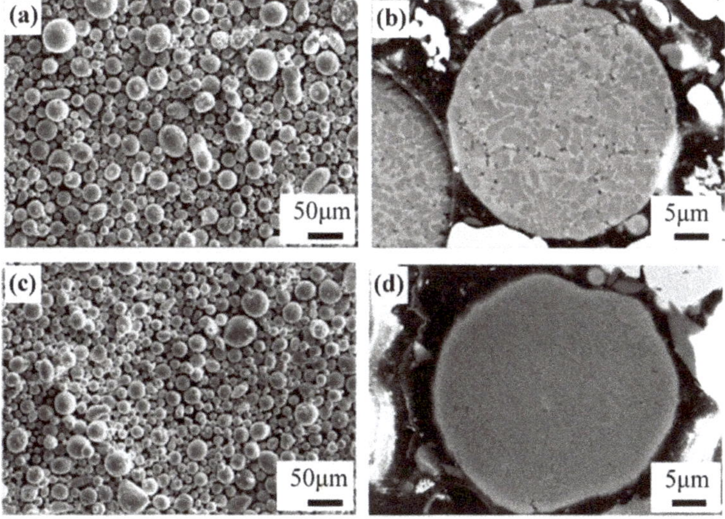

Figure 1. Microstructures of (**a,b**) Al–12Si, and (**c,d**) Al–20Si gas atomized powders.

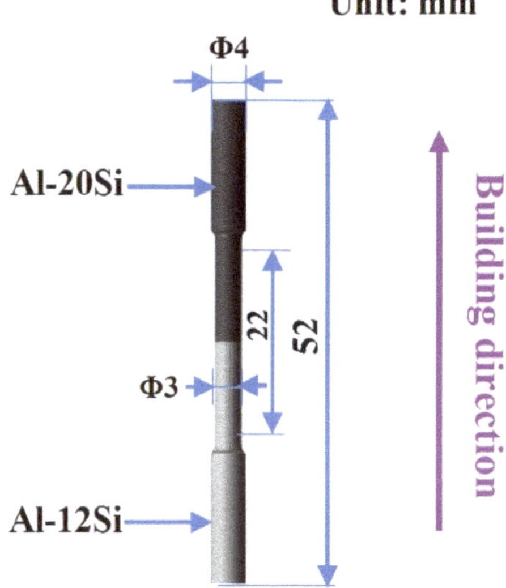

Figure 2. Schematic diagram of tensile samples of Al–(12-20)Si bi-material.

3. Results and Discussion

Figure 3 shows OM and SEM images of the Al–12Si and Al–20Si bi-materials fabricated by SLM with and without heat treatment at different temperatures. The morphology of the as-SLM material is shown in Figure 3a,b. The boundary between Al–12Si and Al–20Si is distinct due to differences in Si content (marked by dotted lines). As can be seen in Figure 3a, the average width of the melt pool is about 160 μm, and the penetration depth is around 90 μm along the Al–12Si side. On the other hand, the average width of the melt pool is approximately 204 μm, and the penetration depth is about 87 μm along the Al–20Si side. This suggests that Al–20Si exhibits elongated melt pools with a typical length-to-width aspect ratio of about 2.35:1. Concerning the geometrical characteristics, a single track of the melt pool dimensions, including the width and depth, is correlated with the energy input. As more Si exists in the Al–20Si alloy than in the Al–12Si alloy and the experimental parameters used are the same for both, the additional melting of 8 wt % more Si in the Al–20Si alloy requires more energy. Upon consumption of additional energy for melting, the depth of the melt pool in Al–20Si is reduced; however, the width is increased compared to the Al–12Si side and the length of the melt pool remains constant, suggesting shallow melt pool characteristics on the Al–20Si side [25,26].

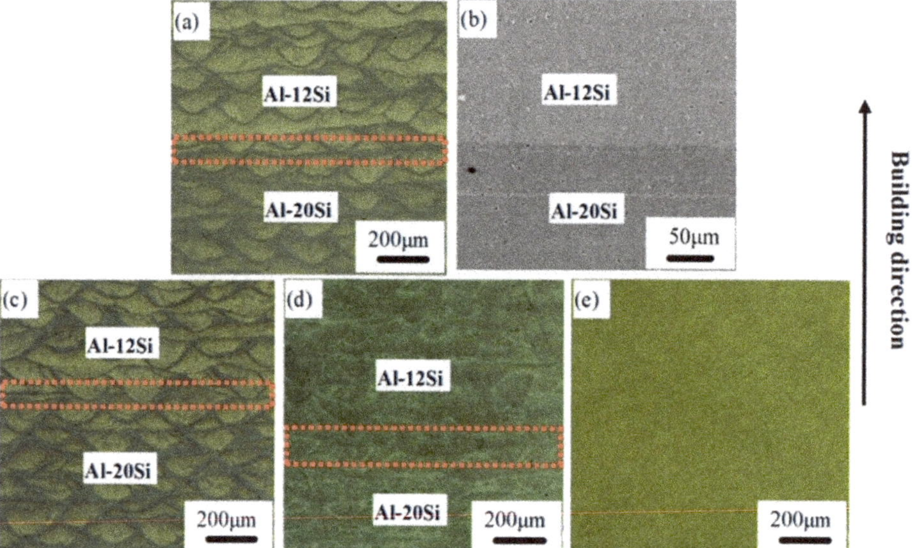

Figure 3. Microstructure of Al–Si bi-materials fabricated by selective laser melting (SLM) and heat treatment: (**a**,**b**) SLM; (**c**) SLM + 473 K/6 h; (**d**) SLM + 573 K/6 h; (**e**) SLM + 673 K/6 h.

Because of Marangoni convection [26], a large amount of Si segregates at the hatch overlaps and/or boundaries are present in the material and this phenomenon is more distinct at the Al–20Si side. The primary silicon phase is driven by the fluid flow in the melt pool and starts to solidify along the low temperature region of the melt pool. Thus, the primary silicon phase is concentrated at the contour regions of the melt pool and along the hatch overlaps. The interfaces are still visible and there are no significant variations in the microstructure after heat treatment at 473 K for 6 h (Figure 3c). Compared with Figure 3a–c, the interface in Figure 3d is rather blurred; moreover, the contour regions also become unclear and it is difficult to distinguish the track cores and the hatch overlaps. Finally, both the heat affected zone and the interface disappear after heat treatment at 673 K for 6 h, as shown in Figure 3e. Therefore, with increasing heat treatment temperature, the interface becomes indistinct and the microstructure becomes uniform. When heat treated at 673 K, the diffusion of Si atoms as well as growth of Si particles are accelerated, both promoting the uniform distribution of Si in the Al matrix [23,25].

Figure 4 displays the hardness profile of the Al–Si bi-material as a function of the distance from the interface after SLM and after heat treatment at different temperatures. As observed in Figure 4a, the average Vickers hardness of the Al–20Si side is 188 $HV_{0.1}$ and 131 $HV_{0.1}$ for the Al–12Si side. The SLM-processed samples demonstrate superior hardness compared to alloys with the same composition fabricated with other methods [27,28]. As expected, the hardness of the Al–20Si alloy is significantly higher than at the Al–12Si side because of its higher Si content [27,28].

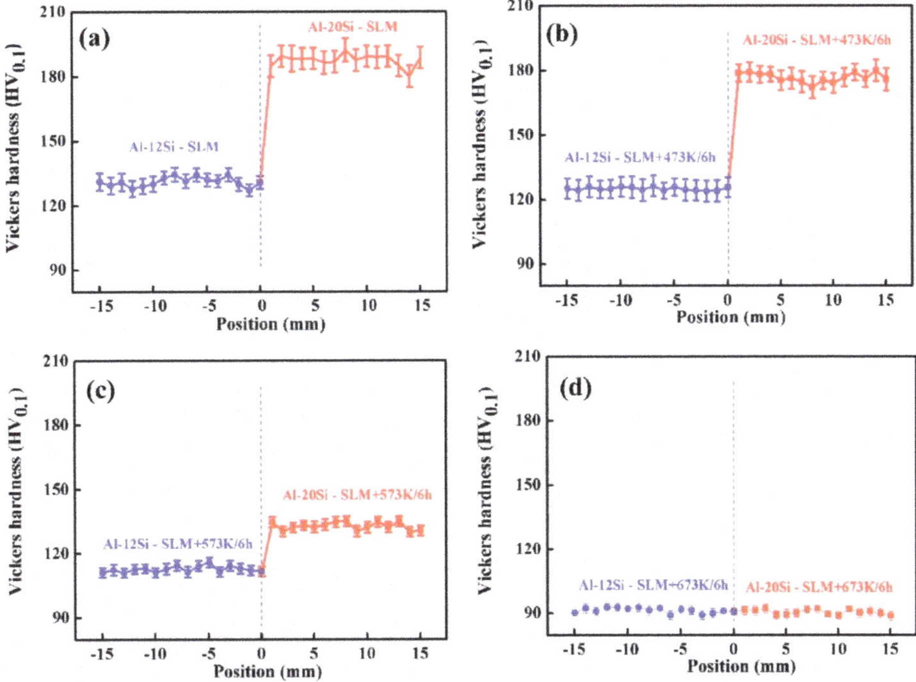

Figure 4. Vickers hardness of Al–(12-20)Si bi-materials: (**a**) SLM; (**b**) SLM + 473 K/6 h; (**c**) SLM + 573 K/6 h; (**d**) SLM + 673 K/6 h.

It is well-known that both the hardness and the strength of Al-based alloys processed by SLM decrease with heat treatment [24,27]. Similarly, in the Al–(12-20)Si bi-material, the hardness decreases after thermal treatment (Figure 4b–d). The hardness of Al–12Si decreases from 131 $HV_{0.1}$ (as-prepared SLM) to 125 $HV_{0.1}$, 113 $HV_{0.1}$ and 91 $HV_{0.1}$ after heat treatment at 473 K, 573 K and 673 K for 6 h, respectively. Meanwhile, along the Al–20Si side, the hardness decreases from 188 $HV_{0.1}$ (as-prepared SLM) to 177 $HV_{0.1}$, 133 $HV_{0.1}$ and 91 $HV_{0.1}$ after heat treatment at 473 K, 573 K and 673 K for 6 h, respectively. Heat treatment at 673 K causes the hardness on both the Al–12Si and the Al–20Si side to become similar, suggesting that there is enough time for the Si atoms to diffuse from one side to the other and to reach an equilibrium state [27,29]. Further heat treatment will further reduce the hardness, but it will remain similar on both sides. Hence, the hardness profile for the 673 K heat treated sample remains flat without any variations from the Al–12Si to the Al–20Si side.

Representative room temperature stress–strain curves of the quasistatic tensile tests of the SLM-fabricated Al–Si alloys and the bi-material are presented in Figure 5. The as-prepared Al–12Si alloy exhibits an ultimate tensile strength (UTS) and yield strength (YS) of approximately 380 MPa and 260 MPa, respectively, with a room temperature plasticity of about 2.5%. Most of the SLM samples have excellent tensile strength, significantly better than their cast counterparts [27]. For instance, as-cast Al–12Si normally has a tensile strength of approximately 200 MPa [27–34] and in the present as-prepared condition, the SLM samples exhibit a strength of approximately 380 MPa. Some studies introduced ultrasonic melt treatments and equal-channel angular extrusions in order to refine the grains and enhance their mechanical properties [31,32]. After such treatment, the UTS was about 309 MPa and 230 MPa, respectively. Moreover, conventional powder metallurgy techniques were used to fabricate Al–12Si alloys, and the tensile strength was observed to be 174 MPa and 220 MPa, respectively, after hot pressing and hot extrusion [33]. Similarly, the strength of Al–20Si SLM parts is higher than the values obtained by most manufacturing methods [34–37]. Yoon et al. [36] reported an ultimate tensile strength of the Al–20Si alloy after extrusion and equal channel angular pressing of about 350 MPa. Furthermore, high pressure solidified Al–20Si samples show an UTS of 365 MPa after high pressure solidification [37].

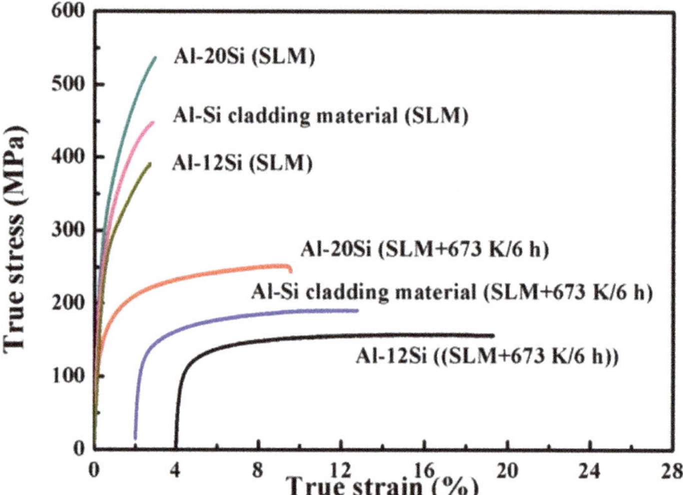

Figure 5. True stress–strain curves of Al–Si alloys and Al–(12-20)Si bi-materials.

Figure 5 reveals that the YS and UTS of the bi-materials are about 450 MPa and 300 MPa, respectively, with approximately 2.2% ductility. Moreover, the Al–(12-20)Si bi-material exhibits mechanical properties that are between those of the Al–12Si and Al–20Si alloys. Since the melt pool is very small (a few micrometers in length), depending on the powder characteristics and on the processing parameters used, the cooling rates during SLM can reach very high values (1×10^3–1×10^6 K/s), which can increase the nucleation rate and suppress grain growth [13,38]. Correspondingly, this high strength can be attributed to the very fine microstructure as well as the formation of a supersaturated solid solution during the process.

The longitudinal microstructure of the SLM bi-materials is shown in Figure 6, revealing that the morphology of the Al–12Si side in the Al–(12-20)Si bi-material is parallel to the building direction. The microstructure is anisotropic and is composed of track cores and hatch overlaps. Fibrous eutectic Si is located around α-Al and presents a columnar morphology in the track cores. In contrast, the α-Al phase is surrounded in circles with Si particles at the relatively narrow hatch overlaps.

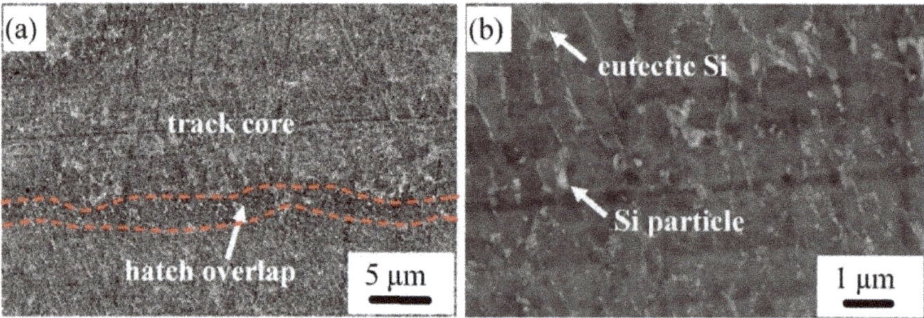

Figure 6. Longitudinal microstructure of the Al–12Si side in the Al–(12-20)Si bi-material (**a**) low magnification and (**b**) high magnification.

The microstructure in Figure 7 along the Al–20Si side is inhomogeneous and shows typical SLM features like track core morphology with hatch overlaps. The track cores consist of fibrous eutectic (Al + Si), Si particles and a small quantity of α-Al phase, as illustrated in Figure 7b. The size of the fibrous eutectic Si is approximately 500 nm. The grain boundaries can be seen clearly in Figure 7. The hatch overlap region (Figure 7c) is composed of Si particles with an average size of 300 nm. Grain refinement leads to reduction of the distance between the Si particles, which can give a considerable contribution to the strength because the increased number of Al–Si interfaces can effectively reduce the movement of dislocations [39,40]. As discussed above, such microstructural modification obtained through SLM processing can enhance the mechanical properties. Similar observations were made for SLM-processed Al–Si based alloys [27,28].

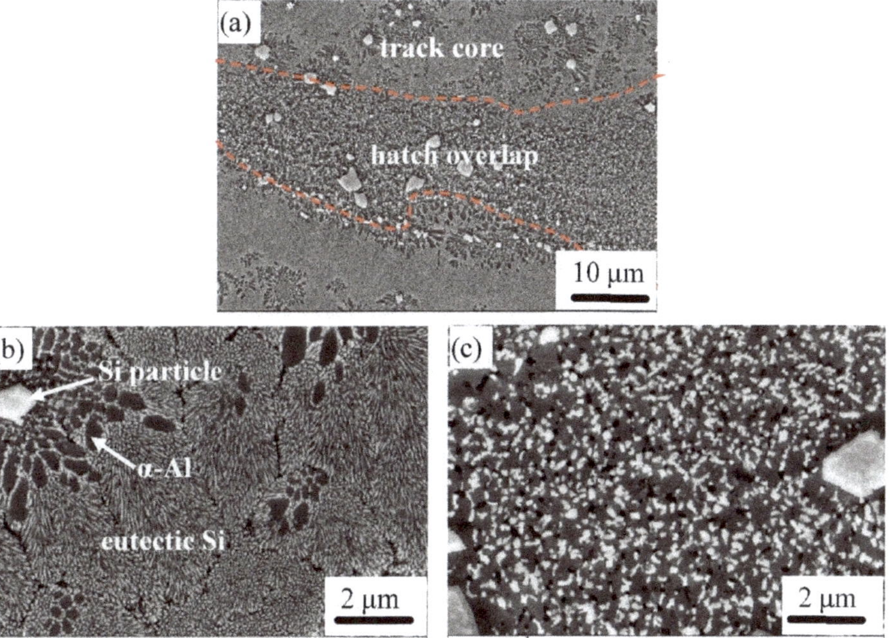

Figure 7. Longitudinal microstructure of the Al–20Si side in the Al–(12-20)Si bi-material, where (**a**) shows the microstructure along with both track cores and hatch overlaps; (**b**) microstructure along with the core of the tracks; and (**c**) microstructure along with the hatch overlaps.

The bi-material heat treated at 673 K for 6 h was further investigated in order to gain insight about the effect of heat treatment on the tensile properties. The UTS decreases to approximately 200 MPa and the YS decreases to approximately 125 MPa. However, the elongation increases to 9.0% for the heat treated bi-material. Table 1 lists the tensile stress–strain data of the Al–12Si alloy, the Al–20Si alloy and the Al–(12-20)Si bi-material after SLM fabrication and after heat treatment at 673 K for 6 h. The comparison reveals that the Al–(12-20)Si bi-material after heat treatment has mechanical properties that are still between the tensile properties of the Al–12Si alloy and the Al–20Si alloy under heat-treated conditions.

Figure 8 presents the morphology of the Al–Si bi-material after heat treatment at 673 K for 6 h. Due to different Si contents, Si particles diffuse from the Al–20Si side to the Al–12Si side during heat treatment. Moreover, with increased heat treatment temperature, the melt pool boundaries and the heat affected zones disappear. In addition, the Si particles become coarse and their average size increases to approximately 1.31 μm.

Table 1. Mechanical properties of Al–Si alloys and the Al–Si bi-material. UTS = ultimate tensile strength; YS = yield strength.

Composition	Status	UTS (MPa)	YS (MPa)	Elongation (%)
Al–20Si	SLM	500 ± 13	340 ± 9	1.8 ± 0.1
	SLM + 673 K/6 h	240 ± 7	150 ± 3	5.0 ± 0.2
Al–Si bi-material	SLM	450 ± 11	300 ± 9	2.2 ± 0.1
	SLM + 673 K/6 h	200 ± 7	125 ± 3	9.0 ± 0.4
Al–12Si	SLM	380 ± 10	260 ± 7	2.5 ± 0.1
	SLM + 673 K/6 h	130 ± 3	100 ± 3	14.1 ± 0.5

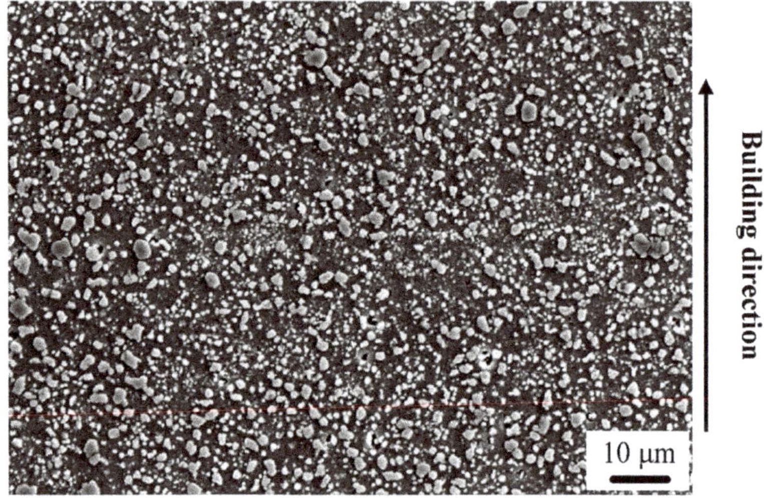

Figure 8. Microstructure of the Al–(12-20)Si bi-material after heat treatment at 673 K for 6 h.

During heat treatment at 673 K, the number of Si particles decreases as a result of diffusion of the Si particles from the Al–20Si to the Al–12Si side as well as due to particle coalescence and growth. The distance between adjacent particles becomes large, and upon loading, the dislocations can move easily for long distances without obstacles. Furthermore, the decrease in the number of Si particles and the increase in size induce a reduction of localized stress or strain. The residual stresses that are built up during the SLM process are also relieved during the thermal treatment. Accordingly, the heat treated bi-material is characterized by decreased strength and large elongation.

Figure 9 depicts the fracture morphologies of the Al–(12-20)Si bi-material both in as-built as well as heat treated conditions. It can be observed that the failure occurs along the Al–20Si side rather than in the interfacial zone, which demonstrates that good bonding is established between Al–12Si and Al–20Si by the SLM process. Figure 9a,b shows typical fracture surfaces of the Al–(12-20)Si bi-material after SLM processing, revealing a layered structure. Some cleavage planes surrounded by tearing ridges are also visible. Moreover, the dimples are not continuous and are mixed with brittle fracture characteristics, indicating a typical quasi-cleavage fracture mode. The fracture morphologies of the bi-materials after heat treatment are shown in Figure 9c,d. A large number of dimples can be observed, and the size of the dimples is considerably larger compared with Figure 9a,b (in the as-prepared SLM material). The observation of more dimples is well correlated with larger elongation after heat treatment and points to a ductile failure mode.

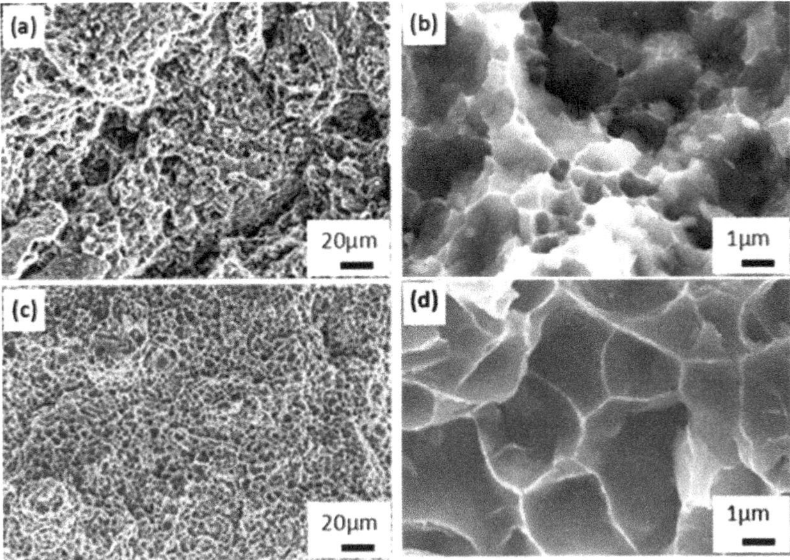

Figure 9. Fracture surface of tensile samples: (**a**,**b**) as-prepared SLM, (**c**,**d**) SLM and heat treated at 673 K for 6 h.

4. Conclusions

Al–(12-20)Si bi-materials were manufactured by selective laser melting and were further heat treated at different temperatures. The microstructure and the mechanical properties were investigated, and the main results can be summarized as follows:

(1) The interfaces are well bonded and can be evidently observed in the bi-material after SLM, whereas they become blurry and finally disappear with increased heat treatment temperature. The microstructure is well refined along both sides (the Al–12Si side and the Al–20Si side). The hardness difference between both sides gradually decreases when the samples are heat treated at 673 K and the hardness values tend to be consistent.

(2) The tensile strength and elongation of the bi-material after SLM are between that of as-prepared SLM Al–12Si and Al–20Si alloys, and failure occurs along the Al–20Si side with quasi-cleavage fracture. The average yield strength, ultimate tensile strength and elongation of the Al–Si bi-material after SLM are 300 MPa, 450 MPa, and 2.2% respectively.

(3) The yield strength and the ultimate tensile strength decrease to 100 MPa and 130 MPa, respectively, and the elongation increases to 14.1% after heat treatment at 673 K for 6 h. Failure is characterized by a ductile fracture mode as a result of increased Si particle size and a decreased number of Si particles.

Author Contributions: Conceptualization, K.G.P. and P.M.; methodology, K.G.P. and Y.J.; experiments, S.Z., Z.Y., R.S., X.S. and P.J.; formal analysis, P.M., and Y.J.; investigation, P.M., Y.J. and K.G.P.; writing—original draft preparation, S.Z., Z.Y., R.S., X.S. and P.J.; writing—review and editing, P.M., Y.J., and J.E.; supervision, K.G.P.

Funding: This work was supported by the National Key Research and Development Program of China (2016YFB0700203), National Natural Science Foundation of China (Grant numbers 51601110, 51601109), the Shanghai Science and Technology Committee Innovation (Grant numbers 17JC1400600, 17JC1400601), and the Talent Project of Shanghai University of Engineering Science. Support from the Estonian Research Council through the MOBERC15 project is also acknowledged.

Acknowledgments: Authors acknowledge Sergio Scudino for simulating discussion.

Conflicts of Interest: The authors declare no conflict of interest.

References

1. Gupta, S.; Taupin, V.; Fressengeas, C.; Jrad, M. Geometrically Nonlinear Field Fracture Mechanics and Crack Nucleation, Application to Strain Localization Fields in Al-Cu-Li Aerospace Alloys. *Materials* **2018**, *11*, 498. [CrossRef] [PubMed]
2. Miller, W.S.; Zhuang, L.; Bottema, J.; Wittebrood, A.J.; Smet, P.D.; Haszler, A.; Vierrgge, A. Recent development in aluminium alloys for the automotive industry. *Mater. Sci. Eng. A* **2000**, *280*, 37–49. [CrossRef]
3. Ferreira, V.; Egizabal, P.; Popov, V.; Cortázar, M.G.; Irazustabarrena, A.; López-Sabirón, A.M.; Ferreira, G. Lightweight automotive components based on nano diamond-reinforced aluminium alloy: A technical and environmental evaluation. *Diam. Relat. Mater.* **2019**, *92*, 174–186. [CrossRef]
4. Maamoun, A.H.; Xue, Y.F.; Elbestawi, M.A.; Veldhuis, S.C. Effect of Selective Laser Melting Process Parameters on the Quality of Al Alloy Parts: Powder Characterization, Density, Surface Roughness, and Dimensional Accuracy. *Materials* **2018**, *11*, 2343. [CrossRef] [PubMed]
5. Prashanth, K.G.; Scudino, S.; Chaubey, A.K.; Loeber, L.; Schimansky, P.; Pyczak, F.; Eckert, J. In processing of Al-12Si-TNM composites by selective laser melting and evaluation of compressive and wear properties. *J. Mater. Res.* **2016**, *31*, 55–65. [CrossRef]
6. Prashanth, K.G.; Scudino, S.; Eckert, J. Tensile properties of Al-12Si fabricated via selective laser melting (SLM) at different temperatures. *Technologies* **2016**, *4*, 38. [CrossRef]
7. Koli, D.K.; Agnihotri, G.; Purohit, R. Advanced Aluminium Matrix Composites: The Critical Need of Automotive and Aerospace Engineering Fields. *Mater. Today Proc.* **2015**, *2*, 3032–3041. [CrossRef]
8. Prashanth, K.G.; Scudino, S.; Chatterjee, R.P.; Salman, O.O.; Eckert, J. Additive Manufacturing: Reproducibility of metallic parts. *Technologies* **2017**, *5*, 8. [CrossRef]
9. Wang, F.; Liu, H.; Ma, Y.J.; Jin, Y.S. Effect of Si content on the dry sliding wear properties of spray-deposited Al-Si alloy. *Mater. Des.* **2004**, *25*, 163–166. [CrossRef]
10. Guo, C.Y.; He, X.B.; Ren, S.B.; Qu, X.H. Effect of (0-40) wt % Si addition to Al on the thermal conductivity and thermal expansion of diamond/Al composites by pressure infiltration. *J. Alloys Compd.* **2016**, *664*, 777–783. [CrossRef]
11. Jung, J.G.; Lee, J.M.; Cho, Y.H.; Yoon, W.H. Combined effects of ultrasonic melt treatment, Si addition and solution treatment on the microstructure and tensile properties of multicomponent Al-Si alloys. *J. Alloys Compd.* **2017**, *693*, 201–210. [CrossRef]
12. Li, R.X.; Liu, L.J.; Zhang, L.J.; Sun, J.H.; Shi, Y.J.; Yu, B.Y. Effect of Squeeze casting on Microstructure and Mechanical Properties of Hypereutectic A$_{1-x}$Si Alloys. *J. Mater. Sci. Technol.* **2017**, *33*, 404–410. [CrossRef]
13. Prashanth, K.G.; Shakur Shahabi, H.; Attar, H.; Srivastava, V.C.; Ellendt, N.; Uhlenwinkel, V.; Eckert, J.; Scudino, S. Production of high strength Al$_{85}$Nd$_8$Ni$_5$Co$_2$ alloy by selective laser melting. *Addit. Manuf.* **2015**, *6*, 1–5. [CrossRef]
14. Dinnis, C.M.; Taylor, J.A.; Dahle, A.K. As-cast morphology of iron-intermetallics in Al-Si foundry alloys. *Scr. Mater.* **2005**, *53*, 955–958. [CrossRef]
15. Prashanth, K.G.; Scudino, S.; Eckert, J. Defining the tensile properties of Al-12Si parts produced by selective laser melting. *Acta Mater.* **2017**, *126*, 25–35. [CrossRef]
16. Yan, X.C.; Li, Q.; Yin, S.; Chen, Z.Y.; Jenkins, R.; Chen, C.Y.; Wang, J.; Ma, W.Y.; Bolot, R.; Lupoi, R.; et al. Mechanical and in vitro study of an isotropic Ti6Al4V lattice structure fabricated using selective laser melting. *J. Alloys Compd.* **2019**, *782*, 209–223. [CrossRef]
17. Prashanth, K.G.; Scudino, S.; Maity, T.; Das, J.; Eckert, J. Is the energy density a reliable parameter for materials synthesis by selective laser melting. *Mater. Res. Lett.* **2017**, *5*, 386–390. [CrossRef]
18. Schwab, H.; Prashanth, K.G.; Loeber, L.; Kuehn, U.; Eckert, J. Selective laser melting of Ti-45Nb alloy. *Metals* **2015**, *5*, 686–694. [CrossRef]
19. Jia, Y.D.; Ma, P.; Prashanth, K.G.; Wang, G.; Yi, J.; Scudino, S.; Cao, F.Y.; Sun, J.F.; Eckert, J. Microstructure and thermal expansion behavior of Al-50Si synthesized by selective laser melting. *J. Alloys Compd.* **2017**, *699*, 548–553. [CrossRef]
20. Scudino, S.; Unterfoerfer, C.; Prashanth, K.G.; Attar, H.; Ellendt, N.; Uhlenwinkel, V.; Eckert, J. Additive manufacturing of Cu-10Sn bronze. *Mater. Lett.* **2015**, *156*, 202–204. [CrossRef]
21. Xi, L.X.; Zhang, H.; Wang, P.; Li, H.C.; Prashanth, K.G.; Lin, K.J.; Kaban, I.; Gu, D.D. Comparative investigation of microstructure, mechanical properties and strengthening mechanisms of Al-12Si/TIB2 fabricated by selective laser melting and hot pressing. *Ceram. Int.* **2018**, *44*, 17635–17642. [CrossRef]

22. Wang, P.; Deng, L.; Prashanth, K.G.; Pauly, S.; Eckert, J.; Scudino, S. Microstructure and mechanical properties of Al-Cu alloys fabricated by selective laser melting of powder mixtures. *J. Alloys Compd.* **2018**, *735*, 2263–2266. [CrossRef]
23. Zhao, X.; Song, B.; Fan, W.R.; Zhang, Y.J.; Shi, Y.S. Selective laser melting of carbon/AlSi10Mg composites: Microstructure, mechanical and electronical properties. *J. Alloys Compd.* **2016**, *665*, 271–281. [CrossRef]
24. Ma, P.; Prashanth, K.G.; Scudino, S.; Jia, Y.D.; Wang, H.W.; Zou, C.M.; Wei, Z.J.; Eckert, J. Influence of Annealing on Mechanical Properties of Al-20Si Processed by Selective Laser Melting. *Meterials* **2014**, *4*, 28–36. [CrossRef]
25. Kang, N.; Coddet, P.; Liao, H.L. Macrosegregation mechanism of primary silicon phase in selective laser melting hypereutectic Al-High Si alloy. *J. Alloys Compd.* **2016**, *662*, 259–262. [CrossRef]
26. Prashanth, K.G.; Eckert, J. Formation of metastable cellular microstructures in selective laser melted alloys. *J. Alloys Compd.* **2017**, *707*, 27–34. [CrossRef]
27. Prashanth, K.G.; Scudino, S.; Klauss, H.J.; Surreddi, K.B.; Loeber, L.; Wang, Z.; Chaubey, A.K.; Kühn, U.; Eckert, J. Microstructure and mechanical properties of Al-12Si produced by selective laser melting: Effect of heat treatment. *Mater. Sci. Eng. A* **2014**, *590*, 153–160. [CrossRef]
28. Ma, P.; Jia, Y.D.; Prashanth, K.G.; Yu, Z.S.; Li, C.G.; Zhao, J.; Yang, S.L.; Huang, L.X. Effect of Si content on the microstructure and properties of Al-Si alloys fabricated using hot extrusion. *J. Mater. Res.* **2017**, *32*, 2210–2217. [CrossRef]
29. Prashanth, K.G.; Damodaram, R.; Scudino, S.; Wang, Z.; Prasad Rao, K.; Eckert, J. Friction welding of Al-12Si parts produced by selective laser melting. *Mater. Des.* **2014**, *57*, 632–637. [CrossRef]
30. Kucukomeroglu, T. Effect of equal-channel angular extrusion on mechanical and wear properties of eutectic Al-12Si alloy. *Mater. Des.* **2010**, *31*, 782–789. [CrossRef]
31. Purcek, G.; Saray, O.; Kul, O. Microstructural Evolution and Mechanical Properties of Severely Deformed Al-12Si Casting Alloy by Equal-Channel Angular Extrusion. *Met. Mater. Int.* **2010**, *16*, 145–154. [CrossRef]
32. Jung, J.G.; Lee, S.H.; Cho, Y.H.; Yoon, W.H.; Ahn, T.Y.; Ahn, Y.S.; Lee, J.M. Effect of transition elements on the microstructure and tensile properties of Al-12Si alloy cast under ultrasonic melt treatment. *J. Alloys Compd.* **2017**, *712*, 277–287. [CrossRef]
33. Wang, Z.; Prashanth, K.G.; Chaubey, A.K.; Löber, L.; Schimansky, F.P.; Pyczak, F.; Zhang, W.W.; Scudino, S.; Eckert, J. Tensile properties of Al-12Si matrix composites reinforced with Ti-Al-based particles. *J. Alloys Compd.* **2015**, *630*, 256–259. [CrossRef]
34. Choi, H.; Konishi, H.; Li, X.C. Al$_2$O$_3$ nanoparticles induced simultaneous refinement and modification of primary and eutectic Si particles in hypereutectic Al-20Si alloy. *Mater. Sci. Eng. A* **2012**, *541*, 159–165. [CrossRef]
35. Kilicaslan, M.F.; Lee, W.R.; Lee, T.H.; Sohn, Y.; Hong, S.J. Effect of Sc addition on the microstructure and mechanical properties of as-atomized and extruded Al-20Si alloys. *Mater. Lett.* **2012**, *71*, 164–167. [CrossRef]
36. Yoon, S.C.; Hong, S.J.; Hong, S.I.; Kim, H.S. Mechanical properties of equal channel angular pressed powder extrudates of a rapidly solidified hypereutectic Al-20 wt % Si alloy. *Mater. Sci. Eng. A* **2007**, *449–451*, 966–970. [CrossRef]
37. Ma, P.; Wei, Z.J.; Jia, Y.D.; Zou, C.M.; Scudino, S.; Prashanth, K.G.; Yu, Z.S.; Yang, S.L.; Li, C.G.; Eckert, J. Effect of high pressure solidification on tensile properties and strengthening mechanisms of Al-20Si. *J. Alloys Compd.* **2016**, *688*, 88–93. [CrossRef]
38. Jung, H.J.; Choi, S.J.; Prashanth, K.G.; Stoica, M.; Scudino, S.; Yi, S.; Kuehn, U.; Kim, D.H.; Kim, K.B.; Eckert, J. Fabricate of Fe-based bulk metallic glass by selective laser melting. *Mater. Des.* **2015**, *86*, 703–708. [CrossRef]
39. Li, W.; Li, S.; Liu, J.; Zhang, A.; Zhou, Y.; Wei, Q.S.; Yan, C.Z.; Shi, Y.S. Effect of heat treatment on AlSi10Mg alloy fabricated by selective laser melting: Microstructure evolution, mechanical properties and fracture mechanism. *Mater. Sci. Eng. A* **2016**, *663*, 116–125. [CrossRef]
40. Wei, P.; Wei, Z.Y.; Chen, Z.; Du, J.; He, Y.Y.; Li, J.F.; Zhou, Y.T. The AlSi10Mg samples produced by selective laser melting: Single track, densification, microstructure and mechanical behavior. *Appl. Surf. Sci.* **2018**, *408*, 38–50. [CrossRef]

Review

New Aluminum Alloys Specifically Designed for Laser Powder Bed Fusion: A Review

Alberta Aversa [1,*], Giulio Marchese [1], Abdollah Saboori [1], Emilio Bassini [1], Diego Manfredi [2], Sara Biamino [1], Daniele Ugues [1], Paolo Fino [1] and Mariangela Lombardi [1]

[1] Department of Applied Science and Technology, Politecnico di Torino, Corso Duca degli Abruzzi 24, 10129 Torino, Italy; giulio.marchese@polito.it (G.M.); abdollah.saboori@polito.it (A.S.); emilio.bassini@polito.it (E.B.); sara.biamino@polito.it (S.B.); daniele.ugues@polito.it (D.U.); paolo.fino@polito.it (P.F.); mariangela.lombardi@polito.it (M.L.)

[2] Center for Sustainable Future Technologies CSFT@PoliTo, Istituto Italiano di Tecnologia, Via Livorno 60, 10144 Torino, Italy; diego.manfredi@iit.it

* Correspondence: alberta.aversa@polito.it; Tel.: +39-011-090-4763

Received: 26 February 2019; Accepted: 22 March 2019; Published: 27 March 2019

Abstract: Aluminum alloys are key materials in additive manufacturing (AM) technologies thanks to their low density that, coupled with the possibility to create complex geometries of these innovative processes, can be exploited for several applications in aerospace and automotive fields. The AM process of these alloys had to face many challenges because, due to their low laser absorption, high thermal conductivity and reduced powder flowability, they are characterized by poor processability. Nowadays mainly Al-Si alloys are processed, however, in recent years many efforts have been carried out in developing new compositions specifically designed for laser based powder bed AM processes. This paper reviews the state of the art of the aluminum alloys used in the laser powder bed fusion process, together with the microstructural and mechanical characterizations.

Keywords: laser powder bed fusion; additive manufacturing; aluminum; composition; mechanical properties

---

1. Introduction

Additive manufacturing (AM) defines a class of production technologies that can create 3D components layer by layer based on 2D patterns defined by slicing the 3D computer aided design (CAD) of the component. Among the different AM technologies, powder bed technologies such as laser powder bed fusion (LPBF) and electron beam melting (EBM) are the most used to process metallic alloys [1]. A schematic representation of the LPBF process is reported in Figure 1.

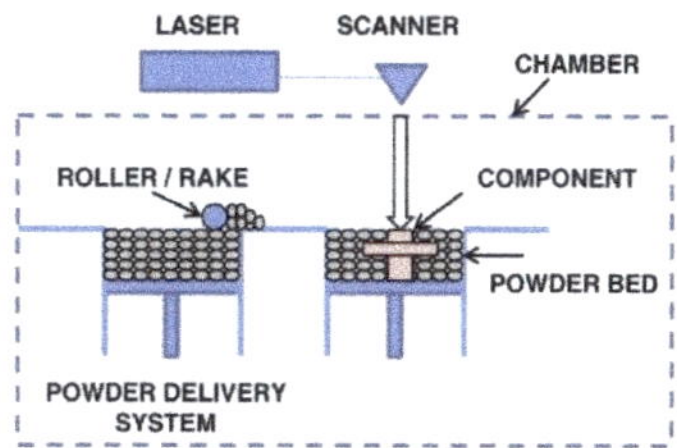

Figure 1. Schematic representation of a laser powder bed fusion (LPBF) process. Adapted from [1], with permission from © 2014 Springer Nature.

The advantages from a design perspective of LPBF and in general of AM processes have been widely addressed in recent years [2,3]. However, it is important to underline that AM brings many interesting aspects also from a material science perspective. The high cooling rate (10^3–10^6 K/s [4]) of the melt pool generated by the successive laser scan tracks is also a key factor of the success of these technologies. From a metallurgical point of view in fact, the rapid cooling generates the solidification of peculiar microstructures characterized by extremely interesting features which will be discussed more in details in the paper.

LPBF showed to be very successful with many alloys such as titanium, aluminum, nickel, steels, and refractory materials [1]. In the case of titanium and nickel alloys, the poor machinability and the issues related to their formability drove the success of AM processes. On the contrary, most of the aluminum alloys can be easily formed by conventional processes. Furthermore, many studies reported that the LPBF process of Al alloys are generally critical. One of the main issues faced when processing Al by AM is related to difficulties in spreading the powder bed due to the poor flowability of the powders [5]. Furthermore the LPBF process of aluminum alloys generally requires high laser power because of the high reflectivity of the powder and the high thermal conductivity of the solidified material [6]. Finally, the presence of a thin oxide film on the gas atomized particles and on the solidified layers might reduce the processability as it reduces the wettability and hinders the remelting of the previous layer, causing some porosities in the built parts [7–9]. The thickness of the oxide layer on both, the melt pool and the particles, strongly depends on the processing conditions [10,11]. Uslan et al. for example showed that on aluminum gas atomized particles the oxide thickness is in the range of 2.75–2.96 nm and it is correlated to the particles size [11]. Notwithstanding these issues, the LPBF processing of Al alloys gained much interest in recent years mainly due to the peculiar microstructure and enhanced mechanical properties it is possible to achieve.

Among commercial aluminum alloys, mainly near eutectic Al-Si alloys, such as AlSi10Mg, Al-12Si, A357, and A356, are generally used in AM processes and in particular in LPBF and, among them, the most studied composition is certainly AlSi10Mg [12–16]. The success of this composition is mainly related to the Si content, which is close to the eutectic one and hinders the solidification cracking phenomenon. It is well-known that this cracking mechanism is related to the solidification range of the alloy, to the fluidity of the molten phase, the solidification shrinkage and to the coefficient of thermal expansion (CTE) of the alloy [17]. In AlSi10Mg, the presence of 10 wt.% Si implies a fine solidification range ($\Delta T = T_{liquidus} - T_{solidus}$) of this alloy, which was calculated to be about 30 °C [18]. This value is extremely low with respect to other high strength aluminum alloys such as the 2024 ($\Delta T = 135$ °C) [19]. In addition, it well-known that Si improves the fluidity of molten aluminum, reduces the solidification shrinkage and the coefficient of thermal expansion [19]. Furthermore, Sercombe et al. suggested that Si is also fundamental for the AlSi10Mg LPBF processing because it is responsible for the laser absorption. Silicon in fact has a low solubility in Al, and it is mainly contained in the alloy as pure particles which are characterized by a high laser absorption (~70%) [20].

The microstructure and properties of AlSi10Mg by LPBF have been widely investigated in recent years [12,21]. It is well-accepted that as-built AlSi10Mg LPBF microstructure is made by large columnar grains with 10–20 microns width and hundreds of microns length (Figure 2a). These grains, which can be detected only by electron backscatter diffraction (EBSD) analyses, are formed during the building process thanks to the epitaxial grain growth along the building direction. It is well known that these large columnar grains contain a fine dendritic structure, due to the rapid and directional cooling that arises during the laser scanning and implies enhanced mechanical properties. The effect of the laser scanning can be detected by optical or scanning electron microscope (SEM) images by a network of interconnected melt pools that contain an extremely fine cellular structure. The size and the morphology of the cells strongly depend on the building parameters and on the position within the melt pool [18,22]. Kim et al. for example distinguished three regions of the melt pool defined as, fine melt pool (FMP), heat affected zone (HAZ) and coarse melt pool (CMP) characterized by cells with different size and morphology (Figure 2b) [22]. EDS analyses showed that this cellular structure is made of Al cells surrounded by Si-rich regions (Figure 2c–e).

It was demonstrated that as-built samples are characterized by higher mechanical properties with respect to cast plus conventional solution-ageing T6 treatment samples with the same composition [23]. Hadadzadeh et al. attributed the high mechanical performances of AlSi10Mg LPBF samples to the Hall-Petch mechanism, due to the eutectic cells boundaries, to the Orowan effect due to the presence of Si and Mg_2Si and to the dislocation hardening [24,25].

Because of this reason, LPBF AlSi10Mg was deeply investigated in terms of post process heat treatments, surface roughness, residual stresses, corrosion and fatigue resistance [26–29].

Despite the results obtained with AlSi10Mg showed the great potentiality of the LPBF process of Al alloys, only other few casting alloys have been processed, without meeting the strict requirements of high strength and ductility of the aerospace industry. In fact, most of the high strength aluminum alloys such as 2000, 6000, and 7000 series are hardly processable by LPBF because of their solidification cracking susceptibility. In addition, 2000, 6000, and 7000 series contain volatile alloying elements such as Zn, Mg, and Li that can easily evaporate during the building process [17,30].

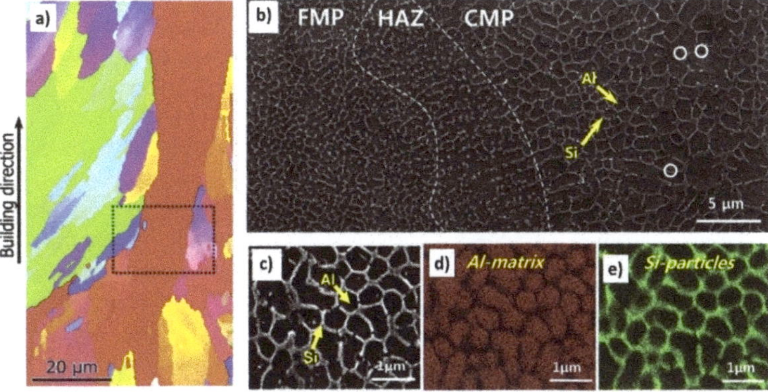

Figure 2. (**a**) Electron backscatter diffraction (EBSD) inverse pole figure (IPF), adapted from [31], with permission from © 2016 Elsevier, (**b,c**) SEM images and (**d**) Al- and (**e**) Si EDS maps of an as-built AlSi10Mg sample, adapted from [22], with permission from © 2016 Elsevier.

Because of these reasons, there is a strong industrial interest in the development of high strength aluminum alloys to be processed by LPBF. Even if the research on this topic started only a few years ago, there has been a significant increase in the number of scientific publications on this topic. Figure 3 reports the number of publications per year on new Al alloys specifically designed for AM processes and clearly demonstrates the increasing interest on this topic.

The purpose of this review is to discuss the approaches used in the selection of the alloy composition and to sum up the main results obtained in recent years. The major outcomes are here discussed as follows. At first the main methodologies used to study new aluminum alloys compositions are discussed. Afterwards, the advantages of the rapid solidification processes to which LPBF belongs will be presented. Then, the limitations related to the LPBF processability of high strength aluminum alloys (e.g., 7075, 2024) will be presented together with the consequent investigated composition adjustments. Finally, the design of new aluminum alloys obtained by the addition of rare earth (RE) and transition metals (TM) will be examined together with Al-based metallic glasses. As a conclusion the mechanical properties of the alloys will be compared and discussed on the basis of the microstructural results and with emphasis on the key findings.

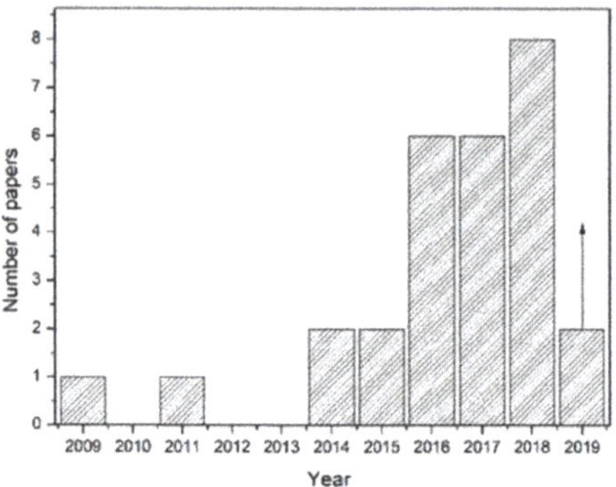

Figure 3. Number of papers published per year on new alloys for LPBF process (author's image).

2. Methodologies

One of the main issues related to the development of a new aluminum alloy for LPBF process is the difficulty to source gas atomized powders with customized compositions. Furthermore, in some cases, gas atomization of specific compositions might be difficult [32]. Because of these reasons, in recent years many researchers mixed commercial powders with different chemical compositions to obtain a batch with the final required alloying elements content [32–38]. The phenomena that arise in the melt pool, such as the Marangoni flow and the recoil pressure effects, generally allow the distribution of the alloying elements within the melt pool [39]. Despite this, some of the published studies reported some alloying elements mixing issues [33,38]. Wang et al., for example, produced Al-xCu alloys using mixed Al-4.5Cu and Cu particles and found some Cu rich areas in as-built samples (Figure 4). The inhomogeneity of the composition is strongly undesired not only because it alters the results, but also because it hinders the production of large samples due to the formation of some areas rich in brittle phases.

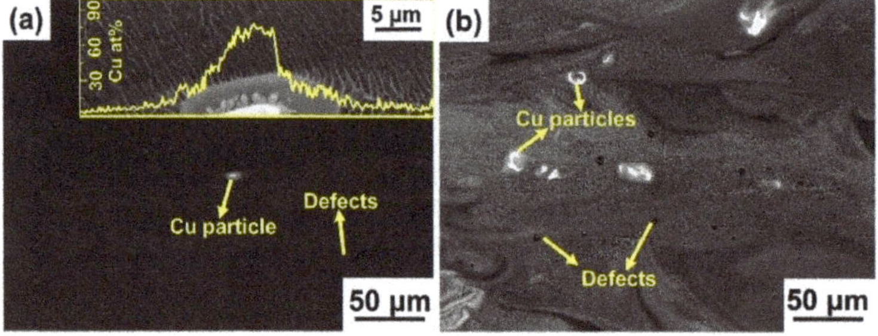

Figure 4. Backscattered Electrons (BSE) SEM micrographs of LBF (**a**) Al-6Cu and (**b**) Al-40Cu alloy (inset: Cu distribution across the Cu-rich zone), adapted from [38], with permission from © 2018 Elsevier.

Furthermore, another critical aspect in developing a new composition is the quantity of powder required for an LPBF build. Accordingly, some researchers used an alternative method to reduce the powder quantity needed to investigate the processability and the properties of a new alloy: laser single scan tracks (SSTs) [40–44]. In these experiments, the laser scans a powder bed with a single track and thereafter the SSTs are generally cut perpendicularly to the scanning direction, polished and analyzed. The morphological observations of the melt pool indicate the processability of the compositions, based on the type melting and solidification phenomena. This method leads to the identification of the most suitable scanning parameters that brings to the formation of a stable melt pool and of a continuous scan track. These types of track are generated when the material melts in conduction melting mode. On the contrary, when high power associated with low scan speed are used the powder bed melts in a keyhole mode, causing a strong evaporation of some alloy elements, the formation of deep melt pools and many undesired effects [45]. The sets of parameters that, because of surface tension effects, cause the rupture of the scan track into separated balls are also undesired as they strongly increase the porosity of the final part. This morphology is generally identified as balling [46].

The main geometrical feature of the cross-section of SSTs can be evaluated in order to select the most suitable hatching distance and layer thickness for the production of dense bulk components. Finally, the melt pool microstructure is generally observed by SEM and EBSD analyses and the mechanical properties are evaluated by nanoindentation measurements [41,42,47].

This approach is very useful to have a first insight on the processability, properties and microstructures of a new alloy by LPBF. However, some aspects have to be taken into account:

- the dimension of the laser scan in a real part might be slightly larger with respect to the SSTs due to the heat accumulated during the scanning of the previous layers;
- the microstructure of a SST might be slightly different from the real part one because it does not undergo the intrinsic heat treatment due to the melting and solidification of the following layers.

Notwithstanding this, if these aspects are considered, the SSTs analysis showed to be a promising methodology for the development of new alloys for LPBF. Nie et al. confirmed the applicability of single scan methods by selecting the building parameters for the production of an Al-Cu-Mg alloy on the basis of single and multi-tracks morphology and crack density [41].

In literature there are different methods and approaches to create the powder surface to be scanned. Aversa et al. for example used a film depositor with a 50 μm gap to spread the powders mixed with 50 vol.% ethanol in order to facilitate the deposition [40]. In this case, in order to simulated as much as possible a scan in a real AM part, specific attention has to be payed to the platform composition. Li et al. slightly consolidated a pre-alloyed powder in order to be able to remove the scan track for further analyses (Figure 5) [42]. This approach may facilitate the analyses as it does not involve the platform, but it must be carefully used for the selection of building parameters because the partially consolidated powder bed has a different thermal behavior with respect to the bulk material. Finally, Jia et al. did not create a powder bed but performed SSTs on cast platforms [48]. This method has many advantages from a homogeneity point of view; furthermore critical alloying elements can be added without too many safety issues. However, the lack of the powder bed may modify the laser absorption and the melting phenomena. It is well-known that the laser absorption depends on the surface quality and strongly changes when a powder bed is used [43]. Notwithstanding this, this method is mainly useful for the understanding of the solidification phenomena and microstructure analyses.

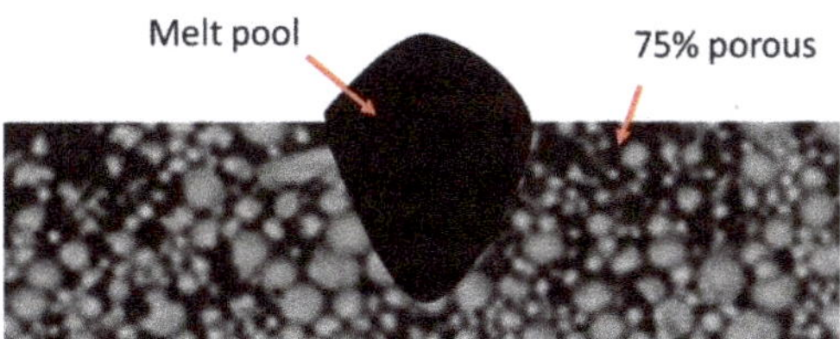

Figure 5. Schematic representation of a melt pool on a partially consolidated powder (This figure was redrawn based on Figure 3 of [42], with permission from © 2014 Elsevier).

3. Rapid Solidification

In recent years many studies modelled the melt pool temperature during LPBF revealing the extraordinary temperature profile and cooling rates of these technologies [49–51]. Li et al. for example calculated that the melt pool maximum temperature for an AlSi10Mg is in the range 1000–1800 °C and the cooling rate is in the range 1.2×10^6–6.1×10^6 K/s, depending on the main building parameters of the LPBF system employed [49].

These results suggested that LPBF may be considered a rapid solidification process (RSP). As reported in literature, RSPs have been deeply investigated in past years because of their many advantages such as [52]:

- the extension of the solid solubility;
- the formation of non-equilibrium and metastable phases;
- the reduction of number and size of segregated phases;
- the changes in grain morphology such as grain refinement, location and distribution of the phases;
- the reduced phase crystallinity.

The main RSPs are the following: melt spinning (or planar flow casting), splat quenching, laser surface melting and several types of atomization [52]. These technologies produce ribbons or powders and have many limitations in terms of the shape and size of the final products. Nevertheless, because of the possibility to obtain peculiar microstructures and interesting mechanical properties, these processes and their thermodynamic phenomena received much attention in past years in metallurgy literature [53,54]. Many studies in fact focused the attention on the selection of the most suitable composition that can be processed by RSPs and that can take advantage of these extremely high cooling rates [52].

Recently, Marola et al. compared the microstructure of LPBF AlSi10Mg samples with AlSi10Mg specimens produced with common RSPs, in particular melt spinning (MS) and copper mold casting (CMC) [55]. The comparison was carried out by means of Field Emission Scanning Electron Microscopy (FESEM) micrographs and consequent evaluation of the eutectic percentage, X-Ray Diffraction (XRD) and differential scanning calorimetry (DSC) analyses. This study demonstrated that the LPBF samples have a higher extension of the Si solid solubility in Al with respect to the conventional RSPs. This result was verified by the released enthalpy during DSC analyses. These findings confirm that LPBF can be fully considered as a RSP and that it can therefore benefit, from a material point of view, of the above discussed advantages of these techniques.

4. Aluminum Alloys for LPBF

4.1. Processability of High Strength Aluminum Alloys

Many studies already revealed that most of the high strength aluminum alloys are hardly processable by LPBF as they suffer from solidification cracking during the laser scanning [17,30]. This cracking mechanism, deeply investigated in welding, arises when, during the melt pool solidification, the thin liquid film that forms on the grain boundaries cannot accommodate the solidification shrinkage, generating a crack [56]. It was demonstrated that solidification cracking is mainly due to some alloy characteristics, i.e., the large solidification range, the solidification shrinkage, the CTE value, and to the poor fluidity of the molten phase [56]. Furthermore, high strength aluminum alloys usually contain volatile alloying elements such as Zn, Mg, and Li. These elements can evaporate during the LPBF process causing a modification of the composition and consequently of the microstructure and the properties. In some cases, the modification of the composition due to the evaporation might even increase the susceptibility to the cracking mechanism [30].

The occurrence of the cracking mechanism during LPBF of high strength aluminum alloys was demonstrated in several studies [57–59]. The processability of an Al-Cu-Mg alloy (with a composition close to 2024) has been investigated by Zhang et al. who observed long cracks formed along the building direction with most of the parameters [57].

Kaufmann et al. studied the LPBF process of the 7075, an aluminum alloy known thanks to its outstanding strength [58,60]. It was demonstrated that with high power (P > 300 W) nearly dense samples (porosity < 1%) can be obtained using a 7075 powder. However, the high cooling rate causes the generation of long cracks, oriented along the building direction also on samples built on a preheated platform (T = 200 °C) [58]. Furthermore, in this study energy-dispersive X-ray spectroscopy (EDS) analyses revealed that LPBF samples contain a lower Zn content with respect to the starting powder, confirming that Zn evaporates during the building process. Qi et al. deeply studied the cracking phenomena that arise during the laser scanning of a 7075 powder [59]. They proved that three different types of cracks could generate in 7075 LPBF samples, all oriented along the grain boundaries. The melting mode was also evaluated on the basis of the geometrical features of the last layers melt pools and was correlated with the crack density. The lowest crack density was observed in samples in which the keyhole melting mode (Figure 6a) arose due to the solidification of fine and irregular grains [45,59]. With high energy density, therefore with high laser power and low scan speed, the keyhole phenomenon takes place due to the evaporation of metal and the consequent creation of a deep melt pool which enhances the laser absorption (Figure 6a) [45]. The EBSD maps of samples built in conduction mode (Figure 6b,c) show that cracks are present along the columnar grain boundaries. On the contrary, the keyhole melting mode causes the formation of irregular grains due to the strong the fluctuation of the melt pool. The different microstructure causes a reduction in the crack density (Figure 6d,e).

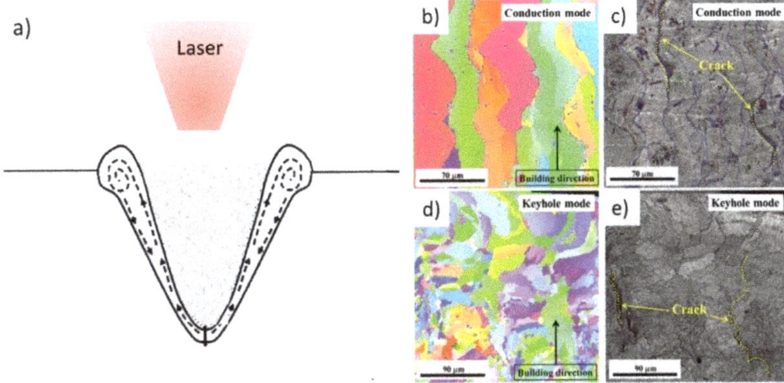

Figure 6. (a) Schematic representation of a keyhole melting and EBSD maps of 7050 samples produced in (**b**,**c**) a conduction melting mode and (**d**,**e**) a keyhole melting mode. (**b**,**d**) orientation image map and (**c**,**e**) grain boundary misorientation angle map. Adapted from [59], with permission from © 2017 Elsevier.

4.2. High Strength Aluminum Alloys Modification

Considering the difficulties in LPBF processing high strength aluminum alloys, many studies tried therefore to modify their composition in order to make it processable by AM [4,32,34,36,37,41,57].

Montero Sistiaga et al. and Aversa et al. modified the composition of the 7075 alloy increasing the Si content, trying in this way to change the solidification range and the fluidity of the molten phase [36,37]. In both cases, the introduction of Si has been obtained by dry mixing the 7075 powders to Si or to a Si-rich aluminum alloy powder. It was demonstrated that Si is very effective in reducing the cracks density of the alloy. DSC analyses confirmed the reduction of the solidification range and EBSD analyses performed on samples with different Si contents showed that silicon also causes a strong grain refinement. This microstructure refinement might also contribute to the improved processability of this alloy. Based on Martin et al. [32] and Qi et al. [59] results, cracks propagated along columnar grains. However, the effect of Si on the laser absorption has also to be taken into account. In fact, the alloy with higher Si content might melt in a keyhole melting mode with a consequent grain refinement [59]. The microstructure of these Si modified 7075 alloys is similar to the AlSi10Mg one, consisting of fine α-Al cells surrounded by the eutectic phase. Thanks to the high cooling rates, the as-built samples are constituted by an α-Al solid solution, whereas a direct ageing heat treatment allows the precipitation of strengthening phases. The tensile tests performed on Si modified samples showed that the as-built alloy has the yield and ultimate tensile strengths of 315 and 387 MPa, respectively, and these values can increase up to 11% and 7%, respectively, after the appropriate ageing treatment [37].

On the other hand, Martin et al. studied a new approach, based on the refinement of the microstructure, to produced crack-free 7075 and 6061 parts by LPBF [32]. The authors introduced hydrogen stabilized Zr nanoparticles to the powder batches. Zr reacts with Al forming Al_3Zr fine particles which act as nucleant for the aluminum alloy. The microstructure of Zr modified alloy consists then in equiaxed grains that can easily accommodate the solidification shrinkage allowing the production of crack-free samples. A similar approach was used by Zhou et al. who added 1wt.% Sc + Zr to an Al-6Zn-2Mg alloy [61]. This composition showed to have a good processability and promising mechanical properties which can be further increased by a T6 heat treatment.

Similar approaches were used for the alloys belonging to the 2000 series. Dense and crack-free Al-Cu-Mg samples were produced by Zhang et al. only with a reduced range of parameters; with high scan speed, long cracks, oriented along the building direction, were observed [57]. In more recent studies, two main ways to enlarge the process window of this alloy have been studied [4,34,35]. On one hand, Zhang et al. added Zr to its composition in order to form Al_3Zr precipitates which act as seeds for the heterogeneous nucleation [34]. The experimental results showed that the introduction of Zr is effective in the production of crack-free sample when high scan speed is used [34]. XRD and EBSD analyses confirmed the precipitation of Al_3Zr which act as a grain refiner for the alloy, allowing to reach extremely high mechanical properties. The EBSD maps of the cross section of Al-Cu-Mg and Zr/Al-Cu-Mg samples processed using different scan speeds are reported in Figure 7. The comparison of the images clearly shows the reduction of the grain size obtained by the introduction of Zr and by using high scan speed values. The reason for the success of the introduction of Zr to the alloy was also clarified by Nie et al. on a similar composition, for which a reduced process windows could be found (Al-4.24Cu-1.97Mg-0.56Mn) [35]. The authors investigated the effect of different Zr contents and different scan speeds on the crack density, revealing that higher scan speed can be used with higher Zr contents [41]. The solidification phenomena have been deeply investigated by EBSD and simulations, confirming that Zr allows the crack reduction thanks to the reduction of the solidification range and the refinement of the microstructure due to heterogeneous nucleation during solidification.

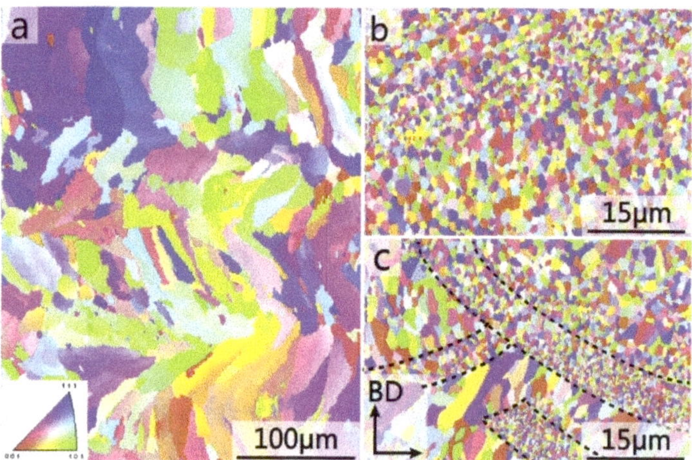

Figure 7. EBSD inverse pole figure (IPF) maps of Al-Cu-Mg fabricated at v = 5 m/min (**a**) and Zr/Al-Cu-Mg sample fabricated at v = 5 m/min (**b**) and v = 15 m/min (**c**), respectively. Adapted from [34], with permission from © 2017 Elsevier.

In another recent study, Wang et al. investigated a different approach to increase the processability of Al-Cu-Mg alloys investigating the effect of the introduction of Si on the consolidation and the microstructure of the starting material [4]. This modification of the composition might reduce the crack density due to the increase in the fluidity of the molten phase and to the reduction in the liquation cracking phenomena. The authors demonstrated that this new Al-3.5Cu-1.5Mg-1Si can be successfully processed by LPBF and that dense and crack-free samples can be produced with the optimized parameters. Concerning the microstructure, it was demonstrated that as-built samples are made of α-Al and strengthening phases and EBSD analyses showed again that large columnar grains are formed upon cooling. XRD and TEM analyses revealed that, in the as-built condition, the strengthening phase is an unknown group identified as Q phase and rich in Cu, Mg, and Si. After the T6 heat treatment, only Mg_2Si and Al_xMn_y were detected.

Finally, the LPBF processability of the alloys belonging to the 6000 series was also investigated and results revealed that these compositions are also very critical [62–64]. Louvis et al. reported a poor consolidation and large delamination issues in 6061 samples processed by laser-based AM processes [65]. Fulcher et al. and Maamoun et al. compared the 6061 and the AlSi10Mg processability showing that, because of the larger solidification range, the 6061 alloy suffers from solidification cracking during LPBF [63,66]. In their studies, it was also demonstrated that the cracking mechanism was related to the building parameters and thanks to the rapid cooling, the as-built 6061 microstructure is characterized by fine Al cells and nanometric Si particles [63]. The processability of the 6061 alloy was firstly improved by Martin et al. following the same approach used for the 7075 alloy [32]. The authors introduced Zr nanoparticles in order to cause the precipitation of Al_3Zr and control the solidification process favoring the heterogeneous nucleation [32]. The 6061 LPBF cracking issues were also recently overcome by Uddin et al. by increasing the building platform temperature at 500 °C and selecting the most suitable building parameters [62]. However, the building platform heating completely altered the solidification mechanism eliminating the melt pool features and leading to a reduced texture, larger grains and precipitates.

All these studies showed that the three main approaches have been studied for the reduction of the crack density of high strength aluminum alloys processed by LPBF. The methods, based on different microstructural mechanisms, are summarized here:

- Control of the solidification process by the formation of nucleant phases (e.g., Al_3Zr);
- Modification of the solidification processes by the reduction of the solidification range;
- Reduction of the thermal gradient by preheating the building platform.

4.3. Effect of Transition Metals and Rare Earth Elements to Aluminum Alloys

A different approach for the development of new Al-based compositions can be defined on the basis of the results obtained with rapid solidified alloys, introducing transition metals (TM) and rare earth elements (RE) to the alloys compositions [67–69].

It is well known that TM and RE elements are extremely suitable alloying elements for rapidly solidified Al alloys. Because of this reason, recently, many works focused on the investigation of the microstructure and the properties of aluminum alloys containing RE and TM processed by LPBF. The main advantages of RE are related to the strongly coherent and stable precipitates they form with Al that allow a grain refinement of the alloy.

The most known application of the approach based on TM and RE addition is certainly the Scalmalloy®, a new composition developed and patented by the Airbus Group and deeply investigated in recent years. In 2011 Schmidtke et al. proofed the applicability of an Al-Mg alloy containing Sc and Zr through LPBF process (Al4.5Mg0.7Sc0.4Zr0.5Mn) [70].

The main idea behind this alloy is related to the obtainment of a supersaturated solid solution due to the rapid cooling and the precipitation of Al_3Sc by direct ageing. Scandium was selected because, together with erbium and zirconium, is a promising alloying element for aluminum as it gives L12

Al$_3$X ordered structures strongly coherent with the Al matrix [71]. This high coherency reduces the tendency of particles to grow, so that these precipitates are stable up to 350 °C. It is worthy to remind that the most common Mg-, Si-, and Cu-based precipitates are stable only up to about 250 °C. Moreover, thanks to their high coherency with aluminum, Al$_3$(Sc,Zr) precipitates act as seed crystals for the Al heterogeneous nucleation implying high degrees of grain refinement [72].

The peculiar microstructure and the properties of Scalmalloy® samples processed by LPBF were investigated in various works by Spierings et al. [72–74]. The authors demonstrated that the as-built Scalmalloy® microstructure is made of fine grain (FG) and coarse grain (CG) regions (Figure 8). This bimodal microstructure is probably due to the formation of MgO and Al$_3$Sc seeds that cause the formation of fine grains. The CG region, on the contrary, is due to the lack of seeds and to the temperature gradient that cause a columnar grain growth [73]. The authors also investigated the effect of LPBF building parameters on Scalmalloy® microstructure and properties, revealing that the scan speed has an important effect on microstructure and in particular on the size of Al$_3$Sc particles. However, this difference only slightly affects the mechanical properties of the as-built alloy in terms of hardness and yield strength [74]. Li et al. also observed the effect of process parameters on the microstructure of as-built LPBF samples with a composition close to the Scalmalloy® (Al-6.2Mg-0.36Sc-0.09Zr). The authors demonstrated that the building parameters have an effect on the texture and on the Mg content of LPBF samples [75]. In particular high energy density caused the evaporation of Mg from the melt pool and a reduction of the Al lattice parameter and the solidification of a structure characterised by a weaker texture [75]. Shi et al. studied the relation between the building parameters and the microstructure and the properties of this composition in the aged condition [76]. It was demonstrated that a balance between the low densification obtained at low energy density and the low supersaturated solid solution obtained at high energy density has to be found in order to reach the highest mechanical performances [76].

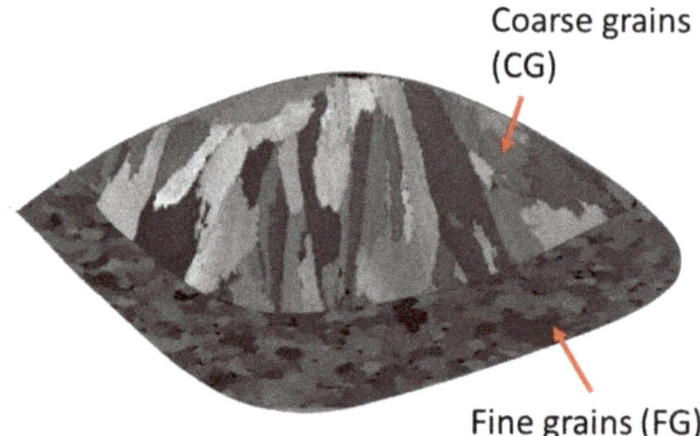

Figure 8. Schematic representation of an Al-Mg-Sc-Zr melt pool microstructure (This figure was drawn based on Figure 2 of [77], with permission from © 2018 Elsevier).

A similar bimodal microstructure was observed by Griffiths et al. in Sc-free Al-Mg-Zr (Addalloy™) LPBF samples [77]. The authors focused the attention on the effect of laser rescanning on the microstructure of Addalloy™ observing a refinement of the microstructure thanks to the remelting of columnar grains and the precipitation of Al$_3$Zr seeds. Croteau et al. also investigated the strengthening mechanisms of Al-Mg-Zr alloys processed by LPBF [78]. The results suggested that the Mg acts as solid solution strengthener while Zr strengthens the alloy based on two main mechanisms, i.e., grain refinement thanks to the precipitation of coherent Al$_3$Zr phase during solidification and by precipitation strengthening.

Because of the extremely promising results obtained with Scalmalloy®, various works followed the same idea. Jia et al., for example, investigated the laser processability and the microstructure of

Al-Sc-Zr and Al-Er-Zr alloys [48]. Sc and Er form similar L12 precipitates with Al, both extremely coherent with the Al lattice structure [71]. Despite this, the microstructure analyses showed that these compositions have an extremely different behavior: Er has, in fact, a lower solubility in Al with respect to Sc and this causes the reduced thermal stability of these precipitates that easily grow with ageing treatments. Furthermore, EBSD maps confirmed that Er has a different effect with respect to Sc on the solidification mechanisms of Al alloys, implying the solidification of larger grains.

Zheng et al. investigated the microstructure and the properties of the FVS0812 alloy (Al-8.5Fe-1.3V-1.7Si) processed by LPBF [79]. This alloy, developed by Allied Signal Inc., was deeply studied in the 1990s by RSP and powder metallurgy, reaching elevated strengths at high temperature due to the nanoscale $Al_{12}(Fe,V)_3Si$ precipitates [80,81]. Zheng et al. stated that dense and crack-free FVS0812 samples can be successfully produced by LPBF and that as-built samples have a hardness value similar to the planar flow casting one. In the as-built sample microstructure three zones can be recognized: the laser melted zone made of α-Al and fine and round $Al_{12}(Fe,V)_3Si$, the melt pool border made of fine $Al_{13}Fe_4$ precipitates and the heat affected zones (Figure 9a) [82].

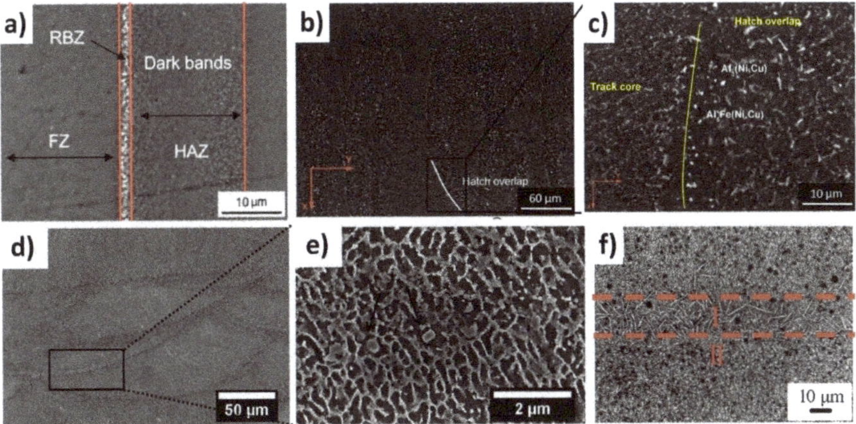

Figure 9. SEM images of (**a**) FVS0812, adapted from [82], with permission from © 2015 Cambridge University Press; (**b**,**c**) Al-Si-Ni-Fe water quenched and aged at 250 °C for 5 h, adapted from [83], with permission from © 2019 Elsevier; (**d**,**e**) Al-Si-Ni, adapted from [33], with permission from © 2017 Elsevier; and (**f**) Al-20Si-Fe-Cu-Mg adapted from [68], with permission from © 2016 Elsevier.

Manca et al. recently proofed the potentiality of a new Al-Si-Ni-Fe alloy containing Cu minor addition, which presented not only a good LPBF processability but also high mechanical performances [83]. The high hardness value of this alloy (186 HV) was attributed to the fine Si, Al_5Fe (Ni,Cu), and $Al_3(Ni,Cu)$ phases (Figure 9b,c).

Aversa et al. studied the processability and the properties of an Al-Si-Ni alloy with a composition close to the ternary eutectic one using AlSi10Mg and Ni particles [33]. From a microstructural point of view, the as-built samples are constituted by α-Al cells surrounded by the eutectic and Al_3Ni particles. Similarly to what was observed on the FVS0812 alloy, slightly larger precipitates were recognized at the melt pool boundary (Figure 9d,e). The hardness measurements revealed that the introduction of 5 wt.% Ni to the AlSi10Mg composition caused an increase of 23% in Vickers hardness. Nanoindentation measurements were then used to demonstrate that the high hardness value is induced by the presence of fine Al_3Ni particles.

On the other hand, Ma et al. studied the effect of Fe, Cu, and Mg addition on the microstructure of an Al-20Si alloy processed by LPBF [68]. It was demonstrated that the microstructure consists of α-Al, Si bulk particles, Al-Si eutectic and coarse acicular δ-Al_4FeSi_2 phase (Figure 9f). The size of these phases is strongly lower with respect to the cast ones thanks to the high cooling rate of the LPBF process. Furthermore, the presence of Fe, Cu, and Mg hinders the growth of Si phase.

Wang et al. investigated the microstructure and the properties of Al-Cu samples with different Cu contents (4.5–40 wt.%) produced by LPBF starting from mixed Al and Cu particles [78]. The microstructures of the low Cu content alloys were made of an Al matrix and Al_2Cu particles. Moreover, the microstructure became fine eutectic as the Cu content increases. The highest mechanical properties, evaluated by compression tests, were recorded in the case of the compositions closest to the eutectic (Al-33Cu). Despite the sensitivity of Al-Cu alloys to cracking mechanisms, no consolidation issues of these alloys were reported in this paper.

These studies showed that the success of the Scalmalloy® composition is related not only to the precipitation of L12 coherent strengthening phases but also to the strengthening effect of Mg which is contained in a supersaturated solid solution in the α-Al matrix. Subsequent studies on the effect of other RE elements such as Er demonstrated that the relatively high solubility of Sc in Al is also a key-point for the success of AlMgScZr compositions.

The introduction of TM to the LPBF alloys compositions can be then considered a promising method to improve the mechanical properties of the alloy by maintaining a good processability. The composition studied up now were mainly based on the most promising alloys processed by RSP. The microstructural and mechanical investigations demonstrated that, due to rapid cooling, fine microstructures and by high mechanical properties can be achieved by LPBF.

4.4. Metallic Glasses and Nanocrystalline Materials

Metallic glasses (MGs) and nanocrystalline materials (NCMs) are two class of materials that received great attention in recent years thanks to their unique properties due to the lack of a conventional metal structure characterized by long-range order [84]. The LPBF processability of a few materials belonging to these class was studied in recent scientific publications.

Prashanth et al. studied the microstructure and the properties of thermally stable $Al_{85}Nd_8Ni_5Co_2$ nanocrystalline material produced by pre-alloyed gas atomized powder. The authors reported a composite like microstructure made of Ni, Nd and Co-rich sub micrometric platelets like intermetallic precipitates [85]. These particles implied an increase in the mechanical properties of the alloys and a reduction of the microstructure coarsening. Furthermore, the elongated particles, characterized by a strong interface with the matrix, were very effective in deflecting cracks and increasing the mechanical strength.

Li et al. investigated the possibility to process an Al-based metallic glass ($Al_{86}Ni_6Y_{4.5}Co_2La_{1.5}$) by LPBF [42]. The authors firstly performed SSTs and observed that some process parameters caused cracks within the SSTs. It was also demonstrated that the microstructure and the hardness strongly depend on the position within the melt pool: some areas crystallize due to a different thermal history. In a second study, the same authors investigated the effect of low energy density rescanning in the processing of bulk metallic glass composites (BMGCs) [86]. It has been demonstrated that for these materials rescanning with low power allows the reduction of residual stresses.

5. Mechanical Properties

The main mechanical properties of the above-described alloys in terms of hardness (HV and HB), yield strength (YS), ultimate tensile strength (UTS) and elongation (ε) are reported in Table 1. Undoubtedly, the heat treated Scalmalloy® samples showed the highest YS and UTS values together with a high elongation. The authors attributed this extraordinary mechanical strength to Sc and Zr which form Al_3Sc and $Al_3(Sc_xZr_y)$, to Zr that creates a shell around Al_3Sc particles, and mainly to the presence of Mg which causes a solid solution strengthening effect [70]. It is important to underline that the post processing heat treatment plays a crucial role in determining the properties of the alloy [76], it can be noticed indeed that the as-built Scalmalloy® properties are similar to the as-built AlSi10Mg ones [76]. The comparison of the Scalmalloy® with the Al-Sc-Zr confirms the relevance of both strengthening effects. Furthermore Jia et al. demonstrated that Er does not imply the same strengthening effect of Sc on LPBF Al alloys mainly because of the large size and reduced volume fraction of the strengthening phase [48].

Table 1. Mechanical properties of modified aluminum alloys processed through LPBF. Hardness (HV and HB), yield strength (YS), ultimate tensile strength (UTS) and elongation (ε).

Composition	Heat Treatment	HV	HB	YS [MPa]	UTS [MPa]	ε (%)	Reference
AlSi10Mg H		135.0 ± 0.9	128.6 ± 1.9	270 ± 10	460 ± 20	9 ± 2	[33]
AlSi10Mg H	S.R.	~95	93 ± 3	230 ± 15	345 ± 50	12 ± 2	[12]
Scalmalloy®	-	-	100–115	276–287	403–427	14–17	[74]
Scalmalloy®	325 °C 4 h	~180	-	520	530	14	[70]
AlScZr L.R.	5 h 300 °C	113 HV0.5	-	-	-	-	[48]
AlErZr L.R.	2 h 300 °C	91 HV0.5	-	-	-	-	[48]
AlCuMgMn	-	-		276.2 ± 41	402.4 ± 9.5	6 ± 1.4	[57]
Zr/AlCuMgMn	-	-		446 ± 4.3	451 ± 3.6	2.7 ± 1.1	[34]
Zr/AlCuMgMn	-	153.6		464.06 ± 2	493.30 ± 10	4.76 ± 1	[35]
2219	-	94 ± 6.6					[87]
2219	T6	147 ± 2.3					[87]
Al-3.5Cu-1.5Mg-1Si	-			223 ± 4	366 ± 7	5.3 ± 0.3	[4]
Al-3.5Cu-1.5Mg-1Si	T6			368 ± 6	455 ± 10	6.2 ± 1.8	[4]
7075+Zr	T6	130–140	-	32–373	383–417	3.8–5.4	[32]
Si mod. 7075	6 h 150 °C	~170	-	-	-	-	[36]
Si mod. 7075	6 h 160 °C		140–150	350	415	-	[37]
Al-8.5Fe-1.3V-1.7Si	-	135–175	-	-	-	-	[79]
AlSiNi	-	158.7 ± 3.0	179.5 ± 3.0	-	-	-	[33]
Al-3.60Mg-1.18Zr	400 °C 8 h	-	-	353 ± 5	386 ± 3	18.6 ± 0.9	[78]
Al-3.66Mg-1.57Zr	400 °C 8 h	-	-	365 ± 11	389 ± 4	23.9 ± 4.4	[78]
6061	-	67–84		246.7	392		[63]
6061 500 °C platform	-	54 ± 2.5		66–75	133–141	11–15	[62]
6061 500 °C platform	T6	119 ± 6		282–290	308–318	3.5–5.4	[62]

* H = Horizontal, S.R. Stress Relieved (300 °C 2 h), L.R. Laser remelted.

The comparison of the YS values of Al-Cu-Mg-Mn and Zr-Al-Cu-Mg-Mn highlights the positive effect of Zr on the mechanical properties of as-built materials. In this case, the increase in the high mechanical properties was attributed to the grain refinements and to the strengthening effect of the precipitates [34,35]. On the other hand, the increase in the processability obtained by adding Si to the Al-Cu-Mg-Mn composition also allows a slight increase in YS and UTS values, especially after a T6 heat treatment [4]. In this case the strengthening effect was mainly attributed to the Q phase, in the as-built state, and to the Mg_2Si and Al_xMn_y precipitation in heat-treated samples [4]. Contrarily, Zr and Si showed to have a similar strengthening effect on the 7075 composition [32,36]. The comparison of the 6061 properties suggests that similar YS and UTS can be obtained by as-built 6061 (T = 200 °C) and T6 6061 built at 500 °C [62,63]. The high mechanical properties obtained by processing materials with lower building platform temperatures are related to the fine microstructure that solidifies thanks to the rapid cooling. However, it must be pointed out that in this case the authors reported the presence of small micro-cracks in as-built samples.

Finally, the strengthening effect of the transition metals can be clearly seen by comparing the Vickers hardness values of AlSi10Mg with the AlSiNi and Al-8.5Fe-1.3V-1.7Si ones. In both cases the high mechanical properties were attributed to the fine Al_3Ni, $Al_{12}(Fe,V)_3Si$, and h-$Al_{13}Fe$ phases.

6. Conclusions

This review aims to demonstrate the strong interest and the great possibilities related to new alloys development specifically designed for AM processes, and in particular Al alloys for LPBF process. It could be stated that, up to now, the several approaches based on the following effects have been used:

1. The reduction of the solidification cracking mechanism of commercial high strength aluminum alloys (e.g., 7075, 2024, and 6061) thanks to the modification of the melting behavior. This effect was obtained mainly by the introduction of Si which increases the fluidity of the molten phase

and reduces the alloy melting range, its coefficient of thermal expansion and its solidification shrinkage. The decisive effect of Si was demonstrated on the 7075 and 2024 compositions.

2. The reduction of the solidification cracking obtained thanks to the reduction of the grain size of commercial high strength aluminum alloys achieved as a result of the precipitation of strongly coherent phases. Zr was mainly employed in order to obtain fine coherent Al_3Zr particles which act as nucleant during the solidification process. This method demonstrated to be promising for the processability of 7075, 2024, and 6061 alloys.

3. The reduction of the solidification cracking by the increase in the building platform temperature. This method implies a reduction of the thermal stresses and therefore of the cracking density. The high temperature of the building platform causes however a reduction in the cooling rate and precludes the solidification of fine microstructures. This method was however successfully applied to the 6061 composition which could achieve high mechanical properties in the T6 condition.

4. The introduction of rare earth and transition metallic elements to standard Al alloys compositions, which implied a strong increase in the mechanical properties of LPBF samples. The most promising composition was undoubtedly produced by the introduction of Sc and Zr to an Al-Mg alloy leading to the patented Scalmalloy® composition. The high mechanical properties of this alloy are mainly due to the precipitation of coherent $Al_3(Sc,Zr)$ particles and to the Mg solid solution strengthening effect. Furthermore, this composition resulted to be stable up to high temperatures thanks to the $Al_3(ScZr)$ poor tendency to grow. On the basis of the success of this composition, many authors focused on the study of similar compositions containing Sc, Er, and Zr and obtained promising results. The introduction of less expensive TM elements seems to be also a promising approach for the production of high strength LPBF Al alloys. The rapid solidification achieved during the laser scanning allows the precipitation of extremely fine strengthening phases and therefore high mechanical properties.

5. The production of metallic glass and nanocrystalline materials. Thanks to the rapid solidification that arises as a consequence of the laser scanning, it seems that the LPBF process might be a promising production technology for these materials. However, because of the peculiar thermal history to which the material undergoes, different microstructures can be obtained. This aspect has to be carefully taken into account if complex parts have to be built.

The analyses of the mechanical properties of newly designed alloys showed that it is possible to produce high strength Al alloys by LPBF. The mechanical performances of LPBF Al alloys are mainly due to the fine microstructure and the supersaturated solid solution obtained by the rapid solidification and by the strengthening effect of fine strongly coherent particles. It must also be underlined that, in most of the cases, non-standard heat treatments were used on these materials because of the peculiar microstructure of as-built components.

Furthermore, the results highlight that the microstructure and the properties of LPBF compositions strongly depend on the building parameters used during the AM process. On the basis of this consideration, it is important to underline that a promising composition must be coupled together with the most suitable parameters that allow the solidification of a fine microstructure and a supersaturated solid solution in order to achieve the highest mechanical properties.

The industrial scalability of these compositions also has to be taken into account. The cost of some elements such as Sc, Er, and Zr, used as grain refiners, has to be considered when evaluating the applicability of these compositions. In AM however, unlike conventional processes, the usage of more expensive materials is admitted thanks to a low buy-to-flight ratio close to 1. Furthermore, as demonstrated by the Scalmalloy® example, the high mechanical performances achieved by these compositions can often justify the high cost of the powder. On the other hand, the introduction of silicon is certainly a less expensive modification; however a complete characterization of this composition is necessaire. Finally, when considering the increase in the building platform temperature for the production of crack-free high strength Al alloy parts, the technological limit of this approach has to be considered. Up to now, in fact, most of the commercial systems do not reach sufficiently high temperatures.

To conclude, it must be pointed out that, up to now, data about advanced characterizations, such as fatigue and high temperature properties, on these materials are not available in scientific publications. This knowledge is extremely important to meet the strict requirements of the aerospace industry. This lack of data might be due, on the one hand, to the novelty of these studies and, on the other hand, to the strong industrial interest in this topic which causes to the non-disclosure of specific materials properties. Only advanced mechanical characterizations will allow the definition of the best approach for improving strength and fatigue values of Al alloys for LPBF.

Author Contributions: Writing—Original Draft Preparation, A.A., G.M. and A.S.; Data Curation, A.A., E.B. and D.U.; Writing—Review and Editing, D.M., M.L. and S.B.; Project Administration, M.L. and P.F.

Funding: This research and the APC was funded by STAMP (Sviluppo Tecnologico dell'Additive Manufacturing in Piemonte).

Conflicts of Interest: The authors declare no conflict of interest.

References

1. Frazier, W.E. Metal Additive Manufacturing: A Review. *J. Mater. Eng. Perform.* **2014**, *23*, 1917–1928. [CrossRef]
2. Adam, G.A.O.; Zimmer, D. Design for Additive Manufacturing—Element transitions and aggregated structures. *CIRP J. Manuf. Sci. Technol.* **2014**, *7*, 20–28. [CrossRef]
3. Chu, C.; Graf, G.; Rosen, D.W. Design for additive manufacturing of cellular structures. *Comput. Aided. Des. Appl.* **2008**, *5*, 686–696. [CrossRef]
4. Wang, P.; Gammer, C.; Brenne, F.; Prashanth, K.G.; Mendes, R.G.; Rümmeli, M.H.; Gemming, T.; Eckert, J.; Scudino, S. Microstructure and mechanical properties of a heat-treatable Al-3.5Cu-1.5Mg-1Si alloy produced by selective laser melting. *Mater. Sci. Eng. A* **2018**, *711*, 562–570. [CrossRef]
5. Aboulkhair, N.T.; Everitt, N.M.; Maskery, I.; Ashcroft, I.; Tuck, C. Selective laser melting of aluminum alloys. *MRS Bull.* **2017**, *42*, 311–319. [CrossRef]
6. Li, X.P.; Ji, G.; Chen, Z.; Addad, A.; Wu, Y.; Wang, H.W.; Vleugels, J.; van Humbeeck, J.; Kruth, J.P. Selective laser melting of nano-TiB2 decorated AlSi10Mg alloy with high fracture strength and ductility. *Acta Mater.* **2017**, *129*, 183–193. [CrossRef]
7. Tang, M.; Pistorius, P.C. Oxides, porosity and fatigue performance of AlSi10Mg parts produced by selective laser melting. *Int. J. Fatigue* **2016**, *94*, 192–201. [CrossRef]
8. Olakanmi, E.O.; Cochrane, R.F.; Dalgarno, K.W. A review on selective laser sintering/melting (SLS/SLM) of aluminium alloy powders: Processing, microstructure, and properties. *Prog. Mater. Sci.* **2015**, *74*, 401–477. [CrossRef]
9. Gu, D.D.; Meiners, W.; Wissenbach, K.; Poprawe, R. Laser additive manufacturing of metallic components: Materials, processes and mechanisms. *Int. Mater. Rev.* **2012**, *57*, 133–164. [CrossRef]
10. Vreeling, J.A.; Ocelik, V.; Pei, Y.T.; Agterveld, V.D.T.L.; de Hosson, J.T.M. Laser Melt Injection in Aluminum Alloys: On the Role of the Oxidaton Skin. *Acta Mater.* **2000**, *48*, 4225–4233. [CrossRef]
11. Uslan, I.; Saritas, S.; Davies, T.J. Effects of variables on size and characteristics of gas atomised aluminium powders. *Powder Metall.* **2013**, *42*, 157–163. [CrossRef]
12. Trevisan, F.; Calignano, F.; Lorusso, M.; Pakkanen, J.; Aversa, A.; Ambrosio, E.P.; Lombardi, M.; Fino, P.; Manfredi, D. On the Selective Laser Melting (SLM) of the AlSi10Mg Alloy: Process, Microstructure, and Mechanical Properties. *Materials* **2017**, *10*, 76. [CrossRef] [PubMed]
13. Aversa, A.; Lorusso, M.; Trevisan, F.; Ambrosio, E.; Calignano, F.; Manfredi, D.; Biamino, S.; Fino, P.; Lombardi, M.; Pavese, M. Effect of Process and Post-Process Conditions on the Mechanical Properties of an A357 Alloy Produced via Laser Powder Bed Fusion. *Metals* **2017**, *7*, 68. [CrossRef]
14. Vora, P.; Mumtaz, K.; Todd, I.; Hopkinson, N. AlSi12 in-situ alloy formation and residual stress reduction using anchorless selective laser melting. *Addit. Manuf.* **2015**, *7*, 12–19. [CrossRef]
15. Kempen, K.; Thijs, L.; van Humbeeck, J.; Kruth, J.-P. Mechanical Properties of AlSi10Mg Produced by Selective Laser Melting. *Phys. Procedia* **2012**, *39*, 439–446. [CrossRef]

16. Rao, J.H.; Zhang, Y.; Zhang, K.; Huang, A.; Davies, C.H.J.; Wu, X. Multiple precipitation pathways in an Al-7Si-0.6Mg alloy fabricated by selective laser melting. *Scr. Mater.* **2019**, *160*, 66–69. [CrossRef]

17. Olabode, M. Weldability of High Strength Aluminium Alloys. Ph.D. Thesis, Lappeenranta University of Technology, Lappeenranta, Finland, 2015.

18. Tang, M.; Pistorius, P.C.; Narra, S.; Beuth, J.L. Rapid Solidification: Selective Laser Melting of AlSi10Mg. *J. Miner. Met. Mater. Soc.* **2016**, *68*, 960–966. [CrossRef]

19. Guo, H.; Yang, X. Preparation of semi-solid slurry containing fine and globular particles for wrought aluminum alloy 2024. *Trans. Nonferrous Met. Soc. China* **2007**, *17*, 799–804. [CrossRef]

20. Sercombe, T.B.; Li, X. Selective laser melting of aluminium and aluminium metal matrix composites: Review. *Mater. Technol.* **2016**, *31*, 77–85. [CrossRef]

21. Manfredi, D.; Ambrosio, E.P.; Calignano, F.; Krishnan, M.; Canali, R.; Biamino, S.; Pavese, M.; Atzeni, E.; Iuliano, L.; Fino, P.; et al. Direct Metal Laser Sintering: An additive manufacturing technology ready to produce lightweight structural parts for robotic applications Memorie. *La Metall. Ital.* **2013**, *10*, 15–24.

22. Kim, D.-K.; Woo, W.; Hwang, J.-H.; An, K.; Choi, S.-H. Stress partitioning behavior of an AlSi10Mg alloy produced by selective laser melting during tensile deformation using in situ neutron diffraction. *J. Alloys Compd.* **2016**, *686*, 281–286. [CrossRef]

23. Tradowsky, U.; White, J.; Ward, R.M.; Read, N.; Reimers, W.; Attallah, M.M. Selective laser melting of AlSi10Mg: Influence of post-processing on the microstructural and tensile properties development. *Mater. Des.* **2016**, *105*, 212–222. [CrossRef]

24. Hadadzadeh, A.; Baxter, C.; Shalchi, B.; Mohammadi, M. Strengthening mechanisms in direct metal laser sintered AlSi10Mg: Comparison between virgin and recycled powders. *Addit. Manuf. J.* **2018**, *23*, 108–120. [CrossRef]

25. Hadadzadeh, A.; Amirkhiz, B.S.; Mohammadi, M. Contribution of Mg_2Si precipitates to the strength of direct metal laser sintered AlSi10Mg. *Mater. Sci. Eng. A* **2019**, *739*, 295–300. [CrossRef]

26. Li, W.; Li, S.; Liu, J.; Zhang, A.; Zhou, Y.; Wei, Q.; Yan, C.; Shi, Y. Effect of heat treatment on AlSi10Mg alloy fabricated by selective laser melting: Microstructure evolution, mechanical properties and fracture mechanism. *Mater. Sci. Eng. A* **2016**, *663*, 116–125. [CrossRef]

27. Brandl, E.; Heckenberger, U.; Holzinger, V.; Buchbinder, D. Additive manufactured AlSi10Mg samples using Selective Laser Melting (SLM): Microstructure, high cycle fatigue, and fracture behavior. *Mater. Des.* **2012**, *34*, 159–169. [CrossRef]

28. Cabrini, M.; Lorenzi, S.; Pastore, T.; Pellegrini, S.; Ambrosio, E.P.; Calignano, F.; Manfredi, D.; Pavese, M.; Fino, P. Effect of heat treatment on corrosion resistance of DMLS AlSi10Mg alloy. *Electrochim. Acta* **2016**, *206*, 346–355. [CrossRef]

29. Salmi, A.; Atzeni, E.; Iuliano, L.; Galati, M. Experimental analysis of residual stresses on AlSi10Mg parts produced by means of Selective Laser Melting (SLM). *Procedia CIRP* **2017**, *62*, 458–463. [CrossRef]

30. Mauduit, A. Study of the suitability of aluminum alloys for additive manufacturing by laser powder bed fusion. *Sci. Bull.* **2017**, *79*, 219–238.

31. Wu, J.; Wang, X.Q.; Wang, W.; Attallah, M.M.; Loretto, M.H. Microstructure and strength of selectively laser melted AlSi10Mg. *Acta Mater.* **2016**, *117*, 311–320. [CrossRef]

32. Martin, J.H.; Yahata, B.D.; Hundley, J.M.; Mayer, J.A.; Schaedler, T.A.; Pollock, T.M. 3D printing of high-strength aluminium alloys. *Nature* **2017**, *549*, 365–369. [CrossRef] [PubMed]

33. Aversa, A.; Lorusso, M.; Cattano, G.; Manfredi, D.; Calignano, F.; Ambrosio, E.P.; Biamino, S.; Fino, P.; Lombardi, M.; Pavese, M. A study of the microstructure and the mechanical properties of an AlSiNi alloy produced via selective laser melting. *J. Alloys Compd.* **2017**, *695*, 1470–1478. [CrossRef]

34. Zhang, H.; Zhu, H.; Nie, X.; Yin, J.; Hu, Z.; Zeng, X. Effect of Zirconium addition on crack, microstructure and mechanical behavior of selective laser melted Al-Cu-Mg alloy. *Scr. Mater.* **2017**, *134*, 6–10. [CrossRef]

35. Nie, X.; Zhang, H.; Zhu, H.; Hu, Z.; Ke, L.; Zeng, X. Effect of Zr content on formability, microstructure and mechanical properties of selective laser melted Zr modified Al-4.24Cu-1.97Mg-0.56Mn alloys. *J. Alloys Compd.* **2018**, *764*, 977–986. [CrossRef]

36. Montero-Sistiaga, M.L.; Mertens, R.; Vrancken, B.; Wang, X.; Van Hooreweder, B.; Kruth, J.P.; Van Humbeeck, J. Changing the alloy composition of Al7075 for better processability by selective laser melting. *J. Mater. Process. Technol.* **2016**, *238*, 437–445. [CrossRef]

37. Aversa, A.; Marchese, G.; Manfredi, D.; Lorusso, M.; Calignano, F.; Biamino, S.; Lombardi, M.; Fino, P.; Pavese, M. Laser Powder Bed Fusion of a High Strength Al-Si-Zn-Mg-Cu Alloy. *Metals* **2018**, *8*, 300. [CrossRef]

38. Wang, P.; Deng, L.; Prashanth, K.G.; Pauly, S.; Eckert, J.; Scudino, S. Microstructure and mechanical properties of Al-Cu alloys fabricated by selective laser melting of powder mixtures. *J. Alloys Compd.* **2018**, *735*, 2263–2266. [CrossRef]

39. Zhou, J.; Tsai, H.-L.; Wang, P.-C. Transport Phenomena and Keyhole Dynamics during Pulsed Laser Welding. *Trans. Am. Soc. Mech. Eng.* **2006**, *128*, 680–690. [CrossRef]

40. Aversa, A.; Moshiri, M.; Librera, E.; Hadi, M.; Marchese, G.; Manfredi, D.; Lorusso, M.; Calignano, F.; Biamino, S.; Lombardi, M.; et al. Single scan track analyses on aluminium based powders. *J. Mater. Process. Technol.* **2018**, *255*, 17–25. [CrossRef]

41. Nie, X.; Zhang, H.; Zhu, H.; Hu, Z.; Ke, L.; Zeng, X. Analysis of processing parameters and characteristics of selective laser melted high strength Al-Cu-Mg alloys: From single tracks to cubic samples. *J. Mater. Process. Technol.* **2018**, *256*, 69–77. [CrossRef]

42. Li, X.P.; Kang, C.W.; Huang, H.; Zhang, L.C.; Sercombe, T.B. Selective laser melting of an $Al_{86}Ni_6Y_{4.5}Co_2La_{1.5}$ metallic glass: Processing, microstructure evolution and mechanical properties. *Mater. Sci. Eng. A* **2014**, *606*, 370–379. [CrossRef]

43. Boley, C.D.; Khairallah, S.A.; Rubenchik, A.M. Calculation of laser absorption by metal powders in additive manufacturing. *Appl. Opt.* **2015**, *54*, 2477–2482. [CrossRef]

44. Bartkowiak, K.; Ullrich, S.; Frick, T.; Schmidt, M. New Developments of Laser Processing Aluminium Alloys via Additive Manufacturing Technique. *Phys. Procedia* **2011**, *12*, 393–401. [CrossRef]

45. King, W.E.; Barth, H.D.; Castillo, V.M.; Gallegos, G.F.; Gibbs, J.W.; Hahn, D.E.; Kamath, C.; Rubenchik, A.M. Observation of keyhole-mode laser melting in laser powder-bed fusion additive manufacturing. *J. Mater. Process. Technol.* **2014**, *214*, 2915–2925. [CrossRef]

46. Gu, D.; Shen, Y. Balling phenomena during direct laser sintering of multi-component Cu-based metal powder. *J. Alloys Compd.* **2007**, *432*, 163–166. [CrossRef]

47. Aboulkhair, N.T.; Maskery, I.; Tuck, C.; Ashcroft, I.; Everitt, N.M. On the formation of AlSi10Mg single tracks and layers in selective laser melting: Microstructure and nano-mechanical properties. *J. Mater. Process. Technol.* **2016**, *230*, 88–98. [CrossRef]

48. Jia, Q.; Rometsch, P.; Cao, S.; Zhang, K.; Huang, A.; Wu, X. Characterisation of AlScZr and AlErZr alloys processed by rapid laser melting. *Scr. Mater.* **2018**, *151*, 42–46. [CrossRef]

49. Li, Y.; Gu, D. Parametric analysis of thermal behavior during selective laser melting additive manufacturing of aluminum alloy powder. *Mater. Des.* **2014**, *63*, 856–867. [CrossRef]

50. Scipioni, U.; Guss, G.; Wu, S.; Matthews, M.J.; Schoenung, J.M. In-situ characterization of laser-powder interaction and cooling rates through high-speed imaging of powder bed fusion additive manufacturing. *Mater. Des.* **2017**, *135*, 385–396. [CrossRef]

51. Masoomi, M.; Thompson, S.M.; Shamsaei, N. Laser powder bed fusion of Ti-6Al-4V parts: Thermal modeling and mechanical implications. *Int. J. Mach. Tools Manuf.* **2017**, *119*, 73–90. [CrossRef]

52. Srivastava, V.K. A Reviev on Advances in Rapid Prototype 3D Printing of Multi-Functional Applications. *Sci. Technol.* **2017**, *7*, 4–24.

53. Tawfik, N.L. Mechanical properties of rapidly solidified ribbons of some Al–Si based alloys. *J. Mater. Sci.* **1997**, *32*, 2997–3000. [CrossRef]

54. Ovecoglu, M.L.; Necip, U.; Niyazi, E.; Eruslu, N.; Genc, A. Characterization investigations of a melt-spun ternary Al–8Si–5.1Cu (in wt.%) alloy. *Mater. Lett.* **2003**, *57*, 3296–3301. [CrossRef]

55. Marola, S.; Manfredi, D.; Fiore, G.; Poletti, M.G.; Lombardi, M.; Fino, P.; Battezzati, L. A comparison of Selective Laser Melting with bulk rapid solidification of AlSi10Mg alloy. *J. Alloys Compd.* **2018**, *742*, 271–279. [CrossRef]

56. Ghaini, F.M.; Sheikhi, M.; Torkamany, M.J.; Sabbaghzadeh, J. The relation between liquation and solidification cracks in pulsed laser welding of 2024 aluminium alloy. *Mater. Sci. Eng. A* **2009**, *519*, 167–171. [CrossRef]

57. Zhang, H.; Zhu, H.; Qi, T.; Hu, Z.; Zeng, X. Materials Science & Engineering A Selective laser melting of high strength Al–Cu–Mg alloys: Processing, microstructure and mechanical properties. *Mater. Sci. Eng. A* **2016**, *656*, 47–54.

58. Kaufmann, F.; Imran, N.; Wischeropp, M.; Emmelmann, T.; Siddique, C.; Walther, S. Influence of process parameters on the quality of aluminium alloy EN AW 7075 using selective laser melting (SLM). *Phys. Procedia* **2016**, *83*, 918–926. [CrossRef]

59. Qi, T.; Zhu, H.; Zhang, H.; Yin, J.; Ke, L.; Zeng, X. Selective laser melting of Al7050 powder: Melting mode transition and comparison of the characteristics between the keyhole and conduction mode. *Mater. Des.* **2017**, *135*, 257–266. [CrossRef]

60. Rajan, K.; Wallace, W.; Beddoes, J.C. Microstructural study of a high-strength stress-corrosion resistant 7075 aluminium alloy. *J. Mater. Sci.* **1982**, *17*, 2817–2824. [CrossRef]

61. Zhou, L.; Pan, H.; Hyer, H.; Park, S.; Bai, Y.; McWilliams, B.; Cho, K.; Sohn, Y. Scripta Materialia Microstructure and tensile property of a novel AlZnMgScZr alloy additively manufactured by gas atomization and laser powder bed fusion. *Scr. Mater.* **2019**, *158*, 24–28. [CrossRef]

62. Uddin, S.Z.; Murr, L.E.; Terrazas, C.A.; Morton, P.; Roberson, D.A.; Wicker, R.B. Processing and characterization of crack-free aluminum 6061 using high-temperature heating in laser powder bed fusion additive manufacturing. *Addit. Manuf. J.* **2018**, *22*, 405–415. [CrossRef]

63. Maamoun, A.H.; Xue, J.; Elbestawi, M.; Veldhuis, S.C. The Effect of Selective Laser Melting Process Parameters on the Microstructure and Mechanical Properties of Al6061 and AlSi10Mg Alloys. *Materials* **2019**, *12*, 12. [CrossRef] [PubMed]

64. Loh, L.E.; Liu, Z.H.; Zhang, D.Q.; Mapar, M.; Sing, S.L.; Chua, C.K.; Yeong, W.Y. Selective Laser Melting of aluminium alloy using a uniform beam profile Selective Laser Melting of aluminium alloy using a uniform beam profile. *Virtual Phys. Prototyp.* **2014**, *9*, 11–16. [CrossRef]

65. Louvis, E.; Fox, P.; Sutcliffe, C.J. Selective laser melting of aluminium components. *J. Mater. Process. Technol.* **2011**, *211*, 275–284. [CrossRef]

66. Fulcher, B.A.; Leigh, D.K.; Watt, T.J. Comparison of AlSi10Mg and Al 6061 processes trough DMLS. In Proceedings of the Solid Freeform Fabrication (SFF) Symposium, Austin, TX, USA, 4–6 August 2014; pp. 404–419.

67. Lee, T.H.; Hong, S.J. Microstructure and mechanical properties of Al–Si–X alloys fabricated by gas atomization and extrusion process. *J. Alloys Compd.* **2009**, *487*, 218–224. [CrossRef]

68. Ma, P.; Jia, Y.; Gokuldoss, K.; Scudino, S.; Yu, Z. Microstructure and phase formation in Al–20Si–5Fe–3Cu–1Mg synthesized by selective laser melting. *J. Alloys Compd.* **2016**, *657*, 430–435. [CrossRef]

69. Rajabi, M.; Vahidi, M.; Simchi, A.; Davami, P. Effect of rapid solidification on the microstructure and mechanical properties of hot-pressed Al-20Si-5Fe alloys. *Mater. Charact.* **2009**, *60*, 1370–1381. [CrossRef]

70. Schmidtke, K.; Palm, F.; Hawkins, A.; Emmelmann, C. Process and Mechanical Properties: Applicability of a Scandium modified Al-alloy for Laser Additive Manufacturing. *Phys. Procedia* **2011**, *12*, 369–374. [CrossRef]

71. Kendig, K.L.; Miracle, D.B. Strengthening mechanisms of an Al-Mg-Sc-Zr alloy. *Acta Mater.* **2002**, *50*, 4165–4175. [CrossRef]

72. Spierings, A.B.; Dawson, K.; Voegtlin, M.; Palm, F.; Uggowitzer, P.J. Microstructure and mechanical properties of as-processed scandium-modified aluminium using selective laser melting. *CIRP Ann. Manuf. Technol.* **2016**, *65*, 213–216. [CrossRef]

73. Spierings, A.B.; Dawson, K.; Heeling, T.; Uggowitzer, P.J.; Schäublin, R.; Palm, F.; Wegener, K. Microstructural features of Sc- and Zr-modified Al-Mg alloys processed by selective laser melting. *Mater. Des.* **2017**, *115*, 52–63. [CrossRef]

74. Spierings, A.B.; Dawson, K.; Uggowitzer, P.J.; Wegener, K. Influence of SLM scan-speed on microstructure, precipitation of Al₃Sc particles and mechanical properties in Sc- and Zr-modified Al-Mg alloys. *Mater. Des.* **2018**, *140*, 134–143. [CrossRef]

75. Li, R.; Wang, M.; Yuan, T.; Song, B.; Chen, C.; Zhou, K.; Cao, P. Selective laser melting of a novel Sc and Zr modified Al-6.2 Mg alloy: Processing, microstructure, and properties. *Powder Technol.* **2017**, *319*, 117–128. [CrossRef]

76. Shi, Y.; Rometsch, P.; Yang, K.; Palm, F.; Wu, X. Characterisation of a novel Sc and Zr modified Al–Mg alloy fabricated by selective laser melting. *Mater. Lett.* **2017**, *196*, 347–350. [CrossRef]

77. Griffiths, S.; Rossell, M.D.; Croteau, J.; Vo, N.Q.; Dunand, D.C.; Leinenbach, C. Effect of laser rescanning on the grain microstructure of a selective laser melted Al-Mg-Zr alloy. *Mater. Charact.* **2018**, *143*, 34–42. [CrossRef]

78. Croteau, J.R.; Griffiths, S.; Rossell, M.D.; Leinenbach, C.; Kenel, C.; Jansen, V.; Seidman, D.N.; Dunand, D.C.; Vo, N.Q. Microstructure and mechanical properties of Al-Mg-Zr alloys processed by selective laser melting. *Acta Mater.* **2018**, *153*, 35–44. [CrossRef]

79. Zheng, L.; Liu, Y.; Sun, S.; Zhang, H. Selective laser melting of Al-8.5Fe-1.3V-1.7Si alloy: Investigation on the resultant microstructure and hardness. *Chin. J. Aeronaut.* **2015**, *28*, 564–569. [CrossRef]

80. Leng, Y.; Porr, W.C.; Gangloff, R.P. Tensile Ddeformation of 2618 and Al-Fe-Si-V Aluminium Alloys at Elevated Temperatures. *Scr. Metall. Mater.* **1990**, *24*, 2163–2168. [CrossRef]

81. Pickens, J.R.; Marietta, M. High-Strength Aluminum P/M Alloys. *Adv. Mater. Process.* **1999**, *155*, 200–215.

82. Sun, S.; Zheng, L.; Liu, Y. Characterization of Al-Fe-V-Si heat-resistant aluminum alloy components fabricated by selective laser melting. *J. Mater. Res.* **2015**, *30*, 1661–1669. [CrossRef]

83. Manca, D.R.; Churyumov, A.Y.; Pozdniakov, A.V.; Ryabov, D.K.; Korolev, V.A.; Daubarayte, D.K. Novel heat-resistant Al-Si-Ni-Fe alloy manufactured by selective laser melting. *Mater. Lett.* **2019**, *236*, 676–679. [CrossRef]

84. Inoue, A. Amorphous, nanoquasicrystalline and nanocrystalline alloys in Al-based systems. *Prog. Mater. Sci.* **1998**, *43*, 365–520. [CrossRef]

85. Prashanth, K.G.; Shahabi, H.S.; Attar, H.; Srivastava, V.C.; Ellendt, N.; Uhlenwinkel, V.; Eckertae, J.; Scudino, S. Production of high strength Al85Nd8Ni5Co2 alloy by selective laser melting. *Addit. Manuf.* **2015**, *6*, 1–5. [CrossRef]

86. Li, X.P.; Kang, C.W.; Huang, H.; Sercombe, T.B. The role of a low-energy-density re-scan in fabricating crack-free Al85Ni5Y6Co2Fe2 bulk metallic glass composites via selective laser melting. *Mater. Des.* **2014**, *63*, 407–411. [CrossRef]

87. Cornelius, M.; Karg, H.; Ahuja, B.; Wiesenmayer, S.; Kuryntsev, S.V.; Schmidt, M. Effects of Process Conditions on the Mechanical Behavior of Aluminium Wrought Alloy EN AW-2219 (AlCu6Mn) Additively Manufactured by Laser Beam Melting in Powder Bed. *Micromachines* **2017**, *2219*, 1–11.

 *materials*

Article

The Martensitic Transformation and Mechanical Properties of Ti6Al4V Prepared via Selective Laser Melting

Junjie He [1], Duosheng Li [1,*], Wugui Jiang [2], Liming Ke [2], Guohua Qin [2], Yin Ye [1], Qinghua Qin [3] and Dachuang Qiu [1]

[1] School of Materials Science and Engineering, Nanchang Hangkong University, Nanchang 330063, China; 233hjj@gmail.com (J.H.); 13627007624@163.com (Y.Y.); Dachuangq@gmail.com (D.Q.)
[2] School of aeronautical manufacturing and engineering, Nanchang Hangkong University, Nanchang 330063, China; jiangwugui@nchu.edu.cn (W.J.); limingke@nchu.edu.cn (L.K.); qghwzx@126.com (G.Q.)
[3] Research School of Engineering, Australian National University, Acton, ACT 2601, Australia; qinghua.qin@anu.edu.au
* Correspondence: ldsnuaa@nuaa.edu.cn; Tel.: +86-181-7008-9973

Received: 14 December 2018; Accepted: 18 January 2019; Published: 21 January 2019

Abstract: This article investigated the microstructure of Ti6Al4V that was fabricated via selective laser melting; specifically, the mechanism of martensitic transformation and relationship among parent β phase, martensite (α′) and newly generated β phase that formed in the present experiments were elucidated. The primary X-ray diffraction (XRD), transmission electron microscopy (TEM) and tensile test were combined to discuss the relationship between α′, β phase and mechanical properties. The average width of each coarse β columnar grain is 80–160 μm, which is in agreement with the width of a laser scanning track. The result revealed a further relationship between β columnar grain and laser scanning track. Additionally, the high dislocation density, stacking faults and the typical $(10\bar{1}1)$ twinning were identified in the as-built sample. The twinning was filled with many dislocation lines that exhibited apparent slip systems of climbing and cross-slip. Moreover, the α + β phase with fine dislocation lines and residual twinning were observed in the stress relieving sample. Furthermore, both as-built and stress-relieved samples had a better homogeneous density and finer grains in the center area than in the edge area, displaying good mechanical properties by Feature-Scan. The α′ phase resulted in the improvement of tensile strength and hardness and decrease of plasticity, while the newly generated β phase resulted in a decrease of strength and enhancement of plasticity. The poor plasticity was ascribed to the different print mode, remained support structures and large thermal stresses.

Keywords: selective laser melting; Ti6Al4V alloy; martensitic transformation; texture evolution; mechanical properties

---

1. Introduction

As opposed to traditional subtractive manufacturing, selective laser melting (SLM) is a layer-by-layer overlapping technology that was used to create three-dimensional (3D) components from 3D model data by using laser as the input heat source. SLM has attracted much attention over the past few years for its immanent advantages, such as high material utilization, short work cycle time and the ignorance of the geometric shape during processing. Furthermore, SLM is an environmentally friendly and advanced manufacturing technology that is widely used in processing titanium, Co-Cr, aluminum and stainless alloys in different fields [1–7].

As a kind of α + β dual-phase titanium alloy, Ti6Al4V has been widely used in national defense, automobile and medicine due to its excellent performance including good anticorrosive, low density, superior weldability and high specific strength. However, it is difficult to machine Ti6Al4V alloy via conventional measures for its high melting point, frangibility to oxidation, large deformation resistance and poor cutting ability. Those characteristics lead to high production costs [8]. In addition, the low surface roughness and poor plasticity of Ti6Al4V alloys that were fabricated by SLM are not ideal; these have caused the property of SLM parts to be of lower quality when compared to traditional manufactured parts.

Recently, many experiments have focused on the mechanical properties and microstructures of the Ti6Al4V alloy. For instance, the morphology, utilization and particle size of additive manufacturing powder has been reviewed [9,10]. The as-built sample and its support structures were investigated to explore the performance of the region in sample and support structures, respectively [11,12]. The surface roughness of as-received and laser polished samples of Ti6Al4V and TC11 alloys were also compared, respectively [13]. Moreover, researches have reported on the relationship between the sample quality and the process parameters as laser power, exposure time, point distance, particle size, layer thickness and hatching distance during SLM [14–16]. The effect of six different kinds of scan patterns on microstructure and mechanical properties has been investigated [17]. The microstructures and mechanical properties of different alloys manufactured by various additive manufacturing (AM) technologies were introduced, respectively [3,18]. While the typical hexagonal close-packed (HCP) diffraction spots were presented and that α′ martensite was the only structure in SLM samples [19].

Nowadays, many studies have focused on the mechanical properties and microstructures of Ti6Al4V alloys, and different heat treatments or different processing technologies have been designed to promote the comprehensive property of Ti6Al4V. Little attention has been paid to the origin of martensitic transformation mechanism during SLM between the parent β phase and the newly generated α′ phase in-depth, owing to its complexity. This study focused on the martensite that is generated from Ti6Al4V by SLM, which is distinct from the traditional subtractive methodologies. The deformation mechanism related to the β phase and α′ phase and the influence of microstructure on macroscopic mechanical properties were also elucidated. Furthermore, the commercial SLM printer was used to build samples with different print mode.

2. Material Preparation and Methods

In the present work, the SLM system (Space Traveler Tr150, Profeta Intelligent Technology, Nanjing, China) was used to fabricate the rectangular and cube parts as shown in Figure 1. The optimal process parameters are listed in Table 1.

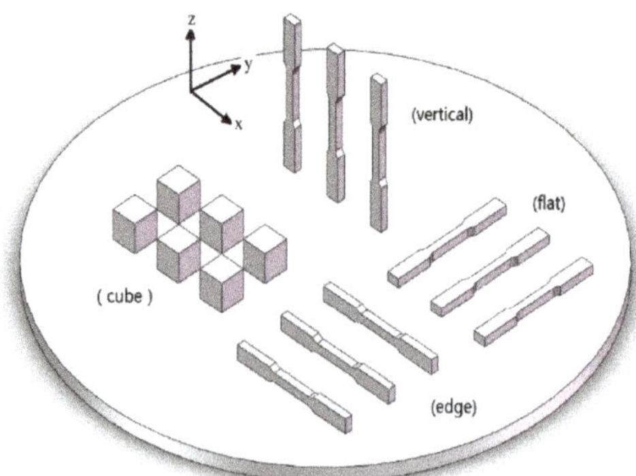

Figure 1. Computer-aided design (CAD) models of cubes and three orthogonal orientations (flat, edge and vertical) tensile samples on the stainless steel substrate.

Table 1. Process parameters of preparing Ti6Al4V specimens.

Process Parameters	Values
Laser power (W)	135
Layer thickness (μm)	30
Exposure time (μs)	400
Scan speed (mm/s)	800
Laser spot size (μm)	52

The plasma atomized Ti6Al4V-ELI powder with an average diameter of 34 μm was used in the present experiments. Figure 2a shows the powder morphology under scanning electron microscope (SEM, FEI Quanta 200, Hillsboro, OR, USA). Figure 2b shows the powder size and composition of the Ti6Al4V powder that were obtained by laser-based particle size analyzer and energy dispersive spectrometer (EDS) respectively. The high sphericity and smooth-faced powder has a narrow size distribution between 10 μm and 50 μm in which the particle diameter of 37 μm occupies highest proportion. Table 2 displays the detail chemical composition of the experimental Ti6Al4V powder.

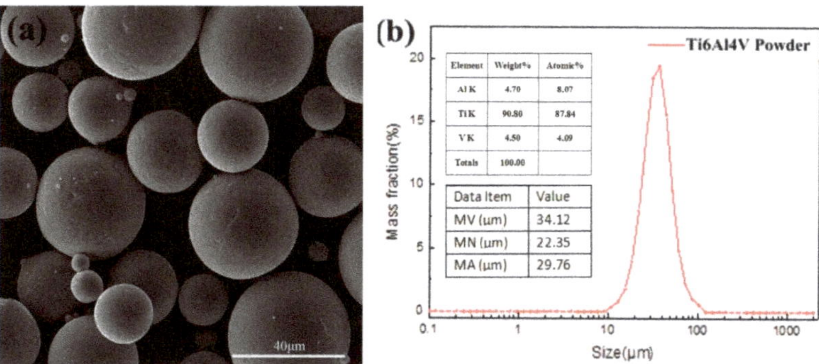

Figure 2. (**a**) The morphology of Ti6Al4V powder; (**b**) particle size and distribution.

Table 2. Chemical composition of Ti6Al4V powder (wt %).

Element	Ti	Al	V	O
Ti6Al4V	89.84	6.25	3.90	<0.1

The laying powder device spread the powder across the building substrate, and then, a laser beam scanned the uniform powder. The laser scanned the area of each layer with parallel scan vectors and the border of each layer was then scanned twice (interior and exterior contour). The scan direction initially fused at a 45° angle to the horizontal line and then rotated by 90° between each new layer as shown in Figure 3. The laser had a power of 135 W; a scan speed of 800 mm/s was applied to the center area of the parts, which were in variance from the border where the laser power was lower. A reduced scan speed resolved the problem of energy accumulation. In this study, the edge area of sample was optimized to be easily removed by post-machining once the print process was completed; however, due to the surface roughness of SLM, certain parts did not satisfy the requirement of the surface roughness national standard. This kind of design pattern is different from the related report in which the print parameters were identical in both the center area and edge area [19]. Additionally, the laser parameters of supporting structure were designed separately on the basis of the principle of easily detachment from the specimens. The height of supports was 2 mm with 0.2 mm-diameter of each fine column. The base plate had been preheated to decrease the occurrence of thermal stress. All the samples were built in the argon atmosphere. Moreover, the porosity of the shaped sample that was measured ranged from 0.33% to 0.69% by using Archimedes method.

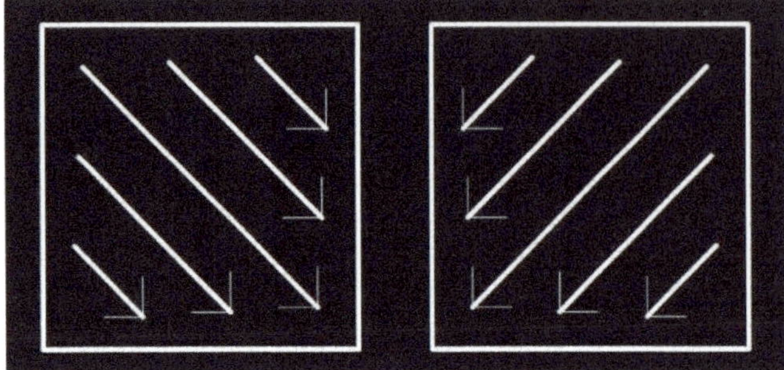

Figure 3. The scan pattern of selective laser melting (SLM).

The hardness of as-built and stress relieving samples was determined with a Vicker's indentation test machine (HX-1000TM, Shanghai optical instrument factory, Shanghai, China) that performed with a load of 500 gf for 5 s at each point. Moreover, the Feature-Scan system (UTF Scan-surface and holes, Nanchang Huali Oil Testing Company, Nanchang, China) was also carried out to detect the defects in as-built and stress relieving samples before the tensile test. The phase composition of as-built and stress relieving samples was analyzed using the software Image-Pro Plus 6.0 (Media Cybernetics, Maryland Co, Rockville, MD, USA), wherein three optical microscopy (OM) (DM1500, Shenzhen mass photoelectric company, Shenzhen, China) images at different position were measured respectively.

The rectangle specimens were fabricated directly into a dog bone shape in three orientations, which were built with a gauge length, width and thickness of 12 mm, 3 mm and 3 mm, respectively. The tensile test was conducted at room temperature with a strain rate of 2 mm/min. Heat treatment was designed at 1003 K for 2 h in N_2 gas conditions. Once the heat preservation process was completed, the specimens were furnace cooled [20–25]. Both as-built and stress relieving specimens were then polished in the support surface and tested by Instron 8827 tensile system (Instron Corporation, Norwood, MA, USA).

The microstructure and morphology of the cube samples were analyzed using X-ray diffraction (XRD, Bruker D8 Advance, Billerica, MA, USA), optical microscopy (OM), scanning electron microscopy (SEM), and transmission electron microscopy (TEM, Talos F200X, FEI Company, Hillsboro, OR, USA). The cube samples for OM were mechanical polished and then etched in kroll solutions consisting of 2 vol.% HF, 5 vol.% HNO_3 and 43 vol.% H_2O [19]. The prepared Ti6Al4V flakes for TEM were polished via twin-jet electropolishing device in mixed solution of 5 vol.% perchloric acid and 95 vol.% anhydrous ethanol at $-30\ ^\circ$C.

3. Results and Discussion

3.1. Microstructure

3.1.1. XRD Analysis

At the beginning of the SLM process, the dual-phase structure at the room temperature transforms rapidly which result in the presence of martensitic α' phase (HCP). The β phase did not have enough time to precipitate α phase in such a large undercooling (even more than one thousand degrees per second); in other words, the composition of prior β phase was almost unchanged, although the crystal lattice did change. As a result, the α phase and the substitutional supersaturated solid solution α' phase generated by this lattice transformation [26].

Commonly, the 100% martensite is impossible to obtain under this large undercooling [26]; thus, the XRD pattern in Figure 4 displays the close-packed hexagonal structure of α/α' martensitic peaks that were similar to other published reports in as-built samples [19,27]. A little deviation of the α/α' phase peak has been observed in comparison with the standard diffraction peak of α-Ti after X-ray diffraction analysis. The phenomenon may derive from the solid strengthening and the other factor is the Ti6AL4V crystal structure distortion. The smaller atomic radius of the V elements dissolved in the Ti lattice, which has a larger atomic radius that led to a crystal structure distortion and then the migration of the diffraction peaks had taken place.

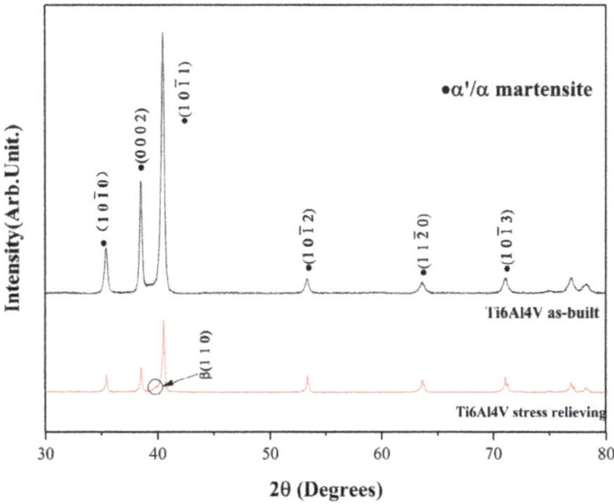

Figure 4. XRD patterns of as-built and stress relieving samples.

Figure 4 shows that the intensity of peaks of the as-built sample were higher than the stress-relieving one, which indicates that the stress relieving samples possessed a finer α/α' phase. The result was attributed to a part of the α' phase transforms into the α phase and β phase, and the other reason is the α phase precipitates the β phase during stress relieving. This can be proven with the stress-relieving XRD pattern. Here, a new β phase was observed because of the decomposition of the α' phase or the precipitation from the α phase. The details were analyzed in the following research.

On the other hand, the decrease of strength and hardness in stress relieving samples were also an important signal of the α' phase transformation. The acicular martensite formed by SLM was a substitutional supersaturated solid solution (α'), which had a positive enhancement in strength and hardness for the whole structure. However, the acicular martensite easily decomposed after stress relieving for its instability. The above two factors resulted in the increase in strength of Ti6Al4V; therefore, the hardness improved a little. Moreover, the driving force for martensitic transformation of Ti alloy was low (approximately -25 J/mol). The thermal hysteresis that was used to characterize the difference between the initial temperature of transformation and start temperature of martensite reverse transformation was rather small. All these phenomena contributed to the precipitation of β phase during the heat treatment.

3.1.2. Microstructural Mechanism

In this research, the martensite transformation was a first-order phase transition; all the parent β phase atoms transformed to an α/α' phase as a lattice reconstruction without diffusion. Thus, the first-order derivative of chemical potential were not equal according to the Equations (1)–(4):

$$\left(\frac{\partial \mu}{\partial P}\right)_T \neq \left(\frac{\partial \mu^\beta}{\partial P}\right)_T, \tag{1}$$

$$\left(\frac{\partial \mu}{\partial P}\right)_T = V, \; V^\alpha \neq V^\beta. \tag{2}$$

$$\left(\frac{\partial \mu^{\alpha}}{\partial T}\right)_P \neq \left(\frac{\partial \mu^{\beta}}{\partial T}\right)_P , \tag{3}$$

$$\left(\frac{\partial \mu}{\partial T}\right)_P = -S, \ S^{\alpha} \neq S^{\beta}. \tag{4}$$

where μ is the chemical potential, P is pressure and T is temperature. The formulas indicate the changes of volume and entropy between the β phase and α/α' phase exist in a first-order phase transition; in other words, volume change and the release (or absorption) of latent heat happened during SLM. The chemical compositions of the new phase and parent phase were identical, while the crystal structure and specific volume were quite different. For these reasons, the huge thermal stress and distortional energy were produced within the as-built alloys [28,29], which brought high dislocation density, stacking faults and twin substructure.

The typical acicular martensitic phase that originated from the prior β boundaries and filled in the β columnar grains are visible in Figure 5a. Each acicular martensite with an average width of 2 ± 0.5 μm, grew spontaneously along the preferred orientation dictated according to the special Burgers relation between the β phase and α/α' phase during the deposition process [30]. Additionally, the average width of β columnar grains was measured at a range of 120 ± 20 μm, which was coincidentally equal to the width of scan track. The similar phenomenon were discussed by Simonelli, M. et al. [31], in which the average width of β columnar grain in Ti6Al4V was 210 ± 50 μm. Curiously, the ratio of the scan track width to the columnar grain width was approximately unitary, owing to the large undercooling once the laser scanned; the content within the β phase transformed to an α' texture by crystal reconstruction instead of the precipitated α phase and the fine acicular martensite formed by preferential growth. Eventually, the morphology of β columnar grains with many acicular martensite emerged.

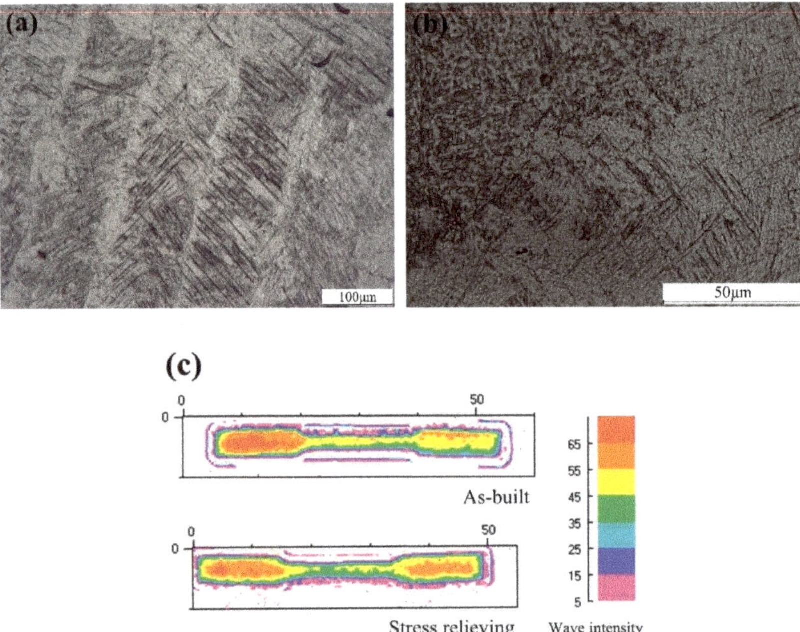

Figure 5. Optical microscopy in top surface: (**a**) the as-built sample; (**b**) middle portion of the stress relieving sample and (**c**) the echo amplitude diagram.

The size of martensite after stress relieving was measured with an average width of 0.4 ± 0.1 μm. These acicular martensite became shallower than that in the Figure 5a. The fine α phase and β phase can be observed from Figure 5b, indicating the decomposition of α' phase. It is generally known that elastic strain energy improved with the martensite growth, and conversely, the excessive energy block the growth of martensite. Comparing with as-built image, the martensite in stress-relieved image is smaller, indicating the reduction of elastic strain energy.

Furthermore, the phase composition of both samples in Table 3 demonstrated that the α' phase decreased after stress relieving, while the proportion of α' phase reduced from 28.50 to 9.90. However, the newly generated β phase rose from 0 to 10.55%. Since the annealing temperature was far from reaching the β transus temperature, there was either no β phase or the number of β phase was too small to be detected by the used methods in as-built samples. The newly generated β phase in the XRD pattern after stress relieving was attributed to two factors. The first reason was the instability of the α' phase. The non-equilibrium α' phase was easy to break down into an $\alpha+\beta$ phase as a metastable phase, which ascribe to its own characterization [21,27]. Moreover, the α' martensite of Ti6Al4V in SLM could be considered as thermal elastic martensitic owing to its low thermal hysteresis, which gives an appropriate explanation about the decomposition of α' phase. Secondly, a 2 h furnace cooling condition during stress relieving was regarded as an aging treatment, which resulted in the appearance of the β phase.

Table 3. The phase composition of as-built and stress relieving samples.

Samples	Phase		
	$\alpha'/\%$	$\alpha/\%$	$\beta/\%$
As-built	28.50	71.50	0
Stress relieving	9.90	79.55	10.55

The acoustic wave detection was applied by a Feature-Scan system to study the microstructure homogeneity and density distribution of the samples in as-built and stress relieving conditions. The echo amplitude diagram is shown in Figure 5c. The wave intensity in red area is higher than in the green area. This phenomenon appeared because of the acoustic attenuation. Once the sound wave propagated in the boundaries (green area), the waves greatly attenuated due to the slag and blowholes. On the other hand, in the center area (red area), there were relatively few defects with fine grains and excellent density. Thus, the intensity of the bounced sound in center area was higher than that of the boundaries. As in the article mentioned before, the boundaries of parts were devised in different process parameters to remove easily to reach the surface roughness standard, while the content in the center of the sample were not affected. Hence, the result reflected the different homogeneous of microstructure or density distribution. In addition, the stress-relieving sample exhibited a better homogeneous density and finer grains in the center area than the as-built sample, indicating the diminish of void ratio, grain refinement and the reduction of residual stress.

The α' acicular martensite transformation of Ti6Al4V during SLM has been introduced in detail in the preceding paragraphs. As it turns out, the newly generated metastable α' phase with close-packed hexagonal structure originated from the parent β phase, which is a body-centered cubic lattice. Generally, a relationship of coherent or semi-coherent interface always existed between the new phase and the prior phase in martensitic transformation. The prior β phase possessed a dominant <100> texture, which is a well know grain growth direction. The newly generated α' phase had a typically crystallographic orientation relationship with the prior β phase $(0001)_{hcp}\|(110)_{bcc}$, $<11\bar{2}0>_{hcp}\|<111>_{bcc}$ [21,32–34].

The apparently higher dislocation density, fine twin substructure and stacking faults are displayed in as-built samples in Figure 6. Since the inner regions subjected uneven heating and cooling cyclically, the large thermal stress occurred during the SLM deposition process, and it was no doubt why so many higher dislocation density and twins were found in the as-built sample. The high amount of twins and the high dislocation density indicated that the high inner stresses certainly existed, which resulted in the improvement of yield strength (YS) and ultimate tensile strength (UTS). Furthermore, based on the mechanism of new martensite transformation [35], the formation of newly emerged α' martensite was preferentially growth through the direction of lower strain energy and as a result, these meta-structures occurred in order to adjust the strain energy during the process of lattice modification.

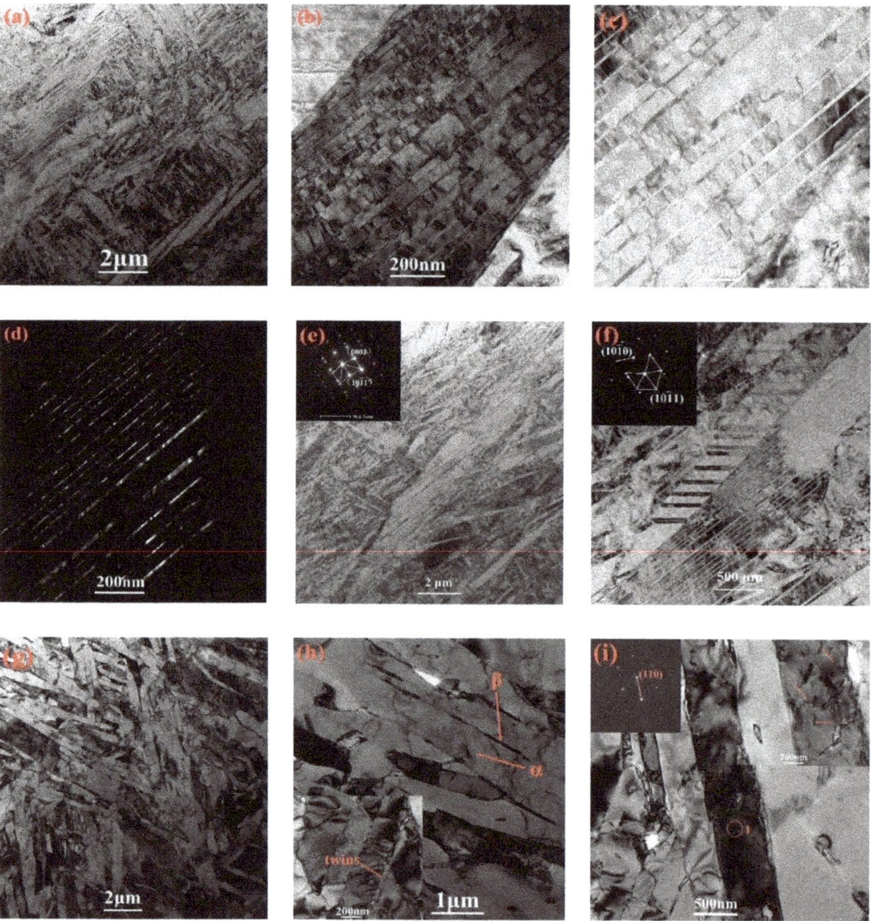

Figure 6. Transmission electron microscopy (TEM) of as-built cube samples: bright field (**a**–**c**); dark field (**d**); twinning and diffraction pattern (**e**,**f**) and stress relieving samples of bright field (**g**–**i**).

The twins in the present experiment were regarded to be transformation twins. The lattice of the twins was modified instead of emerging deformation twins that ascribed to applied stress. The width of twins was in a range distribution from 30–100 nm. The twin planes, for instance, $\{10\bar{1}2\}$ and $\{11\bar{2}1\}$, were observed as extension twins in other reports [36]. In this study, the typical $\{10\bar{1}1\}$ twinning plane were calibrated as the only type of twins by selective area electron diffraction (SAED) data as shown in Figure 6e,f. These types of twins in HCP materials are defined as contraction twins. From Figure 6b,c, it is quite evident that each bamboo-like twin lamellar are glutted with many dislocation lines, which indicates that the slip system of climbing and cross-slip occurred in the deposition process. Beyond that, the average width of neighboring stacking faults was gauged to be about 70 nm wide.

The stress relieving sample in Figure 6g,h shows a disorderly dual phase structure of α + β phase during TEM, which was anticipated due to the martensite decomposition in stress relieving condition differs from the 3D printing process. The reason is that the sample was in a relatively uniform temperature during annealing treatment, while the heat distribution of SLM process was inhomogeneous. Additionally, the residual twins were found in stress-relieved samples, as marked in Figure 6h. Different from the as-built images, the higher density of dislocations did not exist; instead, the fine dislocation lines were discovered in Figure 6i. Thus, the effect of stress relieving was obvious for the high dislocation density, and stacking faults disappeared, twin structure decreased and β phase generated. Such a result differed from the literature [21], where the high dislocation density and stacking faults can also be detected and no β phase in stress-relieved samples. On a macroscopic scale, owing to the reduction of internal stresses and the transformation of α′→α + β, the tensile strength and hardness of stress-relieved samples were decreased and the plasticity was increased.

3.2. Mechanical Properties

3.2.1. Tensile Properties

The performance of three different building orientation samples have been discussed in detail [20]; thus, two batches with six edge oriented samples were selected to study the mechanical properties in as-built and stress-relieved parts by tensile experiment. The stress-strain curve was shown in Figure 7. Additionally, the elastic modulus (E), YS, UTS, fracture stress (FS), fracture elongation (ε fracture) and hardness (H) were also analyzed, as shown in Table 4. The hardness (H) of both samples were in agreement with other reports [20,37].

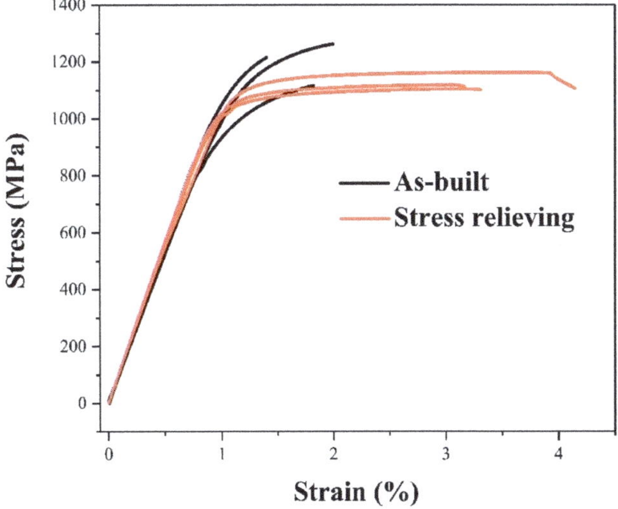

Figure 7. Stress strain curves of as-built and stress relieving Ti6Al4V parts.

Table 4. Tensile properties of SLM Ti6Al4V samples.

Parameters	Elastic Modulus (GPa)	Yield Strength (MPa)	Ultimate Tensile Strength (MPa)	Fracture Stress (MPa)	Fracture Elongation (%)	Hardness (HV)
As-built	107 ± 4	1142 ± 17	1235 ± 37	1235 ± 37	1.3 ± 0.5	395 ± 21
Stress relieving	114 ± 2	1057 ± 25	1130 ± 30	1129 ± 30	2.8 ± 0.4	390 ± 18

The UTS of the as-built sample raised with 9.3% compared to the stress-relieving sample. The elastic modulus of stress-relieving sample was the same as that of the as-built sample. In the as-built sample, the yield strain was less than 2% and the fracture strain was tested to be less than 3% [12]. However, the plasticity improved through stress relieving and the average fracture strain rose from 1.9% to 3.6%. The hardness of stress relieving samples was lower than the as-built sample in the upper surface.

Compared to as-built sample, it was evident that the properties of the stress relieving sample changed. The presence of obvious plastic deformation in stress relieving line was distinct with the as-built curve which almost no plasticity deformation observed. As mentioned in Figure 4, the existence of α' martensitic phase had a prominent improvement in strength and hardness, while it had a negative influence in plasticity, which contributed to the higher strength and hardness before being stress-relieved and better plasticity once the $\alpha' \to \alpha + \beta$ phase transformation took place. As noted in Figure 5a,b, the martensite size of stress-relieved sample was smaller than the as-built one, indicating that the elastic strain energy reduced. Thus, the YS and UTS decreased and plasticity improved.

As a β-phase stabilizing element, the V element was a main ingredient in the $\alpha + \beta$ Ti alloy that caused the morphology of acicular martensite differ from other near α titanium alloys of lower V element and plate martensite. In Ti6Al4V, the V element adopted the mode of displacement to dissolve in β-Ti as a β isomorphous element; thus, the crystal distortion was small and resulted in the low internal stresses and distortional energy. As a result, the strength and hardness of α' martensite were a little higher than the α phase. However, the strength and hardness of this kind of α' martensite did not significantly improve compared with martensite in steel.

In addition, the elastic modulus of stress relieving sample did not vary much compared with as-built one, indicating the total number of the α'/α phase without significant change because of the anisotropic α'/α structure had a marked influence on elastic modulus of samples. Nevertheless, the β texture had no influence on elastic modulus. The details were discussed in the related report [20].

Though the UTS/YS is higher than reported in many previous researches, the fracture strain of both as-built and stress relieving samples was too low compared with other related literature [21]. The three reason are summarized here. Firstly, as mentioned in Section 2, unlike other literature where the print parameters are same in total sample, the present experiment had two sets of parameters applied in the center area and edge area, respectively. The brittle edge area was designed to be easily removed by post-machining for commercial application, which was probably the main reason that resulted in the samples yielded early, and premature fracture during tensile test. Secondly, though the support structure area was mechanically polished, pits and bumps still remained in such a brittle sample surface, which may accelerate the fracture process. Additionally, as shown in Figure 6 as-built images, there were many high densities of dislocation and twins, indicating the larger thermal stresses typically occurred during SLM, which led to the error to come out in the tensile test due to the samples being a little curled.

3.2.2. Tensile Fracture Mechanisms

The transgranular fracture surface of the edge orientation in both the as-built and stress-relieved samples is shown in Figure 8. The microstructure exhibited a typical mixed mode of brittle and ductile fracture that were characterized as the observable cleavage facets, microvoids and shallow dimples [19,20].

Since the fracture strain was less than 3%, the dimple was non-existent. As shown in Figure 8a,c,d, the average diameter of these obvious pores was 38 ± 10 µm in as-built samples and the pores' size in stress relieving samples were reduced slightly: about 32 ± 8 µm wide. Particularly, these lake ripples around the pores in Figure 8c had been considered as reinforcements. However, the EDS image indicated that the compositions of this rippled structure were coincided with elsewhere in the samples. It is generally known that the excess laser power or lower scan speed can cause over-heating (OH), and low heat input or faster scan speed may result in the incomplete melting (IM). Based on this, the multi-layer pores were ascribed to the OH or IM during the SLM process, and each ripple represented one deposition layer. Once the OH or IM happened, the pore of previous layer could not be coated by the following powder layers. Hence, the fused powders of following layers exhibited the sign of ripples.

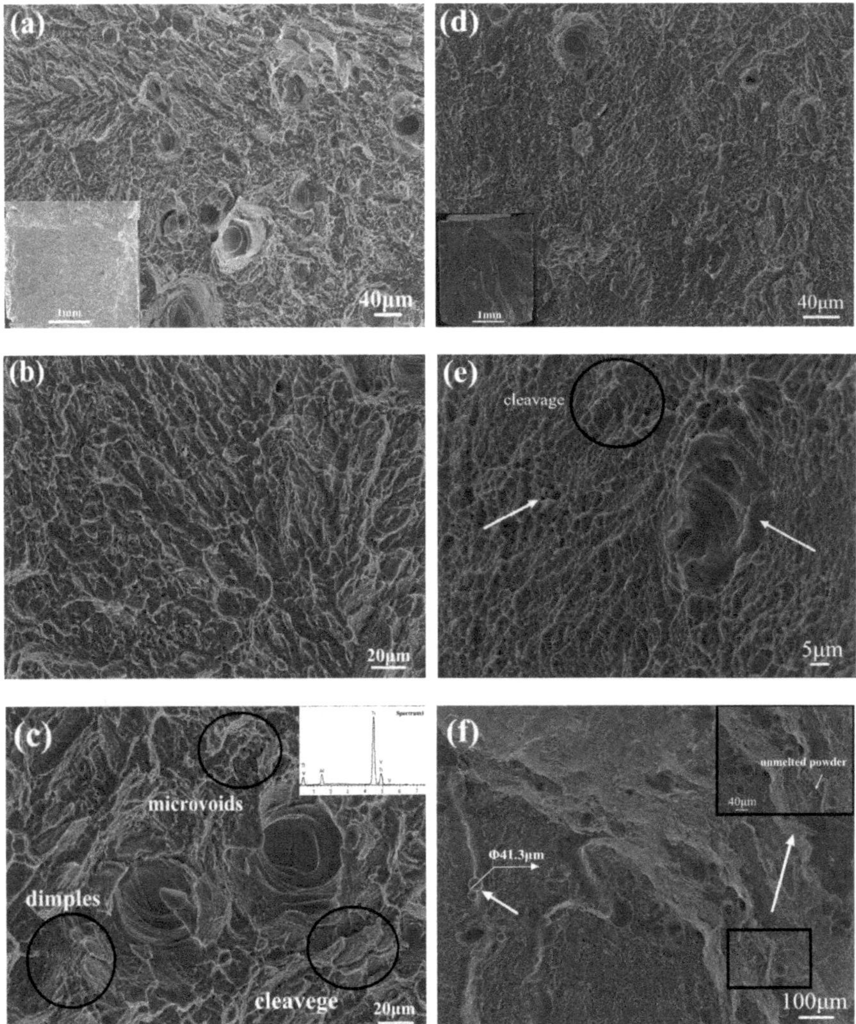

Figure 8. Ti6Al4V fracture surface of the tensile specimens (edge orientation): as-built parts (**a**–**c**); stress-relieved parts (**d**–**f**).

Moreover, a large number of fine grain structures were observed in stress-relieved samples that were marked in Figure 8e, which corroborated that the grains were refined after stress relieving. According to the Hall-Petch strengthening mechanism, smaller grain size will provide more grain boundaries, which can impede the movement of dislocation. Then, the hardness, YS and UTS improved. Nevertheless, the strength of the sample decreased simultaneously for a part of the unstable α' phase transformed into an $\alpha + \beta$ phase, the number of twins significantly decreased and the high dislocation density disappeared after an annealing treatment. Finally, the plasticity of the stress-relieved samples increased, and the YS and UTS were reduced. The result coincided considerably with Figure 7 and the tensile data in Table 4. The unmelted powder particles near the pores are marked in Figure 8f, indicating the IM phenomenon generated [38].

4. Conclusions

The microstructure and mechanical performance of both as-built and stress-relieved Ti6Al4V alloys by SLM technology were investigated in detail, and the main conclusions are as follows:

(1)   After stress relieving, the β phase was observed in XRD pattern, OM metallograph and TEM images, and the composition of α' phase reduced, indicating that the α' phase decomposed during annealing treatment.

(2)   The width of the coarse β phase may have a further relationship with the width of laser scan track.

(3)   In as-built samples, the high dislocation density, twinning and stacking faults were detected, and the typical ($10\bar{1}1$) twinning plane were marked where the clearly dislocation lines were discovered. In stress-relieved samples, a fully disordered α + β phase filled in the TEM images, and there were neither high density of dislocation nor stacking faults. Only fine dislocation lines and a little twinning remained.

(4)   α' phase resulted in an increase in tensile strength and hardness and a decrease in plasticity. The poor plasticity was ascribed to the print mode, remained support structures and large thermal stresses.

Author Contributions: Conceptualization, J.H., D.L. and Q.Q.; methodology, J.H., D.L., Y.Y., L.K. and W.J.; validation, J.H., D.L. and Y.Y.; formal analysis, J.H., D.L., Y.Y. and D.Q.; investigation, J.H., D.L. and Y.Y.; resources, D.L., W.J., L.K., Y.Y. and G.Q.; data curation, D.L., W.J. and L.K.; writing—original draft preparation, J.H.; writing—review and editing, J.H., D.L. and Q.Q.; visualization, J.H., D.L. and Y.Y.; project administration, D.L., W.J., L.K. and G.Q.; funding acquisition, D.L., W.J., L.K. and Q.Q. All authors discussed and participated in writing the manuscript. D.L. initiated and directed this research.

Funding: This research was funded by the National Natural Science Foundation of China (51562027, 11772145 and 51862026), Australian Research Council (DP160102491), the Advantage Technology Innovation Team of Jiangxi Province (20181BCB24007), the Technology Project of Department of Education of Jiangxi Province (GJJ170586), the Jiangsu Key Laboratory of Precision and Micro-Manufacturing Technology (JKL2015001), the Natural Science Foundation of Jiangxi Province (20171BAB216002) and the university science and technology project (KYZXJK123).

Acknowledgments: The authors acknowledge the Profeta Intelligent Technology Company in China and the Department of Materials Science and Engineering of Nanchang Hangkong Universities.

Conflicts of Interest: The authors declare no conflicts of interest.

References

1.   Yadroitsev, I.; Thivillon, L.; Bertrand, P.; Smurov, I. Strategy of manufacturing components with designed internal structure by selective laser melting of metallic powder. *Appl. Surf. Sci.* **2007**, *254*, 980–983. [CrossRef]

2.   Ikeshoji, T.T.; Nakamura, K.; Yonehara, M.; Imai, K.; Kyogoku, H. Selective laser melting of pure copper. *JOM* **2018**, *70*, 396–400. [CrossRef]

3.   Herzog, D.; Seyda, V.; Wycisk, E.; Emmelmann, C. Additive manufacturing of metals. *Acta Mater.* **2016**, *117*, 371–392. [CrossRef]

4.   Demir, A.G.; Previtali, B. Additive manufacturing of cardiovascular CoCr stents by selective laser melting. *Mater. Des.* **2017**, *119*, 338–350. [CrossRef]

5.   Beese, A.M.; Carroll, B.E. Review of mechanical properties of Ti-6Al-4V made by laser-based additive manufacturing using powder feedstock. *JOM* **2016**, *68*, 724–734. [CrossRef]

6.   Azizi, H.; Zurob, H.; Bose, B.; Ghiaasiaan, S.R.; Wang, X.; Coulson, S.; Duz, V.; Phillion, A.B. Additive manufacturing of a novel Ti-Al-V-Fe alloy using selective laser melting. *Addit. Manuf.* **2018**, *21*, 529–535. [CrossRef]

7.   Tan, S.; Li, D.; Liu, H.; Liao, X.; Jiang, L. Microstructure and mechanical properties of 80Ni20Cr alloy manufactured by laser 3D printing technology. *Chin. J. Nonferr. Met.* **2017**, *27*. (In Chinese) [CrossRef]

8. Dutta, B.; Froes, F.S. The additive manufacturing (AM) of titanium alloys. *Met. Powder Rep.* **2017**, *72*, 96–106. [CrossRef]

9. Sutton, A.T.; Kriewall, C.S.; Leu, M.C.; Newkirk, J.W. Powder characterisation techniques and effects of powder characteristics on part properties in powder-bed fusion processes. *Virtual Phys. Prototyp.* **2017**, *12*, 3–29. [CrossRef]

10. Wang, D.; Dou, W.; Yang, Y. Research on selective laser melting of Ti6Al4V: Surface morphologies, optimized processing zone, and ductility improvement mechanism. *Metals* **2018**, *8*, 471. [CrossRef]

11. Bobbio, L.D.; Qin, S.; Dunbar, A.; Michaleris, P.; Beese, A.M. Characterization of the strength of support structures used in powder bed fusion additive manufacturing of Ti-6Al-4V. *Addit. Manuf.* **2017**, *14*, 60–68. [CrossRef]

12. Buchanan, C.; Matilainen, V.P.; Salminen, A.; Gardner, L. Structural performance of additive manufactured metallic material and cross-sections. *J. Constr. Steel Res.* **2017**, *136*, 35–48. [CrossRef]

13. Ma, C.P.; Guan, Y.C.; Zhou, W. Laser polishing of additive manufactured Ti alloys. *Opt. Lasers Eng.* **2017**, *93*, 171–177. [CrossRef]

14. Alfaify, A.Y.; Hughes, J.; Ridgway, K. Critical evaluation of the pulsed selective laser melting process when fabricating Ti64 parts using a range of particle size distributions. *Addit. Manuf.* **2018**, *19*, 197–204. [CrossRef]

15. Gong, H.; Rafi, K.; Gu, H.; Starr, T.; Stucker, B. Analysis of defect generation in Ti-6Al-4V parts made using powder bed fusion additive manufacturing processes. *Addit. Manuf.* **2014**, *1*, 87–98. [CrossRef]

16. Khorasani, A.; Gibson, I.; Awan, U.S.; Ghaderi, A. The effect of SLM process parameters on density, hardness, tensile strength and surface quality of Ti-6Al-4V. *Addit. Manuf.* **2019**, *25*, 176–186. [CrossRef]

17. Kudzal, A.; McWilliams, B.; Hofmeister, C.; Kellogg, F.; Yu, J.; Taggart-Scarff, J.; Liang, J. Effect of scan pattern on the microstructure and mechanical properties of powder bed fusion additive manufactured 17-4 stainless steel. *Mater. Des.* **2017**, *133*, 205–215. [CrossRef]

18. Neikter, M.; Woracek, R.; Maimaitiyili, T.; Scheffzük, C.; Strobl, M.; Antti, M.L.; Åkerfeldt, P.; Pederson, R.; Bjerkén, C. Alpha texture variations in additive manufactured Ti-6Al-4V investigated with neutron diffraction. *Addit. Manuf.* **2018**, *23*, 225–234. [CrossRef]

19. Zhao, X.; Li, S.; Zhang, M.; Liu, Y.; Sercombe, T.B.; Wang, S.; Hao, Y.; Yang, R.; Murr, L.E. Comparison of the microstructures and mechanical properties of Ti-6Al-4V fabricated by selective laser melting and electron beam melting. *Mater. Des.* **2016**, *95*, 21–31. [CrossRef]

20. Simonelli, M.; Tse, Y.Y.; Tuck, C. Effect of the build orientation on the mechanical properties and fracture modes of SLM Ti-6Al-4V. *Mater. Sci. Eng. A* **2014**, *616*, 1–11. [CrossRef]

21. Krakhmalev, P.; Fredriksson, G.; Yadroitsava, I.; Kazantseva, N.; Du Plessis, A.; Yadroitsev, I. Deformation behavior and microstructure of Ti6Al4V manufactured by SLM. *Phys. Procedia* **2016**, *83*, 778–788. [CrossRef]

22. Sallica-Leva, E.; Caram, R.; Jardini, A.L.; Fogagnolo, J.B. Ductility improvement due to martensite α' decomposition in porous Ti-6Al-4V parts produced by selective laser melting for orthopedic implants. *J. Mech. Beha. Biomed. Mater.* **2016**, *54*, 149–158. [CrossRef] [PubMed]

23. Vrancken, B.; Thijs, L.; Kruth, J.P.; Van Humbeeck, J. Heat treatment of Ti6Al4V produced by Selective Laser Melting: Microstructure and mechanical properties. *J. Alloys Compd.* **2012**, *541*, 177–185. [CrossRef]

24. Simonelli, M.; Tse, Y.Y.; Tuck, C. On the texture formation of selective laser melted Ti-6Al-4V. *Metall. Mater. Trans. A* **2014**, *45*, 2863–2872. [CrossRef]

25. Zhao, Z.Y.; Li, L.; Bai, P.K.; Jin, Y.; Wu, L.Y.; Li, J.; Guan, R.G.; Qu, H.Q. The Heat treatment influence on the microstructure and hardness of TC4 titanium alloy manufactured via selective laser melting. *Materials* **2018**, *11*, 1318. [CrossRef] [PubMed]

26. Rafi, H.K.; Karthik, N.V.; Gong, H.; Starr, T.L.; Stucker, B.E. Microstructures and mechanical properties of Ti6Al4V parts fabricated by selective laser melting and electron beam melting. *J. Mater. Eng. Perform.* **2013**, *22*, 3872–3883. [CrossRef]

27. Xu, W.; Brandt, M.; Sun, S.; Elambasseril, J.; Liu, Q.; Latham, K.; Xia, K.; Qian, M. Additive manufacturing of strong and ductile Ti-6Al-4V by selective laser melting via in situ martensite decomposition. *Acta Mater.* **2015**, *85*, 74–84. [CrossRef]

28. Vrancken, B.; Cain, V.; Knutsen, R.; Van Humbeeck, J. Residual stress via the contour method in compact tension specimens produced via selective laser melting. *Scr. Mater.* **2014**, *87*, 29–32. [CrossRef]

29. Yadroitsava, I.; Grewar, S.; Hattingh, D.; Yadroitsev, I. Residual stress in SLM Ti6Al4V alloy specimens. *Mater. Sci. Forum* **2015**, *828*, 305–310. [CrossRef]

30. Tiferet, E.; Rivin, O.; Ganor, M.; Ettedgui, H.; Ozeri, O.; Caspi, E.N.; Yeheskel, O. Structural investigation of selective laser melting and electron beam melting of Ti-6Al-4V using neutron diffraction. *Addit. Manuf.* **2016**, *10*, 43–46. [CrossRef]

31. Simonelli, M.; Tse, Y.Y.; Tuck, C. Microstructure of Ti-6Al-4V produced by selective laser melting. *J. Phys. Conf. Ser.* **2012**, *371*, 012084. [CrossRef]

32. Liu, J.; To, A.C. Quantitative texture prediction of epitaxial columnar grains in additive manufacturing using selective laser melting. *Addit. Manuf.* **2017**, *16*, 58–64. [CrossRef]

33. Mantri, S.A.; Banerjee, R. Microstructure and micro-texture evolution of additively manufactured β-Ti alloys. *Addit. Manuf.* **2018**, *23*, 86–98. [CrossRef]

34. Yamanaka, K.; Saito, W.; Mori, M.; Matsumoto, H.; Chiba, A. Preparation of weak-textured commercially pure titanium by electron beam melting. *Addit. Manuf.* **2015**, *8*, 105–109. [CrossRef]

35. Liu, Z.C.; Ren, H.P.; Ji, Y.P. *The New Theory of Solid State Phase Transformation*; Science Press: Beijing, China, 2015. (In Chinese)

36. Britton, T.B.; Dunne, F.P.E.; Wilkinson, A.J. On the mechanistic basis of deformation at the microscale in hexagonal close-packed metals. *Proc. R. Soc. A Math. Phys. Eng. Sci.* **2015**, *471*, 20140881. [CrossRef]

37. Prashanth, K.G.; Damodaram, R.; Maity, T.; Wang, P.; Eckert, J. Friction welding of selective laser melted Ti6Al4V parts. *Mater. Sci. Eng. A* **2017**, *704*, 66–71. [CrossRef]

38. Hartunian, P.; Eshraghi, M. Effect of Build Orientation on the Microstructure and Mechanical Properties of Selective Laser-Melted Ti-6Al-4V Alloy. *J. Manuf. Mater. Process.* **2018**, *2*, 69. [CrossRef]

Article

The Heat Treatment Influence on the Microstructure and Hardness of TC4 Titanium Alloy Manufactured via Selective Laser Melting

Zhan-Yong Zhao [1,†], Liang Li [1,†], Pei-Kang Bai [1,*], Yang Jin [1], Li-Yun Wu [1], Jing Li [1], Ren-Guo Guan [2] and Hong-Qiao Qu [1]

[1]   School of Materials Science and Engineering, North University of China, Taiyuan 030051, China; syuzzy@126.com (Z.-Y.Z.); nucliliang@126.com (L.L.); nucjinyang@126.com (Y.J.); wuliyunnuc@126.com (L.-Y.W.); jing.li3d@hotmail.com (J.L.); nucqhq1993@126.com (H.-Q.Q.)

[2]   School of Materials Science and Engineering, Northwestern Polytechnical University, Xi'an 710072, China; guanrg@smm.neu.edu.cn

*   Correspondence: baipeikang@nuc.edu.cn

†   These authors contributed equally to this work.

Received: 21 June 2018; Accepted: 28 July 2018; Published: 30 July 2018

Abstract: In this research, the effect of several heat treatments on the microstructure and microhardness of TC4 (Ti6Al4V) titanium alloy processed by selective laser melting (SLM) is studied. The results showed that the original acicular martensite α'-phase in the TC4 alloy formed by SLM is converted into a lamellar mixture of $\alpha + \beta$ for heat treatment temperatures below the critical temperature (T_0 at approximately 893 °C). With the increase of heat treatment temperature, the size of the lamellar mixture structure inside of the TC4 part gradually grows. When the heat treatment temperature is above T_0, because the cooling rate is relatively steep, the β-phase recrystallization transforms into a compact secondary α-phase, and a basketweave structure can be found because the primary α-phase develop and connect or cross each other with different orientations. The residence time for TC4 SLM parts when the treatment temperature is below the critical temperature has little influence: both the α-phase and the β-phase will tend to coarsen but hinder each other, thereby limiting grain growth. The microhardness gradually decreases with increasing temperature when the TC4 SLM part is treated below the critical temperature. Conversely, the microhardness increases significantly with increasing temperature when the TC4 SLM part is treated above the critical temperature.

Keywords: selective laser melting; titanium alloy; heat treatment; microstructure; microhardness measurement

---

1. Introduction

Selective laser melting (SLM) is a layer-based deposition method using a laser to selectively melt successive layers of metal powder in an inert-gas-filled chamber [1,2]. SLM can quickly and accurately produce metallic components of any complex shape on the SLM equipment through 3D CAD data directly [3–5]. In the SLM, the 3D CAD model is imported into the SLM software system and afterwards sliced into layers with a certain thickness [3]. In the manufacturing process, the parameters such as the power of laser previously defined is applied and successive layers of powder is melts and forms a liquid pool by the interaction of a laser beam [6]. The molten pool will solidifies and cools down quickly [6–8]. Then, successive layers of powders are deposited, each one corresponding to a slice of the 3D CAD model [9]. Compared with traditional production technology, it provides a better alternative in certain aspects, such as a reduction of production steps, a high level of flexibility, high material use efficiency, and a near-net-shaped production for the manufacture of complex geometries [10–12]. SLM has been widely used in the fields of medical, military, aerospace, and automobile manufacturing [13].

Titanium alloys such as TC4 have been widely used in both the biomedical and aerospace industries, due to their high weight-to-strength ratio and superior mechanical properties, i.e., high strength, high corrosion resistance, and fracture toughness [14–16]. Extensive previous research has focused on titanium produced by SLM. For instance, Thijs et al. studied the influence of scanning parameters and scanning strategy on the microstructure of TC4 alloy. They found that the microstructure of TC4 alloy processed by SLM is the acicular martensitic phase due to the higher temperature gradient during the SLM process; furthermore, due to the occurrence of epitaxial growth, elongated grains emerge [17]. The direction of grain growth is mainly related to the local heat conduction determined by the scanning strategy; more specifically, the grain direction is parallel with the local conductive heat transfer [17]. Bin Zhou et al. prepared TC4 alloy through SLM under vacuum conditions. Their test results showed that this process could reduce the porosity of Ti6Al4V samples [18]. Investigating the effect of build orientation of TC4 manufactured by SLM, Agius et al. revealed a unique α' martensite microstructure and identified an asymmetric cyclic softening phenomenon [19]. Moreover, the residual stresses presented in the SLM coupons had a significant influence on the cyclic behavior of the material.

TC4 is a typical double-phase titanium alloy. The sharp cycles of steep heating and cooling rates in the SLM process resulted in typical TC4 part internal microstructures that consist of prior β columnar grains filled with acicular a' martensite [3,6,8,17]. Thus, the TC4 titanium alloy prepared by SLM usually displays high yield strength, but limited ductility and low fatigue resistance, therefore making it unsuitable for structural applications [20,21]. Heat treatment has an important effect on the properties of SLM-formed TC4 alloy. Vrancken et al. studied the effect of this process on the alloy's mechanical properties and microstructure. Their results shows that the original martensite α'-phase is converted to a lamellar mixture of α and β following heat treatment temperatures, but features of the original microstructure are maintained [22]. Yao et al. studied the influence of heat treatment on the microstructure evolution of as-built 3D laser-deposited TC4 alloy, and examined the mechanical behaviors in detail; they proved that the heat treatment could improve both the yield stress and the compressive strength of the 3D laser-deposited TC4 alloy [23]. Dai et al. reported the corresponding relationship between the heat treatment temperature and the corrosion resistance of SLM-formed TC4 titanium alloy, and found a negative effect [24], while Young-Kyun Kim et al. improved the creep resistance through heat treatment only [25]. Gerrit et al. studied the effects of extensive heat treatment temperatures on the phase transformation (phase fraction) and tensile properties of SLM-formed TC4 titanium alloys [26]. Furthermore, Wu et al. established the corresponding regular patterns of microstructure and hardness with heat treatment temperature [27]. Their results indicate that the heat treatment can effectively improve the tensile properties and hardness of SLM formed TC4 titanium alloy [26,27]. Therefore, heat treatment can effectively control the microstructure of TC4 alloy after SLM forming. It can eliminate the residual stress and improve its mechanical properties, which is important for improving the fatigue strength and service life of the alloy.

Owing to the $\beta \rightarrow \alpha'$ transformation at rapid cooling rates, TC4 under SLM processing typically results in extensive formation of brittle martensitic microstructures. The appropriate heat treatment process can release the residual stress in the SLM process; control the $\alpha - \beta$ phase transformation; adjust the shape, size and content of the phase; and optimize the microstructure and mechanical properties. Two temperature regions divided by the critical temperature (T_0) are used in this study: below the T_0 region (BTR) and above the T_0 region (ATR). The critical temperature is defined as the temperature above which the titanium alloy forms the α'-phase during rapid cooling (water quenching or air cooling). This temperature T_0 is calculated based on a thermodynamic database by Lu et al. [28] and Ji et al. [29], who identified T_0 as 893 °C and 872 °C, respectively. Therefore, based on previous research, the SLM-formed TC4 titanium alloy was post-treated by a wide range of heat treatment temperatures; the typical microstructure was obtained and the change of hardness was revealed. This study can thus provide recommendations for the preparation of high performance SLM TC4 alloy.

2. Materials and Methods

Cuboidal shape specimens with a side length of 10 mm and height of 5 mm were fabricated on the Ti6Al4V substrate by selective laser melting (SLM). The gas-atomized TC4 powder with spherical particles and a size distribution ranging from 10 to 50 μm was used in this paper, as shown in Figure 1. The composition of the powder is shown in Table 1.

Table 1. Composition of the TC4 alloy powder (wt %).

C	O	N	H	Al	V	Fe	Ti
≤0.03	≤0.1	≤0.01	≤0.002	6.0–6.75	3.5–4.5	≤0.20	Bal.

Figure 1. SEM image of TC4 powders.

All parts were made with Renishaw AM 400 system equipment (Renishaw Plc., Gloucestershire, Britain), which has a 400 W ytterbium fiber laser (Renishaw Plc., Gloucestershire, Britain) with a focused beam diameter of 70 μm. Samples were produced with a scanning speed (v) of 1200 mm/s, a laser power (P) of 400 W, 60 μm hatch spacing (h) (the distance between two adjacent scan vectors), and a 30 μm layer thickness (t). Layers were scanned using a pulsed laser mode according to a zigzag pattern, which was rotated 90° between each layer. This parameter set was determined to obtain fully dense, high-quality Ti6Al4V products.

Heat treatments were executed in a vertical tube furnace, in an argon-protected atmosphere and with a heating rate of approximately 10 °C /min. The parameters of the process were carried out as shown in Table 2. In the table, AC means air cooling and WQ means water quenching. The heat treatment system refers to the current typical wrought titanium alloy heat treatment process [22,26,27].

Table 2. Heat treatments of the TC4 sample produced by SLM.

Sample	Temperature Region	Temperature (°C)	Residence Time (h)	Cooling Rate *
1		As-built	-	-
2		550 °C	4 h	AC
3	BTR	750 °C	4 h	AC
4		850 °C	4 h	AC
5		850 °C	2 h	AC
6	ATR	900 °C	1 h	WQ
7		980 °C	1 h	WQ

* AC = Air cooling; WQ = Water quenching.

Examination of the microstructure occurred after grinding with SiC grinding paper up to a fine 3000-grit size, and polishing with SiO_2 suspension. To reveal the microstructure, samples were etched with a solution containing 50 mL of H_2O, 25 mL of HNO_3 and 5 mL of HF. Because of the anisotropic build process, two cross sections are considered. One is a side view parallel with the build direction, and the other is a top view, perpendicular to the build direction. Micrographs were taken using a conic xip-6A microscope. A Hitachi SEM SU5000 (Hitachi Ltd., Tokyo, Japan) was used for the examination of higher-resolution micrographs. The microhardness of TC4 alloy samples was tested using a DHV-1000Z-type micro-Vickers hardness tester (Shuangxu Ltd., Shanghai, China) with a load of 4.904 N (500 g) and a loading of 15 s (GB/T4340.1–1991). Phase identification of the SLM-formed TC4 alloy specimens before and after heat treatment was conducted with an X-ray diffractometer (XRD; Cu Ka) (Rigaku, Tokyo, Japan) from 30° to 85°, operated at 40 kV and 100 mA, with a step size of 0.02 and a dwelling time of 0.3 s per step.

3. Results and Discussion

3.1. Microstructure

Figure 2a,b show, respectively, a top and side view of non-heat-treated Ti6Al4V, produced by SLM. The microstructures of the top view and side view show completely different morphologies. The top view shows that equiaxed grains with an average diameter of 120 μm were developed in the SLM process. However, the macroscopic structures of the side view consist of columnar grains that grow epitaxially through multiple cladding layers with an average width of 120 μm, as shown in Figure 2b. According to the theory of constitutional supercooling by Chalmers [30], in order to use the alloy orientation to obtain planar crystals, following criteria that is supported to be satisfied,

$$\frac{G_L}{R} \geq \frac{m_0 c_0 (k_0 - 1)}{k_0 D_L} = \frac{\Delta T_0}{D_L} \tag{1}$$

In this equation, G_L is the liquid phase temperature gradient (K/mm) at the front of the solidification interface; R is the interface growth rate (mm/s); m_0 is the liquidus slope; C_0 is the alloy average composition; K_0 is the equilibrium solute distribution coefficient; D_L is the solute diffusion coefficient; ΔT_0 is the equilibrium crystallization temperature interval. The solid–liquid interface morphology in directional solidification is determined by G_L/R. When the liquid phase temperature gradient G_L of the front of the solidification interface is larger, the solid–liquid interface tends to be smoother. In this situation, the crystal grains advance in the opposite direction to the heat flow. Finally, the columnar crystals grow in a certain direction [31].

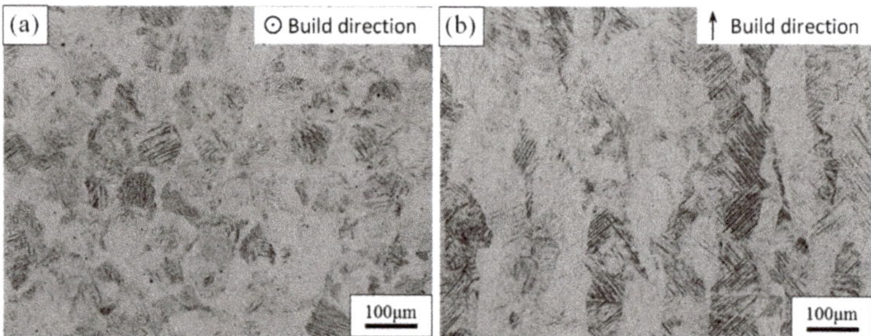

Figure 2. OM image of the top view (**a**) and the side (**b**) view of untreated TC4 produced by SLM.

The main axis of the columnar crystals is perpendicular to the deposition direction, and overlapping tracks in the as-built sample are not obvious. In the side view, the macroscopic structure presents the alternating light and dark growth, which is mainly due to the different crystal orientations. The main axis of the columnar crystal through the cladding layers is perpendicular to the scanning direction of the laser beam. This is mainly due to the temperature gradient of the molten pool being essentially perpendicular to the scanning direction during the SLM process, and the bottom of the molten pool is solidified first. The tops of the columnar crystals that have already solidified are re-melted when the laser beam scans the powder layers and the bottom of the unfused columnar crystal becomes the nucleus of the directionally solidified layer. Consequently, the epitaxial growth of the columnar crystal continues in accordance with the deposition direction [32].

The as-built sample's SEM microstructures of TC4 alloy manufactured by SLM are shown in Figure 3. Figure 3a displays the internal microstructure of the equiaxed grain in the top view observed by SEM. The interior of the equiaxed grains mainly consists of the acicular martensitic α'-phase. The columnar crystals in the side view largely comprise an extremely fine acicular martensite α'-phase with a certain orientation. The β-phase was not detected by XRD in the TC4 titanium alloy fabricated by SLM (Figure 4); moreover, Bey et al. also identified the absence of β-phase in as-built TC4 alloys [22].

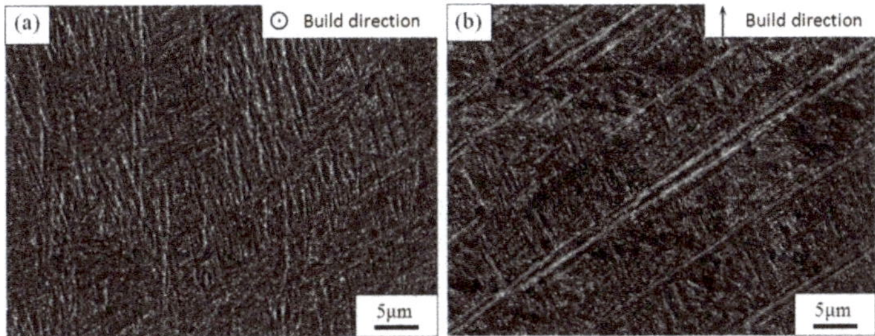

Figure 3. SEM images show the microstructure of untreated TC4 produced by SLM. (**a**) The microstructure of top view; (**b**) the microstructure of side view.

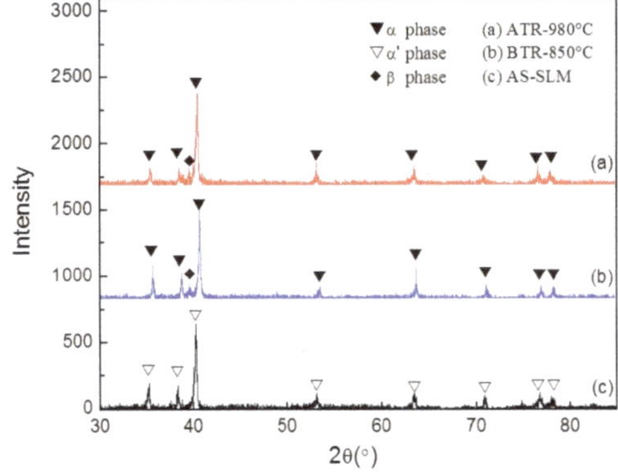

Figure 4. XRD analysis results of SLM formed TC4 before and after heat treatment.

The reason as to why the acicular martensitic α'-phase formed in the TC4 alloy manufactured by SLM is that the transient temperature of the TC4 alloy is extremely high under the action of a laser. This preferentially forms the β-phase, and then forms a non-diffusion shear α'-phase during subsequent rapid cooling. The cooling rate during SLM processing is greater than 410 Ks^{-1}, which has been suggested to be the correct rate for martensite α' formation [33]. The α'-phase is hexagonally close packed (HCP) and maintains a Burgers relationship with the β-phase of the body-centered cubic (BCC): (110) $\beta//(001)$ α; (111) $\beta//(1120)$ [22].

3.2. Influence of Temperature

3.2.1. Sample Annealed in the BTR

The morphology of the SLM material after heat treatment at different temperatures in the BTR is shown in Figure 5. The dark regions indicate the etched β-phase. The long, columnar grains remain visible in the side view of the material at different annealing temperatures, and the width of grain is still approximately 120 μm. Macroscopically, the sample after 550 °C heat treatment has almost no change; however, the sample treated in 750 °C is darker than the OM photograph before heat treatment. This is because a lamellar $\alpha + \beta$ mixture has been formed after treatment at 750 °C but not 550 °C, as can be clearly seen in Figure 6. This observation corresponds to the α-dissolution temperature T_{diss}, defined by Kelly as 705 °C [34], at which the α-phase starts to dissolute into the β-phase under equilibrium heating conditions. After annealing in the BTR, lower hardness but better plasticity are obtained, due to the lamellar $\alpha + \beta$ mixture [35].

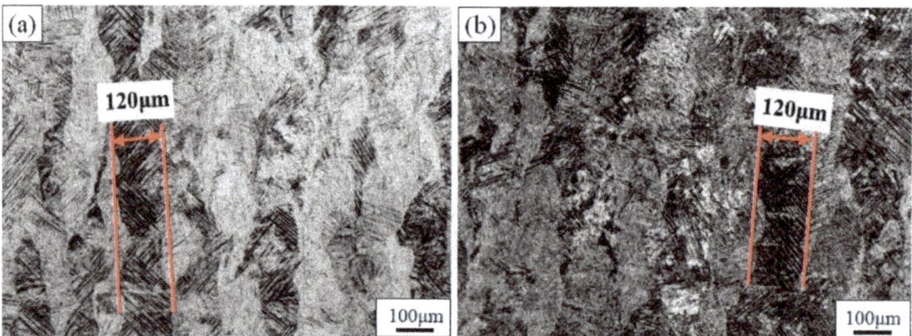

Figure 5. OM images show the etched morphology of TC4 SLM part after heat treating at different temperatures for 4 h, followed by AC. (**a**) 550 °C and (**b**) 750 °C below the T_0.

Microstructures of the TC4 SLM part after heat treatment at different temperatures are shown in Figure 6. As shown in Figure 6a,b, when the residence time is 4 h and the annealing temperature is increased from 550 to 750 °C, the width of the acicular α-phase increases from 0.8 to 2.3 μm. After 4 h at 550 °C, the microstructure does not change significantly compared to the TC4 SLM part. This is because the rapid cooling of the as-built sample during the SLM forming process results in twinning dislocation inside of the alloy, which could be a reason for the poor ductility of the as-received alloy during tension [27]. These defects have somewhat of a hindering effect on the decomposition of the α'-phase. In addition, at 550 °C annealing, the phase decomposition driving-force is insufficient due to the low temperature. In BTR heat treatments, the growth of β-phase at α' grain boundaries, as well as at internal twin boundaries started at T_{diss} and below T_0 [35]. After 4 h at 750 °C, in contrast, the fine martensitic structure has been transformed into a mixture of $\alpha + \beta$, in which the α-phase is present as fine needles (Figure 6b). This is because of the greater phase decomposition driving force at high temperatures, which promotes the decomposition of α'-phase. The width of α-phase in the alloy increases to 2.3 μm, as a result of aggregation and growth during the retaining at 750 °C.

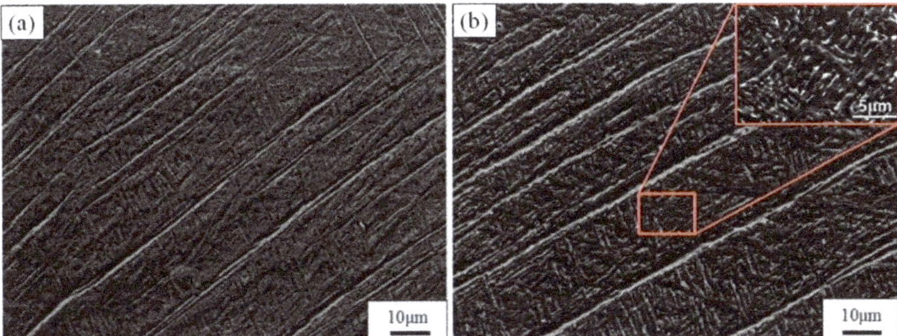

Figure 6. SEM images show the microstructure of TC4 SLM part after heat treating at different temperatures for 4 h, followed by AC. (**a**) 550 °C and (**b**) 750 °C below the T_0.

3.2.2. Sample Annealed in the ATR

Microstructures of the TC4 SLM part after heat treatment at different temperatures in the ATR are shown in Figure 7, in which the light plates indicate primary α-phase. Figure 7a,c show the side view of the SLM material after heat treatment at 900 and 980 °C, respectively. It can be seen that the length of the prior β grains seems unchanged but the width has increased and is now roughly 250 μm wide after 1 h at both 900 and 980 °C. This is because the higher temperature provides more driving force for the α-phase to decompose into the β-phase during the high temperature holding, and the adjacent β grow and connect with each other. The XRD analysis results still show an α + β microstructure, whereas a large amount of the transformed β structure was expected. This suggests that the β-transus of the SLM material is higher than 980 °C.

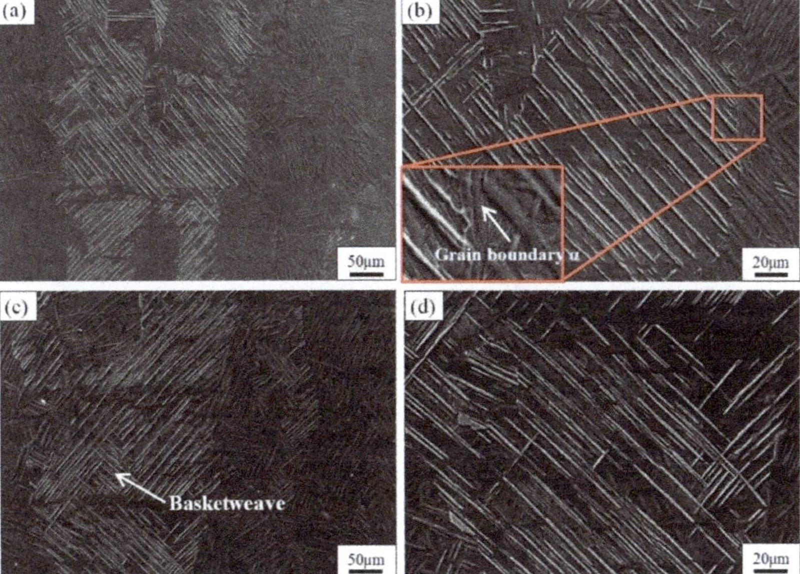

Figure 7. SEM images show the microstructure of TC4 SLM part after heat treatment at different temperatures of recrystallization annealing for 1 h, followed by WQ. The side view of (**a**) 900 °C and (**c**) 980 °C above the T_0; (**b**,**d**) show an enlarged view of the columnar grains of (**a**,**c**), respectively.

As shown in Figure 7b,d, when the temperature is increased from 900 to 980 °C, the plates primary α-phase width inside each of the columnar grains is both about 2.5 μm, because the existence of the β-phase hinders the growth of the primary α-phase [20], which is derived from the martensite α′-phase and surrounded by β-phase. The length of the primary α-phase at 900 °C can reach around 300 μm and at 980 °C it is mostly 200 μm. However, after 1 h at 980 °C (Figure 7d), the number of primary α-phases increases, and some small segments of the primary α-phases with a length of 20–100 μm are also formed between the α-phases. The cause of the above phenomenon is that the driving-force of α′-phase decomposition is stronger during the holding process at 980 °C. During the holding process, the higher the temperature, the smaller the volume fraction of α-phase and the larger the volume fraction of β-phase [26]; then, during the water quenching process, the β-phase precipitates and transformed to α′-phases on the dislocations and along the substructural boundaries of the primary α [36], so that they are separated into small segments. Thus, the long plates shown in Figure 8 are composed of primary α-phase, and spaces between them are filled by α- and α′-phases. This can be seen clearly in Figure 8. When the segments develop with a certain orientation, a fine basketweave structure finally emerges [37]. The mechanical properties of the basket comprise high-temperature properties such as creep behavior deformation and stress rupture properties, better fracture properties and resistance to crack growth, but poor plasticity and thermal stability [19]; therefore, microhardness is significantly increased. As shown in the enlarged view of the red box in the lower left corner of Figure 7b, it can be seen that the α-phase grain boundary is formed outside of the columnar grains after the 900 °C treatment. It is the hindrance of the α-grain boundaries that causes the columnar grains to be connected with each other without limitation. Figure 8 is an enlarged image of the SEM, showing the microstructure of TC4 produced by SLM at different annealing temperatures in the ATR. It can be seen that the secondary α-phase and the coarse primary α-phase are formed after the TC4 SLM part is treated at both 900 and 980 °C. Through a comparison of Figure 8a,b, it is clear that an increased β-phase, and a basketweave microstructure is formed in the β columnar grains of the sample after heat treatment at 980 °C; this is advantageous for increasing the sample's microhardness, but also lowers its plasticity [27].

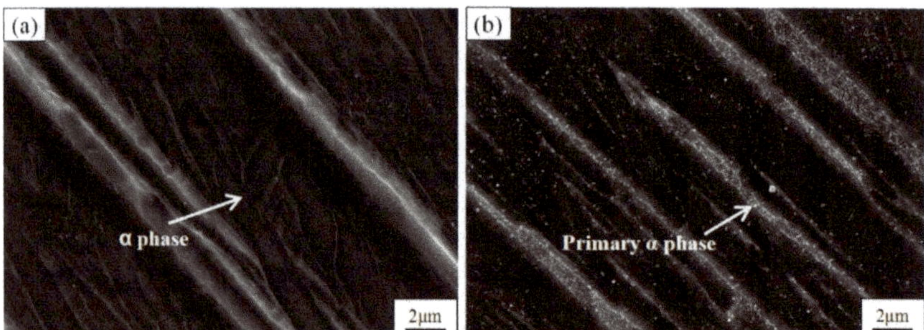

Figure 8. SEM images show the microstructure of TC4 SLM part after heat treating at different temperatures of recrystallization annealing. (**a**) 900 °C/1 h/WQ; (**b**) 980 °C/1 h/WQ.

3.3. Influence of the Residence Time

Figure 9a,b shows, respectively, the OM image of TC4 produced by SLM at 850 °C for 2 h and 850 °C for 4 h, followed by air cooling. When the temperature rises to 850 °C, the β-phase fraction is increased compared with the OM image at 750 °C [26]. With the increase of the residence time from 2 h to 4 h, the β-phase fraction is almost unchanged. This is because the heat treatment temperature is between T_{diss} and T_0, and the coexistence of the α- and β-phases causes them to hinder each other. If the heat treatment temperature is above the β-transus, growth of the β-phase will be without hindrance and rapid [17].

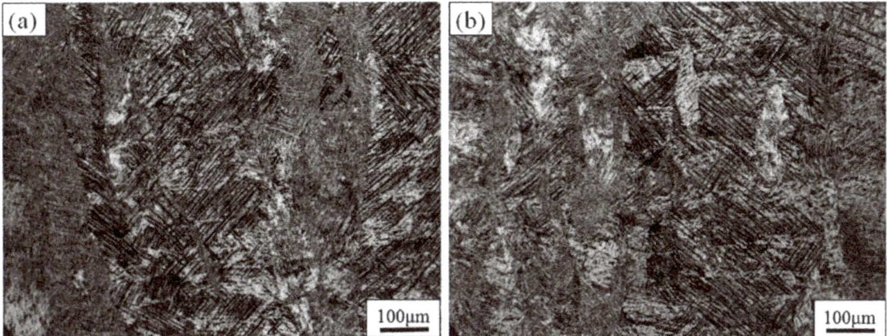

Figure 9. OM images show the etched morphology of TC4 SLM part after heat treating at (**a**) 2 h and (**b**) 4 h under 850 °C, followed by air cooling.

Figure 10a,b show respectively the SEM microstructure of TC4 produced by SLM at 850 °C for 2h and at 850 °C for 4 h, followed by air cooling. The microstructures of the alloys comprise a coarse mixture ($\alpha + \beta$), and the widths of the plate α-phases obtained after 2 h and 4 h of holding at these temperatures are approximately 2.8 and 3.1 μm, respectively. Therefore, the effect of residence time on the growth of the α-phase is not significant, but it is significant compared to the α-phase of TC4 produced by SLM at 750 °C for 4 h, followed by air cooling. The dispersive distribution of β-phase obtained through decomposition of the α'-phase and the α-phase and the β-phase hinder each other; thus, the effect of residence time on the size of grains and precipitated phases is small. Huang et al. similarly found that the α'-phase of TC4 produced by SLM was completely decomposed to the α-phase when held at 800 °C for 2 h and the α-phase will grow by a certain degree when the holding is continued [38]. The microhardness of the sample obtained from different residence times is almost the same due to no significant changes in the microstructure during different heat treatment holding times.

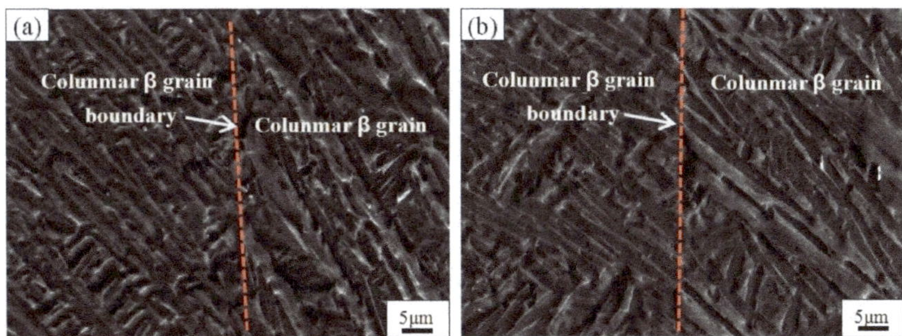

Figure 10. SEM images show the microstructure of TC4 SLM part after heat treatment at different residence time. (**a**) 850 °C/2 h/AC; (**b**) 850 °C/4 h/AC.

3.4. Microhardness

Figure 11 shows the microhardness curves of the TC4 alloy produced by SLM at different heat treatment temperatures. In terms of the annealing temperature in the BTR (left of the dashed line), the microhardness of the TC4 alloy gradually decreases with increasing temperature, but the slope becomes less steep. For the annealing temperature in the ATR (right side of the dashed line), as the heat treatment temperature increases, there is a significant increase in microhardness. The higher cooling rate during the water quenched process after treatment at the ATR temperature causes the hardness to be improved. The microhardness of the as-built sample is 381.3 Hv, which is lower than Point 2, as did work by Wu et al. [27]. In addition, the hardness of samples treated at 850 °C for 2 h with air cooling and for 4 h at the same condition is almost equal, indicating that the effect of holding time on hardness is not significant.

It is evident that the change of hardness with post-heat-treatment temperature is strictly related to the microstructural change of the as-built TC4 specimens. At the BTR temperature, as the temperature exceed T_{diss}, the relatively soft β-phase fraction increases as the temperature increases; this is because the original martensites decompose more thoroughly, thus decreasing the hardness. Because of the insufficient phase decomposition driving-force, α-martensite on twin boundaries and dislocations inside its plates transformed into α-phases partially and formed a refinement substructure, resulting in a high microhardness at Point 2. At Point 4, complete decomposition of $α'$-martensite to an α + β mixture caused the turning point of the microhardness curve. In the ATR heat treatment stage, a martensitic transition will recur in the high-temperature stable β-interlayers during water quenching, and in the obtained basketweave microstructure that also plays a role in strengthening hardness.

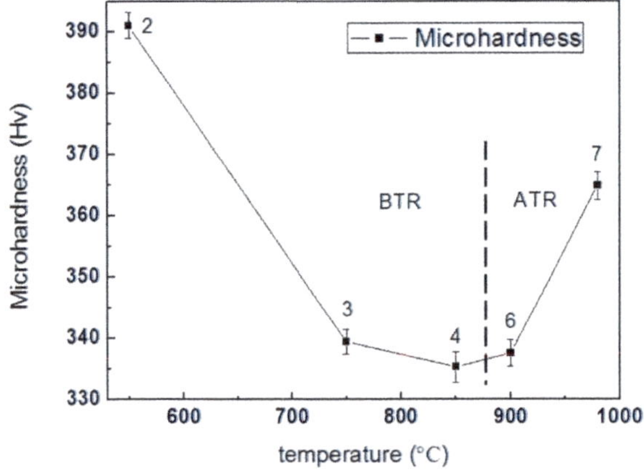

Figure 11. The microhardness curves of TC4 SLM part after heat treating at different temperatures. The number in the curve is equivalent to the sample number in Table 2.

4. Conclusions

1. The side view of non-heat-treated TC4, produced by SLM, reveals long columnar grains which grow through multiple cladding layers and are oriented in the building direction. The extremely high cooling rate during the SLM process leads to the formation of a fine acicular martensite $α'$-phase with a certain orientation inside of the columnar grains.

2. Following heat treatment in the BTR, the internal acicular martensite $α'$-phase of the SLM TC4 part is converted into the α-phase and forms a lamellar α + β mixture, which gradually increases in structure size with increasing temperature. After being annealed in the ATR, the β-phase that formed in the holding process is transformed to martensite α during the cooling process and a fine basketweave structure emerges, which improves the microhardness of the alloy. At the BTR annealing temperature, the holding time has little effect on the microstructure and properties.

3. At the BTR annealing temperature, the microhardness of SLM-formed TC4 alloy gradually decreases with increasing temperature. However, the microhardness increases significantly with increasing temperature when the TC4 SLM part is treated under ATR annealing.

Author Contributions: Z.-Y.Z. designed the original research together with L.L. prepared figures and wrote the manuscript. Z.-Y.Z. and L.L. contributed equally to this work and should be considered co-first authors. Y.J., J.L., and L.-Y.W. performed the experiments and characterization of the alloys. H.-Q.Q. responsible for the written reviewed and edited. R.-G.G. and P.-K.B. collaborated in this study in terms of discussion and useful suggestions. All authors reviewed the manuscript.

Funding: This research was funded by the National Natural Science Foundation of China (Grant No. 51604246 and Grant No. 51775521) and the North University of China for Young Academic Leaders.

Conflicts of Interest: The authors declare no conflict of interest.

References

1. Herzog, D.; Seyda, V.; Wycisk, E.; Emmelmann, C. Additive manufacturing of metals. *Acta Mater.* **2016**, *117*, 371–392. [CrossRef]
2. Prashanth, K.G.; Scudino, S.; Ecker, J. Defining the tensile properties of Al-12Si parts produced by selective laser melting. *Acta Mater.* **2017**, *126*, 25–35. [CrossRef]
3. Li, Z.H.; Xu, R.J.; Zhang, Z.W.; Kucukkoc, I. The influence of scan length on fabricating thin-walled components in selective laser melting. *Int. J. Mach. Tool Manuf.* **2018**, *126*, 1–12. [CrossRef]
4. Maity, T.; Chawake, N.; Kim, J.T.; Ecker, J.; Prashanth, K.G. Anisotropy in local microstructure—Does it affect the tensile properties of the SLM samples? *Manuf. Lett.* **2018**, *15*, 33–37. [CrossRef]
5. Prashanth, K.G.; Scudino, S.; Maity, T.; Das, J.; Eckert, J. Is the energy density a reliable parameter for materials synthesis by selective laser melting? *Mater. Res. Lett.* **2017**, *5*, 386–390. [CrossRef]
6. Kumar, P.; Prakash, O.; Ramamurty, U. Micro-and meso-structures and their influence on mechanical properties of selectively laser melted Ti-6Al-4V. *Acta Mater.* **2018**, *154*, 246–260. [CrossRef]
7. Zhao, Z.Y.; Guan, R.G.; Zhang, J.H.; Zhao, Z.Y.; Bai, P.K. Effects of process parameters of semisolid stirring on microstructure of Mg-3Sn-1Mn-3SiC (wt%) strip processed by rheo-rolling. *Acta Metall. Sin.* **2017**, *30*, 66–72. [CrossRef]
8. Günther, J.; Leudersb, L.; Koppa, P.; Trösterc, T.; Henkel, S.; Biermann, H.; Niendorf, T. On the effect of internal channels and surface roughness on the high-cycle fatigue performance of Ti-6Al-4V processed by SLM. *Mater. Des.* **2018**, *143*, 1–11. [CrossRef]
9. Prashanth, K.G.; Scudino, S.; Klauss, H.J.; Surreddi, K.B.; Löber, L.; Wang, Z.; Chaubey, A.K.; Kühn, U.; Eckert, J. Microstructure and mechanical properties of Al–12Si produced by selective laser melting: Effect of heat treatment. *Mater. Sci. Eng. A* **2014**, *590*, 153–160. [CrossRef]
10. Prashanth, K.G.; Eckert, J. Formation of metastable cellular microstructures in selective laser melted alloys. *J. Alloy Compd.* **2017**, *707*, 27–34. [CrossRef]
11. Kiel-Jamrozik, M.; Szewczenko, J.; Basiaga, M.; Nowinska, K. Technological capabilities of surface layers formation on implants made of Ti-6Al-4V ELI alloy. *Acta Bioeng. Biomech.* **2015**, *17*, 31–37. [PubMed]
12. Kadirgama, K.; Harun, W.S.W.; Tarlochan, F.; Samykano, M.; Ramasamy, D.; Azir, M.Z.; Mehboob, H. Statistical and optimize of lattice structures with selective laser melting (SLM) of Ti6Al4V material. *Int. J. Adv. Manuf. Technol.* **2018**, *97*, 495–510. [CrossRef]
13. Cain, V.; Thijs, L.; Humbeeck, J.V.; Hooreweder, B.V.; Knutsen, R. Crack propagation and fracture toughness of Ti6Al4V alloy produced by selective laser melting. *Addit. Manuf.* **2015**, *5*, 68–76. [CrossRef]
14. Kiel-Jamrozik, M.; Jamrozik, W.; Witkowska, I. The heat treatment influence on the structure and mechanical properties of Ti6Al4V alloy manufactured by SLM technology. *Innov. Biomed. Eng.* **2018**, *623*, 319–327.
15. Jyun-Rong, Z.; Yee-Ting, L.; Wen-Hsin, H.; An-Shik, Y. Determination of melt pool dimensions using DOE-FEM and RSM with process window during SLM of Ti6Al4V powder. *Opt. Laser Technol.* **2018**, *103*, 59–76.
16. Wauthle, R.; Vrancken, B.; Beynaerts, B.; Jorissen, K.; Schrooten, J.; Krutha, J.P.; Humbeeck, J.V. Effects of build orientation and heat treatment on the microstructure andmechanical properties of selective laser melted Ti6Al4V lattice structures. *Addit. Manuf.* **2015**, *5*, 77–84. [CrossRef]
17. Thijs, L.; Verhaeghe, F.; Craeghs, T.; Humbeeck, J.V.; Kruth, J. A study of the microstructural evolution during selective laser melting of Ti-6Al 6Al-4V. *Acta Mater.* **2010**, *58*, 3303–3312. [CrossRef]
18. Zhou, B.; Zhou, J.; Lia, H.X.; Lin, F. A study of the microstructures and mechanical properties of Ti6Al4V fabricated by SLM under vacuum. *Mater. Sci. Eng. A* **2018**, *724*, 1–10. [CrossRef]

19. Agius, D.; Kourousis, K.I.; Wallbrink, C.; Song, T.T. Cyclic plasticity and microstructure of as-built SLM Ti-6Al-4V: The effect of build orientation. *Mater. Sci. Eng. A* **2017**, *701*, 85–100. [CrossRef]

20. Pere, B.; Gussone, J.; Haubrich, J.; Sandlöbes, S.; Silva, J.C.D.; Cloetens, P.; Schell, N.; Requena, G. Inducing stable $\alpha + \beta$ microstructures during selective laser melting of Ti-6Al-4V using intensified intrinsic heat treatments. *Materials* **2017**, *10*, 268.

21. Wu, X.; Brandt, M.; Sun, S.; Elambasseril, J.; Liu, Q.; Latham, K.; Xia, K.; Qian, M. Additive manufacturing of strong and ductile Ti-6Al-4V by selective laser melting via in situ martensite decomposition. *Acta Mater.* **2015**, *85*, 74–84.

22. Vrancken, B.; Thijs, L.; Kruth, J.P.; Humbeeck, J.V. Heat treatment of Ti6Al4V produced by Selective Laser Melting: Microstructure and mechanical properties. *J. Alloy Compd.* **2012**, *541*, 177–185. [CrossRef]

23. Yao, J.; Suo, T.; Zhang, S.Y.; Zhao, F.; Wang, H.T.; Liu, J.B.; Chen, Y.Z.; Li, Y.L. Influence of heat-treatment on the dynamic behavior of 3D laser-deposited Ti-6Al-4V alloy. *Mater. Sci. Eng. A* **2016**, *677*, 153–162. [CrossRef]

24. Dai, N.; Zhang, J.; Chen, Y.; Zhang, L.-C. Heat treatment degrading the corrosion resistance of selective laser melted Ti-6Al-4V alloy. *J. Electrochem. Soc.* **2017**, *164*, C428–C434.

25. Kim, Y.-K.; Park, S.-H.; Yu, J.-H.; AlMangour, B.; Lee, K.-A. Improvement in the high-temperature creep properties via heat treatment of Ti-6Al-4V alloy manufactured by selective laser melting. *Mater. Sci. Eng. A* **2018**, *715*, 33–40. [CrossRef]

26. Gerrit, M.; Ter, H.; Thorsten, H.B. Selective Laser Melting Produced Ti-6Al-4V: Post-Process Heat Treatments to Achieve Superior Tensile Properties. *Materials* **2018**, *11*, 146. [CrossRef] [PubMed]

27. Wu, S.Q.; Lu, Y.J.; Gan, Y.L.; Huang, T.T.; Zhao, C.Q.; Lin, J.J.; Guo, S.; Lin, J.X. Microstructural evolution and microhardness of a selective-laser-melted Ti-6Al-4V alloy after post heat treatments. *J. Alloy Compd.* **2016**, *672*, 643–652. [CrossRef]

28. Lu, S.L.; Qian, M.; Tang, H.P.; Yan, M.; Wang, J.; StJohn, D.H. Massive transformation in Ti-6Al-4V additively manufactured by selective electron beam melting. *Acta Mater.* **2016**, *104*, 303–311. [CrossRef]

29. Ji, Y.; Heo, T.W.; Zhang, F.; Chen, L.Q. Theoretical Assessment on the Phase Transformation Kinetic Pathways of Multi-component Ti Alloys: Application to Ti-6Al-4V. *J. Phase Equilib. Diffus.* **2016**, *37*, 53–64. [CrossRef]

30. Friedrich, J.; Stockmeier, L.; Müller, G. Constitutional Supercooling in Czochralski Growth of Heavily Doped Silicon Crystals. *Acta Phys. Pol. A* **2013**, *124*, 219–226. [CrossRef]

31. Baufeld, B.; Biest, O.V.; Gault, R. Additive manufacturing of Ti-6Al-4V components by shaped metal deposition: Microstructure and mechanical properties. *Mater. Des.* **2010**, *36*, 106–111. [CrossRef]

32. Mok, S.H.; Bi, G.J.; Folkes, J.; Pashby, I. Deposition of Ti-6Al-4V using a high power diode laser and wire, Part I: Investigation on the process characteristics. *Surf. Coat. Technol.* **2008**, *202*, 3933–3939. [CrossRef]

33. Ahmed, T.; Rack, H.J. Phase transformations during cooling in $\alpha + \beta$ titanium alloys. *Mater. Sci. Eng. A* **1998**, *243*, 206–211. [CrossRef]

34. ASTM International. *Standard Specification for Titanium and Titanium-6 Aluminum-4 Vanadium Alloy Powders for Coatings of Surgical Implants*; ASTM F1580–12; ASTM International: West Conshohocken, PA, USA, 2012.

35. Brandl, E.; Schoberth, A.; Leyens, C. Morphology, microstructure, and hardness of titanium (Ti-6Al-4V) blocks deposited by wire-feed additive layer manufacturing (ALM). *Mater. Sci. Eng. A* **2012**, *532*, 295–307. [CrossRef]

36. Katzarov, I.; Malinov, S.; Sha, W. Finite Element Modeling of the Morphology of β to α Phase Transformation in Ti-6Al-4V Alloy. *Metall. Mater. Trans. A* **2002**, *33*, 1027–1040. [CrossRef]

37. Edwards, P.; Ramulu, M. Fatigue performance evaluation of selective laser melted Ti-6Al-4V. *Mater. Sci. Eng. A* **2014**, *598*, 327–337. [CrossRef]

38. Huang, Q.; Liu, X.; Yang, X.; Zhang, R.; Shen, Z.; Feng, Q. Specific heat treatment of selective laser melted Ti-6Al-4V for biomedical applications. *Front. Mater. Sci.* **2015**, *9*, 373–381. [CrossRef]

 *materials* 

Article

The Effect of a Scanning Strategy on the Residual Stress of 316L Steel Parts Fabricated by Selective Laser Melting (SLM)

Di Wang [1,2], Shibiao Wu [1], Yongqiang Yang [1,*], Wenhao Dou [1], Shishi Deng [1], Zhi Wang [1,*] and Sheng Li [2]

[1] Department of Mechatronics Engineering, School of Mechanical and Automotive Engineering, South China University of Technology, Guangzhou 510640, China; mewdlaser@scut.edu.cn (D.W.); siberghost@126.com (S.W.); dwh68@outlook.com (W.D.); shishidengscut@163.com (S.D.)

[2] School of Metallurgy and Materials, The University of Birmingham, Birmingham B152TT, UK; s.li.2@bham.ac.uk

* Correspondence: meyqyang@scut.edu.cn (Y.Y.); wangzhi@scut.edu.cn (Z.W.); Tel./Fax: +86-020-8711-1036 (Y.Y.); Tel./Fax: +86-020-8711-4484 (Z.W.)

Received: 28 August 2018; Accepted: 19 September 2018; Published: 25 September 2018

Abstract: The laser scanning strategy has an important influence on the surface quality, residual stress, and deformation of the molten metal (deformation behavior). A divisional scanning strategy is an effective means used to reduce the internal stress of the selective laser melting (SLM) metal part. In order to understand and optimize the divisional scanning strategy, three divisional scanning strategies and an S-shaped orthogonal scanning strategy are used to produce 316L steel parts in this study. The influence of scanning strategy on the produced parts is verified from the aspects of densification, residual stress distribution and deformation. Experiments show that the 316L steel alloy parts adopted spiral divisional scanning strategy can not only obtain the densification of 99.37%, but they also effectively improve the distribution of residual stress and control the deformation degree of the produced parts. Among them, the spiral divisional scanning sample has the smallest residual stress in plane direction, and its σ_x and σ_y stress are controlled within 204 MPa and 103 MPa. The above results show that the spiral divisional scanning is the most conducive strategy to obtain higher residual stress performance of SLM 316L steel parts.

Keywords: selective laser melting; divisional scanning; residual stress; deformation

---

1. Introduction

Selective laser melting technology adopts single-point high-energy laser beam to melt metal powder layer by layer along the filling path of three-dimensional discrete profile. Theoretically, it is not constrained by the structure of the part and can fabricated any complicated structural part with the densification of nearly 100% [1–4]. In fact, many factors affect the performance of selective laser melting (SLM) parts, not only by the process parameters including laser power, scanning speed, scanning space, layer thickness, and spot size et al. [5–8], but also by the laser scanning strategy and the length of the scanning line [9].

The strategy planning software automatically generates the scanning lines according to the selected filling algorithm, based on the discrete slice data of the three-dimensional model of the produced part. The choice of scanning strategy is very important to the quality improvement of SLM parts [10]. The scanning strategy has a great influence on the heat distribution of the part and plays a decisive role in the residual stress distribution and the deformation trend of the part [11]. In a microscopic view, the scanning strategy affects the grain growth direction and grain size, and it determines the microstructure of the part [12]. The preferred scanning strategy usually controls the length of the scanning lines. Excessive scanning lines can not only decrease the stability of the melt track, but they also reduce the surface quality of the parts. The residual stress can also accumulate in a single direction, and affect the residual stress distribution.

Many scholars conducted a series of studies on the influence of scanning strategy on the performance of SLM parts. These scholars initially studied some simple scanning strategies and adopted finite element simulation methods to study the impact on the performance of parts. Kruth et al. [13] adopted a sub-block scanning method by using a rectangular block of equal length and width and found that the previous scanning region could preheat the adjacent region well and decrease the temperature gradient. Kruth et al. [14] also designed a "bridge curvature method" to study the influence of the scanning line's length, scanning line direction and island scanning strategy on residual stress. It is found that the short scanning line and the optimal orientation of scanning line can effectively reduce the residual stress. Parry et al. [15] simulated the temperature field and the stress field in Z-shaped and S-shaped scanning mode by finite element method and studied the effect of scanning line length and scanning area on temperature field and stress field. It is suggested that too long scanning lines should be avoided during SLM process. With the development of SLM technology, many researchers began to carry out more detailed research on scanning strategies. Lu et al. [16] studied the influence of the size of island shape on the densification, microstructure, and mechanical properties and residual stress of the part, and the optimum width of the region was found to be within 5 mm to 7 mm. Thijs et al. [17] studied the microstructure evolution of SLM parts based on three scanning strategies: zigzag scanning, unidirectional scanning, and orthogonal scanning. Qian et al. [18] studied the residual stress and deformation of SLM parts in helix scanning strategy. Beal et al. [19] used different scanning strategies to statistically analyze the effects with respect to the composition of Cu, and found that the refill strategy produced better results compared to the other scanning strategies. As for the divisional scanning method, Yasa et al. [20] also explored the influence of divisional scanning strategy on densification, surface quality, mechanical properties, and residual stresses formed during SLM, and it was concluded that divisional scanning had some advantages such as lower residual stresses and better surface quality. Rashid et al. [21] also discussed the effect of scan strategy on densification and metallurgical properties of 17-4PH parts fabricated by SLM, and he found that the samples fabricated with double scanning strategy showed an improvement in densification, as compared to that fabricated with a single scanning strategy. Geiger et al. [22] tailored the texture of IN738LC as processed by SLM with specific scanning strategies, and they found that the applied laser scanning strategies allowed the crystallographic texture to be tailored locally.

Many scholars above have studied the influence of the scanning strategy on the quality of SLM parts based on Z-shaped scan, S-shaped scan and divisional scan, and put forward several kinds of divisional scanning methods according to specific problems. However, the effect of different divisional scanning methods on the performance of SLM parts is still not specific. Currently, the mainstream SLM equipment vendors such as Concept Laser, SLM Solutions, and EOS have adopted the divisional scanning strategy on their commercial SLM machines. However, the divisional scanning strategies adopted are all different with each other, and further research on the different divisional scanning strategy is still scarce. Therefore, on the basis of the aforementioned researchers, three kinds of divisional scanning strategies are developed to study the influence on the densification, surface quality, residual stress distribution and deformation of SLM parts in this paper.

2. Research Methods

2.1. Experimental Equipment and Materials

The experiment was carried out using Dimetal-100 SLM equipment, which was developed independently by South China University of Technology, Guangzhou, China, as shown in Figure 1. The equipment is mainly composed of a fiber laser, optical transmission regulation system, gas circulation system, fabricating room, laying powder device, cooling system, and the main control software. The laser was guided by a high-precision optical path transmission system, and it was finally focused by the f-θ lens (f is the focal length) to ensure that the power density of the spot was basically the same throughout the working plane. The maximum fabricating size of the equipment is 100 mm × 100 mm × 100 mm, the scanning speed is 10–7000 mm/s, the focused spot diameter is 70 μm, and the precision can reach ± 0.1 mm. The SLM process was protected by an argon atmosphere (<500 ppm of oxygen), and the purity of argon gas was 99.999%. 316L stainless steel is a common austenitic steel with excellent corrosion resistance, which plays an important role in many fields. Also, SLM has many advantages over traditional machining methods, such as geometric freedom, personalization, and better mechanical properties. Thus, 316L steel was chosen appropriately for the fabricated materials in the study of SLM residual stress. The experimental material was 316L steel, which has a good fluidity, with an average particle size of 30.32 μm. The chemical composition of 316L steel powder is shown in Table 1. In order to ensure the processing quality of produced parts, the optimized process parameters were selected, as shown in Table 2.

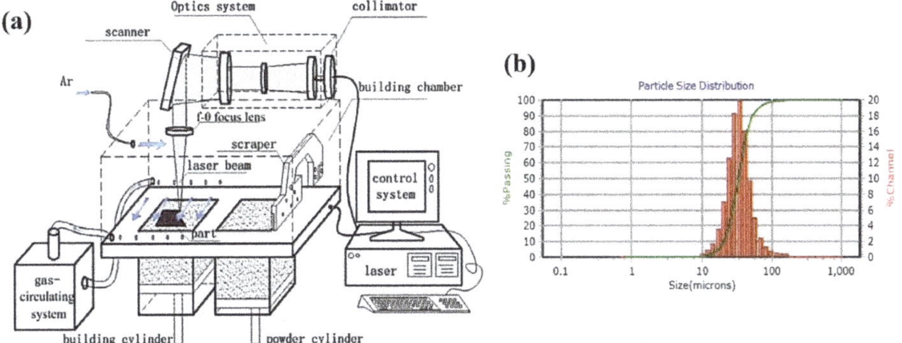

Figure 1. (a) Principle of selective laser melting (SLM) manufacturing equipment; (b) particle size distribution (PSD).

Table 1. Powder composition of 316L (mass fraction, wt %).

Element	Content (%)
C	0.03
Cr	17.5
Ni	12.06
Mo	2.06
Si	0.86
Mn	0.3
O	0.09
S	0.032
P	0.029
Fe	Bal.

Table 2. SLM processing parameters of 316L steel powder.

Laser Power/W	160
Scanning Speed/(mm/s)	600
Layer Thickness/mm	0.03
Hatch Space/mm	0.07
Spot Compensation/mm	0.03
Spot Diameter/μm	80
Oxygen Content/%	≤ 0.01
Shielding Gas	Argon

2.2. Experimental Method

Three kinds of self-developed divisional scanning strategies, normal partition strategy, oblique line, and layer-staggered divisional strategy and spiral divisional strategy, were used in the experiment for the study of residual stress in SLM. At the same time, the commonly used "s" style orthogonal scanning strategy was introduced as a contrast reference for the above-mentioned three divisional scanning strategies. Among them, the "s" style orthogonal scanning strategy is a scanning strategy where the cross section is filled with S-shaped scanning lines, and the upper and lower scanning lines are staggered. The scanning lines of the upper two layers are orthogonal to the x and y directions of the lower two layers, as shown in Figure 2d below. Significantly, the S-shaped scanning lines refer to the mode of the adjacent scanning line; when a scanning line is finished, the origin of the next adjacent scanning line will be next to the destination of the finished scanning line.

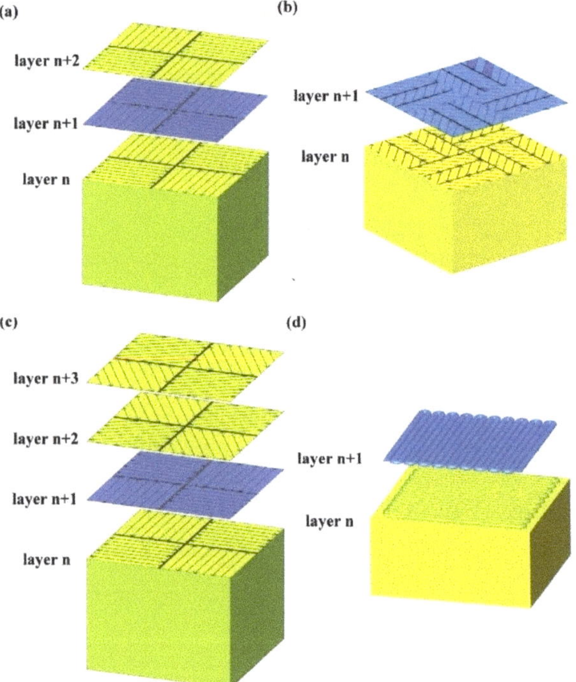

Figure 2. Schematic of three different divisional refilling strategy and "s" style orthogonal scanning strategy: (**a**) normal partition strategy; (**b**) oblique line and layer-staggered divisional strategy; (**c**) spiral divisional strategy; (**d**) "s" style orthogonal scanning strategy.

A normal partition strategy is achieved by dividing the section into the isometrical square sub-areas, and allowing a small amount of overlap between the each two adjacent square areas in the same layer. The overlap area is generally set to equal as a spot diameter. The S-shaped scanning lines follow the rules that the scanning lines in adjacent areas are orthogonal, and that the scanning lines belonging to the same area in the upper and lower layers are orthogonal, as shown in Figure 2a.

A division nested mechanism was applied for the design of oblique lines, and the layer-staggered divisional strategy. In this strategy, the adjacent layers' divisions were orthogonally filled with three parallel rectangular sub-regions respectively. And the adjacent x, y direction division regions are filled with oblique 45° and 135° scanning line. This strategy can not only reduce the number of the areas, but it can also limit the length of the scanning lines inside each area at the same profile shape. This strategy's scanning lines follow the rules as shown in Figure 2b.

The spiral divisional strategy adopted the divisional direction rotation mechanism. Every two adjacent layers are called a group, and the direction between every two adjacent groups is set to $60°$. Besides, the S-shaped scanning lines that belong to the same sub-area in different layers of the same group should be orthogonal, and the scanning lines in the adjacent sub-areas are also orthogonal, as illustrated in Figure 2c.

In order to reduce the negative impact of the width of region segmentation on the part performance, reasonable division size parameters should be chosen. When the width of the sub-area is too small, the area connection is unstable and prone to produce micro-pores. Also, the same cross-section needs to be divided into more areas, and more area connections are introduced, which may affect the densification of the parts. When the sub-area width is far too large, the length of the scanning line is too long, resulting in large residual stress inside the part. The width of the sub-area is suitable for the choice of 5 mm–7 mm [17]. Therefore, the area width in this paper was selected as 5 mm. The animated videos about those divisional strategies are available in the Supplementary Materials.

2.3. Test Methods

2.3.1. Densification Detection and Surface Quality Analysis

Four groups of 10 mm $\times$ 10 mm $\times$ 10 mm cubic samples were experimentally fabricated, and each group consisted of three parts which were group #1 (normal partition strategy), group #2 (oblique line and layer-staggered divisional strategy), group #3 (spiral divisional strategy), and group #4 ("s" style orthogonal scanning strategy). The densification of the SLM parts was measured by Archimedes principle. The weights of the samples in air and water were measured successively, and the calculation formula of the drainage method is as in Equation (1):

$$\rho = \left(\frac{W_{\mathrm{air}} \times \rho_{\mathrm{H_2O}}}{W_{\mathrm{air}} - W_{\mathrm{H_2O}}} \right) / \rho_0, \tag{1}$$

where in $\rho_{\mathrm{H_2O}} = 1.00\ \mathrm{g/cm^3}$ is the density of distilled water, W_{air} is the average weight of the fabricated parts in the air, $W_{\mathrm{H_2O}}$ is the average weight of the fabricated parts in water, and $\rho_0 = 7.98\ \mathrm{g/cm^3}$ is the theoretical density of 316L steel alloy.

The first three groups of the cubic samples did not undergo any surface treatment, and the overall surface quality of each group and the surface area at the junction quality were observed using the VHX-5000 (KEYENCE company, Osaka, Japan) at a magnification of 20 and 200 times.

2.3.2. Measurement of Residual Stress Distribution and Deformation

On the base of relevant references [23], is clear that the accumulated heat in the fabricating part is gradually uniform with an increase of fabrication height. Therefore, the subsequent laser energy input would temper the fabricated layer and release the stress. The residual stress that we discussed in this article was biased toward the distribution in horizontal direction. Thus, two geometric shape samples were designed to test the influence of the scanning strategy on the residual stress: (1) strip samples with a width of 6 mm, a thickness of 10 mm, and a length of 60 mm; (2) square samples with a width and length of 30 mm, and a thickness of 7 mm. Based on those samples, the influence of the different division scanning methods on the stress distribution in the longitudinal direction, and the overall plane stress distribution of SLM parts was discussed. For the two kinds of geometric models, four kinds of samples were processed respectively in group #1 (normal partition strategy), group #2 (oblique line and layer-staggered divisional strategy), group #3 (spiral divisional strategy), and group #4 ("s" style orthogonal scanning strategy). According to ASTM E837, the residual stress of the produced parts is measured by a drilling method. Generally, the drilling method can measure residual stress with a measurement accuracy of 20%. The satisfactory residual stresses in the fabricated parts can be controlled to within 60% of the yield stress. The distribution of the residual stress test points of the strip and square samples are shown in Figure 3a,b. In Figure 3a, the strip sample was designed to discuss the stress distribution in a longitudinal direction with a groove and the typical three-element clockwise stress rosette. The aim of the groove, with a depth of approximately 2 mm, was to relieve residual stress. The holes in Figure 3b were also used to relieve stress, but the difference was that the three-element stress rosette was set as shown in Figure 3b to discuss the overall plane stress distribution of SLM parts. The diameter and depth of hole are 5 mm and 6 mm, respectively.

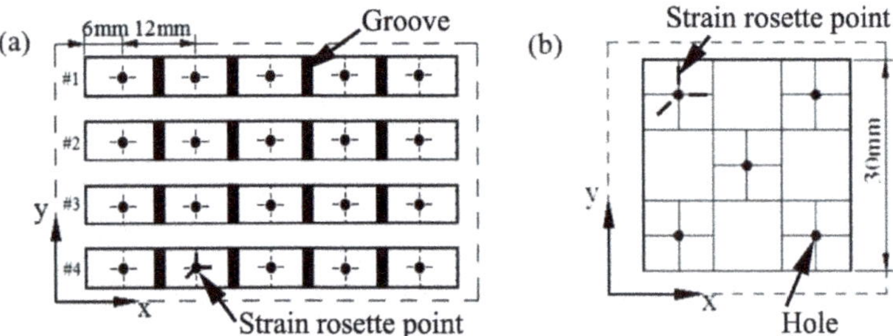

Figure 3. Design model of residual stress testing: (**a**) Residual stress in the direction of length; (**b**) residual stress in the plane distribution.

In addition, the SLM parts were cut from the substrate. The parts were warped upwards at both ends, and they formed a certain arc in the middle due to residual stress. According to the trends and characteristics of this deformation, an arch-like bridge structure was designed. The arch structure is as shown in Figure 4, and the dimensions of the retaining structure W, T and B are constant, with W = 10 mm, T = 2 mm and B = 1 mm. The size L was set to 12 mm, 20 mm, or 28 mm. These three gradients were used to verify the influences of different lengths and scanning methods on the deformation of the parts. For each designated length, four groups of samples were fabricated, which were group #1 (normal partition strategy), group #2 (oblique line and layer-staggered divisional strategy), group #3 (spiral divisional strategy), and group #4 ("s" style orthogonal scanning strategy). According to the deformation characteristics of the specimen, the sample deforms, as shown in Figure 4b, under the joint action of the internal residual stress in a multi-direction. After measuring the distance L of the outer edge of the bottom of the bridge structure, as shown in Figure 4b, the deformation of the bridge structure was measured by $\delta = L' - L$. The deformation of each sample is measured in three positions, and the mean value is the result of measurement.

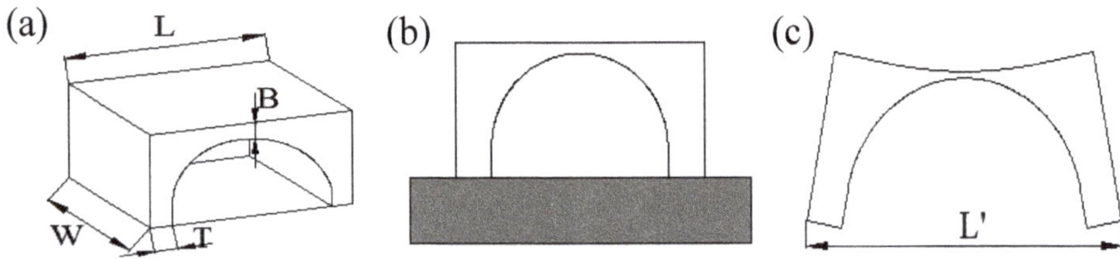

Figure 4. Schematic of bridge structure designing: (**a**) 3D design model of the part; (**b**) placement and manufacturing of parts; (**c**) schematic of parts deformation.

3. Results and Discussion

3.1. Densification and Surface Quality Analysis

Figure 5 shows the fabricated cubic samples. The density of the samples was calculated by the Archimedes principle. The average densities of the (a), (b), (c) and (d) groups were 99.28%,

99.37%, 99.10%, and 98.64% respectively. No big differences existed among the four groups of samples. The density of the produced parts adopting a divisional scanning strategy was up to 99%, which indicates that the parts adopting three kinds of divisional scanning strategies could obtain a dense sample.

Figure 5. Testing sample of densification: (**a**) normal partition strategy; (**b**) oblique line and layer-staggered divisional strategy; (**c**) spiral divisional strategy; (**d**) "s"-style orthogonal scanning strategy.

Figure 6 illustrates the surface morphology of the first three groups of the square blocks at 20 times magnification. As can be seen from the surface morphology in Figure 6, although there were some small holes in the surfaces of the three group samples, the overall surfaces were uniform. The fusion tracks were coherent and clear with a good overlap effect, and the overlap areas between the different zones in the three group samples were also good.

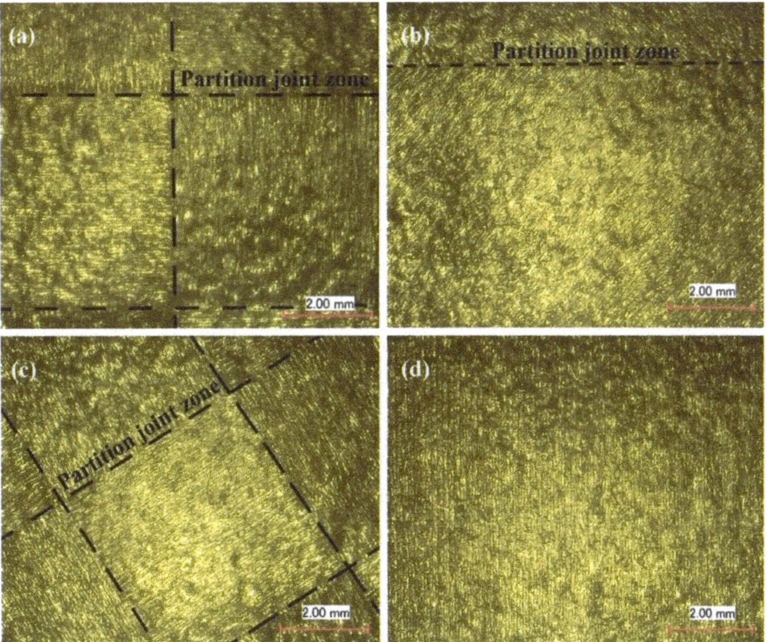

Figure 6. The surface morphology of samples (magnification 20×): (**a**) Normal partition strategy; (**b**) Oblique line and layer-staggered divisional strategy; (**c**) Spiral divisional strategy; (**d**) "s" style orthogonal scanning strategy.

Figure 7 illustrates the morphologies of the overlap area of the first three group blocks at a magnification of 200 times. The results shows that there were obvious microscopic defects, such as micro-pores and micro-bulges. Figure 8 shows the surface measurement of overlap, including the 3D topography of the overlap between two groups of scanning lines, and the overlap between a single scanning line and one group of scanning lines. The Blue line represents the range of measurement, the overlap region. The curves are the measured values of the heights of the overlapped regions. According to the curves from Figure 8a,b, in the oblique line and layer-staggered divisional strategies, there were periodic micro-pores and micro-bulges existing on the surface along the overlap area, with Rz = 40.9 µm. Due to the primitive overlap of the origins of the scanning line, the surface quality of the overlap area was relative poor. The overlap between two groups of scanning lines had a smaller Rz, with 36.8 µm. It could be seen from Figure 7a,c that there was a small amount of micro-pores and micro-bulges in the overlap area of sample groups #1 and #3, but this did not affect the overlapping, and the curve was smoother. It reveals that the micro-pores in sample groups #1 and 3 were shallower than the pores, with a depth of approximately 20 µm in group #2 samples.

The above results show that the adjacent region using a rectangular overlapping mechanism and a spiral divisional mechanism can achieve a better regional overlapping effect. The overlapping effects with one scanning line in the adjacent area and multiple scanning lines in the other area (Figure 9b) were better than another overlapping with multiple scanning lines in the adjacent area. The reason is that in the overlapping mode of multiple Figure 9b, the micro-pores between the semi-circular origin of the parallel track were filled first, or later by a single vertical track. Technically, the micro-pores or even the micro-holes deriving from the "Match head" shape, which were influenced by the delay time of laser on and laser off, and the delay time of the galvo scanning system. This was an inevitable technical feature in laser scanning, but a prominent defect in the divisional scanning strategy.

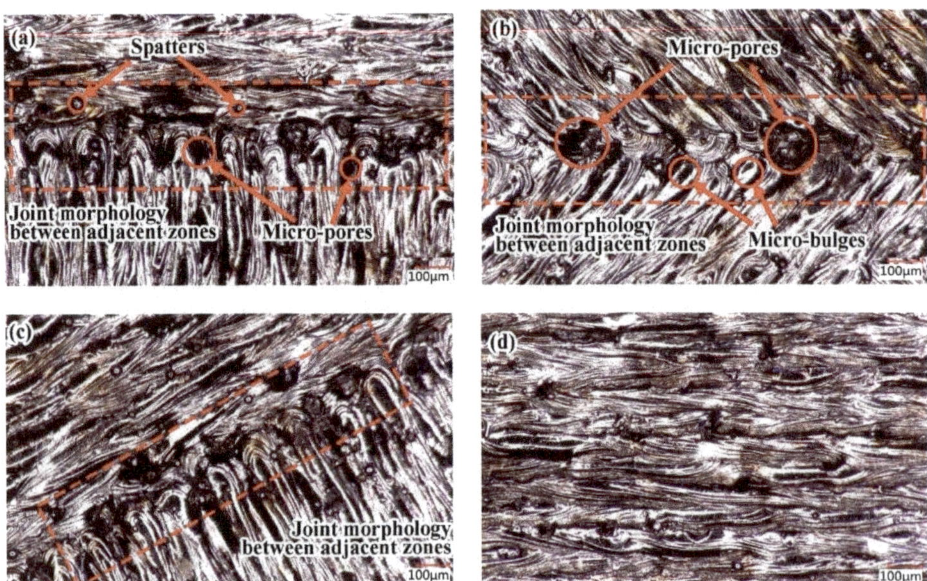

Figure 7. The surface morphology graph of overlap (magnification 200×): (**a**) Normal partition strategy; (**b**) Oblique line and layer-staggered divisional strategy; (**c**) Spiral divisional strategy; (**d**) "s" style orthogonal scanning strategy.

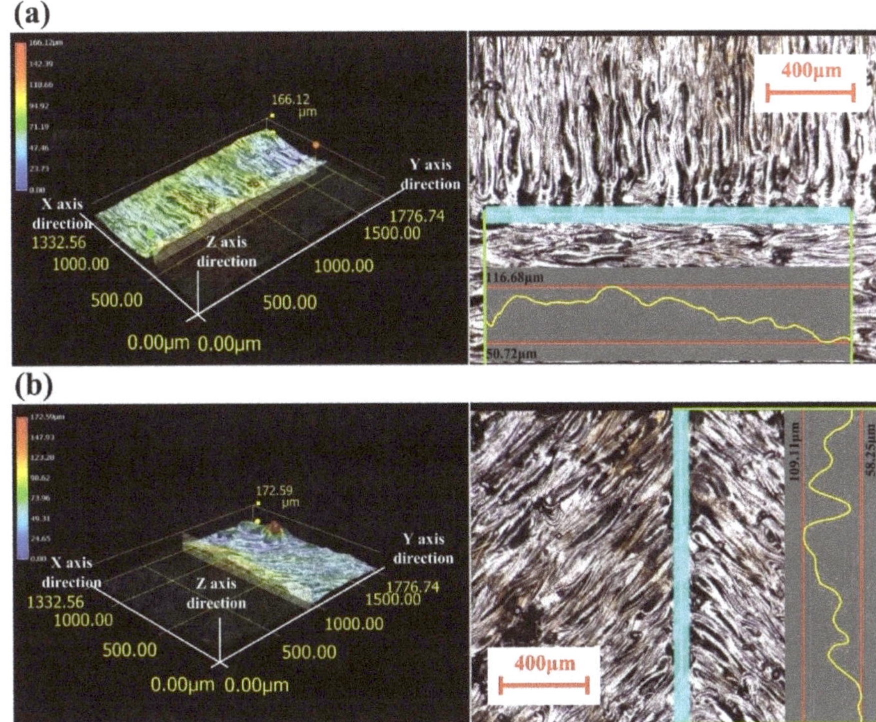

Figure 8. Surface measurement of overlap: (**a**) Overlap between two groups of scanning lines which was used in the normal partition strategy and the spiral divisional strategy; (**b**) Overlap between a single scanning line and one group of scanning lines, which was used in the oblique line and layer-staggered divisional strategy.

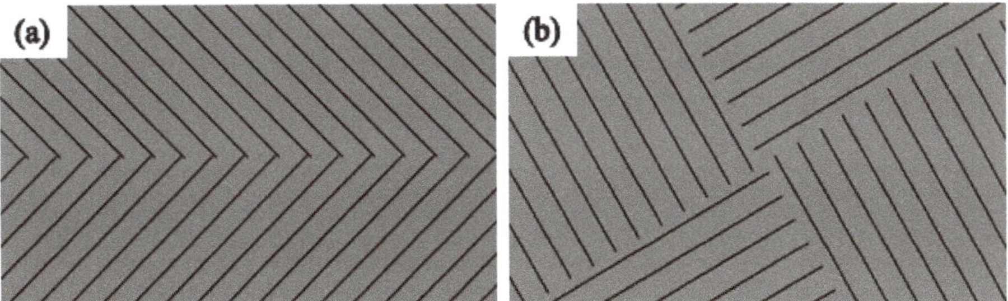

Figure 9. Mechanism of track overlap: (**a**) Overlap between two groups of scanning lines; (**b**) overlap between a single scanning line and one group of scanning lines.

3.2. Residual Stress Distribution Analysis

Figure 10 shows the average results of the first residual stress test model of four group specimens. It can be seen from Figure 10a that the residual stresses of the four group samples are all at a relatively normal level within the range of 98 MPa to 240 MPa, and the σ_x values of all the four samples are larger at the scanning start position and showed a downward trend along the scanning direction. The declining trend of group samples #1, #2 and #3 were more obvious than that of group #4 samples, which was due to the fact that the melting track was surrounded by the powder at the scanning starting position, which thus affected the heat conduction and resulted in greater thermal stress. As the scanning continued forward, the temperature gradient of the melt track tended to be in a state of dynamic equilibrium, and the thermal stress gradually decreased. However, for the group #4 samples, the "s" style orthogonal scanning strategy was used to plan the path, the scanning line was longer, and the scanning direction of each layer was substantially the same. As a result, the thermal stress was accumulated along the scanning direction [24], and the residual stress was large. For #1, #2, and #3 group samples, the divisional scanning method was adopted, the length of the scanning line was set to 5 mm, the control was within the range of 8 mm, and the scanning lines in each adjacent sub-area were orthogonally crossed, reducing the thermal stress accumulation along a single direction, and the stress was decreased along the scanning direction.

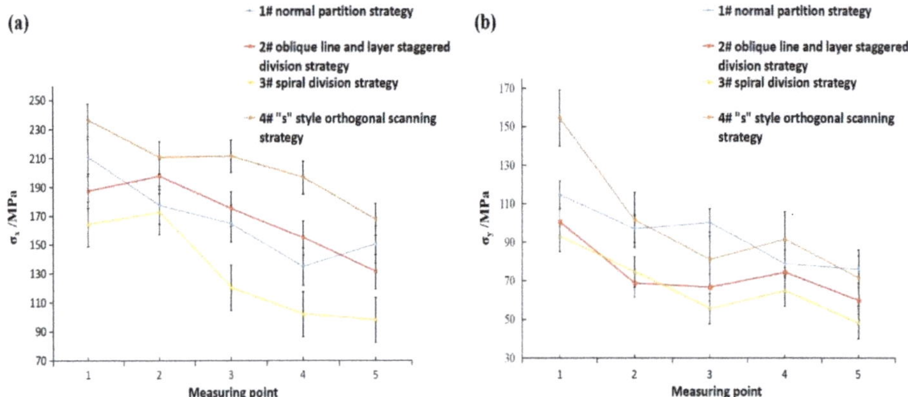

Figure 10. Distribution of residual stress: (**a**) X-direction residual stress distribution curve; (**b**) Y-direction residual stress distribution curve.

Comparing Figure 10a with b, it was shown that the distributions of σ_y of samples from groups #1, #2 and #3 along the x-direction were more stable than σ_x, and they varied from 50 MPa to 115 MPa. The distribution of σ_y of group #4 samples along the length was similar to σ_x. This is because each sub-area scan fabricated by divisional scan strategy planning were all scanned along the x-direction, resulting in a temperature gradient in the x-direction that was greater than the y-direction. Therefore, the residual stress in the x-direction was greater than that in the y-direction [15]. For the "s" style orthogonal scanning strategy, the upper two layers were orthogonal to the x and y directions of the lower two scanning lines, so that the temperature gradient distribution in the x direction and the y direction were almost similar.

Figures 11 and 12 show the results of group #2 of four stress test models, σ_x and σ_y. As can be seen from the table, for the four samples, σ_x values were larger in the lower left corner of the sample, and the stress showed a decreasing trend upward and right, which was due to the scanning start position being located in the left lower corner of the fabricating plane, following the scanning sequence from bottom to top, the left lower part of the produced part presents a larger temperature gradient field, and then it tends to be in a state of dynamic equilibrium. It was also found that the overall stress values of the group #2 and #3 samples were smaller, and the distribution was more uniform. The stress values of the groups #1 and #4 samples were larger, which was due to the normal partition strategy adopted on group #1 samples; the sub-area width of each layer was 5 mm, and the overlapping area between the areas was re-melted, which increased the input energy of the overlapping area, resulting in the accumulation of large amounts of thermal stress in the overlapping area. In the group #4 samples with an "s" style orthogonal scanning strategy, the scanning line was much longer, which could easily intensify the stress accumulation effect. The result showed that the staggered divisional strategy and spiral divisional strategy could effectively reduce the residual stress in the profile of the produced parts and improve the distribution of residual stress to make the residual stress evenly distributed.

Scanning mode / σx / Specimen	Normal partition Strategy			Scanning mode / σx / Specimen	Oblique line and layer staggered divisional strategy		
#1	188.21		160.71	#2	197.20		143.67
		233.48				178.54	
	259.54		201.67		231.43		214.83
Scanning mode / σx / Specimen	**Spiral divisional strategy**			**Scanning mode / σx / Specimen**	**"S" style orthogonal scanning strategy**		
#3	101.12		124.62	#4	212.67		182.75
		153.93				223.08	
	203.73		146.61		274.54		193.43

Figure 11. The result of X-direction residual stress σ_x.

Scanning mode / σy / Specimen	Normal partition strategy			Scanning mode / σy / Specimen	Oblique line and layer staggered divisional strategy		
#1	120.13		125.91	#2	132.46		72.92
		83.91				107.21	
	156.61		113.48		121.59		92.59
Scanning mode / σy / Specimen	**Spiral divisional strategy**			**Scanning mode / σy / Specimen**	**"S" style orthogonal scanning strategy**		
#3	71.23		56.70	#4	140.55		80.68
		84.31				111.30	
	112.73		94.34		173.43		107.37

Figure 12. The result of Y-direction residual stress σ_y.

Overall, most of the measured results were a little bit less than the residual stress value when 316L SLM parts adopt single-direction scanning strategy or other simple scanning strategies [15,23]. Actually, the SLM parts' residual stress relates to many factors, such as materials, part height, processing parameters, and even processing equipment. However, based on the existing related literature [15,23], Figures 10–12 do not just show the distribution characteristics of the residual stress in a longitudinal direction, but they also reveal the tendency effect of the scanning line's direction and length to the residual stress, with three complicated divisional strategies.

3.3. Deformation Analysis

Figure 13a shows a sample of a bridge-shaped structure fabricated by a group #3 scanning strategy (without heat treatment). After the sample was fabricated, a reverse engineering method was used to obtain the actual 3D model of the samples. The theoretical model and the reconstructed model of the sample were then compared and analyzed in Geomagic software, and the deformation diagram of the SLM part was obtained. According to the schematic diagram 13b, it could be seen that the bottom of the bridge structure was obviously yellow, and it appeared to be warped out. The tip of the bridge shaped pillar was obviously blue, and internal contraction appeared. The top beam of the bridge-shaped part also warped up at both ends, while the middle part of the beam was compressed.

Figure 13. The structure and deformation diagram of bridge samples fabricated by a spiral divisional strategy: (**a**) A sample of a bridge-shaped structure; (**b**) the deformation diagram.

As shown in Figure 14, this was the deformation result of the of the four group samples. The curves of samples from groups #1, #2 and #3 group samples in the figure changed gently. With the length of the bridge structure increased from 12 mm to 28 mm, and there was a slight change in the deformation, with a difference of only 0.2 mm. There is no obvious difference among the deformations of the three groups samples, which showed that the deformation of the sample using the divisional scanning strategy had little to do with the length of the bridge structure, and the amount of deformation did not change significantly with the length of the bridge structure. However, the deformation of group #4 samples obviously increased with the length of the bridge, and the deformation of each sample in group #4 was greater than the deformation of the first three groups of samples. When the length was 12 mm, the difference of the deformation was only 0.1 mm. When the length was 28 mm, the difference of the deformation increased to 0.4 mm. Obviously, with a length increase in the bridge structure, the deformation difference between group #4 samples and the first three sample groups became larger and more obvious.

The reason for the above result is related to the length of the scanning line. The first three kinds of divisional scanning strategies all defined area lengths of 5 mm, and the divisional scanning strategy used a sub-area boundary to cut off the scanning lines, so that the scanning line length was controlled to within 8 mm. Therefore, when the bridge structure length becomes longer, the scanning line's lengths in these three kinds of divisional scanning strategies does not increase, and the deformation will not increase. All of the divisional scanning strategies use orthogonal scanning lines in the adjacent sub-areas, so that the deformation in the same direction will not have a cumulative effect along the length of the bridge structure, and the strategies also control the deformation of samples. However, when the "s" style orthogonal scanning strategy is applied to generate the scanning lines, the length of the scanning line is closely related to the size of the sample model. As the size of the model increases, the length of the scanning line and the deformation will also increase.

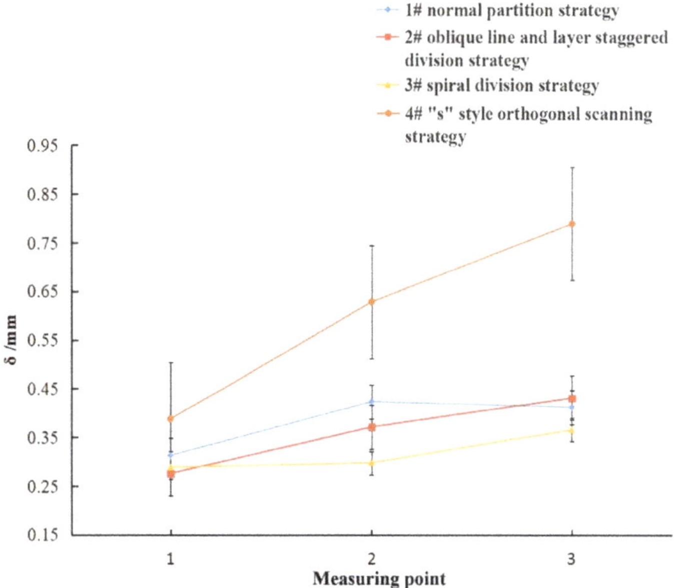

Figure 14. Relationship between the deformation and scanning line lengths under different scanning strategies.

Therefore, the scanning strategy works by indirectly changing the length and direction of the scanning lines, so that the length of the scanning line is not limited by the size and shape of the three-dimensional model. Besides, the deformation of the produced parts can be improved by controlling the length of the scanning lines to within a certain range, and changing the direction of the scanning lines along the length direction. Therefore, it is important to choose the appropriate scanning strategy to control the deformation of the SLM parts.

3.4. Discussions

The experimental results show that all three kinds of divisional scanning strategies can obtain dense parts with a density exceeding 99%. As for the overlap area, the intersection of the slash layer-staggered divisional scanning area, where multiple scanning lines in one area are abutted with multiple scanning lines in another area, is prone to the phenomenon of insufficient powder compensation and micro-pores, which are very unfavorable for the improvement of the overlap quality. Because normal partition strategy adopts a scanning line outward extension mechanism to overlap, the overlap joint areas are re-melted. Thus, the energy input is too large in the overlap of the joint areas, resulting in local excessive residual stress and affecting the residual stress distribution uniformly in the fabricated plane. However, the residual stress of the part plane with an oblique line and layer-staggered divisional strategy and spiral divisional strategy are all at a low level, and the distribution is more uniform. All three kinds of divisional scanning strategies can limit the length of the scanning line to 8 mm, so that the deformation of the part is not affected by the size of the part, and the effects of the three kinds of divisional scanning on the part deformation are not significantly different. The above results might provide a reference for optimizing the scanning strategy of selective laser melting. For example, when using the divisional scanning strategy, the scanning mechanism of docking the scanning lines in the adjacent sub-area should not be adopted, as it will affect the surface quality. Minimizing the size of the in-plane scanning partition, helps to distribute the residual stress evenly and to keep it at a low level. In fact, for the defects in overlap, Yasa et al. also found that the selection of parameters related to sectoral scanning may cause aligned porosity at the edges between sectors or scanned tracks, which is very undesirable in terms of mechanical properties [20]. The authors also made a microscopic analysis on the overlap of the melting track. As shown in Figure 15, it can be found that the microstructure of the samples using three scanning strategies have obvious different track overlapping characterizations.

Figure 15. Microstructures of three different scanning strategies: (a) Normal partition strategy; (b) spiral divisional strategy; (c) "s" style orthogonal scanning strategy.

The above analysis shows that the divisional scanning strategy can inevitably encounter regional overlap problems, and the quality control of overlap area has a crucial influence on the mechanical properties and surface quality of the parts. For example, Dai et al. investigated the influence of the re-melting behavior and scanning strategy on the formation of the "track–track" and "layer–layer" molten pool boundaries (MPBs) [25], and Almangour et al. investigated scanning strategies for texture and anisotropy tailoring during selective laser melting of TiC/316L stainless steel nanocomposites [10]. Carter et al. from Birmingham University also observed the repeating pattern shown in the grain structure, which has been linked to the overlapping of the 'island' pattern used, as is standard in the Concept Laser M2, and they suggested that the formation of this bi-modal grain structure can be linked to heat transfer away from the solidifying melt pool [26].

Of course, the design of this experiment is not well considered. For example, only the three divisional scanning strategies developed based on the current mainstream laser scanning strategies were selected to investigate the influence of the division mechanism on the performance of the SLM part, and other division mechanisms were not fully introduced. The drilling method was adopted to test the residual stress. Due to the constraints of the strain force, the selected test points were limited and the interval was large, and this could only reflect the general rules of residual stress distribution. At the same time, the effects of different divisional scanning strategies on the performance of SLM parts were evaluated only from the aspects of densification, surface quality, residual stress distribution, and deformation. The effect of different divisional scanning modes has not been further explored within the microstructure of parts, and a follow-up remains to be further studied.

4. Conclusions

According to the three kinds of self-developed divisional scanning strategies, the influence of different divisional scanning strategies on the performance of SLM fabricated parts is discussed from the aspects of densification, surface quality, residual stress distribution, and deformation. The main conclusions of this work are as follows:

1. It was found that by using the normal partition strategy, the SLM parts had a large amount of residual stress in the overlap area, which was not conducive to residual stress distribution in the fabricated plane. Using an oblique line and layer-staggered divisional strategy easily forms micro-pores and affects the surface quality of parts, but the overlap regions in the adjacent layers are not located in the same vertical plane, so that the residual stress distribution of the components is more uniform, and the value is smaller.

2. Among the three kinds of divisional scanning strategies, the spiral divisional strategy is the best for obtaining produced parts with better performance. 316L steel alloy parts adopted

a spiral divisional scanning strategy, and this not only obtains a density of 99.37%, but it also effectively improves the distribution of residual stress and controls the deformation degree of the produced parts.

3. The influence of scanning strategy on the deformation of the SLM-produced part is achieved by indirectly changing the length and direction of the scanning lines. The deformation of the produced part can be improved by controlling the length and direction of the scanning lines.

Supplementary Materials: The following are available online at http://www.mdpi.com/1996-1944/11/10/1821/s1: Video S1: Normal partition strategy DEMO; Video S2: Oblique line and layer-staggered divisional strategy DEMO; Video S3: Spiral divisional strategy.

Author Contributions: Conceptualization, D.W. and S.D.; Data curation, D.W., S.W., W.D. and S.D.; Formal analysis, S.W., W.D. and S.L.; Funding acquisition, Y.Y.; Investigation, D.W., S.W., W.D. and Z.W.; Methodology, D.W. and Z.W.; Project administration, Y.Y.; Resources, Y.Y. and W.D.; Software, S.D.; Supervision, D.W. and Y.Y.; Validation, S.L.; Writing—Original draft, D.W., S.W. and S.D.; Writing—Review & Editing, D.W. and S.L.

Funding: This research was funded by [National Natural Science Foundation of China] grant number [51775196]; [Guangdong Province Science and Technology Project] grant number [2017B090912003, 2017A050501058, 2015B010125006]; [High-level Personnel Special Support Plan of Guangdong Province] grant number [2016TQ03X289]; [Guangzhou Pearl River New Talent Project] grant number [201710010064]. [the Fundamental Research Funds for the Central Universities] grant number [2018ZD30], Di Wang thanks the China Scholarship Council for the award to study for one year at the University of Birmingham grant number [201706155082].

Conflicts of Interest: The authors declare no conflict of interest.

References

1. Vandenbroucke, B.; Kruth, J. Selective laser melting of biocompatible metals for rapid manufacturing of medical parts. *Rapid Prototyp. J.* **2007**, *13*, 196–203. [CrossRef]
2. Campbell, I.; Bourell, D.; Gibson, I. Additive manufacturing: Rapid prototyping comes of age. *Rapid Prototyp. J.* **2012**, *18*, 255–258. [CrossRef]
3. Attar, H.; Ehtemam-Haghighi, S.; Kent, D.; Dargusch, M.S. Recent developments and opportunities in additive manufacturing of titanium-based matrix composites: A review. *Int. J. Mach. Tools Manuf.* **2018**, *133*, 85–102. [CrossRef]
4. Thijs, L.; Kempen, K.; Kruth, J.; Van Humbeeck, J. Fine-structured aluminium products with controllable texture by selective laser melting of pre-alloyed AlSi10Mg powder. *Acta Mater.* **2013**, *61*, 1809–1819. [CrossRef]
5. Yan, C.; Hao, L.; Hussein, A.; Raymont, D. Evaluations of cellular lattice structures manufactured using selective laser melting. *Int. J. Mach. Tools Manuf.* **2012**, *62*, 32–38. [CrossRef]
6. Delgado, J.; Ciurana, J.; Rodriguez, C.A. Influence of process parameters on part quality and mechanical properties for DMLS and SLM with iron-based materials. *Int. J. Mach. Tools Manuf.* **2012**, *60*, 601–610. [CrossRef]
7. Attar, H.; Ehtemam-Haghighi, S.; Kent, D.; Wu, X.; Dargusch, M.S. Comparative study of commercially pure titanium produced by laser engineered net shaping, selective laser melting and casting processes. *Mater. Sci. Eng. A* **2017**, *705*, 385–393. [CrossRef]
8. Song, B.; Dong, S.; Liao, H.; Coddet, C. Process parameter selection for selective laser melting of Ti6Al4V based on temperature distribution simulation and experimental sintering. *Int. J. Mach. Tools Manuf.* **2012**, *61*, 967–974. [CrossRef]
9. Gu, D.; He, B. Finite element simulation and experimental investigation of residual stresses in selective laser melted Ti-Ni shape memory alloy. *Comput. Mater. Sci.* **2016**, *117*, 221–232. [CrossRef]
10. AlMangour, B.; Grzesiak, D.; Yang, J. Scanning strategies for texture and anisotropy tailoring during selective laser melting of TiC/316L stainless steel nanocomposites. *J. Alloys Compd.* **2017**, *728*, 424–435. [CrossRef]
11. Bo, C.; Shrestha, S.; Chou, K. Stress and deformation evaluations of scanning strategy effect in selective laser melting. *Addit. Manuf.* **2016**, *12*, 240–251.
12. Miranda, G.; Faria, S.; Bartolomeu, F.; Pinto, E.; Madeira, S.; Mateus, A.; Carvalho, O. Predictive models for physical and mechanical properties of 316L stainless steel produced by selective laser melting. *Mater. Sci. Eng. A* **2016**, *657*, 43–56. [CrossRef]

13. Kruth, J.P.; Froyen, L.; Van Vaerenbergh, J.; Mercelis, P.; Rombouts, M.; Lauwers, B. Selective laser melting of iron-based powder. *J. Mater. Process. Technol.* **2004**, *149*, 616–622. [CrossRef]

14. Kruth, J.P.; Deckers, J.; Yasa, E.; Wauthle, R. Assessing and comparing influencing factors of residual stresses in selective laser melting using a novel analysis method. *Proc. Inst. Mech. Eng. B J. Eng. Manuf.* **2012**, *226*, 980–991. [CrossRef]

15. Parry, L.; Ashcroft, I.A.; Wildman, R.D. Understanding the effect of laser scan strategy on residual stress in selective laser melting through thermo-mechanical simulation. *Addit. Manuf.* **2016**, *12*, 1–15. [CrossRef]

16. Lu, Y.; Wu, S.; Gan, Y.; Huang, T.; Yang, C.; Junjie, L.; Lin, J. Study on the microstructure, mechanical property and residual stress of SLM Inconel-718 alloy manufactured by differing island scanning strategy. *Opt. Laser Technol.* **2015**, *75*, 197–206. [CrossRef]

17. Thijs, L.; Verhaeghe, F.; Craeghs, T.; Van Humbeeck, J.; Kruth, J. A study of the micro structural evolution during selective laser melting of Ti-6Al-4V. *Acta. Mater.* **2010**, *58*, 3303–3312. [CrossRef]

18. Qian, B.; Shi, Y.; Wei, Q.; Wang, H. The helix scan strategy applied to the selective laser melting. *Int. J. Adv. Manuf. Technol.* **2012**, *63*, 631–640.

19. Beal, V.E.; Erasenthiran, P.; Hopkinson, N.; Dickens, P.; Ahrens, C.H. The effect of scanning strategy on laser fusion of functionally graded H13/Cu materials. *Int. J. Adv. Manuf. Technol.* **2006**, *30*, 844–852. [CrossRef]

20. Yasa, E.; Deckers, J.; Kruth, J.P.; Rombouts, M.; Luyten, J. Investigation of sectoral scanning in selective laser melting, biennial conference on engineering systems design and analysis. In Proceedings of the ASME 2010 10th Biennial Conference on Engineering Systems Design and Analysis, Istanbul, Turkey, 12–14 July 2010.

21. Rashid, R.; Masood, S.H.; Ruan, D.; Palanisamy, S.; Rashid, R.R.; Brandt, M. Effect of scan strategy on density and metallurgical properties of 17-4PH parts printed by Selective Laser Melting (SLM). *J. Mater. Process. Technol.* **2017**, *249*, 502–511. [CrossRef]

22. Geiger, F.; Kunze, K.; Etter, T. Tailoring the texture of IN738LC processed by selective laser melting (SLM) by specific scanning strategies. *Mater. Sci. Eng. A* **2016**, *661*, 240–246. [CrossRef]

23. Liu, Y.; Yang, Y.; Wang, D. A study on the residual stress during selective laser melting (SLM) of metallic powder. *Int. J. Adv. Manuf. Technol.* **2016**, *87*, 647–656. [CrossRef]

24. Matsumoto, M.; Shiomi, M.; Osakada, K.; Abe, F. Finite element analysis of single layer forming on metallic powder bed in rapid prototyping by selective laser processing. *Int. J. Mach. Tools Manuf.* **2002**, *42*, 61–67. [CrossRef]

25. Dai, D.; Gu, D.; Zhang, H.; Xiong, J.; Ma, C.; Hong, C.; Poprawe, R. Influence of scan strategy and molten pool configuration on microstructures and tensile properties of selective laser melting additive manufactured aluminum based parts. *Opt. Laser Technol.* **2018**, *99*, 91–100. [CrossRef]

26. Carter, L.N.; Martin, C.; Withers, P.J.; Attallah, M.M. The influence of the laser scan strategy on grain structure and cracking behaviour in SLM powder-bed fabricated nickel superalloy. *J. Alloys Compd.* **2014**, *615*, 338–347. [CrossRef]

 *materials*

Article

Effect of Scanning and Support Strategies on Relative Density of SLM-ed H13 Steel in Relation to Specimen Size

Tomasz Kurzynowski, Wojciech Stopyra *, Konrad Gruber, Grzegorz Ziółkowski, Bogumiła Kuźnicka and Edward Chlebus

Faculty of Mechanical Engineering, Centre for Advanced Manufacturing Technologies (CAMT/FPC), Wroclaw University of Science and Technology, 50-371 Wrocław, Poland;
tomasz.kurzynowski@pwr.edu.pl (T.K.); konrad.gruber@pwr.edu.pl (K.G.);
grzegorz.ziolkowski@pwr.edu.pl (G.Z.); bogumila.kuznicka@pwr.edu.pl (B.K.);
edward.chlebus@pwr.edu.pl (E.C.)
* Correspondence: wojciech.stopyra@pwr.edu.pl; Tel.: +48-71-320-2881

Received: 21 December 2018; Accepted: 8 January 2019; Published: 11 January 2019

Abstract: Standard experimental research works are aimed at optimization of Selective Laser Melting (SLM) parameters in order to produce material with relative density over 99% and possibly the highest scanning speed. Typically, cuboidal specimens with arbitrarily selected dimensions are built. An optimum set of parameters, determined on such specimens, is used for building parts with variable cross-section areas. However, it gives no guarantee that the density of variable-section parts produced with so selected parameters will be as high as that of the specimens measured during the parameters optimization process. The goal of this work was to improve the process of SLM parameter selection according to the criterion of maximum relative density, based on the example of AISI H13 tool steel (1.2344). A selection method of scanning strategy ensuring relative density of parts over 99%, irrespective of their dimensions, was determined. The specimens were produced using several variants of support structures. It was found that proper selection of the support strategy prevents development of columnar pores.

Keywords: selective laser melting; H13 tool steel; process parameters; scanning strategy; support strategy; porosity reduction

---

1. Introduction

The SLM technology belongs to methods of additive manufacturing, where a part is built by adding subsequent portions of material in the form of a thin powder layer that is then selectively melted by a laser beam. This technology can be successfully used in aircraft (fuel injection nozzles), automotive (thermal screens), tooling (injection moulds) and medical (individualized implants) industries [1,2].

By SLM technology, it is possible to manufacture parts of materials like nickel-based superalloys and alloys based on titanium, cobalt, aluminum and iron [3]. Among iron alloys, stainless steels like 316L (1.4404), 15-5PH (1.4540) and 17-4PH (1.4542) are commercially applied for manufacture of parts by SLM. A material commonly used in this method is tool steel H13 (1.2344), applied for injection and casting moulds, worms and cylinders for processing of plastics and tools for extrusion of light-metal profiles. This steel is characterized by high tensile strength, hardness and abrasion resistance at high temperatures. Application of SLM to manufacture of injection moulds of H13 steel gives great possibilities for designing complex geometries including conformal ducts. It should be noted that this complexity means high variability of the cross-section area of the mould along its built direction, which should be considered already at the stage of developing the process window.

By this technology, it is possible to build both thin-walled parts with minimum dimension equal to laser spot diameter (80–200 μm) and solid parts with outside dimensions limited by the size of the working chamber. However, manufacture of high-quality parts with a strongly variable cross-sectional area can cause problems with selection of process parameters precluding occurrence of defects like pores, cracks and deformations.

Cracks and geometry distortions of parts result from creation, during rapid solidification of thin layers of molten metal, significant thermal gradients generating residual stresses. Some attempts are made to reduce these stresses, such as double scanning of each layer, scanning of small areas (chessboard type), application of support structures or heating-up the working platform [4].

In the so-far performed research, attention was mostly paid to defects in the form of pores differing in shape, size and formation mechanism [5,6], with respect to their negative influence on ductility and fatigue strength of parts. The shape of pores (spherical, spheroidal, key-hole, irregular, elongated) depends on their formation mechanism, i.e., due to trapping gases by surface turbulences of the liquid pool and local sinking of the seam, to incomplete melting of powder and to shrinkage [7,8].

Spherical pores (gas pores) are formed during melting powder with increased humidity due to much lower solubility of gases in solid solution in comparison to liquid metal. This is especially visible during processing powders of aluminum alloys [9]. Gas pores can also result from entrapped gas in powder particles and from the Marangoni effect, described as the mass transfer along an interface between two fluids due to a gradient of the surface tension [10].

Key-holes are usually created during the process, when the ratio of the liquid pool depth to half of its width is larger than 1 [11,12]. Then, a tunnel filled with evaporating metal and shield gas is formed in front of the pool face. When the rapidly solidifying metal prevents gases from getting out from under the pool, gas pore type key-holes are created.

Elongated interlayer pores (lack of fusion) occur at the boundary between subsequent travels of a laser beam. The formation of this type of pore is related to too short a lifetime of the liquid metal pool, its wettability, improper selection of the distance between scanning lines or too high scanning speeds. During melting, quickly solidifying liquid metal is unable to fill free spaces, creating later interlayer pores [7,13]. Recent studies and in-process monitoring methods allow prediction of a lack of fusion based on melt pool monitoring data [14].

Columnar pores originate under the conditions of too low energy density, when small discontinuities of the melted layer can develop into defects with their shape similar to tunnels running through many layers [15].

On the grounds of the hitherto carried-out examinations, it was found that the tendency of pore formation in parts manufactured by SLM depends strongly on process parameters, specifically laser power, scanning speed and scanning strategy [5]. Thus, numerous studies focused on investigation of the influence of these parameters on the type, number and arrangement of pores in order to know their formation mechanisms. As was already stated by Gustmann et al. [16] for Cu-Al-Ni-Mn, scanning strategy plays an important role in affecting the quantity of pores, their size and arrangement. Arrangement of pores is dependent on the points characteristic for the given strategy and direction of the laser beam movement. For example, for the strategy "scanning strips in the same direction", Gustmann et al. [16] stated that pores occur along the scanning line and the areas scanned at the beginning show a smaller numbers of pores. In the contour strategy, however, pores are generally arranged in return points of the scanning line, and for the chequered patter, pores occur at the boundaries between fields and at some distance from the scanning initial point. For the considered H13 steel, Beal et al. [17] found a strong influence of the scanning strategy on relative density depending on the applied distances between the lines (hatch distance).

In spite of numerous studies carried-out in this field, the relationships between scanning strategy and the mechanism of formation of pores, especially columnar ones, remain unclear. In addition, there is no research concerning the influence of the cross-sectional area of the built part on relative density. This is significant from the viewpoint of parts with complex shapes, whose manufacture by SLM can be possibly competitive against the traditional technologies.

Support structures are used in laser additive manufacturing to support the parts being built and work as a heat conductor removing heat used in the process, which makes the internal thermal conditions stable during the process. Gan and Wong [18] showed that orientation and distribution of support structures influence levelness of built parts. It seems that a support strategy properly selected to the scanning strategy can significantly affect the creation of pores in the first solidified layers.

The presented work was aimed at determining, based on the example of H13 steel, the influence of the scanning strategy on the amount, shape and arrangement of pores in manufactured specimens depending on their dimensions. The specimens were produced with the additional use of several variants of supporting structures and their inclination angle to the scanning direction, in order to ensure uniform heat dissipation and to prevent columnar pore formation.

2. Materials and Methods

Specimens were made of H13 steel powder, sieved through a screen with a mesh size 63 μm. Next, powder was subjected to chemical analysis, evaluation of particle shapes by Scanning Electron Microscope (SEM, EVO MA25, ZEISS, Oberkochen, Germany) observations and measurement of particle sizes by image computer analysis with use of MicroMeter software (Version 1.0, Warsaw University of Technology, Warsaw, Poland). SEM observations showed spherical or spheroidal shapes of powder particles (Figure 1a) and the absence of closed pores (Figure 1b).

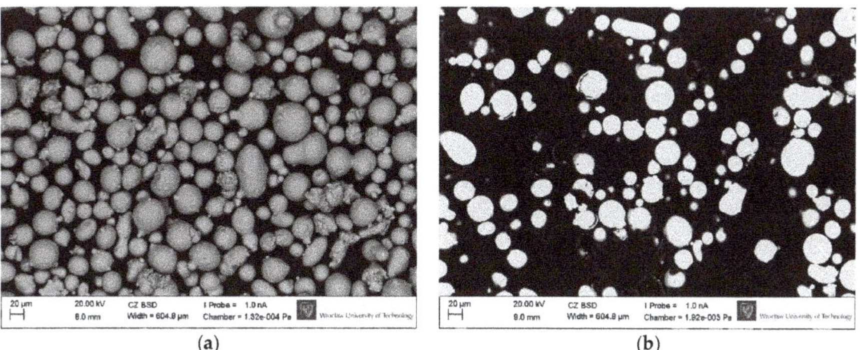

Figure 1. Quality of steel H13 powder: (**a**) spherical and spheroidal shapes, particles surfaces with small amount of satellites; (**b**) cross-sections of particles free from gas pores.

Image analysis of cross-sections of particles (Figure 1b) showed that maximum diameters of particles did not exceed 50 μm (Figure 2). The highest percentage of 45% was found for particles with equivalent diameter of 12 μm. Percentages of particles with equivalent diameters 6 μm and 18 μm were about 20% each. Such distribution of fractions corresponded to the bulk density of powder at the level of 4.6 g/cm^3, i.e., 59% density of solid material (7.8 g/cm^3). As can be seen in Figure 2, the used powder is fine and the distribution of its particles size is favorably skewed towards smaller diameters. The powder with a larger portion of fine particles provides a higher density of the bed, higher density of created parts under low laser energy and generates smoother side surfaces of finished parts [19].

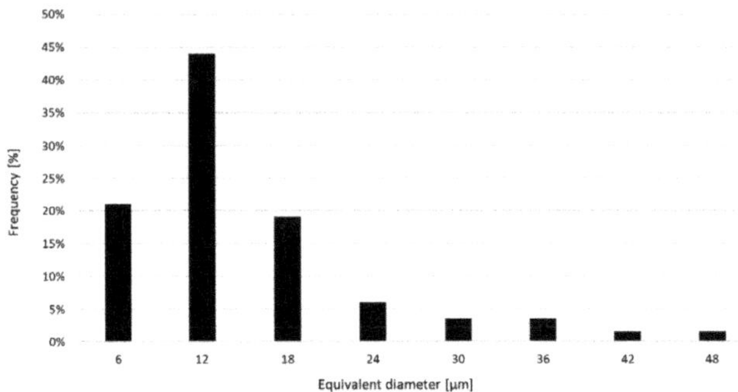

Figure 2. Distribution of fractions of H13 powder used in the research.

The chemical composition of the powder is given in Table 1. Before processing, the powder was soaked at 200 °C for 48 h in order to reduce its humidity thus increasing flowability, and to reduce its tendency to be covered with an oxide layer and, in consequence, to increase its processability.

Table 1. Chemical composition of H13 powder.

Concentration (wt %)	Fe	Cr	Mo	Si	V	Mn	C	P	S
AISI H13	Reminder	4.75–5.50	1.10–1.75	0.8–1.25	0.8–1.20	0.2–0.6	0.32–0.45	0.03	0.03
Used powder	Reminder	5.07	1.72	0.88	1.02	0.44	0.41	0.008	0.009

Specimens were manufactured using an SLM Realizer II 250 (Realizer, Borchen, Germany) machine equipped with a Yb fibre laser with maximum power of 400 W, wave length of 1070 nm and focused laser beam diameter of 200 μm. Material was processed under an argon shield without heating the working platform. All the specimens were manufactured with the use of the optimum set of parameters, given in Table 2. That set of parameters was determined by manufacture, using alternative scanning strategy, of specimens 8 mm × 10 mm × 5 mm (x × y × z) with minimum relative density 99.8 % as the acceptance criterion.

Table 2. Process parameters used for the manufacture of all the specimens.

Laser Power (W)	Exposure Time (μs)	Distance between Scanning Points (μm)	Hatch Distance (μm)	Layer Thickness (μm)	Scanning Speed (mm/s)
200	800	80	180	50	100

Tests were carried out in three stages in that effect of selected factors on relative density of the manufactured specimens was determined, namely:

- stage I—effect of scanning strategy depending on cross-section (x × y) dimensions of specimens (4 mm × 5 mm; 8 mm × 10 mm and 16 mm × 20 mm) and height of 5 mm (z),
- stage II—effect of support structures,
- stage III—effect of intermediate layers (connecting supports with standard layers) built with lower laser power.

At stage I, specimens with various dimensions were manufactured using 2 scanning variants for each size: single (A) and double (B) (re-melting), and 4 strategies (apart from standard strategies with long scan vectors (A0 and B0, see Figure 3), without dividing the scan area into sections) for each of the variants (A) and (B), see Figure 3:

1. Strategy (1) consisted of alternating scanning of stripes, where subsequent stripes are scanned from the edge of that scanning after the previous strip was finished (Figure 4a).
2. Strategy (2) consisted of scanning strips in the same direction, where starting points of scanning individual stripes were located on the same edge (Figure 4b).
3. Strategy (3) consisted of longitudinal scanning of stripes, similar to standard scanning, while scanning of individual stripes took place in a strictly determined order (Figure 5a).
4. Strategy (4), so-called chessboard, consisted of scanning individual square fields relative to those for which the specimen surface was divided. The scanning sequence of the fields is marked in Figure 5b.

Patterns of scanning Strategies 1–4 are shown in Figures 4 and 5 for the variant of single scanning (A). In the variant (B) of double scanning of each layer, Strategies 1–4 are performed in a similar way as strategy 0 (see Figure 3b). The scanning pattern of the solidified and again melted "layer n" is the same as that of "layer n+1" of the single scanning variant (A) (see Figures 4 and 5). This means that the scanning, i.e., powder melting and re-melting with no powder addition, is identical for each layer.

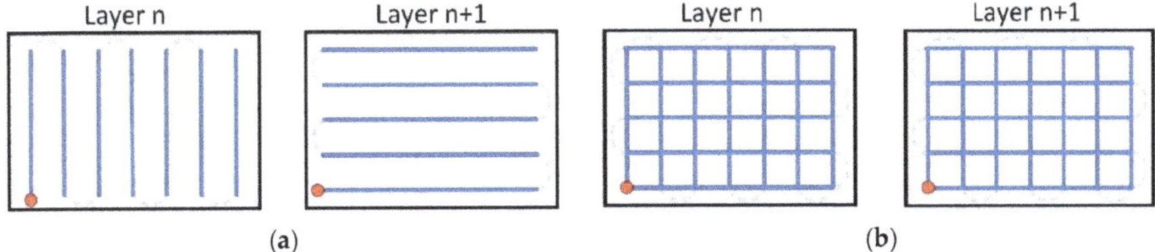

Figure 3. Standard scanning variants: (**a**) single scanning—A0; (**b**) double scanning—B0.

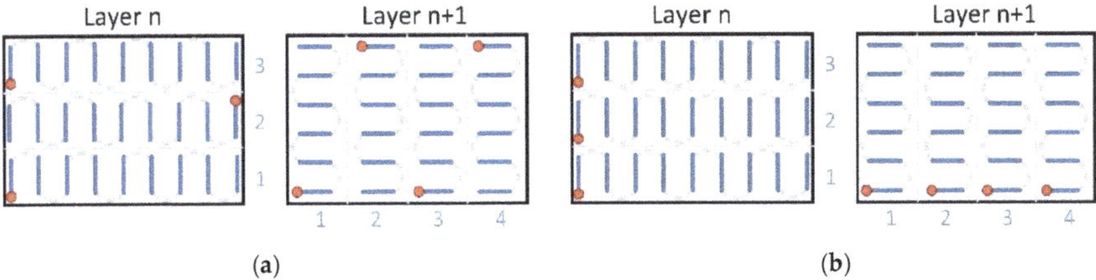

Figure 4. Standard single scanning variants: (**a**) stripes alternating—A1; (**b**) stripes in the same direction—A2.

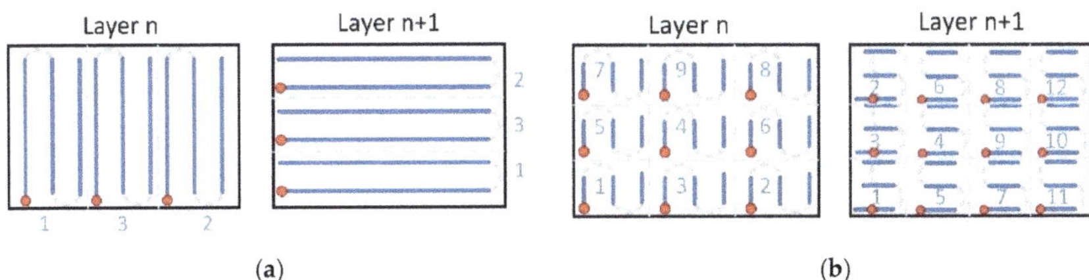

Figure 5. Standard single scanning variants: (**a**) stripes alongside—A3; (**b**) chessboard—A4.

In strategies 1, 2 and 3, various widths of scanning stripes were applied (e.g., A1/2—2 mm, A1/3—3 mm and A1/8—8 mm), but in the chessboard strategy—various sizes of individual fields were applied (e.g., A4/2—2 mm × 2 mm, A4/3—3 mm × 3 mm, A4/8—8 mm × 8 mm).

At stage II, specimens were manufactured using the strategies A3/3 and A4/3, and new supporting structures were designed. The supports used at stage I were located 2 mm away from each other and coincided with the scanning axes X and Y. At the second stage, the distances were reduced to 1.5 mm and the supports were inclined to the axis X at 30° and 45°.

At stage III, two series of specimens were manufactured with use of laser power reduced from 200 W to 150 W and with support structures located 1.5 mm away from each other, inclined to the axis X at 5°, 10° and 30°. In the second series, unlike in the first one, the first 5 layers from the support side were produced at laser power 125 W and the subsequent layers at 200 W. These intermediate layers served as a heat sink for the subsequent standard layers.

Relative density of the specimens manufactured at all the stages was assessed by computer image analysis. Metallographic polished sections were prepared on the planes xy. Surface images obtained with the use of a confocal microscope were subjected to binarization. Porosity was determined as the percentage of pixels falling on pores, i.e., on each kind of empty area irrespective of its shape and origin. Thus, "relative density" in this paper is calculated as the difference between 100% and percentage of pores.

Detection and visualization of three-dimensional distribution of pores within the specimens 16 mm × 20 mm was carried-out by computed tomography (CT, METROTOM 1500, ZEISS, Oberkochen, Germany). Reconstruction was performed using the system ZEISS METROTOM 1500 (Oberkochen, Germany). Resolution of data for steel specimens was 29.81 µm, at the X-ray tube voltage 225 kV and current 130 µA, with integration time of a single projection at the level of 2 s. With regard to the high density of the reconstructed specimens, a copper filter 3 mm thick was applied, and the obtained reconstructions were corrected to minimize measurement artefacts in the form of beam hardening. For the obtained data, porosity was detected with use of the software VG Studio MAX 2.0 (Volume Graphix, Heidelberg, Germany) and relative density was calculated as described above.

3. Results

3.1. Effect of Scanning Strategy on Relative Denisty Depending on Specimen Dimensions

For the research, 22 sets of 3 specimens each were prepared. Results are shown in Figures 6 and 7 for single and double scanning, respectively. It can be seen that, in both cases, the influence of the dimensions of the scanned surfaces xy and scanning strategies on the relative density of the specimens is significant.

All variants of specimens 16 mm × 20 mm did not reach the assumed relative density $\geq$99.8%. This condition was met by all the specimens (8 mm × 10 mm) produced by single scanning irrespective of strategy, and by the specimens produced by single scanning with the strategies A1, A2 and A4, with the dimension (stripe width or field side length) of the scanned field equal to 3 mm.

Double-scanned small specimens (4 mm × 5 mm and 8 mm × 10 mm) did not result in higher relative density, but the influence of scanning strategy was reduced. Nevertheless, similar to single-scanned specimens, a relative density $\geq$99.8% was reached by those manufactured with the strategies B0, B1, B2 and B4, with no clear relationship with dimensions (2, 3 or 8 mm) of scanning fields.

Summarizing the above results, it can be said that, irrespective of the dimensions of the scanned surface xy, relatively higher density was obtained by single scanning of small individual fields with the chessboard strategy (A4).

Microscopic examination of the specimens with surface dimensions x × y = 4 mm × 5 mm and 16 mm × 20 mm showed differences in geometry and arrangement of pores. In the specimens 4 mm ×5 mm, pores were spherical with diameter of 20 to 40 µm, linearly arranged along the scanning direction (Figure 8). Similar pores were also observed in both specimens 8 mm ×10 mm and 16 mm ×20 mm, but their amount was significantly smaller than that of pores with different shapes.

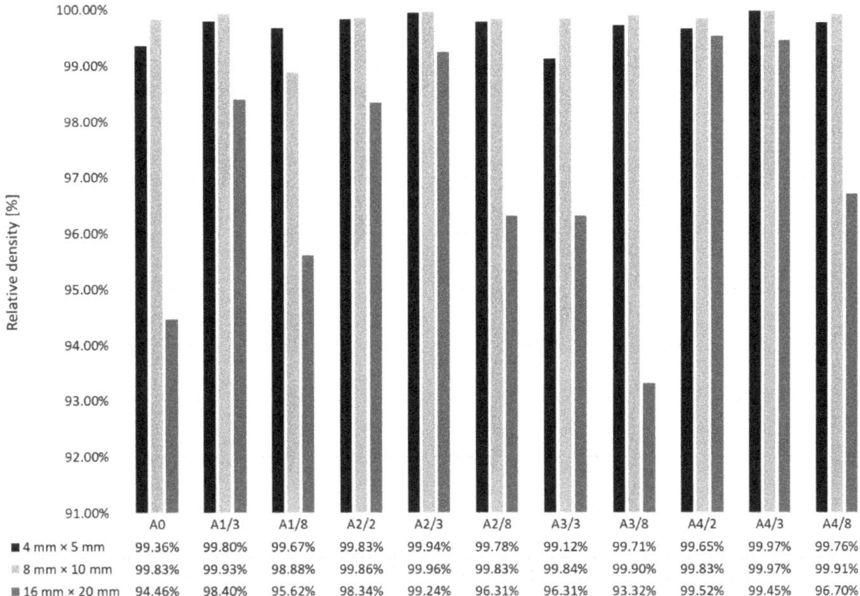

Figure 6. Relative density of specimens manufactured by single scanning with various strategies for various specimen dimensions x × y.

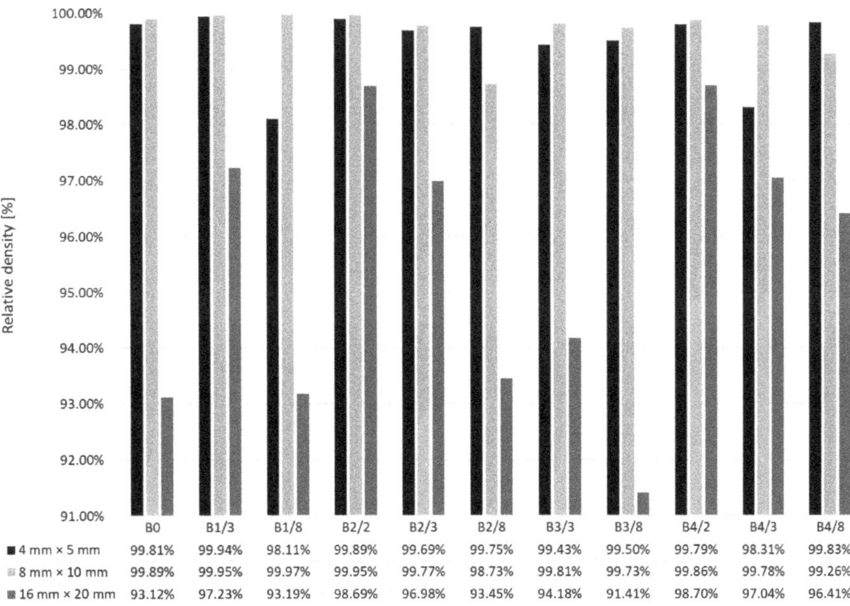

Figure 7. Relative density of specimens manufactured by double scanning with various strategies for various specimen dimensions x × y.

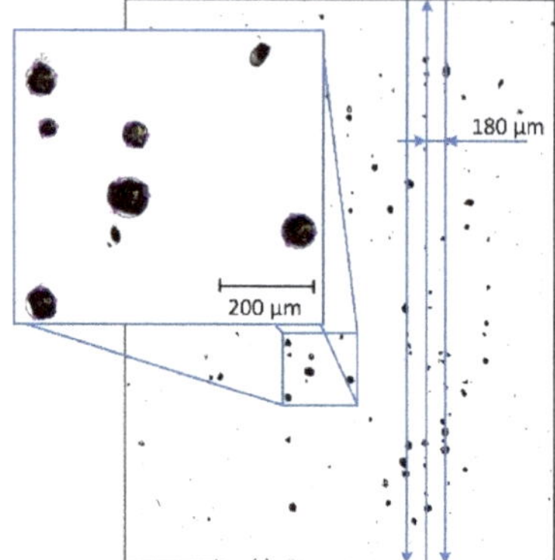

Figure 8. Distribution of pores in specimens manufactured by double scanning with various strategies for various specimen dimensions x × y.

In the specimens 16 mm × 20 mm, pores in the plane xy were mostly elliptical, with substitute diameter from 150 to 1400 μm. A smaller number of pores was visible in the initial scanning areas, and their concentration increased with proceeding scanning of subsequent sections. Figure 9a shows that the pores can merge to create longitudinal tunnels in the plane xy. No similar effects were observed in the specimens 4 mm × 5 mm and 8 mm × 10 mm.

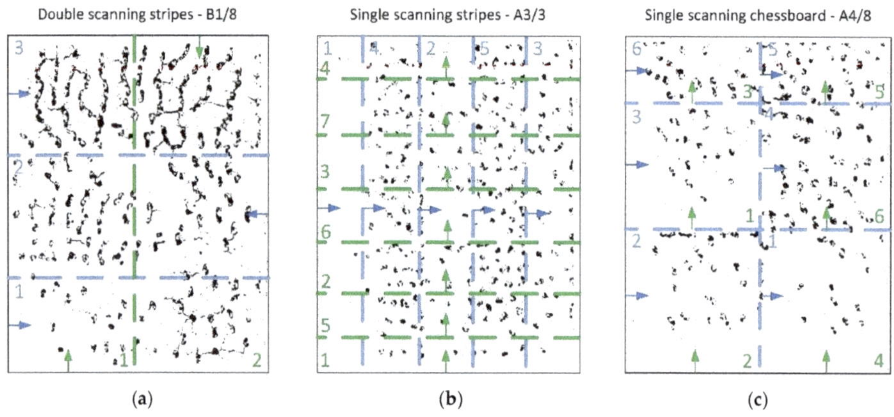

Figure 9. Distribution of pores (black) in the plane xy for the specimens 16 mm × 20 mm produced with scanning strategy: (**a**) B1/8; (**b**) A3/3; (**c**) A4/8. Scanning direction and sequence for even layers (x—green) and odd (y—blue) are presented. During double scanning, the specimen is scanned with the strategy for both even and odd layers.

The result of CT analysis of the specimen 16 mm × 20 mm made with the strategy A4/3, in the form of reconstruction, is shown in Figure 10. It can be seen that the pores shown in Figure 9a have shapes of columns passing through the entire height of the specimen. Visualization 3D of these pores is shown in Figure 11a,b. It was also found that columnar pores occur in the entire volume of the examined specimen. The tunnel-like shape of pores is the cause of the significant difference between results of quantitative 3D analysis of porosity performed by CT, i.e., 1.51%, and of quantitative 2D analysis performed on xy sections of the specimens, i.e., 0.55%.

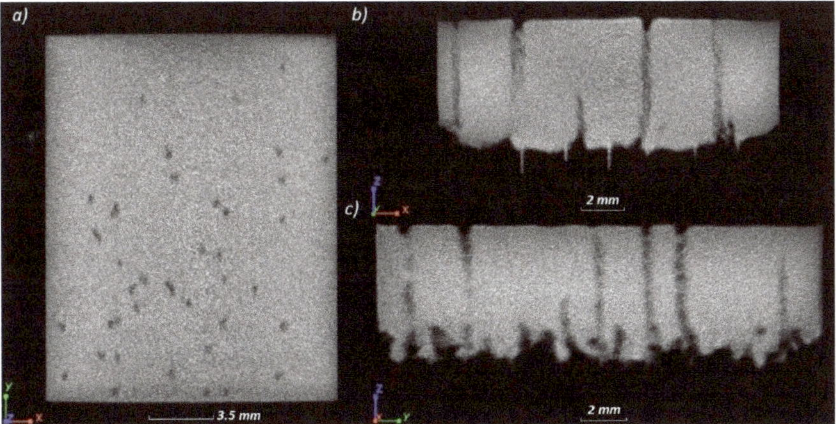

Figure 10. CT image of the specimen A4/3 with marked tunnel-like pores: (**a**) plane xy; (**b**) plane xz; (**c**) plane yz.

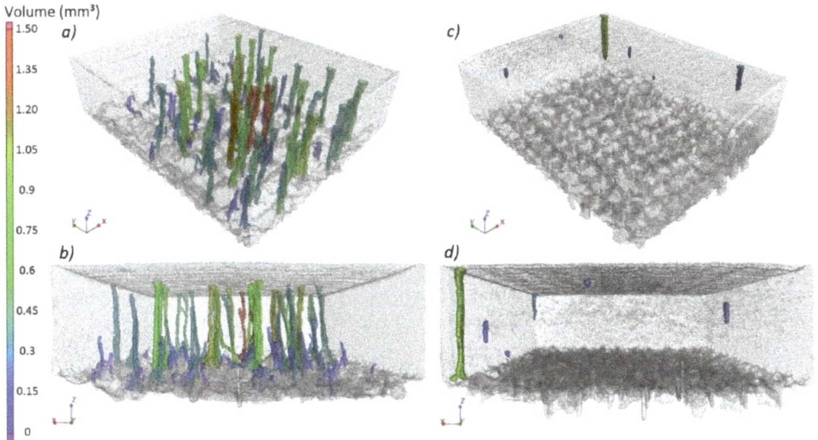

Figure 11. 3D visualization of CT examination of the specimens produced: (**a,b**) with the use of supporting structures 2 mm and the strategy A4/3; (**c,d**) with the use of supporting structures 1.5 mm/5° and the strategy A4/2.

It is visible in Figure 11 that a part of columnar pores (dark blue) disappears with scanning of subsequent layers. Moreover, columnar pores can split and develop in the form of two smaller columnar pores (Figure 12). In the upper part of the specimen, columnar pores expand to create characteristic funnels. Inside the pores, partially re-melted powder particles can be observed, as well as the balling effect on the free surface of solidified material (Figure 13). Due to their character, columnar pores are similar to the lack of fusion type pores. It can be seen in Figures 12 and 13 that columnar pores are created between supporting structures and clear "sinking" of the solidified layers evidences "pouring" of selectively melted metal into the pores.

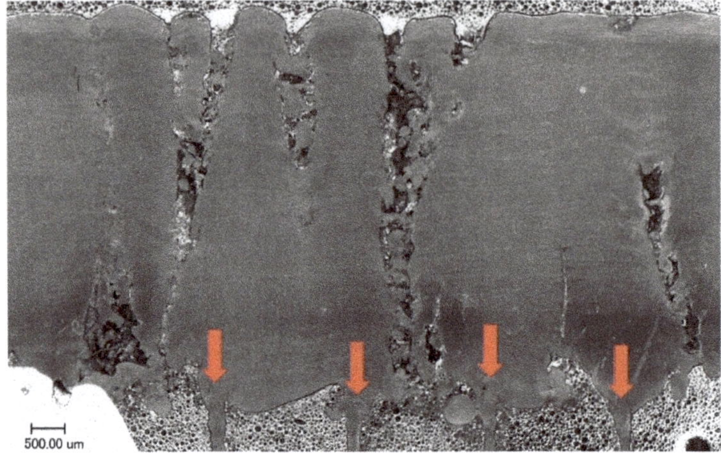

Figure 12. Columnar pores on section xy of the specimen A2/4. These pores are formed between supports shown with red arrows.

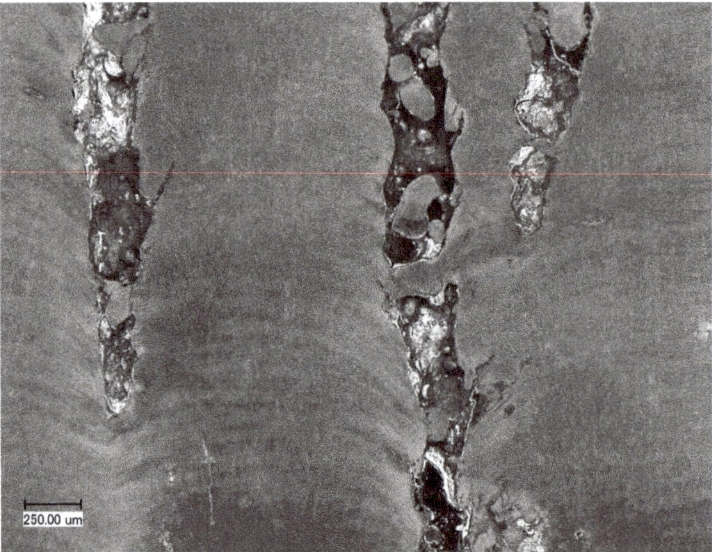

Figure 13. Columnar pores on section xz of the specimen A2/4. Visible sinking of the layer, preceding initiation of pores, and non-melted particles inside pores.

For more precise observation of the phenomenon of disappearance of pores, a series of flat specimens 4 mm × 5 mm, 8 mm × 10 mm and 16 mm × 20 mm were manufactured, the height of each corresponding to 1 to 5 layers. Images of xy surfaces of the specimens made with the scanning strategy A4/2 are shown in Figure 14. It can be seen that, for the specimens 4 mm × 5 mm, pores disappear after 5 layers. For the specimens 8 mm ×10 mm, after scanning 5 layers, columnar pores are still visible with maximum dimension up to ca. 800 μm. However, the fifth layer of the specimens 16 mm × 20 mm is characterized by numerous pores that often merge to create linear discontinuities with a circular cross-section. Measurements of percentage of these pores (open) for all the specimens are shown in Figure 15.

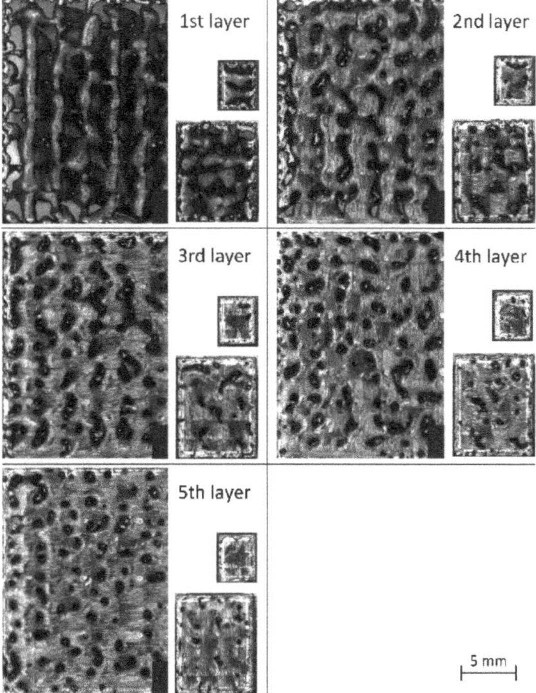

Figure 14. Microscopic images of free surface xy of the specimens with 3 sizes built to the height of 1 to 5 layers with the strategy A4/2.

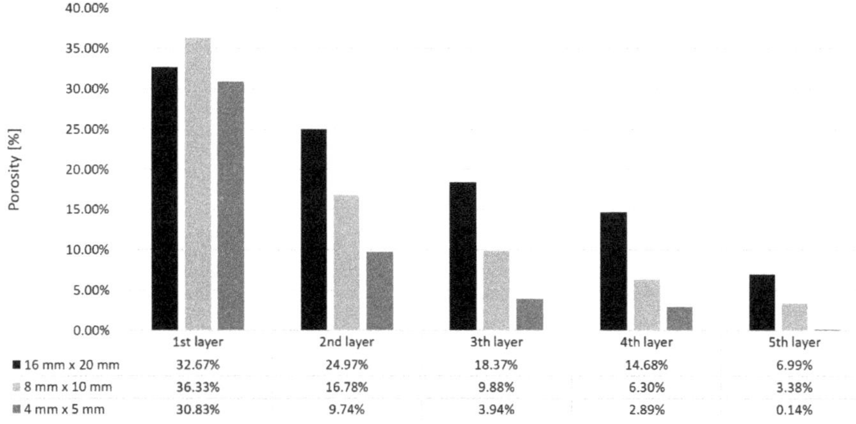

	1st layer	2nd layer	3th layer	4th layer	5th layer
■ 16 mm x 20 mm	32.67%	24.97%	18.37%	14.68%	6.99%
▨ 8 mm x 10 mm	36.33%	16.78%	9.88%	6.30%	3.38%
▧ 4 mm x 5 mm	30.83%	9.74%	3.94%	2.89%	0.14%

Figure 15. Porosity of individual layers for the specimens A4/2, with dimensions x × y = 16 mm × 20 mm, 8 mm × 10 mm and 4 mm × 5 mm.

As can be seen from Figure 15, porosity decreases with the manufacture of subsequent layers. The specimen x × y = 8 mm × 10 mm is characterized by the highest open porosity for the first layer However, with the manufacture of subsequent layers, the highest porosity occurs in the specimens 16 mm × 20 mm. The largest drops of porosity with increasing number of layers were noted for the specimens 4 mm × 5 mm, and the smallest drops—for the specimens 16 mm × 20 mm.

3.2. Effect of Support Structures

Supporting structures are mostly used in order to reduce deformation of the built object. In addition, their task is to remove heat and to support the object. The minimum distance between support structures should be such that the solidifying layers do not sink. However, concentration of supports results in a longer manufacturing time, as well as more difficult removal of the structures and detachment of the model from the working platform.

Considering the fact that columnar pores in Figures 12 and 13 were created between supporting struts, new support structures were used in stage II, with the distance reduced from 2 mm (stage I) to 1.5 mm, with various inclination (0°, 30° and 45°) to the working platform and two scanning strategies: A4/3 and A3/3. These strategies were selected because of very high (A4/3) and very low (A3/3) relative density obtained with their use. Results are shown in Figure 16. Manufacture of the specimens 16 mm × 20 mm with the use of concentrated supporting structures inclined at 30° to the scanning direction resulted in higher relative density, irrespective of scanning strategy. Relative density was improved by 0.29% only for the strategy A4/3 and as much as by 2.37% for the strategy A3/3.

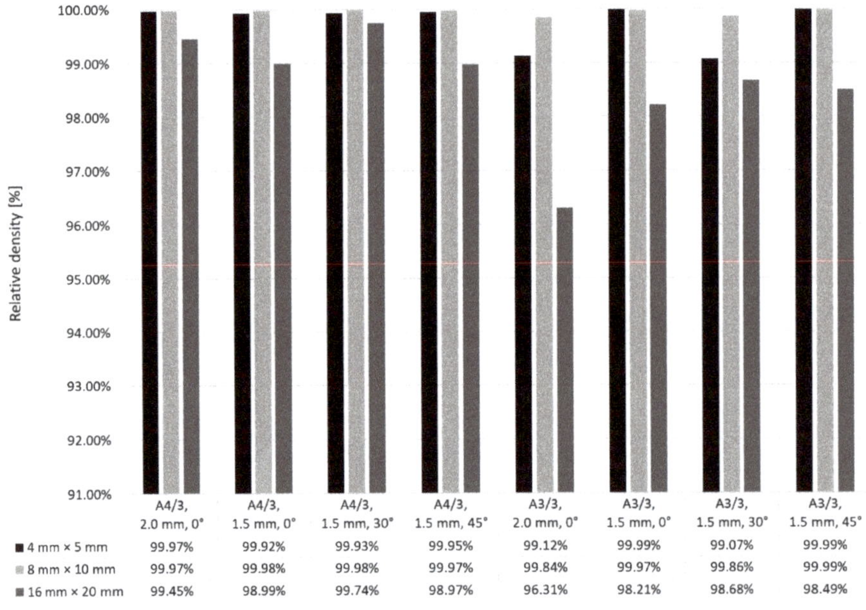

	A4/3, 2.0 mm, 0°	A4/3, 1.5 mm, 0°	A4/3, 1.5 mm, 30°	A4/3, 1.5 mm, 45°	A3/3, 2.0 mm, 0°	A3/3, 1.5 mm, 0°	A3/3, 1.5 mm, 30°	A3/3, 1.5 mm, 45°
4 mm × 5 mm	99.97%	99.92%	99.93%	99.95%	99.12%	99.99%	99.07%	99.99%
8 mm × 10 mm	99.97%	99.98%	99.98%	99.97%	99.84%	99.97%	99.86%	99.99%
16 mm × 20 mm	99.45%	98.99%	99.74%	98.97%	96.31%	98.21%	98.68%	98.49%

Figure 16. Relative density of specimens depending on their scanning and support strategies.

3.3. Effect of Intermediate Layers

Supporting structures are unavoidable in SLM processes; nevertheless, the efforts aimed at minimizing their side effects (increased printing time, cost and impact on surface quality) are necessary. In turn, a restriction for minimization of supporting structures is increasing porosity generated during re-melting of the first layers connected with the supports. According to the authors [20], manufacture (at properly selected laser power) of 20 intermediate layers between the minimized supporting structure and standard layers of the built object prevents an increase of porosity and formation of columnar pores.

Thus, at this stage of the examinations, two series of specimens were manufactured:

- at laser power reduced from 200 W to 150 W in order to decrease the tendency of the first solidifying metal layers to overhang,

- at the same laser power of 200 W but with introduced an intermediate layer, i.e., the first five layers built at lower laser power of 125 W (to decrease volume energy density in order to obtain a stable pool of liquid metal).

Specimens were manufactured on supports with distances 1.5 mm, inclined to the axis X at 5°, 10° and 30°. As can be seen in Figure 17, use of the intermediate layer built at lower laser power influenced relative density of the specimens. In comparison to the specimens with no intermediate layers, with inclination angle of supporting structures equal to 30° and with the scanning strategy A4/3, relative density of the specimens 16 mm × 20 mm increased by 0.16%. For the strategy A3/3, relative density increased by 0.90%. For most of the specimens manufactured at stage III, obtained was relative density over 99.5%. Only one parameter set, i.e. laser power 125 W; strategy A3/3 and supports inclined at 5° for manufacture of specimens 4 mm × 5 mm and 8 mm × 10 mm, did not give such satisfactory results. Very low porosity below 0.1% was obtained by the specimens made at laser power 150 W with supporting structures inclined at 5° and 30°.

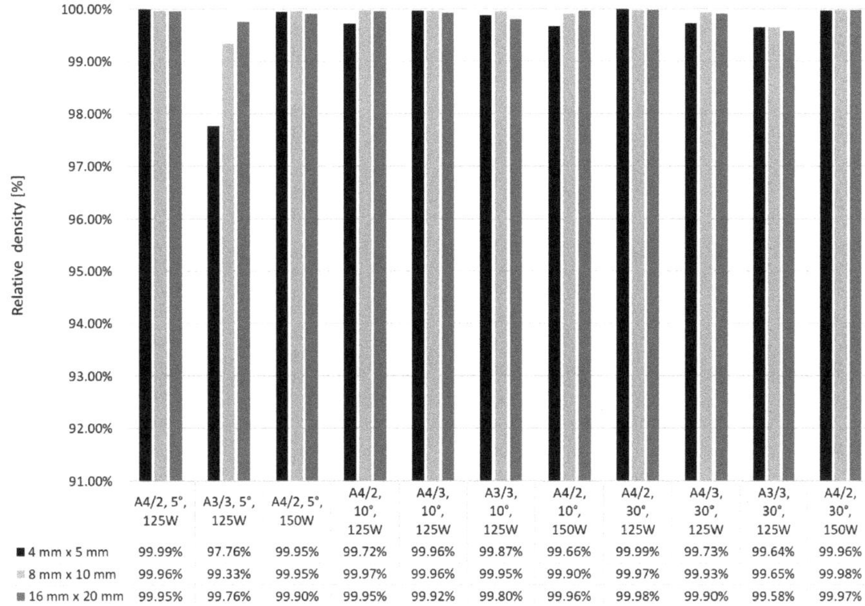

	A4/2, 5°, 125W	A3/3, 5°, 125W	A4/2, 5°, 150W	A4/2, 10°, 125W	A4/3, 10°, 125W	A3/3, 10°, 125W	A4/2, 10°, 150W	A4/2, 30°, 125W	A4/3, 30°, 125W	A3/3, 30°, 125W	A4/2, 30°, 150W
4 mm x 5 mm	99.99%	97.76%	99.95%	99.72%	99.96%	99.87%	99.66%	99.99%	99.73%	99.64%	99.96%
8 mm x 10 mm	99.96%	99.33%	99.95%	99.97%	99.96%	99.95%	99.90%	99.97%	99.93%	99.65%	99.98%
16 mm x 20 mm	99.95%	99.76%	99.90%	99.95%	99.92%	99.80%	99.96%	99.98%	99.90%	99.58%	99.97%

Figure 17. Effect of scanning parameters of intermediate layers and support geometry on relative density.

4. Discussion

In the SLM technology, both constant and variable process parameters influence the quality of manufactured parts. Constant parameters should be fixed in a strictly determined time interval. Variable parameters are controlled by an operator in such a way to ensure a stable run of the process and possibly high quality of the manufactured part. A large number of variable parameters makes it impossible to determine the influence of their possible combinations on product quality. Thus, efforts are made to find such parameters that would consider the total effect of the most important variable parameters. They include laser energy density defined by various formulas given in Table 3. The most often used comparable parameter is volume energy density (VED_H), whose definition considers mostly main variable parameters:

- laser power (W)—P,

- layer thickness (µm)—L,
- hatch spacing (µm)—H,
- scanning speed (mm/s)—V,
- focused beam diameter (µm)—f.

The choice of the definition of energy density depends on the geometry of the manufactured object. For the objects in the form of thin walls, spatial structures and supporting structures, surface energy density is used as a parameter (Equations (2) and (3) in Table 3). In the case of solid parts, the parameter considering hatch spacing is volume energy density (Equations (4) and (5) in Table 3). Most often, the comparative parameter VED_H is applied, since it considers the variable parameters most strongly influencing the efficiency of the melting process. However, this is not an ideal parameter, since it does not consider complex physical phenomena, like thermocapillary convection, hydrodynamic instability and recoil pressure that influence transfer of mass and heat inside the molten metal pool [21,22]. Therefore, for comparative reasons, process windows for the manufacture of parts from H13 powder ensure relative density over 99.9%. Table 4 includes values of energy density calculated according to the formulas cited in Table 3.

Table 3. Definitions of energy density used in powder-bed fusion additive manufacturing.

Item	Designation	Definition	Unit	Reference
1	Surface power density	$P_s = \frac{4P}{\pi f^2}$	$\frac{W}{mm^2}$	-
2	Surface energy density	$E_l = \frac{P}{V \cdot L}$	$\frac{J}{mm^2}$	[13]
3	Surface energy density	$E_f = \frac{P}{V \cdot f}$	$\frac{J}{mm^2}$	-
4	Volume energy density	$VED_H = \frac{P}{V \cdot H \cdot L}$	$\frac{J}{mm^3}$	[13]
5	Volume energy density	$VED_f = \frac{P}{V \cdot f \cdot L}$	$\frac{J}{mm^3}$	-

It can be seen from Table 4 that high density of H13 specimens can be obtained within a wide range of heat input, defined as the ratio of laser power to scanning speed, i.e., 0.25–2.0. High input energy, much higher than that applied by the authors in References [23,24], did not cause the balling effect or any significant increase of surface roughness, similar to results from References [25,26]. In spite of high heat input, which could result in increased width of seams, obtaining high relative density in this research did not require larger values of hatch spacing.

Table 4. Comparison of process parameters and corresponding energy densities used for the manufacture of H13 specimens with relative density >99.8%.

Parameter	This Work A0	This Work A4/3	This Work A4/2, 1.5 mm, 30°, 150 W	Laakso [23]	Mazur VED 80 [24]	Mazur VED 120 [24]
Laser power (W)	200	200	150	251	225	375
Layer thickness (µm)	50	50	50	30	30	30
Scanning speed (mm/s)	100	100	100	994	781	868
Hatch spacing (µm)	180	180	180	100	120	120
Focused beam diameter (µm)	200	200	200	80	80	80
P_s (W/mm²)	6369	6369	4777	49863	44785	74642
E_l (J/mm²)	40	40	30	8	10	14
E_f (J/mm²)	10	10	7.5	3.15	3.6	5.4
VED_H (J/mm³)	222	222	167	84	80	120
VED_f (J/mm³)	200	200	150	105,04	120	180
Strategy	A0	A4/3	A4/2	Rotated 67°	N/A	N/A

Table 4. *Cont.*

Parameter	This Work A0	This Work A4/3	This Work A4/2, 1.5 mm, 30°, 150 W	Laakso [23]	Mazur VED 80 [24]	Mazur VED 120 [24]
Specimen dimensions (mm^3)	8 × 10 × 5	8 × 10 × 5	8 × 10 × 5	10 × 10 × 10	4 × 4 × 4	4 × 4 × 4
Method of relative density measurement	Image analysis without boundary	Image analysis without boundary	Image analysis without boundary	Image analysis without boundary	CT	CT
Relative density (%)	99.83	99.97	99.98	99.91	99.91	99.99

Even though the volumetric energy density parameters VED_H and VED_f consider differences in focused beam diameter, layer thickness and hatch spacing, they cannot be comparative parameters or serve as design parameters in the optimization process. The obtained results indicate that, apart from heat input, selection of scanning strategy and support strategy play important roles in striving for maximum relative density.

The tendency of decreasing density along with increasing dimensions of scanning surfaces, observed in our own research, should be considered in drawing conclusions from a comparison of SLM parameters, like that in Table 4. The specimens manufactured in [24] had smaller dimensions, which reduced the probability of occurrence of tunnel-like pores.

The result of a comparison is also influenced by the method of porosity measurements. Computed tomography makes it possible to determine the quantity, size and arrangement of pores in the entire volume of the specimen. However, for materials with high density, to which H13 steel belongs, limited resolution of laboratory CT restricts detection of small pores. Resolution of the CT can be increased by reducing the thickness of samples (e.g., by cutting out their fragments) or using a system with higher X-ray tube voltage and current. In Reference [24], resolution of measurements is not quoted. In Reference [23], like in the presented work, relative density measurements were analyzed with the use of image analysis of similar cross-sectional areas, i.e., 100 mm^2 and 80 mm^2. No columnar pores occurred in both cases.

Until now, few works have been published in which the occurrence of columnar pores was observed in metal alloys processed by SLM. In one of them, Bauereiß et al. [15], in their experimental work supported by numerical simulation, showed that columnar, channel-like pores in Ti6Al4V occurred in the case when the SLM process was operated at a too low energy density. They also found that, at constant scanning speed and layer thickness, the number and size of channel-like pores decrease with increasing laser power. In this work, however, the opposite tendency of disappearing columnar pores at lower laser power was observed. Nevertheless, it should be stressed that the channel-like defects observed in [15] differed more with more irregular shapes of cross-sections in comparison to more compact, rounded shapes of cross-sections of columnar pores shown in Figure 11 to Figure 13. This can result from various mechanisms of their origin under the conditions of too low energy density (lack of fusion pores) or too high energy density (keyhole induced pores).

Shapes of cross-sections of columnar pores in Figure 10 to Figure 13 are undoubtedly related to geometry of the designed supporting structures and physical factors like wettability, surface tension or gravitation. It can be seen that metal solidifying on the supports tends toward the spherical shape in order to minimize its surface energy. The places most exposed to the occurrence of pores are, first of all, spaces between subsequent supports, as well as between supports and outside edge of the specimen. According to Reference [27,28], geometry and continuity of individual tracks in a single layer are affected by many factors, among others, packing density of powder, scanning speed and ratio of laser power to scanning speed. This applies to building tracks in free powder. When applied, are supports on that solidifying metal "anchors", seams "develop" along with scanning of subsequent layers and liquid metal spreads flowing to the spaces filled with powder. This is why, in microscopic images (Figure 13), single powder particles or splashes inside columnar pores can be observed. Therefore, in the light of the performed research, it seems purposeful to produce a properly selected number of the first densest layers at reduced laser power, in order to maximize density of the entire manufactured part.

5. Conclusions

In this study, the SLM process parameters optimized for H13 steel in relation to the assumed density $\geq 99.8\%$ were used to produce specimens with various dimensions of scanning planes, as well as various scanning and support strategies. The following conclusions can be drawn:

- Application of the same process parameters in the manufacture of specimens with various scanning surface areas results in a decrease of relative density for larger sizes to the degree dependent on the applied scanning strategy.
- Scanning strategy significantly influences quantity, sizes and arrangement of pores. Proper selection of strategy (e.g., chessboard with small dimensions of individual fields) makes it possible to suppress influence of size of the scanning section on relative density of the manufactured parts.
- Double scanning does not significantly reduce porosity of specimens, but reduces the influence of the scanning strategy.
- Distances between supports and their inclination angle to the X axis influence the occurrence of columnar pores. Density of supports and their proper inclination make it possible to reduce the number of columnar pores.
- Application of intermediate layers (first layers from sides of supports, made at properly selected process parameters) makes it possible to reduce the quantity of columnar pores and to maximize density in the entire specimen volume.

Author Contributions: Conceptualization, T.K. and W.S.; methodology, T.K.; validation, K.G. and B.K.; formal analysis, T.K.; investigation, W.S., K.G. and G.Z.; resources, T.K.; data curation, K.G. and G.Z.; writing—original draft preparation, W.S.; writing—review and editing, T.K. and B.K.; visualization, W.S. and K.G.; supervision, E.C.

Funding: This research received no external funding.

Conflicts of Interest: The authors declare no conflict of interest.

References

1. Gausemeier, J.; Echterhoff, N.; Wall, M. Thinking ahead the Future of Additive Manufacturing—Innovation Roadmapping of Required Advancements. Available online: https://dmrc.unipaderborn.de/fileadmin/dmrc/Download/data/DMRC_Studien/DMRC_Study_Part_3.pdf (accessed on 20 December 2018).
2. Gausemeier, J.; Echterhoff, N.; Kokoschka, M.; Wall, M. Thinking ahead the Future of Additive Manufacturing—Analysis of Promising Industries. Available online: https://dmrc.uni-paderborn.de/fileadmin/dmrc/06_Downloads/01_Studies/DMRC_Study_Part_1.pdf (accessed on 20 December 2018).
3. 3D Metals. Available online: https://slm-solutions.com/sites/default/files/downloads/201en171023-01-001-powder_web.pdf (accessed on 20 December 2018).
4. DebRoy, T.; Wei, H.L.; Zuback, J.S.; Mukherjee, T.; Elmer, J.W.; Milewski, J.O.; Beese, A.M.; Wilson-Heid, A.; De, A.; Zhang, W. Additive manufacturing of metallic components—Process, structure and properties. *Prog. Mater. Sci.* **2018**, *92*, 112–224. [CrossRef]
5. Grasso, M.; Colosimo, B.M. Process defects and in situ monitoring methods in metal powder bed fusion: A review. *Measurement Sci. Tech.* **2017**, *28*, 1–25. [CrossRef]
6. Kasperovich, G.; Haubrich, J.; Gussone, J.; Requena, G. Correlation between porosity and processing parameters in TiAl6V4 produced by selective laser melting. *Mater. Des.* **2016**, *105*, 160–170. [CrossRef]
7. Xia, M.; Gu, D.; Yu, G.; Dai, D.; Chen, H.; Shi, Q. Porosity evolution and its thermodynamic mechanism of randomly packed powder-bed during selective laser melting of Inconel 718 alloy. *Int. J. Mach. Tools Manuf.* **2017**, *116*, 96–106. [CrossRef]

8. King, W.E.; Anderson, A.T.; Ferencz, R.M.; Hodge, N.E.; Kamath, C.; Khairallah, S.A.; Rubenchik, A.M. Laser powder bed fusion additive manufacturing of metals; Physics, computational, and materials challenges. *Appl. Phys. Rev.* **2015**, *2*, 041304. [CrossRef]

9. Weingarten, C.; Buchbinder, D.; Pirch, N.; Meiners, W.; Wissenbach, K.; Poprawe, R. Formation and reduction of hydrogen porosity during selective laser melting of AlSi10Mg. *J. Mater. Process. Technol.* **2015**, *221*, 112–120. [CrossRef]

10. Taheri, H.; Shoaib, M.R.B.M.; Koester, L.; Bigelow, T.; Collins, P.C.; Bond, L.J. Powder-based additive manufacturing—A review of types of defects, generation mechanisms, detection, property evaluation and metrology. *Int. J. Addit. Subtractive Mater. Manuf.* **2017**, *2*, 172–209. [CrossRef]

11. Panwisawas, C.; Perumal, B.; Ward, R.M.; Turner, N.; Turner, R.P.; Brooks, J.W.; Basolto, H.C. Keyhole formation and thermal fluid flow-induced porosity during laser fusion welding in titanium alloys: Experimental and modelling. *Acta Materialia* **2017**, *126*, 251–263. [CrossRef]

12. Xiao, R.; Zhang, X. Problems and issues in laser beam welding of aluminum-lithium alloys. *J. Manuf. Processes* **2014**, *16*, 166–175. [CrossRef]

13. Panwisawas, C.; Qiu, C.L.; Sovani, Y.; Brooks, J.W.; Attallah, M.M.; Basoalto, H.C. On the role of thermal fluid dynamics into the evolution of porosity during selective laser melting. *Scr. Materialia* **2015**, *105*, 14–17. [CrossRef]

14. Koester, L.W.; Taheri, H.; Bigelow, T.A.; Collins, P.C.; Bond, L.J. Nondestructive testing for metal parts fabricated using powder-based additive manufacturing. *Mater. Eval.* **2018**, *76*, 514–524. Available online: https://www.researchgate.net/publication/324209458_Nondestructive_Testing_for_Metal_Parts_Fabricated_Using_Powder-Based_Additive_Manufacturing (accessed on 7 January 2019).

15. Bauereiß, A.; Scharowsky, T.; Körner, C. Defect generation and propagation mechanism during additive manufacturing by selective beam melting. *J. Mater. Process. Technol.* **2014**, *214*, 2497–2504. [CrossRef]

16. Gustmann, T.; Neves, A.; Kühn, U.; Gargarella, P.; Kiminami, C.S.; Bolfarini, C.; Eckert, J.; Pauly, S. Influence of processing parameters on the fabrication of a Cu-Al-Ni-Mn shape-memory alloy by selective laser melting. *Addit. Manuf.* **2016**, *11*, 23–31. [CrossRef]

17. Beal, V.E.; Erasenthiran, P.; Hopkinson, N.; Dickens, P.; Ahrens, C.H. Scanning strategies and spacing effect on laser fusion of H13 tool steel powder using high power Nd:YAG pulsed laser. *Int. J. Prod. Res.* **2008**, *46*, 217–232. [CrossRef]

18. Gan, M.X.; Wong, C.H. Practical support structures for selective laser melting. *J. Mater. Process. Technol.* **2016**, *238*, 474–484. [CrossRef]

19. Sames, W.J.; List, F.A.; Pannala, S.; Dehoff, R.R.; Babu, S.S. The metallurgy and processing science of metal additive manufacturing. *Int. Mater. Rev.* **2016**, *61*, 315–360. [CrossRef]

20. Cloots, M.; Spierings, A.B.; Wegener, K. Assessing new support minimizing strategies for the additive manufacturing technology SLM. In Proceedings of the Annual International Solid Freeform. Fabrication Symposium an Additive Manufacturing Conference, Austin, TX, USA, 12–14 August 2013; pp. 631–643. Available online: http://sffsymposium.engr.utexas.edu/Manuscripts/2013/2013-50-Cloots.pdf (accessed on 20 December 2018).

21. Prashanth, K.G.; Scudino, S.; Maity, T.; Das, J.; Eckert, J. Is the energy density a reliable parameter for materials synthesis by selective laser melting? *Mater. Res. Lett.* **2017**, *5*, 386–390. [CrossRef]

22. Scipioni Bertoli, U.; Wolfer, A.J.; Matthews, M.J.; Delplanque, J.P.R.; Schoenung, J.M. On the limitations of volumetric energy density as a design parameter for selective laser melting. *Mater. Des.* **2017**, *113*, 331–340. [CrossRef]

23. Laakso, P.; Riipinen, T.; Laukkanen, A.; Andersson, T.; Jokinen, A.; Revuelta, A.; Ruusuvuori, K. Optimization and simulation of SLM process for high density H13 tool steel parts. *Phys. Procedia* **2016**, *83*, 26–35. [CrossRef]

24. Mazur, M.; Leary, M.; McMillan, M.; Elambasseril, J.; Brandt, M. SLM additive manufacture of H13 tool steel with conformal cooling and structural lattices. *Rapid Prototyp. J.* **2016**, *22*, 504–518. [CrossRef]

25. Lee, J.H.; Jang, J.H.; Joo, B.D.; Yim, H.S.; Moon, Y.H. Application of direct laser metal tooling for AISI H13 tool steel. *Trans. Nonferrous Met. Soc. China* **2009**, *19*, 284–287. [CrossRef]

26. Kruth, J.P.; Froyen, L.; Van Vaerenbergh, J.; Mercelis, P.; Rombouts, M.; Lauwers, B. Selective laser melting of iron-based powder. *J. Mater. Process. Technol.* **2004**, *149*, 616–622. [CrossRef]

27. Körner, C.; Attar, E.; Heinl, P. Mesoscopic simulation of selective beam melting processes. *J. Mater. Process. Technol.* **2011**, *211*, 978–987. [CrossRef]

28. Tan, J.H.; Wong, W.L.E.; Dalgarno, K.W. An overview of powder granulometry on feedstock and part performance in the selective laser melting process. *Addit. Manuf.* **2017**, *18*, 228–255. [CrossRef]

Letter

Effect of Layer-Wise Varying Parameters on the Microstructure and Soundness of Selective Laser Melted INCONEL 718 Alloy

Xiang Wang [1], Jinwu Kang [1,*], Tianjiao Wang [2], Pengyue Wu [3], Tao Feng [3] and Lele Zheng [4]

1   School of Materials Science and Engineering, Key Laboratory for Advanced Materials Processing Technology, Tsinghua University, Beijing 100084, China
2   Department of Mechanical and Energy Engineering, Southern University of Science and Technology, Shenzhen 518055, China
3   Beijing e-Plus 3D Tech. Co., Ltd., Beijing 100027, China
4   School of Mechanical, Electronic and Control Engineering, Beijing Jiaotong University, Beijing 100044, China
*   Correspondence: kangjw@tsinghua.edu.cn; Tel.: +86-10-62784537

Received: 13 June 2019; Accepted: 3 July 2019; Published: 5 July 2019

Abstract: Selective laser melting (SLM) is a promising powder bed fusion additive manufacturing technique for metal part fabrication. In this paper, varying scanning speed in the range of 500 mm/s to 1900 mm/s, and laser power in the range of 100 W to 200 W, were realized from layer to layer in a cycle of 56 layers in a single cuboid Inconel 718 alloy specimen through SLM. Layer-wise variation of microstructure and porosity were acquired, showing the layer-wise controlling capability of microstructural soundness. The melt pool size and soundness are closely linked with the energy input. High energy density led to sound regions with larger, orderly stacked melt pools and columnar grains, while low energy density resulted in porous regions with smaller, mismatched melt pools, un-melted powder, and equiaxed grains with finer dendrites. With the increase of laser energy density, the specimen shifts from porous region to sound region within several layers.

Keywords: selective laser melting; microstructure; defects; Inconel 718; laser energy density

---

1. Introduction

Selective laser melting (SLM) is a type of powder bed fusion additive manufacturing technology, which melts metal powder using tiny a laser spot track-by-track, layer-by-layer to form metal components [1,2]. Due to this unique feature of SLM, it is capable of controlling the microstructure of any location of a part. Laser power (P), scanning speed (v), hatch spacing (d), and layer thickness (t) are the main process parameters. Laser energy density, $E = P/(v \cdot d \cdot t)$, is used to regulate the microstructure and mechanical properties [3,4]. Studies conducted in the past have shown that, with the increase in energy density, soundness and mechanical properties are significantly enhanced [5–7]. Swee et al. [8] conducted experiments on the selective laser melting of TiTa alloy and concluded that increasing laser energy density is needed to achieve fully dense parts and to fully melt the tantalum in the powder. Additionally, Liu et al. [9] revealed that high energy input also leads to higher proportion of columnar grains. Yu et al. [10] applied re-melting on the SLM of AlSi10Mg and found that it reduced surface roughness as well as the porosities in the as-printed specimens. Grain morphologies and texture can also be altered by different scan strategies. Wan et al. [11] found that a 90° rotation of scan directions leads to stronger cubic texture because of the direction change of heat flux. Kurzynowski et al. [12] discovered that perpendicularly re-melting the current layer alters the microstructural texture from cubic to partially fiber. Kirka et al. [13] adopted a point heat source strategy in electron beam melting (EBM) and realized either columnar or equiaxed grain structures upon solidification through changes in the process parameters for nickel-base super-alloy. Studies with post-process treatments were also performed on selective laser melted samples. Mostafa et al. [14] studied the structure and texture of selective laser melted Inconel 718 alloy and concluded that the columnar grains, with strong {100} texture in the as-printed sample, shifted to equiaxed grains after heat treatment. Calandri et al. [15] adopted heat treatment on selective laser melted Inconel 718 alloy in different temperatures and reached maximal hardness value by aging at 714 °C.

However, most studies are limited to one specimen out of one set of parameters. In this paper, specimens with a wide range of varying parameters were fabricated by selective laser melting to gain a thorough understanding of the effect of these parameters on defects and microstructure. This has the potential of bringing out the uniqueness of SLM to achieve a microstructural gradient in one component.

2. Materials and Methods

Nickel-based Inconel 718 alloy was selected for the experiment. Due to its excellent mechanical properties under a wide range of temperature, it has long been considered as one of the most highly applicable materials for gas turbine blades, aero-engines and other parts in the field of aeronautics. Pre-alloyed gas-atomized Inconel 718 powder, produced by AMC Powders Co., Ltd. (Beijing, China), was used. Cuboid specimens of $15 \times 15 \times 60$ mm^3 were fabricated using the SLM based 3D printer EP-M100T manufactured by Beijing e-Plus 3D Tech. Co., Ltd. (Beijing, China). An Al substrate was applied. Highly pure argon gas was introduced to lower oxygen contamination. Raster scan strategy with a rotation of 90° in consecutive layers was adopted. The as-printed specimen with the applied parameters is shown in Figure 1. With the layer thickness (t) and hatch distance (d) held constant at 20 μm, and 100 μm, respectively, laser power (P) and scanning speed (v) varied from layer to layer repeatedly in every 56 layers. The laser power applied increased stepwise from 100 W to 200 W for every 3–6 layers, while scan speed changed back and forth from 500 mm/s to 1900 mm/s. Thus, the volume energy density shifted from 50–150 J/mm^3.

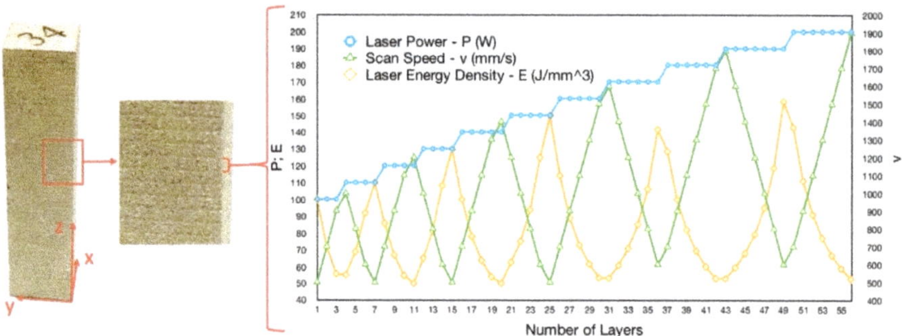

Figure 1. As-printed specimen and the applied parameters of each layer in a cycle.

Specimens were cut and then ground with 240–2000 waterproof abrasive papers and polished with diamond suspension from 2 mm to 0.5 mm and with colloidal silica suspension of 0.05 mm, then electrolytically etched for microstructural analyses. A solution of 70 vol.% phosphoric acid and 30 vol.% water was used as the etchant. Electrolytical etching was performed under the voltage of 5 V for 6–8 s. Microstructural characterization was performed using a Keyence VHX 6000 optical microscope (Osaka, Japan), and a ZEISS Merlin scanning electron microscope (Hallbergmoos, Germany), which was also used for EBSD tests.

3. Results and Discussion

The as-printed specimen presents a macro-morphology of multiple consecutive bands, shown in Figure 1. A total of 53 bands is observed in the specimen, indicating that the average height of one band is approximately 1.13 mm, corresponding to one cycle of 56 20-μm-thick layers. This shows that the morphology is related to the process parameters.

The optical micrograph of x-z plane of the specimen (along the build direction) corresponding to varying laser energy inputs is shown in Figure 2a, which clearly demonstrates that soundness of the specimen directly varied with the laser energy density. High laser energy density resulted in soundness, larger, regularly shaped melt pools, while low laser energy density led to porosity, un-melted powder, smaller, irregularly shaped melt pools. Good stacking and overlap of melt pools was found in regions with high energy input, while significant mismatch of melt pools, between adjacent layers tracks, was found under low energy input. The magnified image of the yellow rectangle in Figure 2a is shown in Figure 2b. As can be seen, the widths of the melt pools are approximately 100–120 μm, and 20–40 μm in depth, well in accordance with the hatch spacing and layer thickness. It is also observed that the overall shape of the melt pools in regions with higher energy inputs are larger in both the width and the depth, which means that increased laser energy will melt more extensively into the metal powder bed both horizontally and vertically.

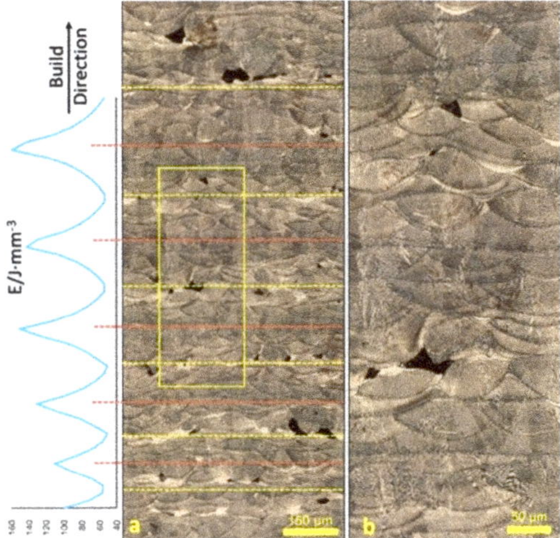

Figure 2. (**a**) Optical micrograph of the x-z plane corresponding to the variation of laser energy input and (**b**) detailed optical micrograph of the region in the yellow rectangle.

During SLM, low energy input cannot fully melt the metal powders, leaving some un-melted or partially melted powders in the as-printed specimen. Meanwhile, low energy input also leads to poor flowability in the molten metal, making it unable to spread out evenly in the present layer. Additionally, this further causes a relatively rugged surface, hindering the swiping of next layer's powder, and leads to possible bridging among powder particles. In layers under continuous low energy inputs, pores can extend to 2–3 layers. However, in layers above those large pores where laser energy input rises, the morphology tends to be smoother, because this increase leads to better flowability of the molten metal, which spreads out more evenly in the current layer, smoothing the surface and improving the spreading of powders in the next layer. Thus, after a few layers with increasing energy inputs, the morphology of the specimen shifts to soundness. Hence, it is possible that the micromorphology of the SLM processed specimens can be altered by increasing or decreasing the laser energy inputs in different layers. This lends insight into controlling the soundness level in selective laser melting by a set of intentionally arranged energy inputs for different layers. Meanwhile, different types of micromorphology were obtained in one specimen, demonstrating the capability of achieving combinations or functional gradients of porous and sound zones in one component using selective laser melting, which would otherwise be improbable with other traditional manufacturing methods.

Inverse pole figure (IPF) of the EBSD test of the x-z section is shown in Figure 3. In sound regions with high laser energy inputs, grains evolved into columnar grains along the build direction. Whereas, in porous regions indicating lower energy input, grains grew in an equiaxed pattern.

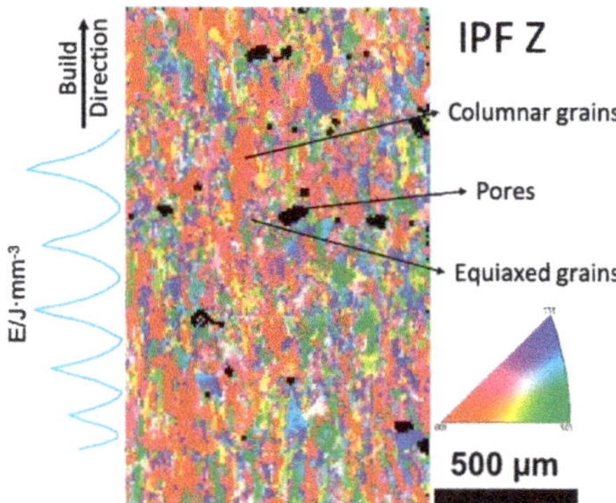

Figure 3. EBSD pattern of the x-z section.

Under high energy input, the temperature gradient was more significant, facilitating columnar grain growth. Also, the stacking of melt pools was in good order and the previous layer was re-melted more deeply under a higher laser energy input. This deeper re-melt caused grains in the previous layers to expose preferable crystal planes, acting as nucleation cites. Grains in the current layer tended to grow along the same orientation in those exposed partially re-melted grains, extending the grain through the melt pool boundary. Thus, grains in the center of melt pools can grow continuously through multiple layers along the build direction. As for regions with low energy input, both the temperature and temperature gradient were smaller. Metals in the previous layer were not re-melted sufficiently. Hence, nucleation cites mostly came from local melt pools, leading to the formation of equiaxed grains. This is in good agreement with other studies of SLM processed Inconel 718 alloys [16], where higher laser power facilitated columnar grain growth. Similar results have also been found in other SLM processed alloys [9].

An assembly of the SEM images of the x-z section along the build direction is shown in Figure 4a. A typical cellular dendritic growth is observed, which is also mentioned in references [7,17], as well as in other SLM processed alloys [3,18]. During a selective laser melting process, molten metal solidifies so rapidly with a cooling rate of up to 10^6 K/s that there is no time for secondary dendrite arms to grow. Thus, only cellular growth appeared. Melt pool boundaries are depicted in the image. It can be seen that the cell sizes varied in different layers. A hardness measurement point was used as a locating spot to match the energy inputs in these layers. As the energy input rose, the size of cellular dendrites coarsened gradually. Shown in magnified images, the widths of the cellular dendrites in regions with high energy inputs are approximately 0.8–1.0 µm (Figure 4b,c), greater than that in regions with low energy inputs (~0.5 µm, Figure 4d).

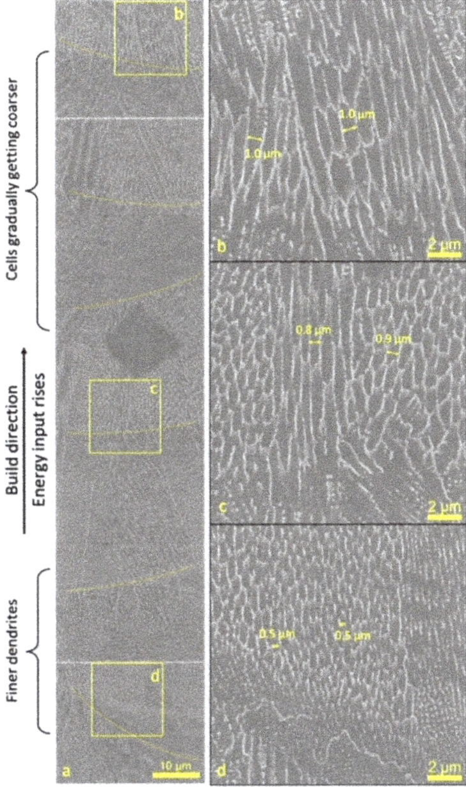

Figure 4. SEM images of the x-z section of the sample, (**a**) an assembly along the build direction showing several layers of melt pools; (**b–d**) magnified images.

Microhardness test results are shown in Figure 5. Overall the average microhardness value of the as-printed specimen is 344.79 ± 15.05 HV, comparable to the values reported in literature with samples fabricated using uniform parameters [6,19,20] (ranging from 325 to 395.8 HV for the as-printed specimens). However, it can be seen that the microhardness values varied slightly in different places. Generally, test points near pores displayed lower microhardness values than those tested in sound regions. This shows that higher energy inputs, corresponding to sound regions, lead to relatively higher hardness. This is because, when the laser beam scans the current layer, there is a heat affected zone in the previous layer where the metal is heated up below its melting point and cooled down. This process can be viewed as a solution heat treatment that enhances the microhardness of the affected zone. Under higher laser energy inputs, the previous layer goes through a more thorough heat treatment, hence, the microhardness tends to be higher.

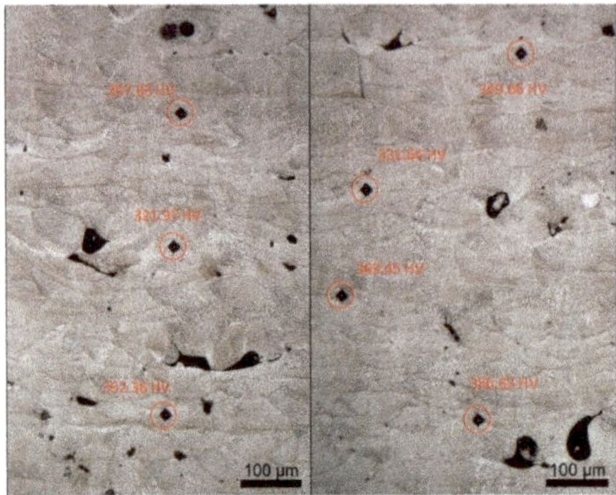

Figure 5. Microhardness test results.

4. Conclusions

To explore the maneuverability of layer-by-layer variation of parameters during selective laser melting, a cycle of 56 layers with varying laser power and scanning speed was realized in an SLM processed Inconel 718 cuboid specimen with 53 cycles. Layer-wise variations of microstructure were acquired, showing the capability of microstructural control. The gradual shifts between soundness and porosity are observed. High energy density leads to sound regions with bigger melt pools and coarser cells, while low energy density results in smaller melt pools and cells, as well as more porosities. Additionally, high energy inputs boost columnar grain growth, while low energy inputs favor equiaxed grain growth. Different types of micromorphology concerning porous and sound regions, as well as different types of grain morphology were realized in a single specimen, which indicates that SLM is capable of achieving combinations or functional gradients of porous and sound zones or different distributions of columnar and equiaxed grains in one component. This provides a huge range of possibilities in the processing design of metal manufacturing to realize gradient structure for gradient function and high throughput microstructure-mechanical properties test.

The as-printed specimens display a comparable microhardness than the ones fabricated using uniform parameters. With the increase of laser energy density, the specimen shifts from porous region to sound region within several layers. This provides insights for the attentive control of microstructures in selective laser melting. By increasing the laser energy input, layers with porosity can gradually transfer into layers with microstructural soundness.

Author Contributions: Conceptualization, X.W. and J.K.; investigation, X.W., T.W., and P.W.; formal analysis, X.W. and L.Z.; resources, T.F. and P.W.; data curation, X.W. and J.K.; writing—original draft preparation, X.W.; writing—review and editing, J.K. and X.W.; supervision, J.K.

Funding: This research was funded by National Science and Technology Major Project of the Ministry of Science and Technology of China, grant number 2016YFB1100703.

References

1. Yap, C.Y.; Chua, C.K.; Dong, Z.L.; Liu, Z.H.; Zhang, D.Q.; Loh, L.E.; Sing, S.L. Review of selective laser melting: Materials and applications. *Appl. Phys. Rev.* **2015**, *2*, 41101. [CrossRef]
2. Gao, W.; Zhang, Y.; Ramanujan, D.; Ramani, K.; Chen, Y.; Williams, C.B.; Wang, C.C.L.; Shin, Y.C.; Zhang, S.; Zavattieri, P.D. The status, challenges, and future of additive manufacturing in engineering. *Comput. Des.* **2015**, *69*, 65–89. [CrossRef]

3. Tucho, W.M.; Lysne, V.H.; Austbø, H.; Sjolyst-Kverneland, A.; Hansen, V. Investigation of effects of process parameters on microstructure and hardness of SLM manufactured SS316L. *J. Alloy. Compd.* **2018**, *740*, 910–925. [CrossRef]

4. Zhang, H.; Zhu, H.; Qi, T.; Hu, Z.; Zeng, X. Selective laser melting of high strength Al–Cu–Mg alloys: Processing, microstructure and mechanical properties. *Mater. Sci. Eng. A* **2016**, *656*, 47–54. [CrossRef]

5. Yi, J.; Kang, J.; Wang, T.; Wang, X.; Hu, Y.; Feng, T.; Feng, Y.; Wu, P. Effect of laser energy density on the microstructure, mechanical properties, and deformation of Inconel 718 samples fabricated by selective laser melting. *J. Alloy. Compd.* **2019**, *786*, 481–488. [CrossRef]

6. Jia, Q.; Gu, D. Selective laser melting additive manufacturing of Inconel 718 superalloy parts: Densification, microstructure and properties. *J. Alloy. Compd.* **2014**, *585*, 713–721. [CrossRef]

7. Deng, C.; Kang, J.; Feng, T.; Feng, Y.; Wang, X.; Wu, P. Study on the Selective Laser Melting of CuSn10 Powder. *Materials* **2018**, *11*, 614. [CrossRef] [PubMed]

8. Sing, S.L.; Wiria, F.E.; Yeong, W.Y. Selective laser melting of titanium alloy with 50 wt% tantalum: Effect of laser process parameters on part quality. *Int. J. Refract. Met. Hard Mater.* **2018**, *77*, 120–127. [CrossRef]

9. Liu, X.; Zhao, C.; Zhou, X.; Shen, Z.; Liu, W. Microstructure of selective laser melted AlSi10Mg alloy. *Mater. Des.* **2019**, *168*, 107677. [CrossRef]

10. Yu, W.; Sing, S.L.; Chua, C.K.; Tian, X. Influence of re-melting on surface roughness and porosity of AlSi10Mg parts fabricated by selective laser melting. *J. Alloy. Compd.* **2019**, *792*, 574–581. [CrossRef]

11. Wan, H.; Zhou, Z.; Li, C.; Chen, G.; Zhang, G. Effect of scanning strategy on grain structure and crystallographic texture of Inconel 718 processed by selective laser melting. *J. Mater. Sci. Technol.* **2018**, *34*, 1799–1804. [CrossRef]

12. Kurzynowski, T.; Gruber, K.; Stopyra, W.; Kuźnicka, B.; Chlebus, E. Correlation between process parameters, microstructure and properties of 316 L stainless steel processed by selective laser melting. *Mater. Sci. Eng. A* **2018**, *718*, 64–73. [CrossRef]

13. Kirka, M.M.; Lee, Y.; Greeley, D.; Okello, A.; Goin, M.J.; Pearce, M.T.; Dehoff, R.R. Strategy for Texture Management in Metals Additive Manufacturing. *JOM* **2017**, *69*, 523–531. [CrossRef]

14. Mostafa, A.; Rubio, I.P.; Brailovski, V.; Jahazi, M.; Medraj, M. Structure, Texture and Phases in 3D Printed IN718 Alloy Subjected to Homogenization and HIP Treatments. *Metals* **2017**, *7*, 196. [CrossRef]

15. Calandri, M.; Yin, S.; Aldwell, B.; Calignano, F.; Lupoi, R.; Ugues, D. Texture and Microstructural Features at Different Length Scales in Inconel 718 Produced by Selective Laser Melting. *Materials* **2019**, *12*, 1293. [CrossRef] [PubMed]

16. Popovich, V.; Borisov, E.; Popovich, A.; Sufiiarov, V.; Masaylo, D.; Alzina, L. Functionally graded Inconel 718 processed by additive manufacturing: Crystallographic texture, anisotropy of microstructure and mechanical properties. *Mater. Des.* **2017**, *114*, 441–449. [CrossRef]

17. Amato, K.; Gaytan, S.; Murr, L.; Martinez, E.; Shindo, P.; Hernandez, J.; Collins, S.; Medina, F.; Murr, L. Microstructures and mechanical behavior of Inconel 718 fabricated by selective laser melting. *Acta Mater.* **2012**, *60*, 2229–2239. [CrossRef]

18. Chen, B.; Moon, S.; Yao, X.; Bi, G.; Shen, J.; Umeda, J.; Kondoh, K. Strength and strain hardening of a selective laser melted AlSi10Mg alloy. *Scr. Mater.* **2017**, *141*, 45–49. [CrossRef]

19. Deng, D.; Peng, R.L.; Brodin, H.; Moverare, J. Microstructure and mechanical properties of Inconel 718 produced by selective laser melting: Sample orientation dependence and effects of post heat treatments. *Mater. Sci. Eng. A* **2018**, *713*, 294–306. [CrossRef]

20. Wang, Z.; Guan, K.; Gao, M.; Li, X.; Chen, X.; Zeng, X. The microstructure and mechanical properties of deposited-IN718 by selective laser melting. *J. Alloys Compd.* **2012**, *513*, 518–523. [CrossRef]

 *materials*

Article

Fabrication of Crack-Free Nickel-Based Superalloy Considered Non-Weldable during Laser Powder Bed Fusion

Oscar Sanchez-Mata, Xianglong Wang, Jose Alberto Muñiz-Lerma, Mohammad Attarian Shandiz, Raynald Gauvin and Mathieu Brochu *

Department of Mining and Materials Engineering, McGill University, 3610 University Street, Wong Building, Montreal, QC H3A 0C5, Canada; oscar.sanchezmata@mail.mcgill.ca (O.S.-M.); xianglong.wang@mail.mcgill.ca (X.W.); jose.muniz@mail.mcgill.ca (J.A.M.-L.); mohammad.attarianshandiz@mail.mcgill.ca (M.A.S.); raynald.gauvin@mcgill.ca (R.G.)
* Correspondence: mathieu.brochu@mcgill.ca; Tel.: +1-514-398-2354

Received: 5 July 2018; Accepted: 23 July 2018; Published: 25 July 2018

Abstract: Crack-free Hastelloy X fabricated through laser powder bed fusion (LPBF) from powder with a standard chemical composition is reported. Electron backscatter diffraction (EBSD) analysis evidenced columnar grains parallel to the building direction. The typical LPBF columnar dendrite microstructure was found to be finer than reported elsewhere. Mo-enriched carbides (~50 nm), presumed to play an important role in the cracking behavior of the alloy, were confirmed along interdendritic regions. Crack-free condition was maintained after heat treatment at 1177 °C for 1 h followed by water quenching, and the resulting microstructure was analyzed.

Keywords: nickel alloys; Hastelloy X alloy; additive manufacturing; microstructure; scanning electron microscopy (SEM); laser powder bed fusion (LPBF)

---

1. Introduction

Laser Powder Bed Fusion (LPBF) of Ni-based Superalloys is finding applications in many fields such as aerospace and land-based turbine engines [1]. This process involves the rastering of a focused beam to selectively melt powder in a layer-by-layer methodology [2]. Unfortunately, LPBF of many of these alloys face the important issue of cracking through a variety of mechanisms, such as liquation and solidification cracking or strain-age and ductility-dip cracking [3]. The welding field has developed criteria to evaluate the cracking susceptibility of alloys [4,5]. Such concepts are often referred in the context of additive manufacturing (AM), but recent experimental evidences have shown that the cracking susceptibility is more severe in the context of LPBF, i.e., a crack-free alloy during welding can develop crack during AM [6,7]. An interesting example is Hastelloy X, a Ni-Cr-Fe-Mo solid solution strengthened alloy [8] that although categorized as a 'weldable' alloy [9], has been shown to be prone to hot cracking during LPBF [10–14].

Various explanations and strategies to eliminate cracking have been reported. Wang, et al. [10] attributed the formation of cracks in samples built using a powder containing 0.78% Si, 0.69% Mn, and 0.009% C to the residual stresses developing during processing, but were not able to recommend an optimal process window to fabricate crack-free samples. Another approach to eliminate cracking included the control of the chemical composition of the feedstock powder. Tomus, et al. [12,13] highlighted that the microsegregation of Si and C contribute to hot cracking and suggested that a composition of 0.11% Si, <0.01% Mn, and 0.01% C would inhibit crack formation, as opposed to an alloy possessing a higher concentration of these three elements. Harrison, et al., suggested that increasing the concentration of substitutional solid solution strengthening elements like Mo, Co and

W helps to increase the resistance to cracking. Their experiments were carried out using a powder containing 0.31% Si, 0.22% Mn, and 0.054% C, and results showed crack density reduction compared to the original composition of 11.6 ± 2.4 to 1.6 ± 0.9 cracks per mm^2 [14]. Recently, Marchese proposed that the formation of intergranular Mo-enriched carbides, coupled with high thermal residual stresses, resulted in hot cracking during LPBF of Hastelloy X [15].

Based on these results, the AM community has been focusing on tailoring the chemical composition of Hastelloy X as a viable avenue to mitigate cracking. A consensus in literature points towards intergranular low melting point eutectics to be responsible for the cracking susceptibility of this alloy. This has led to the commercial availability of Hastelloy X alloys with lower contents of minor elements such as Mn, Si and C, and in some cases offering alloys with a C content below specification. This approach can have, however, detrimental effects on the product's properties; for instance, it has been shown that C content on a Ni-Cr-Fe-Mo alloy has a significant effect on its thermal fatigue properties [16]. This present work reports the fabrication of crack-free Hastelloy X samples made from powder with a standard chemical composition, and where the sound microstructure survived high temperature heat treatment.

2. Materials and Methods

Commercially available gas-atomized Hastelloy X powder provided by LPW Technology Ltd. with a composition displayed in Table 1 was processed using a Renishaw AM400. The powder particles were spherical in morphology, having a particle size distribution D10 = 19 μm, D50 = 41 μm, and D90 = 62 μm. Cubic samples of 1 cm^3 were produced under Ar atmosphere and analyzed in the as-built and annealed condition. For protection of proprietary information, specific process parameters are not included. Cross-sections of samples were prepared for microstructure evaluation using traditional metallography techniques. Electroetching was performed using a 10% oxalic acid solution and 6 V for 5–10 s. The density of the as-built sample was analyzed using a Nikon light optical microscope (OM) equipped with a Clemex Vision System. Microstructures were observed using a Hitachi SU3500 scanning electron microscope (SEM) equipped with an electron backscatter diffraction (EBSD) detector for texture and grain morphology analysis. High magnification micrographs and electron dispersive spectrometer (EDS) line scans were obtained using a Hitachi SU8000 field emission scanning electron microscope (FE-SEM), prior to which the sample's surface was further prepared using ion beam milling. Heat treatment was carried out under an Ar atmosphere at 1175 °C for 1 h followed by water quenching.

Table 1. Comparison of the composition of Hastelloy X from literature.

Reference	Ni	Cr	Fe	Mo	Co	W	Mn	Si	C
Wang, et al. [10]	Bal.	20.6	18.4	8.8	1.3	0.62	0.69	0.78	0.009
Harrison, et al. [14]	Bal.	21.8	18.6	9.4	1.8	1.05	0.22	0.31	0.054
Tomus, et al. [13]	Bal.	21.4	18.4	8.8	1.8	0.86	<0.01	0.11	0.01
Marchese, et al. [15]	Bal.	21.7	18.6	9.2	1.8	0.90	-	0.36	0.056
This work	Bal.	21.2	17.6	8.8	2.0	NM	<0.1	0.20	0.06
ASTM B435 [17]	Bal.	20.5–23	17–20	8–10	0.5–2.5	0.2–1	<1	<1	0.05–0.15

3. Results and Discussion

A mosaic stitched from optical micrographs of the X-Z cross-section of a cubic sample in the as-built condition taken from its polished state is shown in Figure 1. As depicted, crack-free specimens with optical density of 99.76 ± 0.11 were obtained. The remaining porosity, with irregular and spherical morphology, is homogeneously distributed within the sample and attributed to lack of fusion and gas entrapment.

In order to demonstrate the absence of cracks in a larger building volume, a cylindrical sample of radius 1 cm and 5 cm in height was built, vertically cross-sectioned, polished and imaged. An SEM micrograph is shown in Figure 2a, and higher magnification observation from the (b) top, (c) middle, and (d) bottom of the sample further demonstrate a crack-free cross-section.

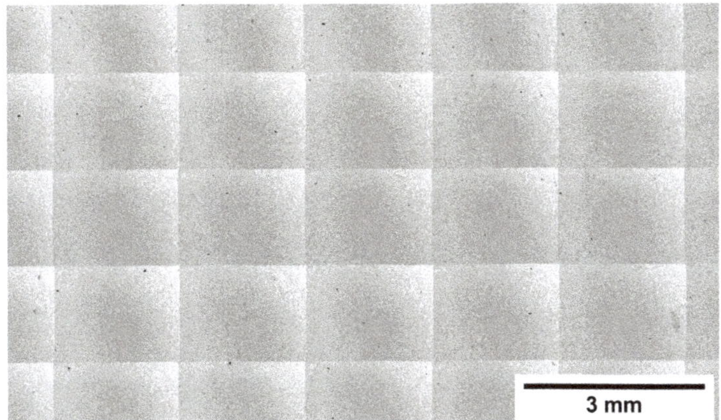

Figure 1. Representative optical micrograph of stitched 50× images from the cross-section of an as-built cubic sample from its polished state.

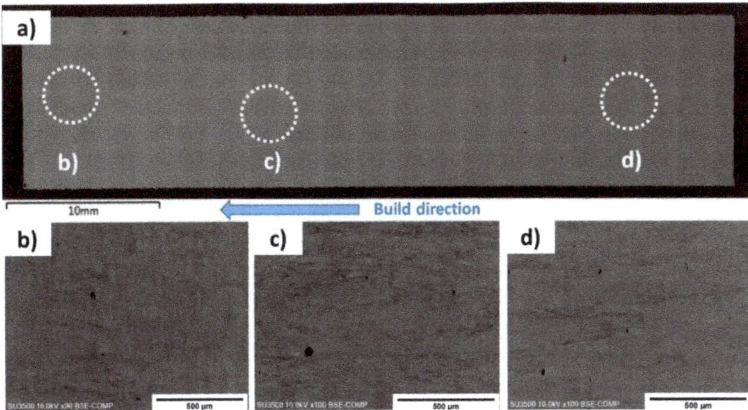

Figure 2. (**a**) SEM montage of the cylindrical sample and higher magnification images of the (**b**) top; (**c**) middle; and (**d**) bottom of the sample. The regions where (**b–d**) were taken from are highlighted on the montage.

Figure 3 shows the EBSD inverse pole figure (IPF) maps along with the corresponding pole figures collected from the cross-section of the cubic sample parallel to the build direction of the as-built condition. Columnar grains with measured widths of 9.76 ± 8.8 µm oriented towards the build direction are evident. The grains grow across several layers without changing their crystallographic orientations, indicating epitaxial growth. As shown in Figure 3a, most columnar grains show either <001> or <110> orientations. It is noted that some of the grains are not perfectly aligned with build direction. As depicted in the {001} pole figure in Figure 3b, the clusters of {001} poles near the center of the pole figure are not concentrated, indicating some angular deviations between the <001> oriented grains and build directions. On the other hand, the grains oriented in the <110> direction exhibit a weak partial fiber type feature, as observed in the {110} pole figure in Figure 3b.

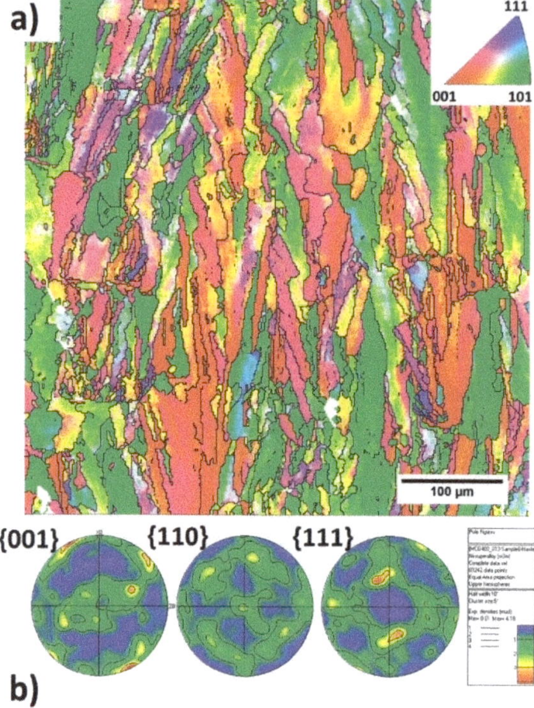

Figure 3. (**a**) Electron backscatter diffraction (EBSD) map taken from the central region of the as-built sample and (**b**) its corresponding pole figure.

Figure 4a illustrates the typical fine columnar microstructure observed in the XZ plane of the sample in the as-built condition. This microstructural morphology is commonly found in the additive manufacturing processes and has been widely reported in LPBF of nickel-based superalloys, including Hastelloy X [13–15] as well as Inconel 625 [18–21], Inconel 718 [22,23], CM247 [7] and Inconel 738 [24,25].

The primary dendrite arm spacing (PDAS) obtained by image analysis of high magnification micrographs, corresponds to 0.4 ± 0.06 μm. An estimation of the cooling rate of the process based on the equation PDAS = $a\varepsilon^{-b}$, where a ≈ 50 (K/s) and b ≈ 0.33 [21], indicates that cooling conditions of 1~2 × 10^6 K/s were achieved. When comparing with the literature data, the PDAS of the current work was found to be smaller. For example, Harrison, et al. [14] observed a similar solidification structure with PDAS ranging between 0.9 and 1.2 μm, Marchese, et al. [15] reported a PDAS of 0.65 ± 0.25 μm, while Tomus showed fine parallel dendrites with a width of 0.53 ± 0.09 μm, calculated from the reported micrographs [26].

The microstructure at higher magnification obtained through FE-SEM is shown in Figure 4b. Boundaries of the cellular structure, which corresponds simply to a different cross-section of the previously described fine columnar features, are clearly visible with nanometric precipitates dispersed along the cell boundaries. Nanometric black dots are also distinguishable in some cell cores and intercellular regions, which are attributed to gas entrapment pores.

Figure 4c displays the EDS line scan spectra crossing from left to right, a precipitate at a cell boundary, three complete cells and their boundaries, and a precipitate lying on the last boundary. Ni and Fe, which are γ matrix elements, are distinctly depleted on the precipitate. On the other hand, the presence of Mo on the same regions causes a distinguishable high-count region on the spectrum, evidencing the fact that these are Mo-enriched precipitates. Furthermore, Si, which is another element that has been suggested to segregate, appears to have no clear concentration gradient throughout the scan.

The size of the Mo enriched precipitates was measured to be 55.4 ± 4.4 nm. Marchese, et al. [15] report the formation of globular and elongated M_6C or M_nC_m Mo-enriched carbides with sizes around 100–500 nm located on interdendritic regions, along with smaller (<100 nm) unidentified precipitates within the dendrites. The authors indicated that these intercellular carbides are present within cracks and thus play a significant role in this alloy's cracking mechanism. In welding of Hastelloy X, previous studies have shown that M_6C carbides are abundant and enriched with Mo, and that they form a liquid film of this lower melting eutectic which is linked to hot cracking of the alloy [27].

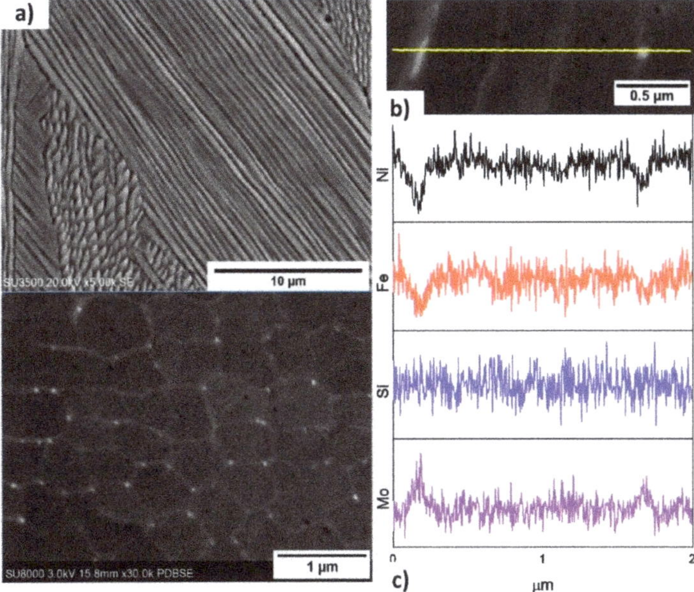

Figure 4. (a) SEM micrograph showing the typical columnar microstructure obtained from as-built LPBF Hastelloy X; (b) High magnification SEM micrograph; (c) Electron dispersive spectrometer (EDS) line scans corresponding to the region displayed above.

Heat treatment was performed on the samples to evaluate any possible change in microstructure. Annealing of Hastelloy X has been done at 1177 °C for an equivalent of 1 h per inch of thickness [28]. After heat treatment, the resulting microstructure is shown in Figure 5. From Figure 5a, the morphology and size of the grains is not significantly altered compared with the as-built sample. Columnar grains with a measured width of 9.54 ± 8.5 µm have been observed. As for the texture, it has been observed from the {001} pole figure (Figure 5b) that the {001} poles at the center become less scattered compared with those in the as-built condition. The texture maxima close to the center of the pole figure indicates a stronger <001> texture towards build direction after heat treatment. Meanwhile, the <110> partial fibre type texture along build direction is further weakened and becomes more diffused compared with that in the as-built condition, suggesting the decreasing <110> grains along build direction after heat treatment. No clear signs of alterations to the as-built columnar grain structure are visible from the EBSD analysis; furthermore, the microstructure depicted in Figure 5c, shows signs of twinning within the sample. Twin formation has been observed in annealing of other additively manufactured superalloys, and attributed to residual stresses from the as-built condition being partially released by forming twin grains during this heat treatment [29]. At higher magnification (Figure 5d), it is clear that the columnar structure visible in the as-built microstructure are almost homogenized. Although no conclusions about segregation in the interdendritic regions can be made from the line scans in Figure 4c, the FE-SEM micrograph in Figure 3b is a backscattered electrons (BSE) image, in which contrast is obtained by differences in composition; thus, it is possible that segregating elements diffuse into the matrix during heat treatment, causing the dendritic structure to disappear.

Etter et al. [30], reported significant grain coarsening and the presence of equiaxed grains after heat treatment at 1200 °C for 4 h, and the complete disappearance of the columnar grain structure from the as-built microstructure. Marchese [31], from a fine LPBF columnar microstructure, obtained a microstructure composed of equiaxed grains and twins after a solution treatment at 1175 °C for up to an hour. He also reported that a heat treatment at 1066 °C for 1 h was ineffective in modifying the microstructure. Finally, a heat treatment performed by Tomus [26] 1175 °C for 2 h followed by air cooling managed to dissolve the as-built dendrites, but columnar grains elongated along the building direction were still observed and attributed to the absence of recrystallization occurring after heat treatment. The heat treatment schedule used here corresponds to the solutionizing treatment time and temperature used for Hastelloy X [28], so any microsegregation present would be expected to homogenize through the microstructure. Although there is a lack of agreement in literature on whether this heat treatment will fully recrystalize and eliminate the as-built columnar grains, and data obtained from EBSD (Figure 4a) does not clearly demonstrate the presence of recrystalized grains, the appearance of annealing twins visible on the studied sample is an indication that recrystallization may have occurred at least partially, since twins have been shown to form mostly during the recrystallization process [32], and have been observed in heat treated samples of IN718 processed through laser metal deposition with different degrees of recrystallization [33].

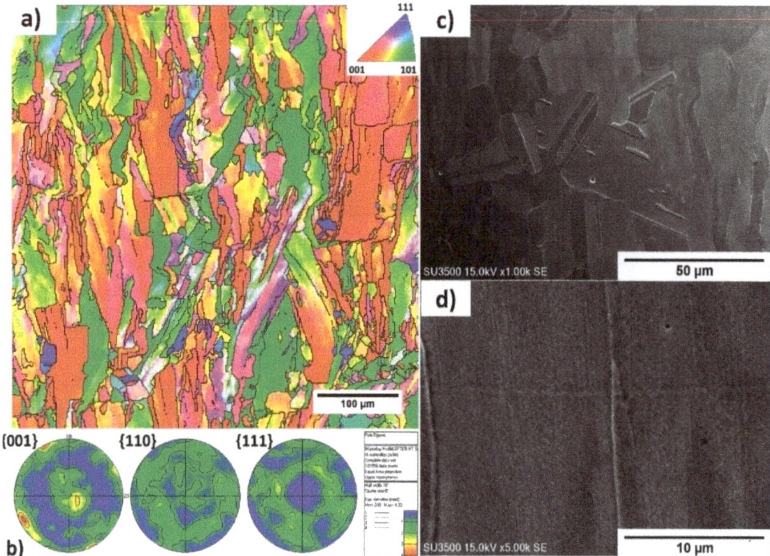

Figure 5. (**a**) EBSD map taken from the central region of the sample after heat treatment; and (**b**) its corresponding pole figure; SEM micrographs taken at (**c**) 5000× and (**d**) 1000× showing the microstructure of the sample after heat treatment.

4. Conclusions

This work has reported the fabrication of crack-free Hastelloy X samples made from powder with a standard chemical composition through LPBF, observable on optical micrographs. Columnar grains parallel to the building directions were evidenced through EBSD in the as-built condition. Likewise, a typical microstructure for LPBF Ni-based superalloys was found, with smaller columnar structure size than reported elsewhere in literature. The presence of Mo-enriched carbides (~50 nm) was confirmed through EDS line scans obtained using an FE-SEM; these Mo-enriched carbides are presumed to play an important role in the cracking mechanism of this alloy. A subsequent heat treatment at 1177 °C for 1 h followed by water quenching was performed, after which the crack-free microstructure was sustained. After EBSD, a columnar grain structure was demonstrated to be present, and SEM micrographs displayed the presence of annealing twins and a disappearance of the as-built columnar microstructure.

Author Contributions: Conceptualization, O.S.-M. and M.B.; Methodology, O.S.-M.; X.W.; J.A.M.-L. and M.A.S.; Validation, O.S.-M.; X.W. and M.A.S.; Formal Analysis, R.G., M.A.S. and X.W.; Investigation, O.S.-M.; X.W. and J.A.M.-L.; Resources, R.G.; M.B.; Data Curation, O.S.-M.; X.W. and J.A.M.-L.; Writing-Original Draft Preparation, O.S.-M.; Writing-Review & Editing, M.B.; J.A.M.-L. and R.G.; Visualization, O.S.-M.; J.A.M.-L. and M.B.; Supervision, M.B. and J.A.M.-L.; Project Administration, M.B.; Funding Acquisition, M.B.

Funding: This research received no external funding.

Acknowledgments: The authors would like to acknowledge the McGill Engineering Doctoral Award (MEDA) and Consejo Nacional de Ciencia y Tecnología (CONACYT, Mexico) scholarships granted to Sanchez-Mata.

Conflicts of Interest: The authors declare no conflict of interest.

References

1. DebRoy, T.; Wei, H.L.; Zuback, J.S.; Mukherjee, T.; Elmer, J.W.; Milewski, J.O.; Beese, A.M.; Wilson-Heid, A.; De, A.; Zhang, W. Additive manufacturing of metallic components—Process, structure and properties. *Prog. Mater. Sci.* **2018**, *92*, 112–224. [CrossRef]

2. Frazier, W.E. Metal Additive Manufacturing: A Review. *J. Mater. Eng. Perform.* **2014**, *23*, 1917–1928. [CrossRef]

3. Chauvet, E.; Kontis, P.; Jägle, E.A.; Gault, B.; Raabe, D.; Tassin, C.; Blandin, J.-J.; Dendievel, R.; Vayre, B.; Abed, S.; et al. Hot cracking mechanism affecting a non-weldable Ni-based superalloy produced by selective electron Beam Melting. *Acta Mater.* **2018**, *142*, 82–94. [CrossRef]

4. DuPont, J.N.; Lippold, J.C.; Kiser, S.D. *Welding Metallurgy and Weldability of Nickel-Base Alloys*; Wiley: Hoboken, NJ, USA, 2009.

5. Kou, S. Solidification and Liquation cracking issues in welding. *JOM* **2003**, *55*, 37–42. [CrossRef]

6. Attallah, M.M.; Jennings, R.; Wang, X.; Carter, L.N. Additive manufacturing of Ni-based superalloys: The outstanding issues. *MRS Bull.* **2016**, *41*, 758–764. [CrossRef]

7. Carter, L.N.; Attallah, M.M.; Reed, R.C. Laser Powder Bed Fabrication of Nickel-Base Superalloys: Influence of Parameters; Characterisation, Quantification and Mitigation of Cracking. *Superalloys* **2012**, *2012*, 577–586.

8. Zhao, J.-C.; Larsen, M.; Ravikumar, V. Phase precipitation and time–temperature-transformation diagram of Hastelloy X. *Mater. Sci. Eng. A* **2000**, *293*, 112–119. [CrossRef]

9. Catchpole-Smith, S.; Aboulkhair, N.; Parry, L.; Tuck, C.; Ashcroft, I.A.; Clare, A. Fractal scan strategies for selective laser melting of 'unweldable' nickel superalloys. *Addit. Manuf.* **2017**, *15*, 113–122. [CrossRef]

10. Wang, F.; Wu, X.H.; Clark, D. On direct laser deposited Hastelloy X: Dimension, surface finish, microstructure and mechanical properties. *Mater. Sci. Technol.* **2011**, *27*, 344–356. [CrossRef]

11. Wang, F. Mechanical property study on rapid additive layer manufacture Hastelloy®X alloy by selective laser melting technology. *Int. J. Adv. Manuf. Technol.* **2011**, *58*, 545–551. [CrossRef]

12. Tomus, D.; Jarvis, T.; Wu, X.; Mei, J.; Rometsch, P.; Herny, E.; Rideau, J.F.; Vaillant, S. Controlling the Microstructure of Hastelloy-X Components Manufactured by Selective Laser Melting. *Phys. Procedia* **2013**, *41*, 823–827. [CrossRef]

13. Tomus, D.; Rometsch, P.A.; Heilmaier, M.; Wu, X. Effect of minor alloying elements on crack-formation characteristics of Hastelloy-X manufactured by selective laser melting. *Addit. Manuf.* **2017**, *16*, 65–72. [CrossRef]

14. Harrison, N.J.; Todd, I.; Mumtaz, K. Reduction of micro-cracking in nickel superalloys processed by Selective Laser Melting: A fundamental alloy design approach. *Acta Mater.* **2015**, *94*, 59–68. [CrossRef]

15. Marchese, G.; Basile, G.; Bassini, E.; Aversa, A.; Lombardi, M.; Ugues, D.; Fino, P.; Biamino, S. Study of the Microstructure and Cracking Mechanisms of Hastelloy X Produced by Laser Powder Bed Fusion. *Materials* **2018**, *11*, 106. [CrossRef] [PubMed]

16. An, N.; An, Y.; Fan, Q.; Fu, Z.M.; Li, Z.R.; Zhang, Y.Y. Effect of Carbon on the Microstructural Evolution and Thermal Fatigue Behavior of a Ni-Base Superalloy. *Mater. Sci. Forum* **2016**, *849*, 497–502. [CrossRef]

17. Standard Specification for UNS N06002, UNS N06230, UNS N12160, and UNS R30556 Plate, Sheet, and Strip. Available online: http://www.htpipe.com/d/files/plate-material-grade/astm-b435.pdf (accessed on 8 June 2011).

18. Marchese, G.; Garmendia Colera, X.; Calignano, F.; Lorusso, M.; Biamino, S.; Minetola, P.; Manfredi, D. Characterization and Comparison of Inconel 625 Processed by Selective Laser Melting and Laser Metal Deposition. *Adv. Eng. Mater.* **2017**, *19*, 1600635. [CrossRef]

19. Lass, E.A.; Stoudt, M.R.; Williams, M.E.; Katz, M.B.; Levine, L.E.; Phan, T.Q.; Gnaeupel-Herold, T.H.; Ng, D.S. Formation of the Ni3Nb δ-Phase in Stress-Relieved Inconel 625 Produced via Laser Powder-Bed Fusion Additive Manufacturing. *Metall. Mater. Trans. A* **2017**, *48*, 5547–5558. [CrossRef]

20. Li, C.; White, R.; Fang, X.Y.; Weaver, M.; Guo, Y.B. Microstructure evolution characteristics of Inconel 625 alloy from selective laser melting to heat treatment. *Mater. Sci. Eng. A* **2017**, *705*, 20–31. [CrossRef]

21. Li, S.; Wei, Q.; Shi, Y.; Zhu, Z.; Zhang, D. Microstructure Characteristics of Inconel 625 Superalloy Manufactured by Selective Laser Melting. *J. Mater. Sci. Technol.* **2015**, *31*, 946–952. [CrossRef]

22. Jia, Q.; Gu, D. Selective laser melting additive manufacturing of Inconel 718 superalloy parts: Densification, microstructure and properties. *J. Alloys Compd.* **2014**, *585*, 713–721. [CrossRef]

23. Tian, Y.; Muñiz-Lerma, J.A.; Brochu, M. Nickel-based superalloy microstructure obtained by pulsed laser powder bed fusion. *Mater. Charact.* **2017**, *131*, 306–315. [CrossRef]

24. Cloots, M.; Kunze, K.; Uggowitzer, P.J.; Wegener, K. Microstructural characteristics of the nickel-based alloy IN738LC and the cobalt-based alloy Mar-M509 produced by selective laser melting. *Mater. Sci. Eng. A* **2016**, *658*, 68–76. [CrossRef]

25. Kunze, K.; Etter, T.; Grässlin, J.; Shklover, V. Texture, anisotropy in microstructure and mechanical properties of IN738LC alloy processed by selective laser melting (SLM). *Mater. Sci. Eng. A* **2015**, *620*, 213–222. [CrossRef]

26. Tomus, D.; Tian, Y.; Rometsch, P.A.; Heilmaier, M.; Wu, X. Influence of post heat treatments on anisotropy of mechanical behaviour and microstructure of Hastelloy-X parts produced by selective laser melting. *Mater. Sci. Eng. A* **2016**, *667*, 42–53. [CrossRef]

27. Savage, W.F.; Krantz, B.M. Microsegregation in Autogenous Hastelloy X Welds. *Weld. Res. Suppl.* **1971**, *50*, 292s.

28. Donachie, M.J.; Donachie, S.J. *Superalloys: A Technical Guide*, 2nd ed.; ASM International: Materials Park, OH, USA, 2002.

29. Dinda, G.P.; Dasgupta, A.K.; Mazumder, J. Laser aided direct metal deposition of Inconel 625 superalloy: Microstructural evolution and thermal stability. *Mater. Sci. Eng. A* **2009**, *509*, 98–104. [CrossRef]

30. Etter, T.; Kunze, K.; Geiger, F.; Meidani, H. Reduction in mechanical anisotropy through high temperature heat treatment of Hastelloy X processed by Selective Laser Melting (SLM). *IOP Conf. Ser. Mater. Sci. Eng.* **2015**, *82*, 012097. [CrossRef]

31. Marchese, G.; Biamino, S.; Pavese, M.; Ugues, D.; Lombardi, M.; Vallillo, G.; Fino, P. Heat Treatment Optimization of Hastelloy X Superalloy Produced by DMLS. Available online: https://iris.polito.it/handle/11583/2646135 (accessed on 5 September 2016).

32. Rollett, A.D.; Rohrer, G.S.; Humphreys, F.J. *Recrystallization and Related Annealing Phenomena*; Elsevier: Amsterdam, The Netherlands, 2017.

33. Sui, S.; Zhong, C.; Chen, J.; Gasser, A.; Huang, W.; Schleifenbaum, J.H. Influence of solution heat treatment on microstructure and tensile properties of Inconel 718 formed by high-deposition-rate laser metal deposition. *J. Alloys Compd.* **2018**, *740*, 389–399. [CrossRef]

Article

Texture and Microstructural Features at Different Length Scales in Inconel 718 Produced by Selective Laser Melting

Michele Calandri [1], Shuo Yin [2], Barry Aldwell [2,3], Flaviana Calignano [4], Rocco Lupoi [2] and Daniele Ugues [1,*]

[1] Dipartimento di Scienza Applicata e Tecnologia (DISAT), Politecnico di Torino, Corso Duca degli Abruzzi, 24, 10129 Turin, Italy; michele.calandri@polito.it
[2] Department of Mechanical and Manufacturing Engineering, Trinity College Dublin, The University of Dublin, Parsons Building, Dublin 2, Ireland; yins@tcd.ie (S.Y.); aldwellb@tcd.ie (B.A.); lupoir@tcd.ie (R.L.)
[3] Nammo Ireland, 15–17 Northwest Business Park, Ballycoolin, Dublin 15, Ireland
[4] Dipartimento di Ingegneria Gestionale e della Produzione (DIGEP), Politecnico di Torino, Corso Duca degli Abruzzi, 24, 10129 Turin, Italy; flaviana.calignano@polito.it
* Correspondence: daniele.ugues@polito.it; Tel.: +39-0110904705

Received: 20 March 2019; Accepted: 16 April 2019; Published: 19 April 2019

Abstract: Nickel-based Inconel 718 is a very good candidate for selective laser melting (SLM). During the SLM process, Inconel 718 develops a complex and heterogeneous microstructure. A deep understanding of the microstructural features of the as-built SLM material is essential for the design of a proper post-process heat treatment. In this study, the microstructure of as-built SLM Inconel 718 was investigated at different length scales using optical microscopy (OM), scanning electron microscopy (SEM), and transmission electron microscopy (TEM). Electron backscatter diffraction (EBSD) was also used to analyze the grain morphology and crystallographic texture. Grains elongated in the build direction and crossing several deposited layers were observed. The grains are not constrained by the laser tracks or by the melt pools, which indicates epitaxial growth controls the solidification. Each grain is composed of fine columnar dendrites that develop along one of their <100> axes oriented in the direction of the local thermal gradient. Consequently, prominent <100> crystallographic texture was observed and the dendrites tend to grow to the build direction or with occasional change of 90° at the edge of the melt pools. At the dendrite length scale, the microsegregation of the alloying elements, interdendritic precipitates, and dislocations was also detected.

Keywords: selective laser melting; Inconel 718; crystallographic texture; subgranular dendrites; epitaxial growth

---

1. Introduction

Among all the nickel-based superalloys, Inconel 718 is one of the most studied alloys, which has been used in a wide range of industrial sectors, such as aeronautical, aerospace, and energy production. Inconel 718 possesses excellent mechanical properties, creep- and fatigue-resistance performance at high temperature [1,2], and hot-oxidation resistance [3,4]. Components manufactured from Inconel 718 often operate in highly aggressive environments at temperatures higher than 700 °C, or at cryogenic temperatures, such as in jet engines, gas turbine engines, chemical and nuclear plants, heat exchangers, and cryogenic tanks [5–8]. Inconel 718 is a ternary Ni-Cr-Fe system and is composed of a solid solution austenitic γ strengthened by very small and finely-dispersed precipitates of intermetallic phases γ′-Ni$_3$(Ti,Al) and γ″-Ni$_3$Nb [9]. This alloy is conventionally produced through casting followed by cold working [5,10], and is one of the most weldable nickel superalloys due to its low strain-age cracking sensitivity during the post-weld heat treatment [11–13]. In particular, this low susceptibility to cracking makes Inconel 718 a good candidate [14] for additive manufacturing (AM).

During the last decade, the fabrication of Inconel 718 alloy through AM has gained more and more attention [15–20] from both the scientific and industrial communities. AM techniques allow the customized production of near net-shape dense metal parts with complex geometries. Therefore, the use of expensive tools, dies, or casting molds can be avoided [21,22]. Among all AM processes, selective laser melting (SLM) is an interesting technique that uses high power-density laser as a heat source to induce local melt of a metallic powder bed. The melt pool then consolidates through rapid solidification, and the fabrication of SLM components is achieved through the addition and melting of several layers of metal powders [23,24]. SLM can produce close to fully dense Inconel 718 components with mechanical properties, after appropriate heat treatment, comparable to and even better than cast or wrought counterparts [25,26].

During the SLM process, the material being built will be rapidly melted by the laser and will then solidify under very high cooling rates and directional heat fluxes, which have a significant effect on the microstructure evolution and mechanical properties [27–31]. In addition, the solidified material is subject to repeated thermal cycles during the deposition of the following layers [32], which further complicates the microstructure evolution because of the partial dissolution of the second phases formed during the solidification, the redistribution of the solutes, the precipitation of new phases [33], and the increased heterogeneity of the microstructure that results from the aforementioned phenomena [34]. Therefore, SLM Inconel 718 alloy in the as-built state will often possess a very complex microstructure characterized by anisotropy [35], strong crystallographic texture [34–36], microsegregation, and the formation of second phases [17,37,38]. These microstructural features can significantly affect the material mechanical properties before and after heat treatments [25,38–40]. Therefore, a comprehensive understanding of the evolution of the microstructure during SLM is essential for the application of SLM technology to the fabrication of Inconel 718 and other nickel-based superalloys.

Although Inconel 718 alloy produced through SLM process has been extensively studied [41,42], a complete and detailed study of the microstructure and the pre-existing second phases in the as-built state focusing on all different length scales is still lacking. In the present study, the microstructure of Inconel 718 produced through a SLM process was investigated at different length scales, starting from 10^{-3} to 10^{-5} m, within which laser-related features (i.e., laser track and melt pool boundaries) and grains with characteristic crystallographic texture could be observed, to 10^{-5}–10^{-6} m, within which intragranular substructures consisting of fine cellular dendrites could be observed, and finally to 10^{-6}–10^{-8} m, within which single dendrite and intercellular precipitates could be observed. Furthermore, the relationship between microstructure and solidification conditions is also discussed in this work. The aim of this study is to clearly identify, at all length scales, the main microstructural features derived directly from the SLM process without any post-process heat treatments. This basic knowledge is fundamental to determine if the as-built condition can eventually be suitable in principle to guarantee mechanical performance and high temperature hot oxidation resistance, and how the as-built microstructure can be corrected to optimize these properties with dedicated subsequent thermal treatments.

A preliminary study of the microstructural evolution of the SLM Inconel 718 alloy during the post-process heat treatment cycle is also reported in this paper. The effect of the solution annealing step on the as-built microstructure has been studied in past work by the authors, with an optimal homogenization/solution recipe of 1065 °C for 2 h identified [43]. Starting from this solutioned condition, analysis of the most important thermodynamic transformations that can occur during the aging process is provided.

2. Materials and Methods

2.1. Manufacturing Procedure

The SLM Inconel 718 samples studied in this work were manufactured using an EOSINT M270 dual mode machine (EOS GmbH, Krailling, Germany) equipped with 200W Yb fiber continuous laser beam. The feedstock was gas atomized Inconel 718 powders (EOS GmbH). The chemical composition of the Inconel 718 powder used in this work, according to the powder supplier's specification, is listed in Table 1. Figure 1 shows the surface morphology and size distribution of the Inconel 718 feedstock powders. The powder was spherical in shape with small particles agglomerated on the surface of larger ones. A bidirectional scanning strategy was adopted for producing the Inconel 718 samples. The scanning trajectory was rotated by an angle of 67° between adjacent layers in order to reduce the anisotropy of the samples.

Table 1. Chemical composition in wt.% of the Inconel 718 EOS GmbH (Krailling, Germany) powders used in this work [44].

Ni	Cr	Nb	Mo	Ti	Al	Fe
50–55	17–21	4.75–5.5	2.8–3.3	0.65–1.15	0.2–0.8	
Co	**Cu**	**C**	**Si + Mn**	**P + S**	**B**	balance
<1	<0.3	<0.08	<0.35	<0.015	<0.006	

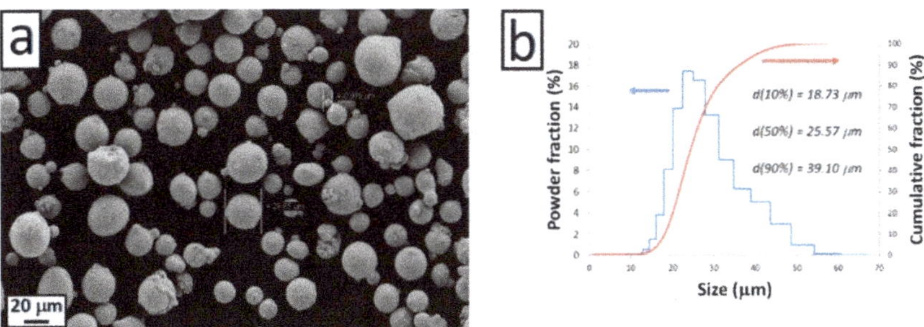

Figure 1. Surface morphology (**a**) and size distribution (**b**) of the Inconel 718 powders used in this work.

The samples were manufactured on the build platform directly without any supports. After manufacture, the samples were removed from the build platform by using electrical discharge machining.

2.2. Processing Parameters Optimization Procedure

Before microstructural characterization, an optimization study of the main process parameters (i.e., laser power, scan speed, and hatching distance) was performed in order to identify the optimum parameters to produce a fully dense part. For this scope, a full factorial design of the experiment involving three levels for each parameter was used. A list of the experiments carried out and the parameter levels used in each is shown in Table 2. For each experiment a 15 mm × 15 mm × 10 mm cubic sample was produced. The spot size and the layer thickness were set to fixed values of 100 and 20 µm, respectively. In Table 2, the volumetric energy density (VED) value associated to each set of process parameters is also reported. VED (J/mm^3) is as shown in Equation (1):

$$VED = \frac{P}{vh_d d} \tag{1}$$

where P is laser power (W), v the scan speed (mm/s), h_d the hatching distance (mm), and d is the layer thickness (mm).

Table 2. Investigated combinations for the optimization study of the process parameters and relative volumetric energy density (VED) values.

Sample	Power [W]	Scan Speed [mm/s]	Hatching Distance [mm]	VED [J/mm^3]
1	175	600	0.09	162.04
2	195	600	0.09	180.56
3	175	900	0.07	138.89
4	175	1200	0.11	66.29
5	175	1200	0.07	104.17
6	185	600	0.09	171.30
7	185	1200	0.11	70.08
8	185	900	0.07	146.83
9	175	1200	0.09	81.02
10	185	900	0.11	93.43
11	195	900	0.07	154.76
12	185	600	0.07	220.24
13	195	600	0.07	232.14
14	195	1200	0.09	90.28
15	175	600	0.07	208.33
16	195	1200	0.11	73.86
17	195	900	0.09	120.37
18	185	900	0.09	114.20
19	185	1200	0.07	110.12
20	185	1200	0.09	85.65
21	175	900	0.09	108.02
22	175	600	0.11	132.58
23	195	600	0.11	147.73
24	195	1200	0.07	116.07
25	185	600	0.11	140.15
26	195	900	0.11	98.48
27	175	900	0.11	88.38

The density level of each sample was evaluated through apparent density measurements, porosity evaluation through image analysis of optical micrographs (Leica DMI 5000 M optical microscope, Wetzlar, Germany), and Brinell hardness measurements. The apparent density was measured through the Archimedes method [45] using a precision balance (model bc, Orma s.r.l., Milan, Italy, resolution: 0.1 mg), with all samples being prepared for testing by polishing of all surfaces with SiC paper, in order to avoid measurement errors related to air trapping on the highly rough surface after the SLM production. The optical micrographs for the porosity evaluation were collected on samples previously polished with 1 μm diamond suspension, without any chemical etching. A total of 28 optical images were collected for each sample, each of them had an area of 1.8 mm × 1.4 mm (total area $\approx$ 73.6 mm^2). The porosity level was calculated as a percentage of the analyzed surface area through the binarization of the micrographs by setting a threshold gray value; for each image, two threshold gray values were chosen in order to obtain a low estimate and a high estimate of porosity.

The average Brinell hardness of each sample was calculated from 5 indentations using an EMCO TEST M4U durometer (EMCO-TEST Prüfmaschinen GmbH, Kuchl, Austria), each indentation was performed by imposing a load of 62.5 kgf ($\approx$ 613 N) for 15 s.

2.3. Microstructural Characterization of the As-Built State

For the microstructural characterization, cylindrical test samples of 15 mm in diameter and 125 mm in length were produced by adopting the optimal set of parameters previously identified and reported in Table 3. No stress relieving heat treatment was applied after the fabrication of these samples.

Table 3. Optimized selective laser melting (SLM) parameters.

Parameter	Value
Laser power (W)	195
Scan speed (mm/s)	1200
Hatching distance (mm)	0.09
Spot size (μm)	100
Layer thickness (μm)	20

Metallographic samples were cut from the cylindrical bars along the direction perpendicular and parallel to the build direction (BD) to characterize the sample microstructure in the horizontal and vertical planes, respectively. The samples were prepared first with SiC grinding paper, and then polished with 1 μm diamond solution, and finally with 0.05 μm alumina suspension. Two etching techniques were used to reveal the microstructure: electrochemical etching using a solution of 100 mL HNO_3 and 10 mL water with an imposed voltage of 1–2 V for about 3–4 s, and normal immersion etching by using waterless Kalling's reagent (5 g $CuCl_2$ in 100 mL HCl and 100 mL ethanol). The etched samples were then observed with an optical microscope (OM) equipped with a digital camera (Leica, Wetzlar, Germany) and with a scanning electron microscope (SEM, Carl Zeiss ULTRA, Oberkochen, Germany).

The crystallographic texture was studied using electron backscattered diffraction (EBSD, Zeiss SUPRA 40, Oberkochen, Germany) equipped with a Bruker detector (Bruker Nano GmbH, Berlin, Germany). The samples for EBSD characterization were polished using standard polishing procedure followed by a long final polishing step with 0.05 μm alumina suspension. The EBSD analysis was carried out with a voltage of 20 kV and an exposure time of 150 to 200 ms. A total area of 215 μm $\times$ 170 μm and 150 μm $\times$ 190 μm was analyzed on the horizontal and vertical planes with a step size of 1.41 μm, respectively. A local area of 30 μm $\times$ 25 μm and 60 μm $\times$ 90 μm was also analyzed, respectively, on the horizontal and vertical planes with high-resolution (step size: 0.35 μm for the former and 0.7 μm for latter). The microstructure at the length scale of 10^{-6}–10^{-8} m was characterized with scanning/transmission electron microscope (S/TEM, FEI Titan, Hillsboro, OR, USA) equipped with energy dispersive X-ray spectrometry (EDS) detector. The thin lamella for TEM observation was prepared at the vertical plane using focused ion beam (FIB, Carl Zeiss Auriga, Oberkochen, Germany) system.

2.4. Microstructural Evolution Investigation during Aging

Differential scanning calorimetry (DSC) was performed to determine the temperature ranges where the most important microstructural modifications occur. Cylindrical samples of 3.5 mm in diameter and 10 mm in height were produced through SLM, using the processing parameters reported in Table 3. A thermal analyzer, Setaram DSC/TGA 92 16.18 (Caluire, France), was used for the DSC analysis, with a heating rate of 20 °C/min, from room temperature to 1200 °C, in order to detect all the solid state transformations that occur both at the as-built condition and after a solution annealing at 1065 °C for 2 h.

On the basis of the DSC analysis, some temperatures of interest were identified. The microstructural modifications of the material were then investigated through progressive exposure to these temperatures. Small round plates (diameters: 13 mm, height: 3 mm) produced through SLM were used for this study: the samples were first treated by solution annealing at 1065 °C for 2 h and then aged following the recipes reported in Table 4. The samples were treated by inserting them into a preheated furnace room. At the end of the thermal exposure they were removed and cooled in still air; the small dimensions of the plate minimize the effects of the heating and cooling transients.

Table 4. List of the post-process heat treatment recipes and total aging exposure time applied for the study on the aging response of selective laser melting (SLM) Inconel 718.

Sample	Heat Treatment Recipe	Total Aging Exposure
Disk 1	1065 °C/2 h, 565 °C/4 h	4 h
	+565 °C/20 h	24 h
Disk 2	1065 °C/2h, 740 °C/2 h	2 h
	+740 °C/6 h	8 h
	+740 °C/16 h	24 h
Disk 3	1065 °C/2 h, 800 °C/4 h	4 h
	+800 °C/20 h	24 h
Disk 4	1065 °C/2 h, 870 °C/4 h	4 h
	+870 °C/4 h	8 h
	+870 °C/16 h	24 h

The Vickers microhardness of the aged samples was measured using a Vickers hardness indenter (Mitutoyo, Kawasaki, Japan) through a total of 10 microindentations for each sample. For comparison, the Vickers microhardness at the as-built state (14 microidentations) and after the 1065 °C/2 h solution annealing (10 microindentations) was also measured.

X-Ray diffraction (XRD) analysis was performed on the aged samples using a X-Pert Philips diffractometer (Amsterdam, The Netherlands) in Bragg–Brentano configuration emitting Cu Kα ration and scanning between 30° and 100° with a step size of 0.013°. Field emission scanning electron microscopy (FESEM) (Merlin Zeiss, Oberkochen, Germany and Carl Zeiss ULTRA) analysis was also used to evaluate the microstructural changes and the formed second phases.

3. Results

3.1. Optimization of the Process Parameters

The samples obtained from the experiments listed in Table 2 were characterized through the measurement of the apparent density, porosity fraction, and Brinell hardness. The results of this preliminary study, reported in Figure 2 as functions of the VED value, show that the process was very robust, with no large variations in the porosity or hardness being observed within the investigated ranges. In particular, for all the combinations of process parameters, the density value fell within the range of 8.17 and 8.22 g/cm^3 (theoretical maximum value) [8]. Furthermore, the porosities evaluated through image analysis of optical micrographs were always lower than 0.5%. Closer examination of the trends reported in Figure 2 shows that intermediate VED values tended to provide slightly better densification and hardness levels and lower variation of the data (i.e., higher process stability). The dotted boxes in the plots of Figure 2 indicate the optimum process window identified by this study.

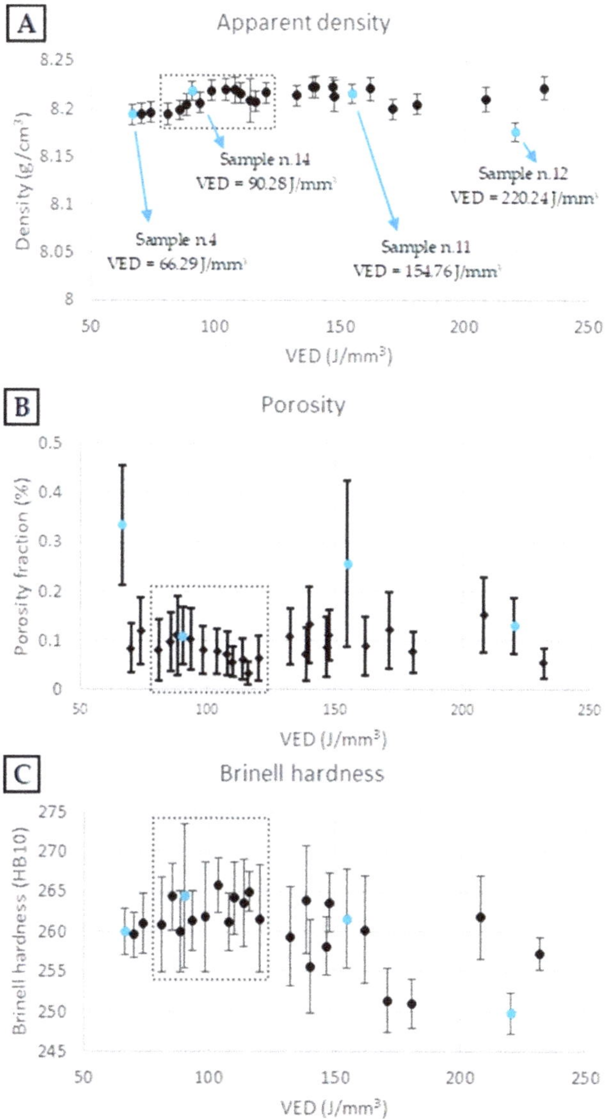

Figure 2. Apparent density (**A**), measured with the Archimedes method; porosity fraction (**B**), evaluated through optical images analysis; and Brinell hardness (**C**) of the selective laser melting (SLM) Inconel 718 samples, produced with different combinations of process parameters as functions of the volumetric energy density (VED) value. In each plot, the blue data points indicate samples n.4, n.14, n.11, and n.12; polished cross sections of which are shown in Figure 3 as examples. The error bars in plots A and C indicate 95% confidence intervals, while the bars in plot B indicate the ranges between the lower and higher estimates of the pore coverage ratio. Adapted from [46].

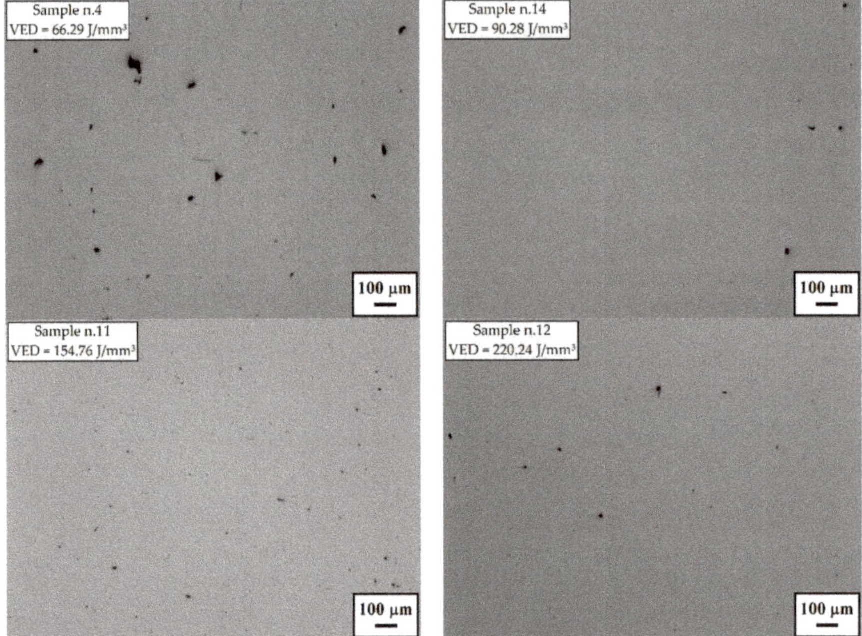

Figure 3. Some example optical micrographs showing the porosity of selective laser melting (SLM) Inconel 718 samples obtained with different volumetric energy density (VED) values. Sample n.14 was obtained with a VED value within the optimum processability window (dotted boxes in the plots of Figure 2).

Selective examples of the optical micrographs used for the porosity evaluation are shown in Figure 3. Sample n.14 is representative of the result that can be obtained with intermediate values of VED. Although some small pores of approximately 20 μm were still present, the evaluated porosity level was lower than 0.12% and the Brinell hardness fell into the range between 260 and 266 HB10.

From the micrographs shown in Figure 3, it is also possible to observe that for the sample manufactured with the lowest VED value the largest pores were observed (40–60 μm) with irregular shape due to lack of fusion. Conversely, higher VED values tend to form a slightly larger amount of spherical gas porosities compared to the ones in the optimum processability window. Furthermore, very high VED tends to reduce the Brinell hardness of the produced material.

The set of parameters used for sample n.14 (Table 3) were chosen for the microstructural characterization presented in the following subsections.

3.2. Grains and Laser-Related Microstructure (10^{-3}–10^{-4} m)

Figure 4a–d shows the OM micrographs of the electrochemically etched SLM Inconel 718 at the length scale of 10^{-3}–10^{-4} m. As can be seen, the microstructure of the SLM Inconel 718 was characterized by two kinds of boundaries, which are laser related boundaries and grain boundaries. The traces of the laser passes are usually referred to as the track–track molten pool boundaries (MPBs) on the horizontal plane and the layer–layer MPBs on the vertical plane [17,47]. The former is related to the laser scanning strategy and hatch distance, while the latter has an arc shape and are created by the local melt boundaries during each laser pass. The grain boundaries were more clearly revealed after etching with Kalling's reagent, as can be observed from the OM micrograph on the vertical plane shown in Figure 4e,f. It is interesting to observe that the grain boundaries were independent of the laser related boundaries. The grains tended to form an elongated shape oriented along the build direction. The grains spanned across several powder layers in the build direction. This shows that re-melting of subsurface layers, during subsequent laser passes, allows for the grain growth process to restart and span multiple layers. This creates strong bonds between the layers, reducing the risk of delamination and formation of inter-layer cracks [20].

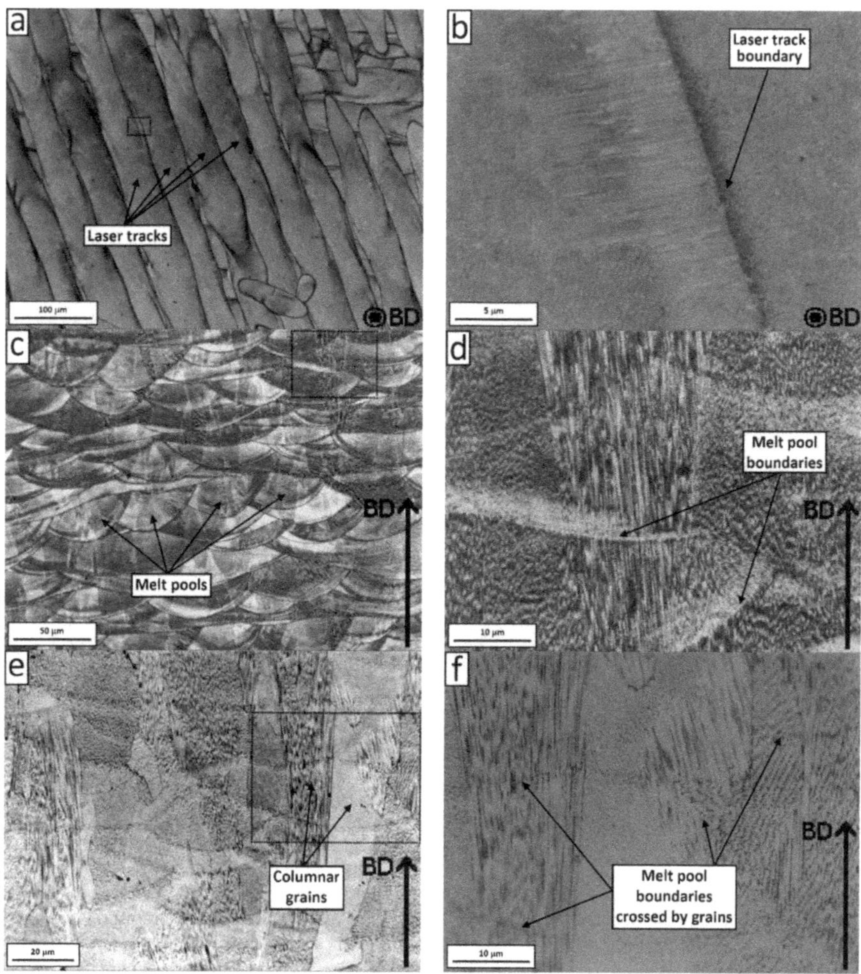

Figure 4. Optical Microscope (OM) micrographs of the electrochemically-etched SLM Inconel 718 on the horizontal plane with track–track MPBs and high magnification view (**a**,**b**), and vertical plane with arc-shape MPBs and high magnification view (**c**,**d**). OM micrographs of the Kalling solution etched Inconel 718 on the vertical plane showing columnar grains across several layers and high magnification view (**e**,**f**).

For further characterization of the grain structure of the SLM Inconel 718, Figure 5 shows EBSD maps of the horizontal and vertical planes. Note that only the matrix γ phase was detected with EBSD analysis. In the as-built state, the grains were characterized by the presence of subgranular domains, which are delimited by low-angle boundaries. In the grain maps shown in Figure 5, the high-angle boundaries between grains, which are defined as the boundaries with a maximum misorientation angle of 10° or higher, are pictured in dark blue; the boundaries with a maximum misorientation angle between 4° and 10° are depicted in lighter blue. Furthermore, the subgranular domains of each grain, surrounded by low-angle boundaries, are shown with different color shades. The largest grains contained five to 11 or even more subgranular domains. Note that the presence of preferred crystallographic orientations, which will be discussed in detail in Section 3.3, leads to some difficulties in identifying a clear misorientation threshold value to define the grain boundaries. The threshold values of 10° and 4° were chosen during the data processing step because they were found to best describe the grain structure.

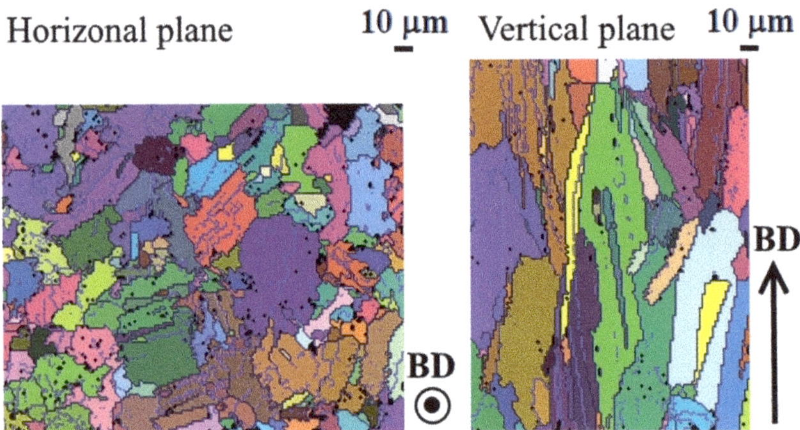

Figure 5. Grain maps of the horizontal and vertical planes of the selective laser melting (SLM) Inconel 718. The boundaries with a maximum misorientation angle of 10° or above are shown in dark blue, while the boundaries with a maximum misorientation angle between 4° and 10° are shown in lighter blue. The colors are used to distinguish the grains, with different shades of the same color indicating sub-granular crystalline domains within a single grain.

It is clear that most of the grains sectioned on the horizontal plane appeared equiaxed, which means that there is no preferential growth direction in this plane (i.e., orthogonal to the build direction).

The average grain size (equivalent diameter) on the analyzed area of the horizontal plane was 10.9 μm, but the grains were very heterogeneous in size, with a standard deviation of 8.7 μm. There were also some large grains of up to 50 μm or greater. On the vertical plane, the grains appeared elongated with the major axis aligned with the build direction. The average grain height, calculated using only the grains that are completely contained in the analysis area, was 28.5 μm, but grains with length of about 180 μm could also be observed. There were a number of grains which have grown across more than one layer (20 μm) and even up to ten layers for the largest grains. The average grain aspect ratio, weighted on the grain sizes, was 5.4.

In addition, the EBSD mapping on the vertical plane also demonstrates that no crystallographic changes occurred between layers (i.e., at each deposition the crystallographic orientation of the underlying material was maintained).

3.3. Crystallographic Texture (10^{-4}–10^{-5} m)

The inverse pole figure (IPF) charts obtained from the EBSD analysis on the horizontal and vertical planes of the SLM Inconel 718 are shown in Figure 6, where a clear crystallographic texture was recognizable. On the horizontal plane, the detected points accumulated at the [001] vertex of the Z axis, with few detected points close to the [111] vertex of the X and Y axes. Similarly, on the vertical plane, the detected points accumulated at the [001] vertex of the Y axis, with few detected points close to the [111] vertex of the X and Z axes. Note that the build direction to the Z axis on the horizontal plane, and the Y axis on the vertical plane. The IPF charts can be interpreted as the preferential orientation of crystals, which tend to align their [001] direction along the build direction. Once the [001] axis is fixed, there still exists one degree of freedom for the orientation of the crystals, which is the Euler angle (φ_2) of rotation around this axis. No preferential orientation related to φ_2 can be deduced from the IPFs; therefore, no clear crystallographic anisotropy existed on the plane of the added layers.

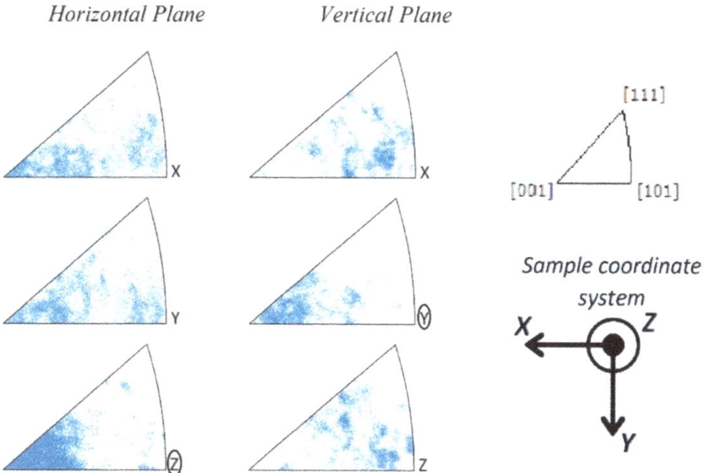

Figure 6. Inverse pole figure (IPF) charts obtained from electron backscatter diffraction (EBSD) analysis on the horizontal and vertical planes of SLM Inconel 718. [001] texture was detected on the IPF Z axis on the horizontal plane and on the IPF Y axis on the vertical plane.

Figure 7 shows the IPF maps of the horizontal plane (vertical to the Z axis of the SCS) and vertical plane (vertical to the Y axis of the SCS). It is clear that most of the grains had their [001] crystallographic direction orientated close to or along the build direction. Figure 8 shows the pole figures at the {100} planes of the horizontal and vertical planes. Each large square indicates the average orientation of one of the grains shown in the maps in Figure 5. The small circles indicate the average orientation of the subgranular domains. It is clear that the [001] axis of most grains was close to the build direction, with no evident orientation around this axis.

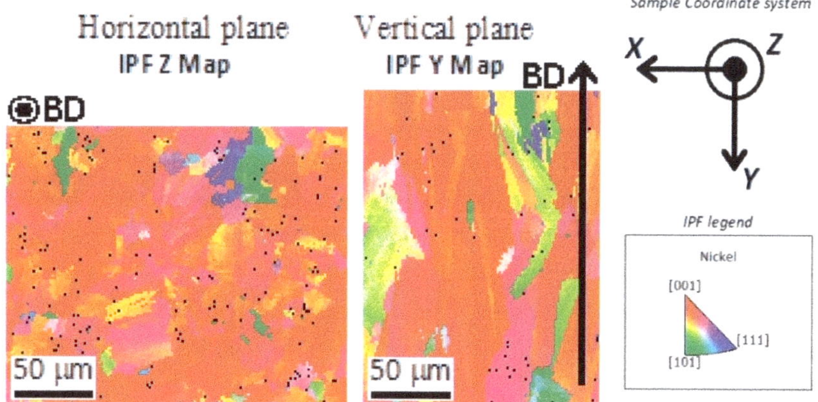

Figure 7. Inverse Pole Figure (IPF) maps of the Z axis for the horizontal plane and Y axis for the vertical plane showing grains with [001] orientation.

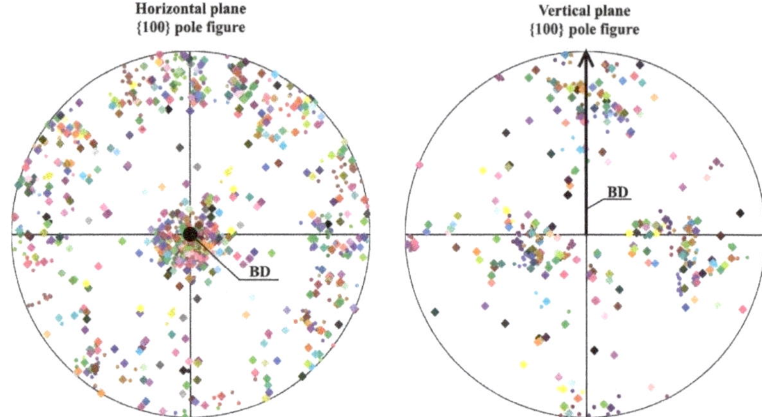

Figure 8. Pole figures of the horizontal and vertical planes of selective laser melting (SLM) Inconel 718. Each square represents the average orientation of the {100} crystallographic planes of the grains, as shown in the electron backscatter diffraction (EBSD) mapping in Figure 5, with the same colors used for each grain. The small circles represent the average orientation of the subgranular domains.

High-resolution EBSD analysis was also conducted on areas of interest in the horizontal and vertical planes. Figure 9 shows the relative IPF maps. The grains on the vertical plane had an interesting texture characterized by two subgrains, with zig-zag shape and a low angle (approximately 1°) boundary between each other. Dendrite substructure marked with a black circle can be observed in a grain on the vertical plane (Figure 9b), which will be further examined in the following section. No substructure was evident on the horizontal plane, with uniform crystallographic orientation inside each grain (Figure 9a).

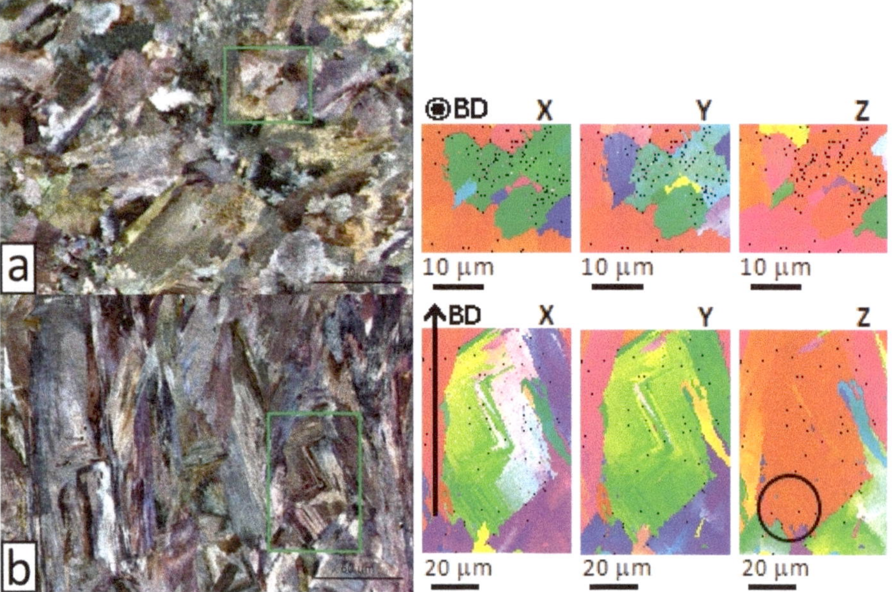

Figure 9. Forescattered/backscattered electron images on the horizontal plane (**a**) and vertical plane (**b**) with relative inverse pole figure (IPF) maps at higher resolution of the areas marked by the green boxes.

3.4. Intragranular Dendrites (10^{-5}–10^{-6} m)

The internal microstructure of the grain was further investigated through SEM observation. At the length scale of 10^{-5}–10^{-6} m, it is possible to observe the microstructure inside a single grain. Figures 10 and 11 show SEM micrographs of the horizontal and vertical planes, respectively. Dendrites with short arm spacing and without secondary branching can be observed inside the grain. These dendrites had the same crystallographic orientation with very little to no misorientation between each other. In addition, it is also found that the dendrite arm spacing and direction were not homogeneous within the grain. This phenomenon can be clearly observed from Figures 11 and 12, which show high-resolution SEM images of the boundary of a melt pool. As can be seen, abrupt changes of arm spacing and dentrite direction occured at the laser related boundaries. In particular, the dendrite size tended to be larger at the top of the melt pools and smaller at the bottom (Figure 12). Furthermore, the growth direction of the dendrites did not change when crossing a melt pool boundary in some cases, but rotateed by 90 degrees in other cases (Figure 11b). At the center of the laser tracks the dendrites developed along the build direction, while at the laser track boundaries they tended to rotate (this is also visible in Figure 4b).

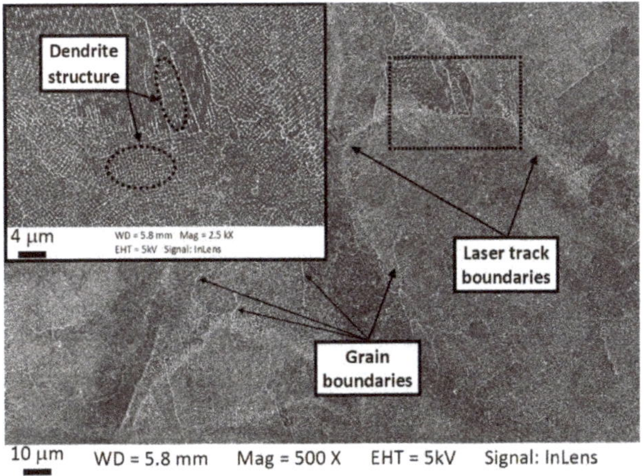

Figure 10. Scanning electron microscopy (SEM) micrographs of the horizontal plane showing laser tracks, grains and the subgranular dendrite structures.

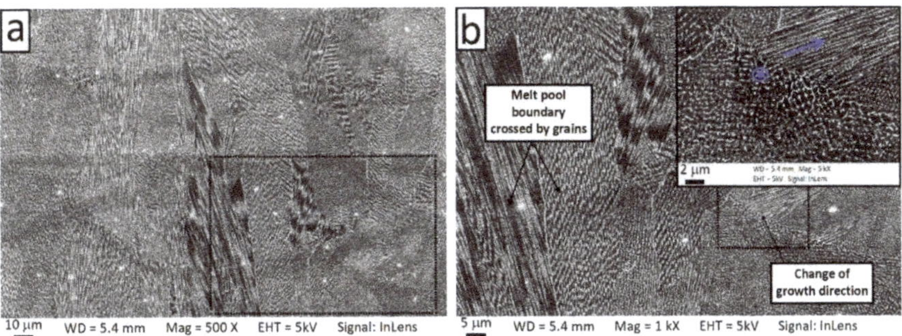

Figure 11. Scanning electron microscopy (SEM) micrographs of the vertical plane showing the columnar dendrites crossing melt pool boundaries (**a**) and the change of dendrite direction at the melt pool boundary (**b**). Blue arrows in (**b**) indicate the growth direction of the dendrites.

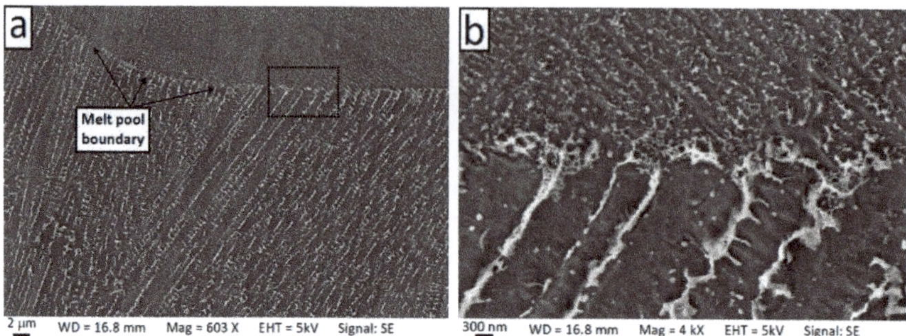

Figure 12. Scanning electron microscopy (SEM) micrograph of the vertical plane showing the change of the dendrite arm spacing and the growth direction when crossing a melt pool boundary. Panel (**b**) contains a higher magnification of the zone indicated by the dashed box in panel (**a**).

3.5. Microsegregation, Interdendrite, and Intradendrite Phases (10^{-6}–10^{-8} m)

Figure 13 shows SEM micrographs of the cross-section of dendrites (also known as the cellular structure) on the horizontal plane. It is clear that the dendrites exhibited a hexagonal pattern with an interspacing of 0.5–1 µm. In addition, a number of precipitates of second phases with a bimodal size distribution were visible. Irregularly-shaped precipitates of approximately 100–120 nm form at the boundaries of adjacent columnar dendrites. Furthermore, a large number of finer precipitates (25–50 nm) were also present in the intradendritic area. The density of the fine precipitates was higher near the edge of the dendrites. Figure 14 shows SEM micrographs of the columnar dendrites across a melt pool boundary on the vertical plane. Precipitates were clearly found around the dendrite boundaries.

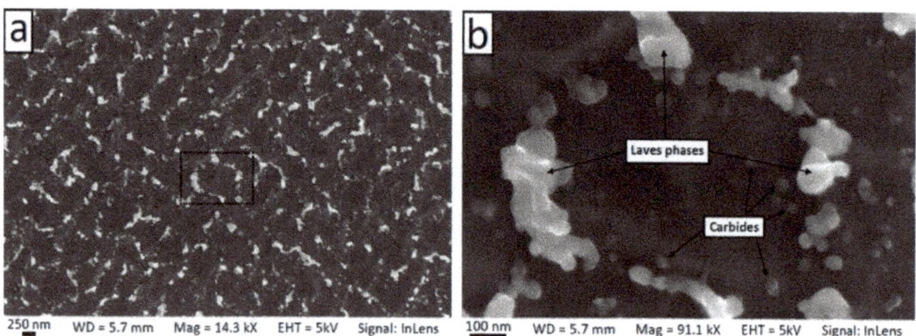

Figure 13. Scanning electron microscopy (SEM) micrographs showing the cellular microstructure on the horizontal plane (**a**) and high magnification view of a single dendrite (**b**).

Figure 15 shows a scanning/transmission electron microscope (STEM) image of a single dendrite. High density dislocations were observed at the edge of the dendrite and around large interdendritic particles. Figure 16 shows an EDS line scan across the dendrite. Microsegregations of Nb, Mo, and Ti were detected at the edges of the dendrite. Ni, Cr, and Fe significantly decreased at one side of the dendrite where Nb content reacheed a peak. This suggests that different second phases were formed at each side of the dendrite. The EDS point analysis also indicates that, as compared with the matrix (point 3), point 1 was richer in Nb, point 2 was richer in Nb, Mo, and Ti, and point 4 was richer in Nb and Ti. Figure 17 shows the TEM image at the interface between the γ dendrite and an interdendritic particle. The selected area electron diffraction (SAED) patterns were taken along the [001] zone axis of the γ phase. It can be observed that the pattern detected on area 2 (i.e., on the interdendritic zone) was made by the superposition of the γ pattern, marked with red circles, and by a second pattern, marked with green circles, which can be interpreted as the $\left[1\bar{1}2\right]$ zone axis of the Laves phase, which has a hexagonal closely-packed structure and lattice parameters a = b = 0. 49 nm and c = 0. 78 nm [48,49]. The result of the SAED analysis suggests that a Laves/γ eutectic mixture was present in the interdendritic zone.

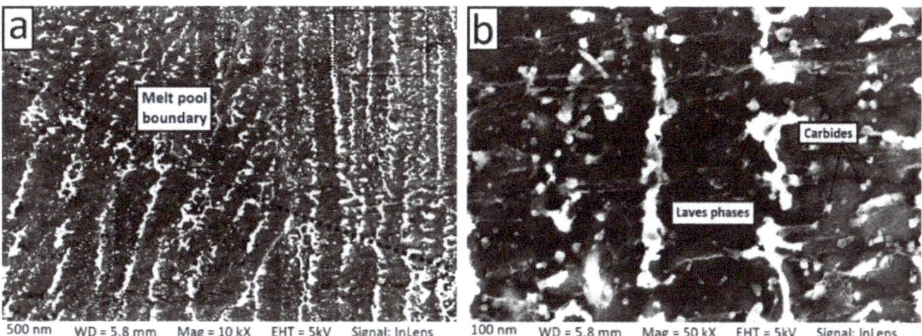

Figure 14. Scanning electron microscopy (SEM) micrographs showing the columnar dendrites on the vertical plane and the change in arm spacing across the melt pool boundary (**a**) and high magnification view (**b**).

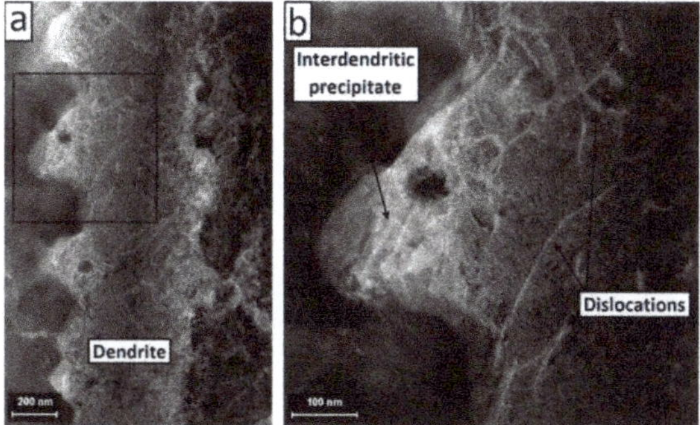

Figure 15. Scanning/transmission electron microscope (STEM) images of a dendrite (**a**) showing the dislocations and the interdendritic phases (**b**).

3.6. Aged State Microstructure

The DSC curves obtained at the as-built and solutioned (1065 °C/2 h) conditions are reported in Figure 18, with the relative marked thermal phenomena occurred during the ramps. In both heating ramps, two exothermal signals (EXO1 and EXO2 peaks) were detected at 500–620 °C and 670–790 °C. Then, a wide endothermal signal (ENDO1) could be observed between 790 and 950 °C. In the ENDO1 temperature range, a third exothermal peak (EXO3) was detected at 850–910 °C. Finally, an endothermal signal (ENDO2) was present between 980–1070 °C. After the solution annealing, the detected signal was similar with respect to the as-built state; however, it was noted that the ENDO2 peak is much weaker.

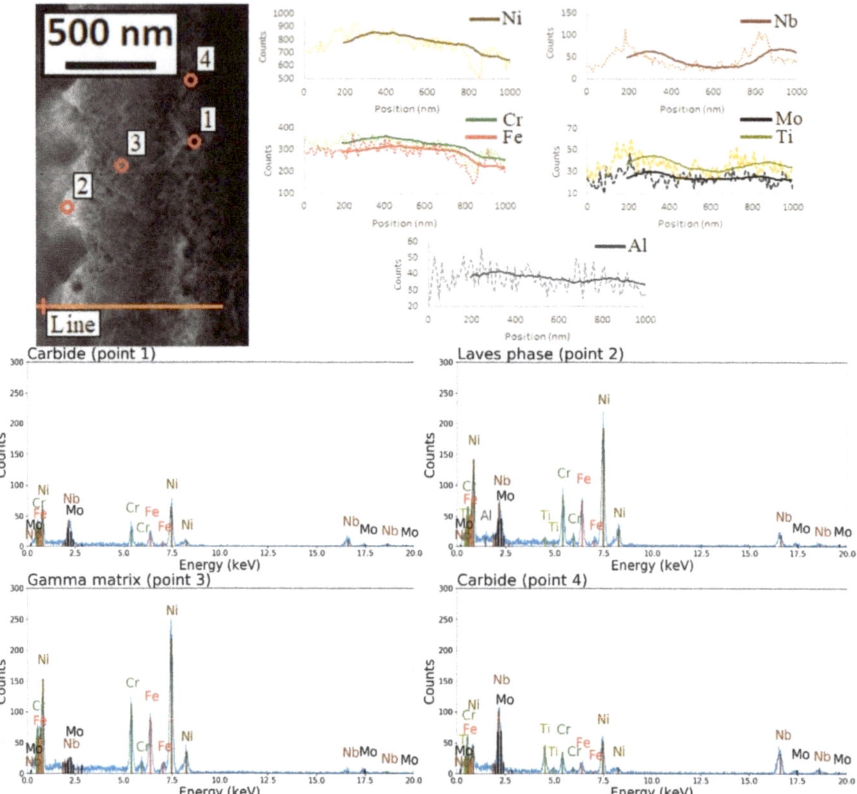

Figure 16. Energy dispersive X-ray spectrometry (EDS) line analysis across the dendrite showing microsegregation of the alloying elements and EDS analysis of interdendritic particles in comparison with the γ intradendritic phase (point 3). Point 2 was compatible with intermetallic Laves phase, while points 1 and 4 were likely carbides.

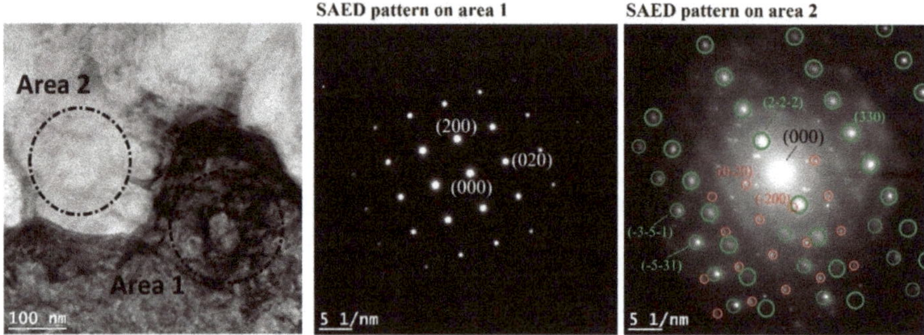

Figure 17. Transmission electron microscope (TEM) image showing the interface between γ phase (dark contrast) and an interdendritic precipitate (bright contrast) with respective selected area electron diffraction (SAED) patterns, both taken along the [001] zone axis of the γ phase. On the SAED pattern on area 2, the green circles underline diffraction points of Laves phase along its $[1\bar{1}2]$ zone axis, instead red circles underline diffraction points of the γ phase.

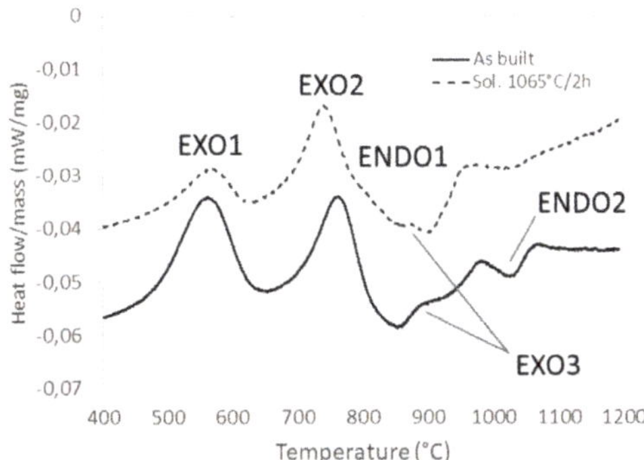

Figure 18. Differential scanning calorimetry (DSC) curves recorded on selective laser melting (SLM) Inconel 718 in the as-built state (full line) and solution heat treated at 1065 °C for 2 h (dotted line). The labels indicate the peaks related to exothermal (EXO) and endothermal (ENDO) phenomena.

Based on the DSC results from the solution heat-treated sample, the following temperatures were considered for the microstructural evolution during the aging step: 565 °C (EXO1 peak), 740 °C (EXO2 peak), 800 °C (EXO2 offset), and 870 °C (EXO3 peak). The mean Vickers microhardness measured on the aged samples are shown in Figure 19 in comparison with the as-built state and the 1065 °C/2 h solution heat-treated state. The solution heat treatment caused a 12.4% reduction of the Vickers microhardness due to the dissolution of most of the pre-existing second phases and the relieving of the residual stresses [50]. The aging treatment increased the Vickers microhardness of the alloy. Aging at 565 °C caused a slight hardening with respect to the solutioned condition; however, the mean Vickers microhardness was lower when compared to the as-built state, even after 24 h of aging. The greatest increase in hardness was obtained when aging at 740 °C, where a 46.6% increase of the Vickers microhardness, with respect to the as-built state, could be reached. The hardness reduced when the sample is aged for 24 h due to over aging. Aging at 800 °C was still able to improve the hardness of the as-built state, although the obtainable Vickers microhardness was lower as compared to that achieved after aging at 740 °C, especially when the treatment was prolonged to 24 h. After aging at 870 °C for 4 h, the Vickers microhardness was comparable to the as-built condition, with a slight decrease with longer treatment durations.

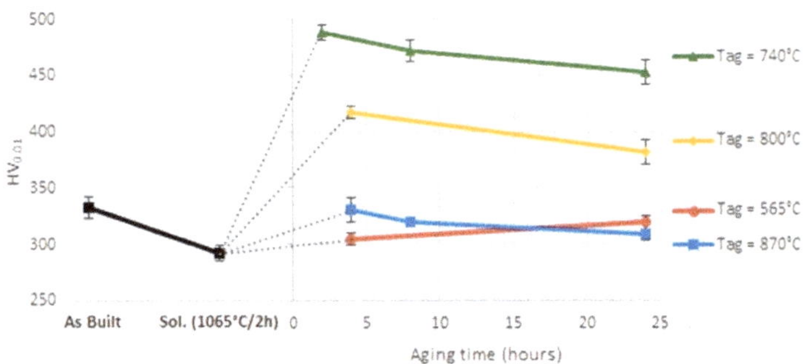

Figure 19. Vickers microhardness at different heat treatment conditions (as-built, solutioning, and aging at different temperatures and times).

The XRD spectra are shown in Figure 20. In the majority of cases, only the γ matrix peaks were detected because there were insufficient amounts of the second phases to be detected by the XRD apparatus (carbides and Laves phases) or their peaks are overlapped with the γ peaks, as in the case of γ′ and γ″ [19]. The only exception was the δ (211) peak observed in the samples aged at 870 °C, which was also reported by Cao et al. in their study [51].

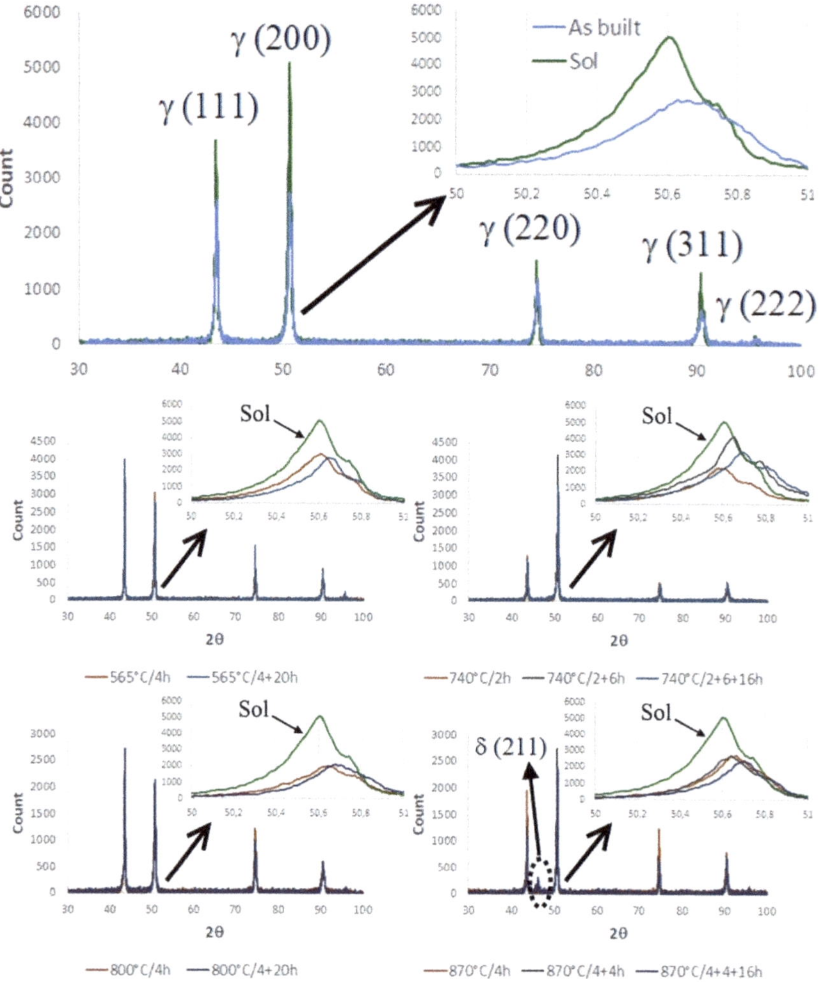

Figure 20. X-Ray diffraction (XRD) spectra acquired from SLM Inconel 718 samples at different heat treatment conditions (as-built, solutioning, and aging at different temperatures and times). The γ(200) peak is shown in higher resolution.

However, the effect of the thermal treatments (solution and aging) can be indirectly measured though the shift of the γ peaks. For example, in Figure 20 the shift of the (200) peak is shown in the higher magnification panels. The shift of the 2θ position was due to a slight variation of the γ matrix lattice parameter, which was related to the amount of the solute dissolved in it. A similar variation of the lattice parameter after heat treatment has previously been reported by Zhang et al. [52].

The lattice parameters can be calculated from the peak 2θ positions through the Bragg equation:

$$n\lambda = 2d \sin\theta \tag{2}$$

where λ is the wavelength used, d is the interplanar distance and θ is the reflection angle.

The plot in Figure 21 shows the value of the γ lattice parameters after solution annealing at 1065 °C for 2 h followed by aging treatment, compared to the as-built condition. For each sample, the lattice parameter is the average of the interplanar distance values obtained from the $\gamma(111)$, $\gamma(200)$, $\gamma(220)$, and $\gamma(311)$ peaks using Equation (2). The cell parameter was obtained by averaging the values obtained from the first four peaks (the fifth is usually too weak).

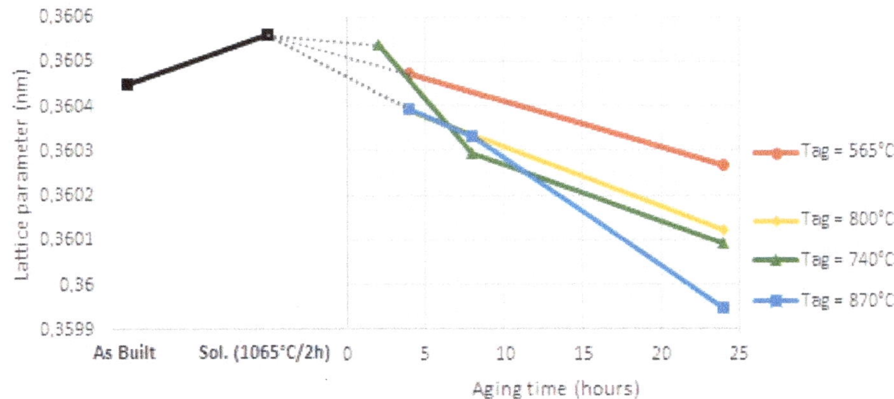

Figure 21. Lattice parameter of the γ matrix calculated from the X-Ray diffraction (XRD) spectra of Figure 20 with the Bragg equation. Trends at different heat treatment conditions (as-built, solutioning and aging at different temperatures and times).

Examples of FESEM micrographs collected for the solutioned and aged samples are shown in Figure 22. All micrographs show the horizontal plane. After the thermal treatment cycle, the as-built microstructure was significantly modified: The laser-related boundaries vanished, relatively large elongated or blocky precipitates (likely carbides [53]) formed at the grain boundaries, and the interdendritic Laves phases were dissolved so that only the smaller eutectic carbides remained as residuals. These observations are in agreement with Brenne et al. [50].

After aging at 565 °C for 4 h, no new second phases could be detected from the FESEM observation (not shown here for the sake of brevity); however, after 24 h very small particles (γ' phase, see discussion) appeared in the intradendritic zone (see panel b of the 565 °C/24 h condition micrograph in Figure 22). Small intradendritic γ' particles could be detected after aging at 740 °C for 2 h. Furthermore, a film-like precipitation occurred at the interdendritic boundaries where the Nb content was higher due to microsegregation. For longer durations, a high density of discoidal precipitates of about 35–65 nm (γ'' phase, see discussion) was observed.

During aging at 800 °C, the grain boundaries displayed plate-like δ precipitates and coarser discoidal γ'' precipitates formed in the intragranular zone. After 24 h aging at 800 °C, the intergranular plate-like precipitates were coarser and smaller plate-like δ precipitates were also formed within the grain.

The intergranular δ plates grew rapidly at 870 °C, furthermore a lot of plate-like intragranular precipitates were formed. The intragranular precipitates were 200–250 nm in length after 4 h of aging and 300–450 nm in length after 8 h (not included in the micrographs of Figure 22 for the sake of brevity). A large amount of very large plates (4–7 μm) uniformly dispersed across the metallographic surface was clearly visible after 24 h of aging at 870 °C.

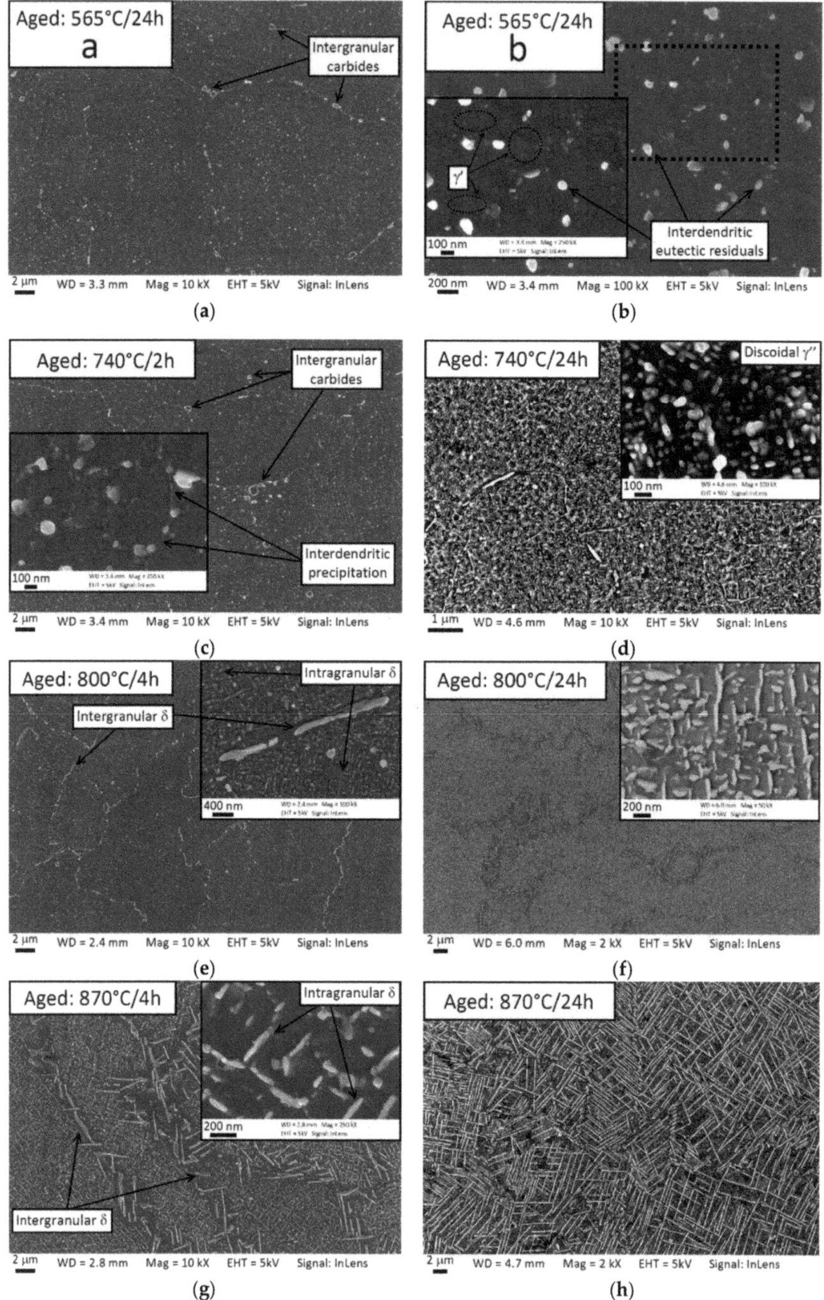

Figure 22. Field emission scanning electron microscopy (FESEM) micrographs showing the microstructure and the formation of precipitates after aging treatments of varying temperature and duration. Aging at 565 °C for 24 h (**a**) and higher magnification (**b**), aging at 740 °C for 2 h (**c**) and 24 h (**d**), aging at 800 °C for 4 h (**e**) and 24 h (**f**), aging for 870 °C for 4 (**g**) and 24 h (**h**).

4. Discussion

4.1. Considerations on the Grain Structure and Texture

During the SLM process, materials are built through sequential solidification steps. At each pass, laser radiation causes an extremely fast melting of a local area of the powder layer and also part of the material that has already solidified during the previous passes [35], causing the formation of liquid volume. When the laser beam leaves the melt pool, the heat is rapidly released to the liquid–Ar atmosphere interface by convection and radiation and to the underlying substrate by conduction [54]. During this rapid cooling of the molten pool, solidification occurs predominantly through two competing phenomena: heterogeneous nucleation of new grains and epitaxial growth [55]. The epitaxial growth of the partially re-melted grains during the laser pass is a phenomenon widely reported in the literature [16,39,56,57] and causes the newly solidified material to inherit the crystallographic orientation of grains in contact with the liquid. Conversely, when heterogeneous nucleation occurs, a new grain with a random crystallographic orientation forms at the solid–liquid interface, interrupting the growth of the underlying grains.

The grain evolution during solidification depends on the crystallographic orientation with respect to the local thermal flux direction in the molten pool. The cubic crystals grow preferentially along the <100> directions [55,58]; therefore, those grains with a <100> axis oriented at a low angle with the local heat flux direction are favored (i.e., they grow faster and prevail over the others). Therefore, if some favorably oriented grains are present at the solid–liquid interface, these grains develop quickly, holding their crystallographic orientation through epitaxial growth and the frequency of formation of new randomly oriented grains through heterogeneous nucleation is low. Conversely, if the existing grains are not favorably oriented, their epitaxial growth is inhibited and thus the solidification of the molten volume occurs mainly through heterogeneous nucleation. When the epitaxial growth prevails, it leads to the formation of a small number of large grains, which are not confined by melt pools (i.e., independent from the laser related features) and a strong crystallographic texture is observed. When heterogeneous nucleation is the dominant solidification mechanism, many small grains confined in the melt pools and in the laser tracks are expected to form. The balance between epitaxial growth and heterogeneous nucleation depends on the thermal flux field in the molten pool, which in turn is influenced by the SLM process parameters and the adopted scan strategy. The exact form of the thermal flux field is usually very complex because of the influence of convective phenomena and the Marangoni effect [59,60]; however, at the solid–liquid interface the thermal fluxes tend to be aligned in a direction normal to the solidification front [61–63]. Therefore, at the boundaries between the laser tracks, the thermal gradients are oriented toward the center of the molten pool rather than along the build direction due to the arc shape of the melt pool [21,47,57].

When a bidirectional scan mode without rotation of the scan direction between successive layers is adopted, an alternated bands granularity is usually reported; this band structure is made of narrow columnar grains with a <100> axis predominately aligned to the build direction and elongated grains along the direction perpendicular to the scan direction, and with <100> and <110> texture along the scan direction and the build direction, respectively [63,64]. The band structure is related to the scan strategy, in particular to the repetition of the scanning pattern that makes the shape of the thermal flux field similar during deposition of each new layer. At the center of the laser scan path, the thermal fluxes have a strong component along the build direction that drives the epitaxial growth of the columnar grains, with [001] orientation created during the deposition of the previous layers. Instead, at the boundaries between the scan lines, the thermal fluxes are at an angle with respect to the build direction; therefore, the grains develop transversally in these zones through epitaxial growth, while heterogeneous nucleation prevails along the build direction. The elongated grains at the boundaries between the laser tracks tend to have a <100> axis aligned to the scan direction and the other two orthogonal <100> axes tilted at 45° respect to the build direction; in fact, this is the configuration that aligns the thermal flux field at the borders of the scan lines, and so grains nucleated with similar crystallographic orientations can grow faster and prevail over the others. If the scan direction is rotated by 90° between each new layer, the resultant bands structure is weaker and the formation of columnar grains along the build direction prevails [63]. A bimodal grain structure has been observed when this 90° rotation scan strategy is adopted due to the periodicity of the scanning pattern; this bimodal grain structure is made by columnar grains with a strong cubic texture, in which the orthogonal <100> axes tend to be oriented along the build direction and along the edges of the squares on the horizontal plane, and by fine grains, which are slightly elongated in the direction perpendicular to the edges of the squares and with a more random crystallographic orientation [34,65,66]. Wang et al. also [65] reported the frequent nucleation of randomly oriented grains at the solidification front that interrupts the epitaxial growth of the columnar grains of the previous layer. In accordance with the above discussion, the prevalence of the heterogeneous nucleation on the epitaxial growth leads to the formation of grains that are more confined within the thickness of the deposited layer and by the laser related features (i.e., the squared scan islands in that case).

In the current study, a bimodal grain structure was not observed, with most of the grains being columnar in shape and aligned along the build direction crossing multiple melt pools (Figures 4e,f and 11). Furthermore, a marked preferential [001] orientation along the build direction was detected from the EBSD analysis (Figures 6 and 8), but no preferential orientation of the other two orthogonal <100> axes was evident. The development of the grains during the SLM process was not confined to the melt pools and the laser tracks, with the laser related boundaries not being detectable in the EBSD images (Figures 5 and 7). This shows the prevalence of the epitaxial growth of the columnar grains across different layers on the nucleation of new randomly oriented grains during the advancement of the solidification front.

Based on the comparison between our results and the aforementioned studies, it can be inferred that the grains which can grow across more deposition steps have a crystallographic orientation that better fits the global shape of the thermal flux field resulting from the adopted scan strategy. The repetitive scan strategies, or those which involve a periodic repetition of the scanning pattern after a small number of deposited layers, lead to a more heterogeneous and anisotropic thermal flux field, to which the grains are aligned. For example, as discussed above, the simple scan strategy without rotation selects the grains with <100> texture along the scan direction and <110> texture along the build direction, while the island scan strategy selects the grains with the <100> axes parallel to the build direction and to the edges of the squared islands. In both of these examined cases, the ideal crystallographic orientation is rigid and so few grains are expected to be selected.

The non-repeating nature of the 67° rotation scan strategy adopted for this study led to a more isotropic situation because the effects of the horizontal components of thermal fluxes (i.e., lying on the plane of the added layers) offset each other, thus only the component parallel to the build direction had a strong role in driving the growth at the length scale of the grains. The result of this scan strategy was the observed strong and quite homogeneous <100> texture along the build direction (Figure 7), but another important consequence is that the distribution of the favored crystallographic orientations was symmetric around the build direction, thus the selection of the grains was much less severe. In these conditions, the epitaxial growth of the preexisting grains prevailed over the heterogeneous nucleation of new grains and, for this reason, small randomly oriented grains at the laser related boundaries were not observed. Additionally, Thijs et al. [67] have observed how the choice of the scan strategy in the SLM process has an important effect on the competitive growth of the grains; however, they state that the rotation of the scan direction of an angle of 90° or 60° after each deposition step leads to a more severe competition, with the consequent development of a stronger texture. Furthermore, Wan et al. [63] reported a stronger texture in an Inconel 718 alloy obtained with a 90° rotation scan strategy with respect to those obtained without any rotation. We state that the 67° rotation has the effect of reducing the harshness of the competition between grains by relaxing one constraint (i.e., the angle around the build direction) on the ideal crystallographic orientation. The disagreement with the above-mentioned studies [63,67] can be explained by considering that the 90° and 60° rotation scan strategies both involve a repetition of the scan pattern after two and three layers, respectively, which are probably not enough to guarantee an effective symmetry along the build direction.

4.2. Development of the Intragranular Dendrites

The morphology of the solidification front is controlled by the G/V ratio [68], where G is the thermal gradient in the liquid and V is the solidification velocity. A high G/V value leads to a stable planar interface, while a low value results in the formation of cells and dendrites. The destabilization of the planar interface is related to the formation of constitutional undercooled liquid in front of it. The essential condition for this to occurs for a binary alloy under steady-state solidification is given by [58]:

$$\frac{G}{V} < \frac{T_L - T_S}{D} \tag{3}$$

where $T_L - T_S$ is the range between the liquidus and the solidus temperatures and D is the diffusivity of the solute in the liquid. It is known from Equation (3) that, at a given value of G, a constitutional undercooling region forms in front of the solid–liquid interface when the solidification velocity exceeds a critical value. As a consequence, the planar interface becomes unstable and the observed cellular dendrites can develop and grow along the direction of the thermal gradient [58,60,69]. Wei et al. calculated the G/V value of the SLM process [61]: In each layer, the G/V value falls in the range between 20 and 100 K s mm^{-2}, and a threshold value for the stabilization of the planar front is in the order of 7000 K s mm^{-2}. Therefore, the columnar dendrite formation is significantly favored as observed. No evidence was found to suggest nucleation in the core of the liquid volume, far from the solid–liquid interface, or the subsequent formation of equiaxed dendrites.

The melt pool solidification is also characterized by very high cooling rates. Hooper [70] used a coaxial high-speed temperature imaging system to monitor the temperature field in the melt pool of a Ti6Al4V powder bed and reported an average thermal gradient and cooling rate of 5–20 K/μm and 1–40 K/μs, respectively. Li et al. [71] and Song et al. [72] used numerical modeling to obtain the cooling rate at different zones of the melt pool of Inconel 718 during directed energy deposition (DED) and reported an average cooling rate range of 2300–6800 K/s. The high cooling rate of the melt pool led to strong microsegregation and the development of an extremely fine subgranular cellular microstructure characterized by small dendrite arm spacing. The mean dendrite arm spacing observed in this study was in the order of 1 μm (Figure 13), which is comparable to that reported by Amato et al. [19] and other studies [16,17,20,47,53]. However, Popovich et al. [35,40] reported coarser dendrites with a dendrite arm spacing of approximately 2–3 μm; in their work they used higher laser power and lower scan speed when compared to those used this study, which favor the formation of larger dendrite sizes due to the formation of larger melt pools and thus lower cooling rates [73]. The primary dendrite arm spacing obtained in SLM process is usually much lower when compared to the DED techniques, for comparison Tian et al. observed a dendrite size of 5–7 μm with peaks of 20 μm [33].

Prominent [001] texture and fine cellular substructure along the build direction were also reported by Mostafa et al. [36], who have also adopted the 67° rotation scan strategy. In addition, morphological differences were found in the laser overlapping regions, where the authors observed a larger dendrite arm spacing and changes in the dendrite growth direction, which are explained with a different experienced cooling rate respect to the zone within the laser tracks [36]. The dendrite growth direction and size were found to be non-uniform over the entire grain also in the current study, although the observed variations seem to us more related to the laser related boundaries: Abrupt changes of 90° in the growth direction were sometimes observed at the boundaries between adjacent laser tracks (Figures 4b, 9b and 11b); the dendrite size changed across the arc-shaped boundaries of the melt pools (Figures 12 and 14).

As discussed above, the growth direction and length scale of the dendrites are related to the direction of the heat fluxes in the melt pool and the cooling rate, respectively. Therefore, it is inferred that the aforementioned inhomogeneities are related to the dynamic temperature field that evolves during the solidification of the melt pool [60,70–72]. At the beginning of solidification, the local cooling rate of the liquid in contact with the solid substrate is high leading to the formation of narrow dendrites, but after this first solidification stage the cooling rate is reduced [70,72] and so a slight increase in the dendrite size occurs in the top part of the melt pool which experiences solidification last.

A 90° rotation of the dendrite growth direction was observed to occur at the boundary of the laser tracks due to local variations in the heat fluxes direction toward the center of the melt pool. The schematic shown in Figure 23 clarifies this process. As can be seen, the dendrites of favorably oriented grains develop along their [001] axis, and can grow epitaxially in the last deposited layer through the melt pool boundary if the thermal gradient direction does not vary across it. However, the thermal gradient direction changes abruptly close to the borders of laser scans, nevertheless the dendrites can still grow epitaxially from the new layer if they are favorably oriented with respect to the new direction of the local thermal gradient (i.e., if one of their [100] or [010] axes is set at a small angle with respect to the local heat flux). Whenever this condition is satisfied, the dendrites start to grow at a 90° angle relative to the previous growth direction. This phenomenon leads to the zig-zag path of grain growth, as observed in Figure 9b. A similar zig-zag epitaxial growth of the grains across subsequent layers was observed in the deposition through DED techniques when a scan strategy of parallel scan lines with altered direction is adopted [37,61,74]. In the current study, where a powder-bed technique and a more complex scan strategy were adopted, the zig-zag growth was not a global feature, but it concerned only the grains that are coincidentally in a favorable position and with the correct crystallographic orientation, as represented in Figure 23. The 90° rotation of the dendrites at the melt pool boundaries in an Inconel 718 alloy produced through SLM is also reported by Deng et al. [75].

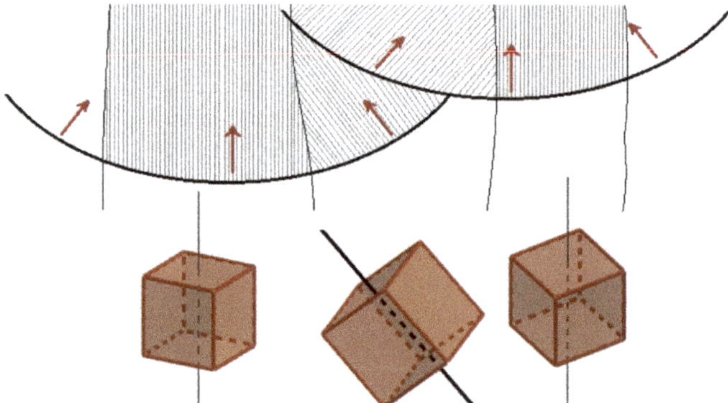

Figure 23. Schematic model of the solidification of columnar grains with different crystallographic orientation. The dendrites of the middle grain change their growth direction of 90° at the melt pool boundary following the local thermal gradients (red arrow) resulting in a zig-zag shape similar to that observed in Figure 9b.

Theoretically, it is possible to assert that the dendrites in each grain should have the same crystallographic orientation. However, the high-resolution EBSD maps in Figure 9b shows some elongated features with slight reciprocal misorientation inside one grain that developed with abrupt change of 90° in the direction. Although the resolution of 0.71 μm/pixel in the EBSD analysis (Figure 9b) is too low to obtain a sharp orientation map at the dendrite length scale, our observations suggest that it is possible that a subgranular structure caused by slightly misoriented colonies of dendrites formed inside some grains. Intragranular dendrites with low-angle misorientation were also reported by Chlebus et al. [38] and Choi et al. [64]. Furthermore, Divya et al. [30] reported a high-resolution EBSD analysis in a single grain of SLM CM247LC Ni alloy, showing a misorientation of lower than 1° between intragranular dendrites that leads to a gradual variation of the orientation across the grain.

The intragranular misorientation can explain the features observed in the grain maps reported in Figure 5, where the largest grains were constituted by subgranular domains surrounded by low-angle boundaries. The slight misorientation between the forming intragranular dendrites led to a crystallographic orientation gradient inside the grain; then a rearrangement of the atoms occured, driven by the reduction of the misorientation energy, with consequent formation of the observed subgranular domains. The rearrangement can be triggered by the thermal cycles to which the as deposited material is subjected at each deposition of a new layer. The presence of the subgranular domains can make the unambiguous identification of the grains in the as-built material difficult.

4.3. Formation of the Dendrite's Features: Microsegregation, Eutectic Phases, and Dislocations

Microsegregation and microstructural inhomogeneities in a single dendrite as shown in Figures 13b, 14b and 15 were due to the path of solidification during the SLM process. The high solidification rate caused the elements to distribute in the solid and remaining liquid at a ratio that reflects their own partition coefficient k (i.e., the ratio between the equilibrium solute concentrations in the solid and in the liquid phase; approximatively constant with the temperature). Elements with a value of k lower than 1 tended to segregate in the last solidifying zone (i.e., the interdendritic boundaries), while the dendrite cores remained more enriched in the elements when k is higher 1. In the Inconel 718 system, the alloying elements with a $k > 1$ are Ni (1.03), Cr (1.09), and Fe (1.20), while those with $k < 1$ are Nb (0.28), Ti (0.41), Mo (0.73), and Al (0.79) [76]. As expected, the EDS analysis reported in Figure 16 indicates that Nb was the most dominant segregation among all the other alloying elements due to its very low partition coefficient, large atomic radius and the consequent low diffusivity in the γ phase which prevents solute redistribution. The interdendritic liquid is progressively enriched by Nb during the formation of the primary γ phase, then the solidification ceases through two non-invariant eutectic transformations that occurs at the interdendritic boundaries [77–80]: $L \rightarrow L + \gamma + NbC$ and L $\rightarrow \gamma +$ Laves phases.

The microstructure of the dendrite observed in the current study (Figures 13–15) was in good agreement with the solidification path described above. Carbides and intermetallic Laves phases are expected near and in correspondence to the interdendritic boundaries. The eutectic products were present in divorced form [38]. Carbides were identified as 25–50 nm sized blocky or rounded particles whose density increases approaching the interdendritic edges (they appear in dark contrast in the STEM images in Figure 15 and are significantly enriched by Nb). Laves phases were placed mostly at the triple points between cells on the horizontal plane (Figure 13b) and disposed along the boundaries between dendrites on the vertical plane (as shown in Figure 14b, in bright contrast in the STEM images in Figures 15 and 17 as confirmed by the SAED analysis). A Laves phase is a metastable Topologically Close Packed (TCP) phase [81] with general formula $(Ni,Cr,Fe)_2(Nb,Mo,Ti)$. Usually, Laves phases are undesirable due to its embrittlement effect and the decrease of the availability of Nb for the formation of the γ'' strengthening phase which can results in the reduction of the mechanical properties [82]. A post homogenization heat treatment is often required to dissolve Laves phases.

Dislocations were also present due to internal stress and consequent plastic deformation caused by the high thermal gradients and consecutive thermal cycles during the SLM process [53]. Dislocations tended to accumulate at the interdendritic boundaries (Figure 15) in order to accommodate the misorientation between cells [30]. Furthermore, higher dislocation density was observed around the interdendritic second phases, in particular the relatively coarse Laves particles, suggesting that the presence of these precipitates, which are able to block the dislocation motion, contributed to the accumulation of the line defects at the interdendritic boundaries.

4.4. Evolution of the Microstructure during the Aging Treatment

In the collected DSC curves (Figure 18), the exothermal peaks indicate the precipitation of second phases, while the dissolution of the second phases is an endothermic phenomenon.

The first two exothermal peaks were due to the precipitation of the strengthening γ' and γ'' phases, respectively. ENDO1 was due to the dissolution of the previously formed γ' and γ'' precipitates. EXO3 was related to the formation of the δ phase.

The last endothermic peak ENDO2 was caused by the dissolution into solid solution of the previously formed δ phase and the partially dissolution of the pre-existing metastable phases, as mainly the Laves compounds observed in the as-built state.

The variation in the lattice parameters after thermal treatment was ascertained from the XRD analysis (Figures 20 and 21). The lattice parameter increased after the solution step because of the greater amount of solute in the γ matrix, then it reduced during the aging because of the precipitation of γ', γ'', and δ second phases. Increasing aging duration led to higher levels of solute transfer from the solid solution to the second phases.

Although the γ' particles after aging at 565 °C are hardly visible at the FESEM images (Figure 22), the DSC analysis (Figure 18), the Vickers microhardness measurements (Figure 19), and the lattice shrinkage (Figure 21) detected through XRD indicate the formation of this phase at this aging temperature.

The greatest contribution to the hardness came from the γ'' formation. At 740 °C the discoidal γ'' particles were formed, preventing the formation of δ phase. The loss of Vickers microhardness after an over aging of 24 hours was likely due to coarsening of the strengthening phases [9,83,84] and by a reduction of the solid solution strengthening caused by the Nb dissolved in the γ matrix.

At 800 °C the δ phase started to form firstly at the grain boundaries, while γ'' discoidal precipitates formed inside the grain and undergo rapid coarsening. The stacking faults of γ'' are nucleation sites for the δ phase [85,86]; therefore, the presence of γ'' particles favor the formation of intragranular δ plates after long aging exposure. The γ'' and δ phase have the same Ni_3Nb stoichiometry, but δ is the thermodynamically stable form; therefore, γ'' transforms progressively into δ during aging with consequent reduction in the strengthening level [87,88]. Furthermore, the δ phase is usually unfavorable because of its plate morphology that causes stress concentrations, although it is sometimes reported as beneficial for creep resistance [86,89].

Consistent with the DSC analyses, no trace of γ' and γ'' were detected at the FESEM after aging at 870 °C, with the rapid formation and growth of the δ plates occurring both at the grain boundaries and inside the grains. The intragranular δ plates were oriented in a regular pattern relative to each other based on a parallelepiped grid due to the well-known crystallographic relationship between the δ phase and the γ matrix [51,90,91]: $(010)_\delta \parallel \{111\}_\gamma$, $[100]_\delta \parallel \langle 1\bar{1}0 \rangle_\gamma$.

The δ precipitates provided a strengthening of the alloy, which is sufficient to recover the decrease in hardness caused by the solution annealing step. However, the hardness level was equal or even slightly lower than the one measured for the as-built sample.

5. Conclusions

The microstructure of as-built SLM Inconel 718 superalloy was deeply investigated at different length scales following an optimization study of processing parameters. The as-built microstructure was characterized by high complexity, which can be described at different length scales. A systematic study of the microstructure of as-built SLM Inconel 718 was carried out and the following main features were detected:

- Length scale from 10^{-3} to 10^{-4} m: Laser related features, columnar grains developed mainly along the build direction and not confined within the melt pools or laser tracks, and presence of subgranular domains separated by low-angle boundaries. The bimodal grain structure, which is usually reported when a periodic scanning strategy is adopted, was not observed in the current study. The 67° rotation scan strategy has a role in impeding the formation of the bimodal grain structure because it removes one constraint in the selection of the most favorably oriented grains; therefore, it is assumed to reduce harshness of the competitive growth and thus to favor epitaxial growth at the expense of heterogeneous nucleation;

- Length scale from 10^{-4} to 10^{-5} m: Predominant <100> crystallographic texture of the grains and substantial isotropy of the crystallographic orientation around the build direction. The lack of observable texture on the horizontal plane is further evidence of the less severe selection of the growing grains;
- Length scale from 10^{-5} to 10^{-6} m: Columnar intragranular dendrites mainly oriented along the build direction or grown following a zig-zag path along the melt pools due to abrupt changes of 90° in the growth direction. Contrary to what was found at the grain length scale, the microstructure at the length scale of dendrites was affected by the laser related boundaries, with abrupt changes in growth direction and dendrite size observed due to the nonuniform solidification conditions and the complex thermal field in the melt pool;
- Length scale from 10^{-6} to 10^{-8} m: Microsegregation of the alloying elements inside the dendrite and presence of extremely fine particles (i.e., carbides and Laves phases) and dislocations at the interdendric boundaries.

Based on the characterization reported in this paper, it can be concluded that the as-built microstructure is not suitable for an immediate application of the material because of the observed heterogeneities, the microsegregation of the alloying elements, and the uneven distribution of a large amount of brittle precipitates. Therefore, a post-process heat treatment is required to correct the microstructure.

The temperature ranges at which the most important precipitation and solutioning phenomena of the second phases can occur, and their effect on the microhardness, were determined in this work. The formation of γ' particles at 565 °C resulted in a slight increase in hardness, with the peak hardness being reached at 740 °C because of large precipitation of discoidal γ''; however, these precipitates underwent coarsening after prolonged aging and tended to transform to plate-like δ phase with consequent decrease in hardness. At the interdendritic boundaries the small eutectic carbides persisted after the aging process and the precipitation of γ'' was more concentrated along them because of the locally high Nb content.

This work provides a complete framework of the Inconel 718 microstructure in the as-built state following the SLM process and how it can be modified through thermal treatment, and it can be used as a base for the development of a post-process heat treatment cycle specifically designed for the specifications required by the final application.

Author Contributions: Conceptualization, M.C., S.Y. and D.U.; formal analysis, M.C.; funding acquisition, R.L. and D.U.; investigation, M.C., S.Y., B.A., and F.C.; methodology, M.C. and F.C.; resources, R.L. and D.U.; supervision, R.L. and D.U.; visualization, M.C.; writing—original draft, M.C.; writing—review and editing, S.Y. and B.A.

Funding: This research received no external funding.

Acknowledgments: This work was supported by the European Space Agency (ESA Frame Contract number 4000112844/14/NL/FE), and was enabled by the CRANN Advanced Microscopy Laboratory (AML). The authors want to express special thanks to Manfredi Diego (Center for Sustainable Futures Technologies—CSFT@POLITO, Istituto Italiano di Tecnologia, Via Livorno, 60, 10144, Turin, Italy) and Ambrosio Elisa Paola (Research Support Department, Politecnico di Torino, Corso Duca degli Abruzzi 24, 10129, Turin, Italy) for their contribution in the preliminary stage of the present study, in particular for the production, through SLM, of all the analyzed Inconel 718 samples.

Conflicts of Interest: The authors declare no conflict of interest.

References

1. Thomas, A.; El-Wahabi, M.; Cabrera, J.M.; Prado, J.M. High temperature deformation of Inconel 718. *J. Mater. Process. Technol.* **2006**, *177*, 469–472. [CrossRef]
2. Pieraggi, B.; Uginet, J.F. Fatigue and Creep Properties in Relation with Alloy 718 Microstructure. *Superalloys* **1994**, *718*, 625–706. [CrossRef]
3. Greene, G.A.; Finfrock, C.C. Oxidation of Inconel 718 in Air at High Temperatures. *Oxid. Met.* **2001**, *55*, 505–521. [CrossRef]

4. Al-Hatab, K.A.; Al-Bukhaiti, M.A.; Krupp, U.; Kantehm, M. Cyclic oxidation behavior of in 718 superalloy in air at high temperatures. *Oxid. Met.* **2011**, *75*, 209–228. [CrossRef]

5. Akca, E.; Gürsel, A. A Review on Superalloys and IN718 Nickel-Based INCONEL Superalloy. *Period. Eng. Nat. Sci.* **2015**, *3*, 15–27. [CrossRef]

6. Barker, J.F. The Initial Years of Alloy 718–A GE Perspective. *Superalloy 718-Metall. Appl.* **1989**, 269–277. [CrossRef]

7. Paulonis, D.F.; Schirra, J.J. Alloy 718 at Pratt & Whitney-Historical Perspective and Future Challenges. *Superalloys* **2001**, *718*, 13–23. [CrossRef]

8. INCONEL Alloy 718. In Special Metals Company, Publication Number SMC-045. 2007. Available online: http://www.specialmetals.com/assets/smc/documents/inconel_alloy_718.pdf (accessed on 14 March 2019).

9. Cozar, R.; Pineau, A. Morphology of Y' and Y" precipitates and thermal stability of Inconel 718 type alloys. *Metall. Trans.* **1973**, *4*, 47–59. [CrossRef]

10. Radavich, J.F. The Physical Metallurgy of Cast and Wrought Alloy 718. *Superalloys* **1989**, *718*, 229–240. [CrossRef]

11. Lingenfelter, A. Welding of Inconel Alloy 718: A Historical Overview. *Superalloys* **1989**, *718*, 673–683. [CrossRef]

12. DuPont, J.N.; Lippold, J.C.; Kiser, S.D. *Welding Metallurgy and Weldability of Nickel-Base Alloys*, 1st ed.; John Wiley & Sons, Inc.: Hoboken, NJ, USA, 2009; ISBN 978-0-470-08714-5.

13. Radhakrishna, C.; Rao, K.P. The formation and control of Laves phase in superalloy 718 welds. *J. Mater. Sci.* **1997**, *32*, 1977–1984. [CrossRef]

14. Carter, L.N.; Attallah, M.M.; Reed, R.C. Laser Powder Bed Fabrication of Nickel-Base Superalloys: Influence of Parameters; Characterisation, Quantification and Mitigation of Cracking. In Proceedings of the Superalloys 2012: 12th International Symposium on Superalloys, Champion, PA, USA, 9–13 September 2012; Huron, E.S., Reed, R.C., Hardy, M.C., Mills, M.J., Montero, R.E., Portella, P.D., Telesman, J., Eds.; The Minerals, Metals & Materials Society: Pittsburgh, PA, USA; pp. 577–586. [CrossRef]

15. Murr, L.E. Metallurgy of additive manufacturing: Examples from electron beam melting. *Addit. Manuf.* **2015**, *5*, 40–53. [CrossRef]

16. Jia, Q.; Gu, D. Selective laser melting additive manufacturing of Inconel 718 superalloy parts: Densification, microstructure and properties. *J. Alloy. Compd.* **2014**, *585*, 713–721. [CrossRef]

17. Kuo, Y.L.; Horikawa, S.; Kakehi, K. The effect of interdendritic δ phase on the mechanical properties of Alloy 718 built up by additive manufacturing. *Mater. Des.* **2017**, *116*, 411–418. [CrossRef]

18. Gong, X.; Chou, K. Microstructures of Inconel 718 by Selective Laser Melting. In Proceedings of the TMS 2015 144th Annual Meeting & Exhibition, Orlando, FL, USA, 15–19 March 2015; The Minerals, Metals & Materials Society: Pittsburgh, PA, USA; pp. 461–468. [CrossRef]

19. Amato, K.N.; Gaytan, S.M.; Murr, L.E.; Martinez, E.; Shindo, P.W.; Hernandez, J.; Collins, S.; Medina, F. Microstructures and mechanical behavior of Inconel 718 fabricated by selective laser melting. *Acta Mater.* **2012**, *60*, 2229–2239. [CrossRef]

20. Wang, Z.; Guan, K.; Gao, M.; Li, X.; Chen, X.; Zeng, X. The microstructure and mechanical properties of deposited-IN718 by selective laser melting. *J. Alloy. Compd.* **2012**, *513*, 518–523. [CrossRef]

21. DebRoy, T.; Wei, H.L.; Zuback, J.S.; Mukherjee, T.; Elmer, J.W.; Milewski, J.O.; Beese, A.M.; Wilson-Heid, A.; De, A.; Zhang, W. Additive manufacturing of metallic components–Process, structure and properties. *Prog. Mater. Sci.* **2018**, *92*, 112–224. [CrossRef]

22. Murr, L.E.; Martinez, E.; Amato, K.N.; Gaytan, S.M.; Hernandez, J.; Ramirez, D.A.; Shindo, P.W.; Medina, F.; Wicker, R.B. Fabrication of metal and alloy components by additive manufacturing: Examples of 3D materials science. *J. Mater. Res. Technol.* **2012**, *1*, 42–54. [CrossRef]

23. Simchi, A. Direct laser sintering of metal powders: Mechanism, kinetics and microstructural features. *Mater. Sci. Eng. A* **2006**, *428*, 148–158. [CrossRef]

24. Yap, C.Y.; Chua, C.K.; Dong, Z.L.; Liu, Z.H.; Zhang, D.Q.; Loh, L.E.; Sing, S.L. Review of selective laser melting: Materials and applications. *Appl. Phys. Rev.* **2015**, *2*, 1–21. [CrossRef]

25. Zhang, D.; Niu, W.; Cao, X.; Liu, Z. Effect of standard heat treatment on the microstructure and mechanical properties of selective laser melting manufactured Inconel 718 superalloy. *Mater. Sci. Eng. A* **2015**, *644*, 32–40. [CrossRef]

26. Qi, H.; Azer, M.; Ritter, A. Studies of Standard Heat Treatment Effects on Microstructure and Mechanical Properties of Laser Net Shape Manufactured INCONEL 718. *Metall. Mater. Trans. A* **2009**, *40*, 2410–2422. [CrossRef]

27. Maamoun, A.H.; Xue, Y.F.; Elbestawi, M.A.; Veldhuis, S.C. The Effect of Selective Laser Melting Process Parameters on the Microstructure and Mechanical Properties of Al6061 and AlSi10Mg Alloys. *Materials* **2019**, *12*, 12. [CrossRef]

28. Uzan, N.E.; Shneck, R.; Yeheskel, O.; Frage, N. Fatigue of AlSi10Mg specimens fabricated by additive manufacturing selective laser melting (AM-SLM). *Mater. Sci. Eng. A* **2017**, *704*, 229–237. [CrossRef]

29. Qiu, C.; Adkins, N.J.E.; Attallah, M.M. Microstructure and tensile properties of selectively laser-melted and of HIPed laser-melted Ti-6Al-4V. *Mater. Sci. Eng. A* **2013**, *578*, 230–239. [CrossRef]

30. Divya, V.D.; Muñoz-Moreno, R.; Messé, O.M.D.M.; Barnard, J.S.; Baker, S.; Illston, T.; Stone, H.J. Microstructure of selective laser melted CM247LC nickel-based superalloy and its evolution through heat treatment. *Mater. Charact.* **2016**, *114*, 62–74. [CrossRef]

31. Kanagarajah, P.; Brenne, F.; Niendorf, T.; Maier, H.J. Inconel 939 processed by selective laser melting: Effect of microstructure and temperature on the mechanical properties under static and cyclic loading. *Mater. Sci. Eng. A* **2013**, *588*, 188–195. [CrossRef]

32. Roberts, I.A.; Wang, C.J.; Esterlein, R.; Stanford, M.; Mynors, D.J. A three-dimensional finite element analysis of the temperature field during laser melting of metal powders in additive layer manufacturing. *Int. J. Mach. Tools Manuf.* **2009**, *49*, 916–923. [CrossRef]

33. Tian, Y.; McAllister, D.; Colijn, H.; Mills, M.; Farson, D.; Nordin, M.; Babu, S. Rationalization of microstructure heterogeneity in INCONEL 718 builds made by the direct laser additive manufacturing process. *Metall. Mater. Trans. A* **2014**, *45*, 4470–4483. [CrossRef]

34. Wang, X.; Chou, K. Effects of thermal cycles on the microstructure evolution of Inconel 718 during selective laser melting process. *Addit. Manuf.* **2017**, *18*, 1–14. [CrossRef]

35. Popovich, V.A.; Borisov, E.V.; Popovich, A.A.; Sufiiarov, V.S.; Masaylo, D.V.; Alzina, L. Functionally graded Inconel 718 processed by additive manufacturing: Crystallographic texture, anisotropy of microstructure and mechanical properties. *Mater. Des.* **2017**, *114*, 441–449. [CrossRef]

36. Mostafa, A.; Rubio, I.P.; Brailovski, V.; Jahazi, M.; Medraj, M. Structure, Texture and Phases in 3D Printed IN718 Alloy Subjected to Homogenization and HIP Treatments. *Metals* **2017**, *7*, 196. [CrossRef]

37. Dinda, G.P.; Dasgupta, A.K.; Mazumder, J. Texture control during laser deposition of nickel-based superalloy. *Scr. Mater.* **2012**, *67*, 503–506. [CrossRef]

38. Chlebus, E.; Gruber, K.; Kuźnicka, B.; Kurzac, J.; Kurzynowski, T. Effect of heat treatment on microstructure and mechanical properties of Inconel 718 processed by selective laser melting. *Mater. Sci. Eng. A* **2015**, *639*, 647–655. [CrossRef]

39. Smith, D.H.; Bicknell, J.; Jorgensen, L.; Patterson, B.M.; Cordes, N.L.; Tsukrov, I.; Knezevic, M. Microstructure and mechanical behavior of direct metal laser sintered Inconel alloy 718. *Mater. Charact.* **2016**, *113*, 1–9. [CrossRef]

40. Popovich, V.A.; Borisov, E.V.; Popovich, A.A.; Sufiiarov, V.S.; Masaylo, D.V.; Alzina, L. Impact of heat treatment on mechanical behaviour of Inconel 718 processed with tailored microstructure by selective laser melting. *Mater. Des.* **2017**, *131*, 12–22. [CrossRef]

41. Li, J.; Zhao, Z.; Bai, P.; Qu, H.; Liu, B.; Li, L.; Wu, L.; Guan, R.; Liu, H.; Guo, Z. Microstructural evolution and mechanical properties of IN718 alloy fabricated by selective laser melting following different heat treatments. *J. Alloy. Compd.* **2019**, *772*, 861–870. [CrossRef]

42. Holland, S.; Wang, X.; Chen, J.; Cai, W.; Yan, F.; Li, L. Multiscale characterization of microstructures and mechanical properties of Inconel 718 fabricated by selective laser melting. *J. Alloy. Compd.* **2019**, *784*, 182–194. [CrossRef]

43. Calandri, M.; Manfredi, D.; Calignano, F.; Ambrosio, E.P.; Biamino, S.; Lupoi, R.; Ugues, D. Solution Treatment Study of Inconel 718 Produced by SLM Additive Technique in View of the Oxidation Resistance. *Adv. Eng. Mater.* **2018**, *20*, 1800351. [CrossRef]

44. EOS NickelAlloy IN718, In EOS Electro Optical System: Industrial 3D Printing, Material Data Sheet. 2014. Available online: http://ip-saas-eos-cms.s3.amazonaws.com/public/4528b4a1bf688496/ff974161c2057e6df56db5b67f0f5595/EOS_NickelAlloy_IN718_en.pdf (accessed on 14 March 2019).

45. *ASTM B962-17, Standard Test Methods for Density of Compacted or Sintered Powder Metallurgy (PM) Products Using Archimedes' Principle*; Astm International: West Conshohocken, PA, USA, 2017.

46. Bassini, E.; Calandri, M.; Parizia, S.; Marchese, G.; Biamino, S.; Ugues, D. Ottimizzazione dei parametri di processo e cenni sui trattamenti termici per la realizzazione di componenti in Inconel 625 e 718 via Selective Laser Melting (SLM). In Proceedings of the 37° Convegno Nazionale AIM, Bologna, Italy, 12–14 September 2018.

47. Shifeng, W.; Shuai, L.; Qingsong, W.; Yan, C.; Sheng, Z.; Yusheng, S. Effect of molten pool boundaries on the mechanical properties of selective laser melting parts. *J. Mater. Process. Technol.* **2014**, *214*, 2660–2667. [CrossRef]

48. Krakow, R.; Johnstone, D.N.; Eggeman, A.S.; Hünert, D.; Hardy, M.C.; Rae, C.M.F.; Midgley, P.A. On the crystallography and composition of topologically close-packed phases in ATI 718Plus®. *Acta Mater.* **2017**, *130*, 271–280. [CrossRef]

49. Manikandan, S.G.K.; Sivakumar, D.; Rao, K.P.; Kamaraj, M. Laves phase in alloy 718 fusion zone-microscopic and calorimetric studies. *Mater. Charact.* **2015**, *100*, 192–206. [CrossRef]

50. Brenne, F.; Taube, A.; Pröbstle, M.; Neumeier, S.; Schwarze, D.; Schaper, M.; Niendorf, T. Microstructural design of Ni-base alloys for high-temperature applications: Impact of heat treatment on microstructure and mechanical properties after selective laser melting. *Prog. Addit. Manuf.* **2016**, *1*, 141–151. [CrossRef]

51. Cao, G.H.; Sun, T.Y.; Wang, C.H.; Li, X.; Liu, M.; Zhang, Z.X.; Hu, P.F.; Russell, A.M.; Schneider, R.; Gerthsen, D.; et al. Investigations of γ', γ'' and δ precipitates in heat-treated Inconel 718 alloy fabricated by selective laser melting. *Mater. Charact.* **2018**, *136*, 398–406. [CrossRef]

52. Zhang, F.; Levine, L.E.; Allen, A.J.; Stoudt, M.R.; Lindwall, G.; Lass, E.A.; Williams, M.E.; Idell, Y.; Campbell, C.E. Effect of heat treatment on the microstructural evolution of a nickel- based superalloy additive-manufactured by laser powder bed fusion. *Acta Mater.* **2018**, *152*, 200–214. [CrossRef]

53. Tucho, W.M.; Cuvillier, P.; Sjolyst-Kverneland, A.; Hansen, V. Microstructure and hardness studies of Inconel 718 manufactured by selective laser melting before and after solution heat treatment. *Mater. Sci. Eng. A* **2017**, *689*, 220–232. [CrossRef]

54. Yan, Z.; Liu, W.; Tang, Z.; Liu, X.; Zhang, N.; Li, M.; Zhang, H. Review on thermal analysis in laser-based additive manufacturing. *Opt. Laser Technol.* **2018**, *106*, 427–441. [CrossRef]

55. Kou, S. *Welding Metallurgy*, 2nd ed.; Chapman & Hall: London, UK, 2003; ISBN 0-471-43491-4.

56. Liu, F.; Lin, X.; Huang, C.; Song, M.; Yang, G.; Chen, J.; Huang, W. The effect of laser scanning path on microstructures and mechanical properties of laser solid formed nickel-base superalloy Inconel 718. *J. Alloy. Compd.* **2011**, *509*, 4505–4509. [CrossRef]

57. Kuo, Y.-L.; Horikawa, S.; Kakehi, K. Effects of build direction and heat treatment on creep properties of Ni-base superalloy built up by additive manufacturing. *Scr. Mater.* **2017**, *129*, 74–78. [CrossRef]

58. Porter, D.A.; Easterling, K.E. *Phase Transformations in Metals and Alloys*, 2nd ed.; Chapman & Hall: London, UK, 1992; ISBN 978-1-4899-3051-4.

59. Xia, M.; Gu, D.; Yu, G.; Dai, D.; Chen, H.; Shi, Q. Selective laser melting 3D printing of Ni-based superalloy: Understanding thermodynamic mechanisms. *Sci. Bull.* **2016**, *61*, 1013–1022. [CrossRef]

60. Acharya, R.; Sharon, J.A.; Staroselsky, A. Prediction of microstructure in laser powder bed fusion process. *Acta Mater.* **2017**, *124*, 360–371. [CrossRef]

61. Wei, H.L.; Mazumder, J.; DebRoy, T. Evolution of solidification texture during additive manufacturing. *Sci. Rep.* **2015**, *5*, 1–7. [CrossRef] [PubMed]

62. Wei, H.L.; Elmer, J.W.; Debroy, T. Origin of grain orientation during solidification of an aluminum alloy. *Acta Mater.* **2016**, *115*, 123–131. [CrossRef]

63. Wan, H.Y.; Zhou, Z.J.; Li, C.P.; Chen, G.F.; Zhang, G.P. Effect of scanning strategy on grain structure and crystallographic texture of Inconel 718 processed by selective laser melting. *J. Mater. Sci. Technol.* **2018**, *34*, 1799–1804. [CrossRef]

64. Choi, J.P.; Shin, G.H.; Yang, S.; Yang, D.Y.; Lee, J.S.; Brochu, M.; Yu, J.H. Densification and microstructural investigation of Inconel 718 parts fabricated by selective laser melting. *Powder Technol.* **2017**, *310*, 60–66. [CrossRef]

65. Wang, X.; Chou, K. Electron Backscatter Diffraction Analysis of Inconel 718 Parts Fabricated by Selective Laser Melting Additive Manufacturing. *JOM* **2017**, *69*, 402–408. [CrossRef]

66. Carter, L.N.; Martin, C.; Withers, P.J.; Attallah, M.M. The influence of the laser scan strategy on grain structure and cracking behaviour in SLM powder-bed fabricated nickel superalloy. *J. Alloy. Compd.* **2014**, *615*, 338–347. [CrossRef]

67. Thijs, L.; Sistiaga, M.L.M.; Wauthle, R.; Xie, Q.; Kruth, J.-P.; Van Humbeeck, J. Strong morphological and crystallographic texture and resulting yield strength anisotropy in selective laser melted tantalum. *Acta Mater.* **2013**, *61*, 4657–4668. [CrossRef]

68. Lee, J.S.; Gu, J.H.; Jung, H.M.; Kim, E.H.; Jung, Y.G.; Lee, J.H. Directional solidification microstructure control in CM247LC superalloy. *Mater. Today Proc.* **2014**, *1*, 3–10. [CrossRef]

69. Stefanescu, D.M.; Ruxanda, R. Fundamentals of Solidification. In *ASM Handbook, Metallography and Microstructures*; Vander Voort, G.F., Ed.; ASM International: Materials Park, OH, USA, 2004; Volume 9, pp. 71–92. ISBN 978-0-87170-706-2.

70. Hooper, P.A. Melt pool temperature and cooling rates in laser powder bed fusion. *Addit. Manuf.* **2018**, *22*, 548–559. [CrossRef]

71. Li, S.; Xiao, H.; Liu, K.; Xiao, W.; Li, Y.; Han, X.; Mazumder, J.; Song, L. Melt-pool motion, temperature variation and dendritic morphology of Inconel 718 during pulsed- and continuous-wave laser additive manufacturing: A comparative study. *Mater. Des.* **2017**, *119*, 351–360. [CrossRef]

72. Song, J.; Chew, Y.; Bi, G.; Yao, X.; Zhang, B.; Bai, J.; Moon, S.K. Numerical and experimental study of laser aided additive manufacturing for melt-pool profile and grain orientation analysis. *Mater. Des.* **2018**, *137*, 286–297. [CrossRef]

73. Ali, H.; Ghadbeigi, H.; Mumtaz, K. Effect of scanning strategies on residual stress and mechanical properties of Selective Laser Melted Ti6Al4V. *Mater. Sci. Eng. A* **2018**, *712*, 175–187. [CrossRef]

74. Parimi, L.L.; Ravi, G.; Clark, D.; Attallah, M.M. Microstructural and texture development in direct laser fabricated IN718. *Mater. Charact.* **2014**, *89*, 102–111. [CrossRef]

75. Deng, D.; Peng, R.L.; Brodin, H.; Moverare, J. Microstructure and mechanical properties of Inconel 718 produced by selective laser melting: Sample orientation dependence and effects of post heat treatments. *Mater. Sci. Eng. A* **2018**, *713*, 294–306. [CrossRef]

76. Kang, S.-H.; Deguchi, Y.; Yamamoto, K.; Ogi, K.; Shirai, M. Solidification Process and Behavior of Alloying Elements in Ni-Based Superalloy Inconel718. *Mater. Trans.* **2004**, *45*, 2728–2733. [CrossRef]

77. Knorovsky, G.A.; Cieslak, M.J.; Headley, T.J.; Romig, A.D.; Hammetter, W.F. INCONEL 718: A Solidification Diagram. *Metall. Trans. A* **1989**, *20*, 2149–2158. [CrossRef]

78. DuPont, J.N.; Robino, C.V.; Michael, J.R.; Notis, M.R.; Marder, A.R. Solidification of Nb-Bearing Superalloys: Part, I. Reaction Sequences. *Metall. Mater. Trans. A* **1998**, *29*, 2785–2796. [CrossRef]

79. DuPont, J.N.; Robino, C.V.; Marder, A.R.; Notis, M.R. Solidification of Nb-bearing superalloys: Part II. Pseudoternary solidification surfaces. *Metall. Mater. Trans. A* **1998**, *29*, 2797–2806. [CrossRef]

80. Dupont, J.N.; Robino, C.V.; Marder, A.R. Modeling solute redistribution and microstructural development in fusion welds of Nb-bearing superalloys. *Acta Mater.* **1998**, *46*, 4781–4790. [CrossRef]

81. Wilson, A.S. Formation and effect of topologically close-packed phases in nickel-base superalloys. *Mater. Sci. Technol.* **2017**, *33*, 1108–1118. [CrossRef]

82. Schirra, J.J.; Caless, R.H.; Hatala, R.W. The effect of laves phases on the mechanical properties of wrought and cast + HIP Inconel 718. *Superalloys* **1991**, *718*, 375–388. [CrossRef]

83. Slama, C.; Abdellaoui, M. Structural characterization of the aged Inconel 718. *J. Alloy. Compd.* **2000**, *306*, 277–284. [CrossRef]

84. Rao, G.A.; Kumar, M.; Srinivas, M.; Sarma, D.S. Effect of standard heat treatment on the microstructure and mechanical properties of hot isostatically pressed superalloy inconel 718. *Mater. Sci. Eng. A* **2003**, *355*, 114–125. [CrossRef]

85. Sundararaman, M.; Mukhopadhyay, P.; Banerjee, S. Precipitation and Room Temperature Deformation Behaviour of Inconel 718. *Superalloys* **1994**, *718*, 419–440. [CrossRef]

86. Azadian, S.; Wei, L.Y.; Warren, R. Delta phase precipitation in Inconel 718. *Mater. Charact.* **2004**, *53*, 7–16. [CrossRef]

87. Anderson, M.; Thielin, A.-L.; Bridier, F.; Bocher, P.; Savoie, J. Phase precipitation in Inconel 718 and associated mechanical properties. *Mater. Sci. Eng. A* **2017**, *679*, 48–55. [CrossRef]

88. Slama, C.; Servant, C.; Cizeron, G. Aging of the Inconel 718 alloys between 500 and 750 °C. *J. Mater. Res.* **1997**, *12*, 2298–2316. [CrossRef]

89. Sjoberg, G.; Ingesten, N.-G.; Carlson, R.G. Grain Boundary delta-phase Morphologies, Carbides and Notch Rupture Sensitivity of Cast Alloy 718. *Superalloys* **1991**, *718*, 603–620. [CrossRef]

90. Nunes, R.M.; Pereira, D.; Clarke, T.; Hirsch, T.K. Delta Phase Characterization in Inconel 718 Alloys Through X-ray Diffraction. *ISIJ Int.* **2015**, *55*, 2450–2454. [CrossRef]

91. Rong, Y.; Chen, S.; Gengxiang, H.U.; Gao, M.; Wei, R.P. Prediction and characterization of variant electron diffraction patterns for γ″ and δ precipitates in an INCONEL 718 alloy. *Metall. Mater. Trans. A* **1999**, *30*, 2297–2303. [CrossRef]

Article

Structural and Mechanical Characteristics of Cu$_{50}$Zr$_{43}$Al$_7$ Bulk Metallic Glass Fabricated by Selective Laser Melting

Xiaoyang Lu [1,2], Mussokulov Nursulton [1], Yulei Du [1,*] and Wenhe Liao [1,*]

[1] School of Mechanical Engineering, Nanjing University of Science and Technology, Nanjing 210094, China; njust_luxiaoyang@163.com (X.L.); ac_milan_barsa@mail.ru (M.N.)

[2] Luoyang Ship Material Research Institute, Luoyang 471023, China

* Correspondence: yldu_njust@njust.edu.cn (Y.D.); cnwho@njust.edu.cn (W.L.)

Received: 9 December 2018; Accepted: 31 January 2019; Published: 6 March 2019

Abstract: In this work, the structural and mechanical characteristics of Cu$_{50}$Zr$_{43}$Al$_7$ bulk metallic glass (BMG) fabricated by selective laser melting (SLM) are studied and the impacts from the SLM process are clarified. Cu$_{50}$Zr$_{43}$Al$_7$ alloy specimens were manufactured by the SLM method from corresponding gas-atomized amorphous powders. The as-built specimens were examined in terms of phase structure, morphologies, thermal properties and mechanical behavior. The x-ray diffraction and differential scanning calorimetry results showed that structural relaxation and partial crystallization co-exist in the as-fabricated Cu$_{50}$Zr$_{43}$Al$_7$ glassy samples. The nano- and micro- hardness and the elastic modulus of the SLM-fabricated Cu$_{50}$Zr$_{43}$Al$_7$ BMG were higher than CuZrAl ternary BMGs with similar compositions prepared by conventional mold casting, which can be attributed to the structural relaxation in the former sample. However, the macro compressive strength of the SLM-fabricated Cu$_{50}$Zr$_{43}$Al$_7$ BMG was only 1044 MPa mainly due to its porosity. This work suggests that the SLM process induced changes in structural and mechanical properties are significant and cannot be neglected in the fabrication of BMGs.

Keywords: bulk metallic glasses; selective laser melting; Cu$_{50}$Zr$_{43}$Al$_7$; mechanical properties

---

1. Introduction

Bulk metallic glasses (BMGs) are promising structural materials because of their unique mechanical, physical and chemical properties [1–3]. However, due to their limited ability to form glass, it is difficult to fabricate BMGs with large cross sections or complex shapes by traditional mold casting methods. To overcome such size and shape restrictions, powder metallurgy methods, such as hot-pressing sintering and spark plasma sintering techniques, have been adopted to prepare BMGs [4,5]. However, the consolidation of amorphous powders into bulk samples by these methods requires high pressure and molds. So, the manufacturing of large BMGs with complex shapes remains a problem. Selective laser melting (SLM), a laser additive manufacturing technique, can obtain a cooling rate as high as 10^3–10^4 K/s during the point-by-point and layer-by-layer forming period. It can thus realize the freeform fabrication of metallic materials and has recently been adopted in various attempts to produce BMGs [6]. To date, some Fe- [7–9], Zr- [10–12], Al- [13,14] and Ti-based [15] glass–forming alloys have been fabricated by SLM, preliminarily verifying its capabilities in producing BMGs in intricate shapes and large sizes. It is known that Cu-based BMGs usually possess a higher strength, better ductility and are relatively lower in cost [16,17]. However, the application of SLM and infrared lasering in the manufacturing Cu-based BMGs is restricted by the high thermal conductivity and the high reflectivity of Cu-based alloys, resulting in greater heat loss and an inadequate melting of the powders during the SLM process.

This work seeks to uncover the structural and mechanical characteristics of SLM-fabricated Cu-based BMGs and sort out the related impact of SLM processing. The $Cu_{50}Zr_{43}Al_7$ BMG, a typical ternary Cu-based bulk glass–forming alloy, was chosen for the study and fabricated by SLM. Its structural and mechanical characteristics were analyzed.

2. Materials and Methods

$Cu_{50}Zr_{43}Al_7$ powders were prepared by argon gas atomization. The as-atomized powders with a particle size between 10 and 50 μm were selected for SLM experiments. $Cu_{50}Zr_{43}Al_7$ cylindrical and cubic samples were fabricated using a 3D printer (YLM-120 SLM, manufactured by Jiangsu Yongnian Tech. Co. Ltd., Suzhou, China). As listed in Table 1, the SLM fabrication parameters for these cylinders were as follows: the laser power P = 150 and 190 W; the scanning speed v = 2000, 2200, and 2400 mm/s; the hatch distance h = 0.1 and 0.12 mm; the layer thickness d = 0.03 mm; and the laser energy density E was in the range of 20.8–31.7 J/mm^3 calculated using the formula $E = P/(h \times v \times d)$ (J/mm^3). The SLM fabrication process was performed under an argon atmosphere with an oxygen content of below 100 ppm. The rotation angle of the laser scanning direction of adjacent layers was 67° as shown in Figure 1a. The size of the SLM-fabricated cylinder samples was 4 mm in diameter and 8 mm in height. The samples for each experimental condition were fabricated for several repetitions at a time (as shown in Figure 1b).

Table 1. Selective laser melting (SLM) experimental parameters.

E (J/mm^3)	P (W)	v (mm/s)	h (mm)	d (mm)
31.7	190	2000	0.10	0.03
28.8	190	2200	0.10	0.03
26.4	190	2400	0.10	0.03
25.0	150	2000	0.10	0.03
24.0	190	2200	0.12	0.03
22.7	150	2200	0.10	0.03
20.8	150	2400	0.10	0.03

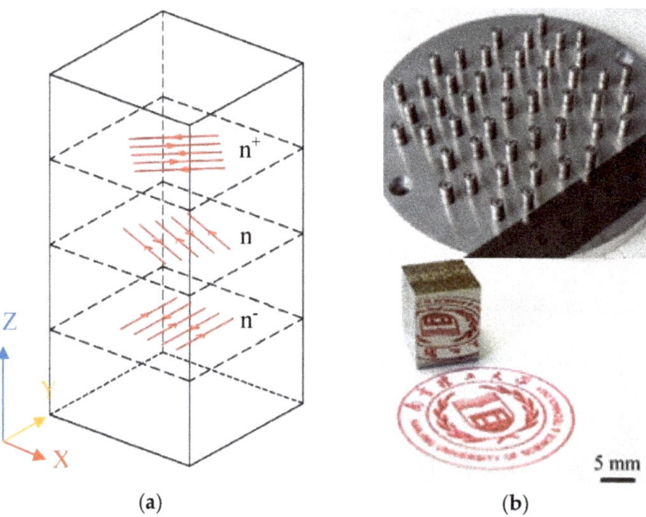

Figure 1. (a) Scanning strategy of the present selective laser melting; (b) The morphology and length of the SLM-fabricated specimens.

3. Results and Discussion

Figure 2 shows the XRD pattern of the as-atomized $Cu_{50}Zr_{43}Al_7$ powders. The pattern exhibits a characteristic broad diffraction without any detectable crystalline Bragg peaks, indicating that fully amorphous structures were obtained for the $Cu_{50}Zr_{43}Al_7$ particles with dimensions smaller than 50 μm. Obviously, the cooling rate during the gas atomization process was sufficient to suppress the formation of the crystalline phases. As shown in the inset of Figure 2, most of the as-atomized $Cu_{50}Zr_{43}Al_7$ powders exhibited spherical or near spherical shape with a smooth surface, which was conducive to the smooth running of the SLM process

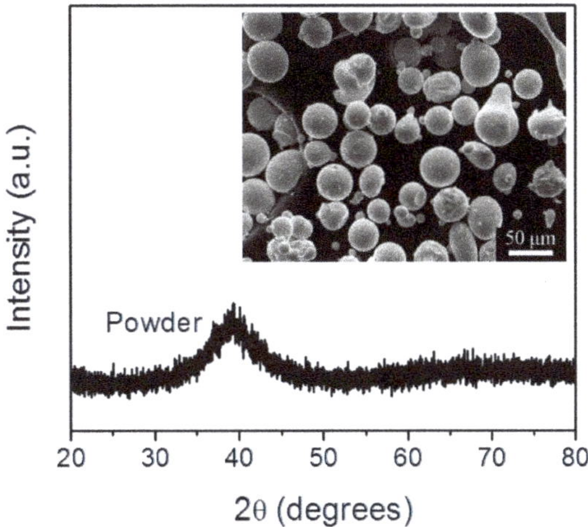

Figure 2. X-ray diffraction (XRD) pattern of the as-atomized $Cu_{50}Zr_{43}Al_7$ powders. The inset depicts a scanning electron microscopy (SEM) image of the powders.

In the present SLM process, the laser energy density was varied in order to study its influence on the amorphization of the SLM-fabricated $Cu_{50}Zr_{43}Al_7$ samples. As can be seen from Figure 1b, when the applied laser energy density was in the range of 20.8–31.7 J/mm^3, all the samples were successfully produced by SLM. A cubic sample with a polished lateral surface was also shown in Figure 1b. As can be seen, a mirror-like surface can be obtained by polishing, indicating the high quality of the SLM-fabricated Cu-based BMG. No obvious cracking or layer delamination was observed by visual inspection of the exterior surface. Figure 3 shows the XRD patterns taken on the cross sections of the SLM-fabricated samples under different laser energy densities. It can be seen that for the samples fabricated with the laser energy density of 31.7, 28.8 and 26.4 J/mm^3, some crystalline phases occurred within the amorphous matrix, which can be attributed to crystallization at heat affect zones under relatively high laser energy. Based on previous reports [18,19], these crystalline phases can be identified as CuZr, $Cu_{10}Zr_7$ and Al_2Zr phases, respectively. Interestingly, for the samples fabricated under a low laser energy density of 20.8, 22.7 and 24 J/mm^3, obvious crystalline phases could also be detected, which might have been due to the crystallization of the un-melted powders under relatively low laser energy. For the samples fabricated with a laser energy density of 25 J/mm^3, a nearly fully amorphous structure was obtained. However, a poor superimposed peak on the main amorphous maximum indicated that slight crystallization was inevitable in the SLM-fabricated BMG samples. In order to further investigate the crystalline phases in the SLM-fabricated Cu-based BMG sample, TEM observations were performed. Figure 4 shows the TEM images of the SLM-fabricated $Cu_{50}Zr_{43}Al_7$ BMG under a laser energy density of 25 J/mm^3. As can be seen from Figure 4a, many crystalline phases dispersed in the amorphous matrix. These crystalline phases were irregular in shape and their size ranged from several to dozens of nanometers. Figure 4b shows the enlarged HRTEM (high resolution transmission electron microscopy) image of the SLM-fabricated Cu-based BMG. It is evident that the nano-sized crystalline phases precipitated from the glassy matrix. As indicated in the figure, the crystalline phases were identified as CuZr phase, consistent with the XRD result. The samples SLM-fabricated under a laser energy density of 25 J/mm^3 were used in the following studies, because of their nearly fully amorphous structure.

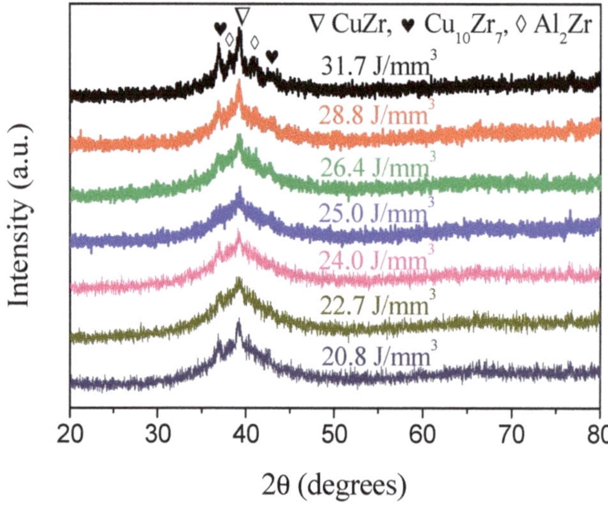

Figure 3. XRD patterns of the SLM-fabricated Cu$_{50}$Zr$_{43}$Al$_7$ specimens under different laser energy densities.

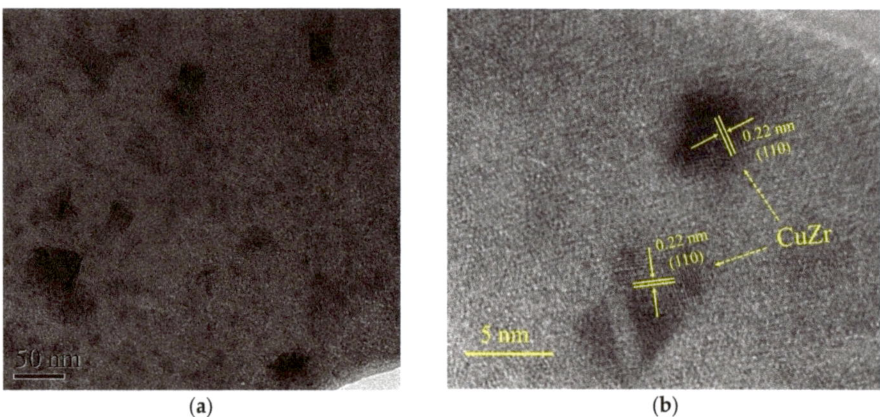

Figure 4. (a) Transmission electron microscopy (TEM) morphology and (b) high resolution transmission electron microscopy (HRTEM) image of SLM-fabricated Cu$_{50}$Zr$_{43}$Al$_7$ bulk metallic glasses (BMG) under a laser energy density of 25 J/mm3.

Figure 5 shows the DSC curves of as-atomized Cu$_{50}$Zr$_{43}$Al$_7$ powders and an SLM-fabricated sample with a laser energy density of 25 J/mm^3. Both of them showed the typical features of the amorphous phase. The values of glass transition temperature (T_g) and relaxation enthalpy (ΔH_r) were 733.1 K, 1.3 J/g and 733.4 K, 1.9 J/g, respectively for the as-atomized Cu$_{50}$Zr$_{43}$Al$_7$ powders and the SLM-fabricated BMG samples. Apparently, the higher relaxation enthalpy of the SLM-fabricated BMG samples shows that they experienced structural relaxation during the SLM process. The existence of structural relaxation in the SLM-fabricated sample can be also proved by the fact that the values of the onset temperature of crystallization (T_x, 777.7 K) for the as-atomized metallic glass powders were significantly higher than those (768.3 K) for the SLM-fabricated BMG samples. As reported previously [20], structural relaxation may lead to a decrease in the T_x of BMGs. In addition, the values of the crystallization enthalpy (ΔH_c, 49.6 J/g) for the as-atomized $Cu_{50}Zr_{43}Al_7$ powders were significantly higher than those (ΔH_c, 40.2 J/g) for the SLM-fabricated BMG sample, which indicates that partial crystallization occurred in the latter sample. SLM is a layer by layer deposition process, during which there exists a fusion zone in the current layer, a remelting zone and a heat-affected zone in the former solidified layers. The specific heat transfer status in different zones led to the occurrence of structural relaxation and partial crystallization in the SLM-fabricated BMGs. The present work implies that structural relaxation and partial crystallization co-exist in the SLM-fabricated BMGs.

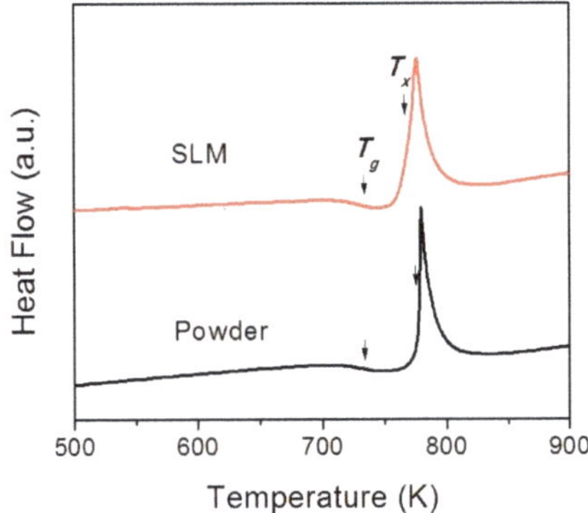

Figure 5. Differential scanning calorimetry (DSC) curves of as-atomized $Cu_{50}Zr_{43}Al_7$ metallic glass powders and the SLM-fabricated $Cu_{50}Zr_{43}Al_7$ BMG sample.

Figure 6 shows the nanoindentation load-displacement (P-h) curves measured on the cross and longitudinal sections of the SLM-fabricated $Cu_{50}Zr_{43}Al_7$ BMG cubic samples with a side length of 10 mm. Following the methods proposed by Oliver and Pharr [20], the nanoindentation hardness (H_{IT}/MPa) and elastic modulus (E_{IT}) values were calculated from the load-displacement curves. The results are listed in Table 2, together with the microscopic Vickers hardness and those from the cast Cu-based BMGs with similar compositions in the literature [22,23]. As can be seen, the hardness and elastic modulus of the SLM-fabricated $Cu_{50}Zr_{43}Al_7$ BMGs were much higher than those of the CuZrAl ternary BMGs prepared by conventional mold casting. This enhancement can mainly be attributed to structural relaxation. As reported previously [24,25], the annealing induced structural relaxation in BMGs resulting from free volume annihilation and atomic rearrangement, hindered the formation and propagation of shear bands. The hardness and elastic modulus of the relaxed BMGs were thus higher than that of the as-cast BMGs. It should be noted that the present SLM-fabricated BMG sample contained crystalline phases embedded in an amorphous matrix, as shown in Figure 3. In the previous work [26], it was revealed that the crystalline phases were softer than the amorphous matrix, thus, it is reasonable to conclude that the increased hardness in the present SLM-fabricated BMG sample was mainly due to the structural relaxation during the SLM process. Figure 7 shows the compressive engineering stress–strain curve of the SLM-fabricated $Cu_{50}Zr_{43}Al_7$ BMG rods. The compressive strength was 1044 MPa, lower than that of the as-cast Cu-based BMGs in previous reports [16,17]. The decrease in compressive strength was also found in other SLM-fabricated BMGs [15]. Obviously, the SLM-fabricated BMGs exhibited higher nanohardness and lower macro compressive strength compared to the corresponding BMGs prepared by traditional casting methods. It is worthy to reveal the reason for this phenomenon. Since SLM is a powder bed fusion additive manufacturing process, the SLM-fabricated samples are usually not completely dense and have many structural defects [7–11,15]. It is known that the macro–mechanical properties of metallic materials are strongly affected by the presence of structural defects. The relative density of the SLM-fabricated $Cu_{50}Zr_{43}Al_7$ BMG sample under a laser energy density of $25 \, J/mm^3$ was measured to be 0.97 by the Archimedes method. It is known that the relative density can be increased by increasing the laser energy density in the SLM process, however, in our case, a higher laser energy density than $25 \, J/mm^3$ will lead to the obvious crystallization of the CuZrAl BMG, as shown in Figure 3.

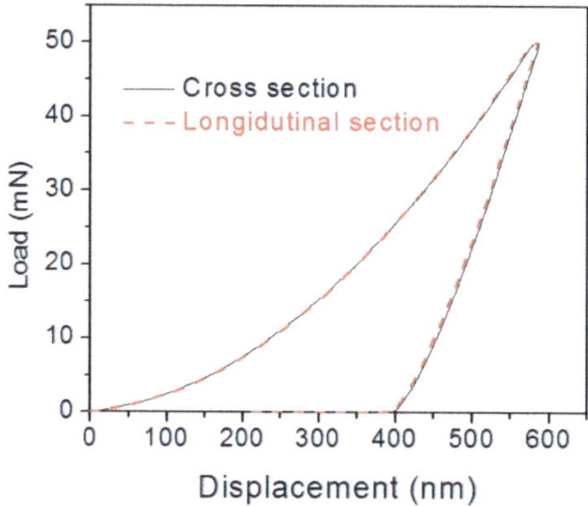

Figure 6. The fitted load-displacement (P-h) curves for nanoindentation on the cross and longitudinal sections of SLM-fabricated $Cu_{50}Zr_{43}Al_7$ BMG samples.

Table 2. The microscopic Vickers hardness (HV0.5), nanoindentation hardness (H_{IT}), elastic modulus (E_{IT}) of SLM-fabricated $Cu_{50}Zr_{43}Al_7$ BMG, as well as some casting CuZrAl ternary BMGs.

Sample	Method	HV0.5 (3σ)	H_{IT} (3σ, MPa)	E_{IT} (3σ, GPa)	Ref.
$Cu_{50}Zr_{43}Al_7$ (Cross section)	SLM	550.1 ± 10.9	9313.7 ± 90.6	129.1 ± 3.6	Present
$Cu_{50}Zr_{43}Al_7$ (Longitudinal section)	SLM	555.8 ± 10.6	9415.6 ± 112.5	129.6 ± 4.6	Present
$(Cu_{50}Zr_{50})_{100-x}Al_x$, $x = 0\sim10$	as-cast	$449.0\sim541.0$	$7300.0\sim8700.0$	$100.5\sim117.3$	[20]
$Cu_{46.5}Zr_{46.5}Al_7$	as-cast	-	7430.0	118.9	[21]

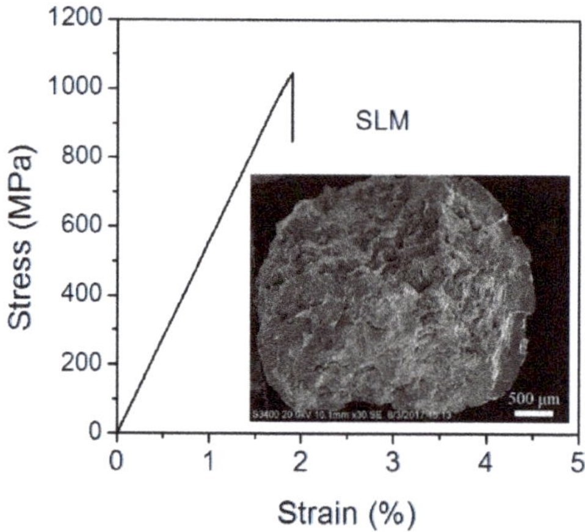

Figure 7. The stress–strain curve in the compression of SLM-fabricated $Cu_{50}Zr_{43}Al_7$ BMG rod samples. The inset depicts an SEM image of the fractured surface.

It is known that the SLM-fabricated component is formed by overlapping multi-track and multi-layer molten pools [27]. Thus, the macro–mechanical properties of SLM-fabricated component will be strongly affected by the solidified molten pools. Figure 8a,b show the morphology of the solidified molten pools in different zones on the uncorroded section parallel to the build direction of the as-fabricated $Cu_{50}Zr_{43}Al_7$ BMG. The arc-shape of the molten pool is due to the Gaussian energy distribution of the laser beam, and can be clearly seen in Figure 8a,b. In addition to the densely packed layers, a number of pores were found in the solidified molten pool boundaries of the local zones on the SLM-fabricated BMG sample, which may be the main reason for the low relative density of the SLM-fabricated $Cu_{50}Zr_{43}Al_7$ BMG. Apparently, formation of pores was much easier in the SLM-fabricated BMGs than in the SLM-produced crystalline metallic materials. This can be attributed to the highly viscous nature of the BMG melts [28–30]. Compared with the SLM-produced crystalline metallic materials, the BMGs are multicomponent alloys with large atomic size mismatches and a composition near deep eutectic. They show high viscosities that are several orders of magnitude higher than pure metal and common alloy melts [31]. The high viscosity of BMG melts causes difficulties in the spread of the melts, leading to the formation of many pores in the SLM-fabricated BMGs. The presence of pores leads to a decrease of the macro compressive strength of the SLM-fabricated BMGs.

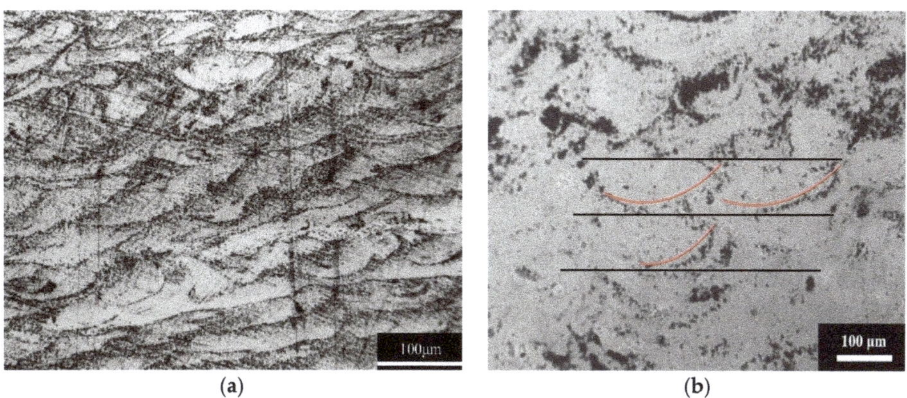

(a)   (b)

Figure 8. Morphology of solidified molten pools in different zones on the uncorroded section parallel to the build direction: (**a**) zones with densely packed layers; (**b**) zones with a number of pores.

In our work, some $Cu_{50}Zr_{43}Al_7$ BMG specimens with complex and delicate geometries were also fabricated by SLM, as shown in Figure 9, in order to evaluate the processability of the present as-atomized BMG powders. It can be seen that all the specimens have a relatively smooth surface without any macroscopic defects, which indicates that $Cu_{50}Zr_{43}Al_7$ BMG specimens can be well manufactured by the SLM method. The present work implied that SLM is a promising processing avenue for fabricating BMGs with large sizes and complex shapes that may be impossible to obtain by traditional casting methods.

Figure 9. Some large and/or complex glassy $Cu_{50}Zr_{43}Al_7$ parts produced by SLM: (**a**) a disk with diameter of 20 mm; (**b**) a micro gear; (**c**) a ring with hollowed-out structure; (**d**) a part with labyrinth-like structure.

4. Conclusions

In conclusion, mainly amorphous $Cu_{50}Zr_{43}Al_7$ BMG specimens are fabricated by SLM. It was found that many nano-sized crystalline phases precipitated from the glassy matrix during the SLM process. The as-prepared BMG specimens without major external cracks had a number of micro-scale pores in the solidified molten pool boundaries due to the high viscosity of the BMG melts, which may be the main reason for the low relative density and high porosity of the SLM-fabricated $Cu_{50}Zr_{43}Al_7$ BMG. The hardness and elastic modulus of the SLM-fabricated $Cu_{50}Zr_{43}Al_7$ BMG was higher than those of the CuZrAl ternary BMGs prepared by conventional mold casting. This can be mainly attributed to structural relaxation during the SLM process. The presence of lots of pores resulted in a decrease in the macro compressive strength of the SLM-fabricated BMGs. Our work proved that SLM is a promising technique for the development of BMGs with complex geometries and large cross sections, and SLM processing brings significant changes to the structural and mechanical properties of BMGs.

Author Contributions: X.L., Y.D. and W.L. designed the experiments; X.L. and M.N. performed the experiments; X.L. and Y.D. analyzed the data and wrote the paper.

Funding: This work was supported by the National Natural Science Foundation of China under Project No. 51571116 and the Fundamental Research Funds for the Central Universities (No. 30915014101).

References

1.   Schroers, J. Processing of Bulk Metallic Glass. *Adv. Mater.* **2010**, *22*, 1566–1597. [CrossRef] [PubMed]
2.   Inoue, A.; Takeuchi, A. Recent development and application products of bulk glassy alloys. *Acta Mater.* **2011**, *59*, 2243–2267. [CrossRef]
3.   Li, H.F.; Zheng, Y.F. Recent advances in bulk metallic glasses for biomedical applications. *Acta Biomater.* **2016**, *36*, 1–20. [CrossRef] [PubMed]

4. Itoi, T.; Takamizawa, T.; Kawamura, Y.; Inoue, A. Fabrication of $Co_{40}Fe_{22}Nb_8B_{30}$ bulk metallic glasses by consolidation of gas-atomized powders and their soft-magnetic properties. *Scr. Mater.* **2001**, *45*, 1131–1137. [CrossRef]

5. Choi, P.P.; Kim, J.S.; Nguyen, O.T.H.; Kwon, Y.S. $Ti_{50}Cu_{25}Ni_{20}Sn_5$ bulk metallic glass fabricated by powder consolidation. *Mater. Lett.* **2007**, *61*, 4591–4594. [CrossRef]

6. Olakanmia, E.O.; Cochranea, R.F.; Dalgarno, K.W. A review on selective laser sintering/melting (SLS/SLM) of aluminium alloy powders: Processing, microstructure, and properties. *Prog. Mater. Sci.* **2015**, *74*, 401–477. [CrossRef]

7. Pauly, S.; Löber, L.; Petters, R.; Stoica, M.; Scudino, S.; Kühn, U.; Eckert, J. Processing metallic glasses by selective laser melting. *Mater. Today* **2013**, *16*, 37–41. [CrossRef]

8. Jung, H.Y.; Choi, S.J.; Prashanth, K.G.; Stoica, M.; Scudino, S.; Yi, S.; Kühn, U.; Kim, D.H.; Kim, K.B.; Eckert, J. Fabrication of Fe-based bulk metallic glass by selective laser melting: A parameter study. *Mater. Des.* **2015**, *86*, 703–708. [CrossRef]

9. Mahbooba, Z.; Thorsson, L.; Unosson, M.; Skoglund, P.; West, H.; Horn, T.; Rock, C.; Vogli, E.; Harrysson, O. Additive manufacturing of an iron-based bulk metallic glass larger than the critical casting thickness. *Appl. Mater. Today* **2018**, *11*, 264–269. [CrossRef]

10. Li, X.P.; Roberts, M.P.; O'Keeffe, S.; Sercombe, T.B. Selective laser melting of Zr-based bulk metallic glasses Processing, microstructure and mechanical properties. *Mater. Des.* **2016**, *112*, 217–226. [CrossRef]

11. Ouyang, D.; Li, N.; Xing, W.; Zhang, J.J.; Liu, L. 3D printing of crack-free high strength Zr-based bulk metallic glass composite by selective laser melting. *Intermetallics* **2017**, *90*, 128–134. [CrossRef]

12. Ouyang, D.; Li, N.; Liu, L. Structural heterogeneity in 3D printed Zr-based bulk metallic glass by selective laser melting. *J. Alloys Compd.* **2018**, *740*, 603–609. [CrossRef]

13. Li, X.P.; Kang, C.W.; Huang, H.; Zhang, L.C.; Sercombe, T.B. Selective laser melting of an $Al_{86}Ni_6Y_{4.5}Co_2La_{1.5}$ metallic glass Processing, microstructure evolution and mechanical properties. *Mater. Sci. Eng. A* **2014**, *606*, 370–379. [CrossRef]

14. Li, X.P.; Roberts, M.; Liu, Y.J.; Kang, C.W.; Huang, H.; Sercombe, T.B. Effect of substrate temperature on the interface bond between support and substrate during selective laser melting of Al–Ni–Y–Co–La metallic glass. *Mater. Des.* **2015**, *65*, 1–6. [CrossRef]

15. Deng, L.; Wang, S.H.; Wang, P.; Kühn, U.; Pauly, S. Selective laser melting of a Ti-based bulk metallic glass. *Mater. Lett.* **2018**, *212*, 346–349. [CrossRef]

16. Kim, Y.C.; Lee, J.C.; Cha, P.R.; Ahn, J.P.; Fleury, E. Enhanced glass forming ability and mechanical properties of new Cu-based bulk metallic glasses. *Mater. Sci. Eng. A* **2006**, *437*, 248–253. [CrossRef]

17. Du, Y.L.; Xu, H.W.; Chen, G.; Deng, Y. Structural and mechanical properties of a Cu-based bulk metallic glass with two oxygen levels. *Intermetallics* **2012**, *30*, 90–93. [CrossRef]

18. Zhang, Y.; Chen, J.; Chen, G.L.; Liu, X.J. Influence of yttrium addition on the glass forming ability in Cu–Zr–Al alloys. *Mater. Sci. Eng. A* **2008**, *483–484*, 235–238. [CrossRef]

19. Ning, Z.; Liang, W.; Zhang, M.; Li, Z.; Sun, H.; Liu, A.; Sun, J. High tensile plasticity and strength of a CuZr-based bulk metallic glass composite. *Mater. Des.* **2016**, *90*, 145–150. [CrossRef]

20. Zhuang, Y.X.; Wang, W.H. Effects of relaxation on glass transition and crystallization of ZrTiCuNiBe bulk metallic glass. *J. Appl. Phys.* **2000**, *87*, 8209–8211. [CrossRef]

21. Oliver, W.C.; Pharr, G.M. An improved technique for determining hardness and elastic modulus using load and displacement sensing indentation experiments. *J. Mater. Res.* **1992**, *7*, 1564–1583. [CrossRef]

22. Cheung, T.L.; Shek, C.H. Thermal and mechanical properties of Cu–Zr–Al bulk metallic glasses. *J. Alloys Compd.* **2007**, *434–435*, 71–74. [CrossRef]

23. Limbach, R.; Kosiba, K.; Pauly, S.; Kühn, U.; Wondraczek, L. Serrated flow of CuZr-based bulk metallic glasses probed by nanoindentation: Role of the activation barrier, size and distribution of shear transformation zones. *J. Non-Cryst. Solids* **2017**, *459*, 130–141. [CrossRef]

24. Ramamurty, U.; Lee, M.L.; Basu, J.; Li, Y. Embrittlement of a bulk metallic glass due to low-temperature annealing. *Scr. Mater.* **2002**, *47*, 107–111. [CrossRef]

25. Song, M.; Li, Y.Q.; Wu, Z.G.; He, Y.H. The effect of annealing on the mechanical properties of a ZrAlNiCu metallic glass. *J. Non-Cryst. Solids* **2011**, *357*, 1239–1241. [CrossRef]

26. Wu, Y.; Xiao, Y.; Chen, G.; Liu, C.T.; Lu, Z. Bulk metallic glass composites with transformation-mediated work-hardening and ductility. *Adv. Mater.* **2010**, *22*, 2770–2773. [CrossRef] [PubMed]

27. Wen, S.; Li, S.; Wei, Q.; Yan, C.; Zhang, S.; Shi, Y. Effect of molten pool boundaries on the mechanical properties of selective laser melting parts. *J. Mater. Process. Technol.* **2014**, *214*, 2660–2667.

28. Lee, K.S.; Kim, S.; Lim, K.R.; Hong, S.H.; Kim, K.B.; Na, Y.S. Crystallization, high temperature deformation behavior and solid-to-solid formability of a Ti-based bulk metallic glass within supercooled liquid region. *J. Alloys Compd.* **2016**, *663*, 270–278. [CrossRef]

29. Li, C.Y.; Zhu, F.P.; Zhang, X.Y.; Ding, J.Q.; Yin, J.F.; Wang, Z.; Zhao, Y.C.; Kou, S.Z. The rheological behavior and thermoplastic deformation of Zr-based bulk metallic glasses. *J. Non-Cryst. Solids* **2018**, *492*, 140–145. [CrossRef]

30. Hong, S.H.; Kim, J.T.; Mun, S.C.; Kim, Y.S.; Park, H.J.; Na, Y.S.; Lim, K.R.; Park, J.M.; Kim, K.B. Influence of spherical particles and interfacial stress distribution on viscous flow behavior of Ti-Cu-Ni-Zr-Sn bulk metallic glass composites. *Intermetallics* **2017**, *91*, 90–94. [CrossRef]

31. Busch, R.; Schroers, J.; Wang, W. Thermodynamics and kinetics of bulk metallic glass. *MRS Bull.* **2007**, *32*, 620–623. [CrossRef]

Article

Estimation of Residual Stress in Selective Laser Melting of a Zr-Based Amorphous Alloy

Wei Xing, Di Ouyang *, Ning Li * and Lin Liu

School of Materials Science and Engineering, and State key Lab for Materials Processing and Die & mold Technology, Huazhong University of Science and Technology, Wuhan 430074, China; wxing@hust.edu.cn (W.X.); lliu2000@mail.hust.edu.cn (L.L.)
* Correspondence: diouyang@hust.edu.cn (D.O.); hslining@hust.edu.cn (N.L.); Tel.: +86-27-8755-9606 (N.L.)

Received: 19 July 2018; Accepted: 17 August 2018; Published: 20 August 2018

Abstract: An accurate estimation of residual stresses is crucial to ensure dimensional accuracy and prevent premature fatigue failure of 3D printed components. Different from their crystalline counterparts, the effect of residual stress would be worse for amorphous alloys owing to their intrinsic brittleness with low fracture toughness. However, the generation of residual stress and its performance in 3D printed amorphous alloy components still remain unclear. Here, a finite element method combined with experiments and theoretical analyses was introduced to estimate the residual stress in selective laser melting of a Zr-based amorphous alloy. The results revealed that XY cross scanning strategy exhibits relatively low residual stress by comparison with X and Y strategies, and the residual stress becomes serious with increasing bar thickness. The residual stress, on the other hand, could be tuning by annealing or preheating the substrate. The above scenario is thoroughly understood according to the temperature gradient mechanism and its effect on microstructure evaluation.

Keywords: selective laser melting; amorphous alloy; finite element analysis; residual stress

---

1. Introduction

Amorphous alloys (also called metallic glasses, MGs) are a unique class of materials that possess an amorphous atomic-level structure and display a plethora of desirable mechanical, chemical, and physical properties, which makes them one of the most promising engineering materials [1–6]. However, the poor processability, combined with ambient-temperature brittleness, and limited size have been the Achilles' heel to structural applications of BMGs [7,8]. Selective laser melting (SLM) as an additive manufacturing (AM) process, has been demonstrated as an effective technique to break through the bottleneck in processing amorphous alloys in recent years [9–13]. For instance, previous literature [14–16] reported that a full amorphous structure could be obtained by controlling parameters such as fast scanning rates, to avoid the risk of crystallization in the heat affected zone (HAZ). Therefore, SLM provides an opportunity to fabricate amorphous alloy components without dimensional or geometric limitations. Furthermore, the 3D printed amorphous alloy specimens also revealed excellent mechanical properties such as combined high strength and fracture toughness [13], exhibiting alluring prospects for near-future applications. However, steep temperature gradients caused by high heating and cooling rates during laser processing generally cause inhomogeneous thermal distribution, induce heterogeneous thermal expansions and contractions, and inevitably result in serious thermal and residual stresses [17]. The high stresses will trigger cracking, delamination, distortion and fatigue-failure in 3D printed components [18], especially for these amorphous alloys with intrinsic brittleness and low fracture toughness. Therefore, it is essential to understand the origin, distribution and evolution of residual stress in 3D printing amorphous alloys.

Residual stress has been widely investigated in laser processing crystalline metals, and various experimental methods [19–21] have been introduced to characterize residual stress in these 3D printed specimens. For example, mechanical methods including sectioning, contour, hole-drilling, ring-core, curvature, etc., were used to evaluate macro-stresses, while these methods only reflect residual stress in some local regions. Non-destructive "physical" methods, such as diffraction analysis, are more relevant for assessing residual stress at the grain level or at the atomic scale, but this "physical" method requires a very thin sample that is hard to print. Different from the above methods, the curvature method measures the deflection or curvature of a part caused by residual stresses, reflecting thermal stresses within layers. Therefore, this method is more suitable for 3D printed components, because SLM is based on the melting of successive layers, and the variation of processing parameters (such as scanning strategy, layer thickness, preheating, etc.) has a significant effect on residual stresses [22]. Accordingly, some researchers [23] used an overhang "cantilever" to investigate the influence of process parameters on distortion or to validate the distortion prediction model. Furthermore, the finite element method (FEM) was introduced to predict and reduce residual stress of SLM-fabricated components in recent years [24–26]. Alvarez et al. [27] established an FEM-model for divine distortion in a Ni-based alloy cantilever structure and to investigate the influence of meshing, layer activation and equivalent thermal loads on prediction capability and computational cost. Li et al. [28] developed a temperature-thread multi-scale modelling approach to predict residual stress and distortion of an $AlSi_{10}Mg$ twin cantilever manufactured by SLM. Parry et al. [29] built a thermo-mechanical model to analyze the effect of laser scanning strategy on the generation of residual stress in SLM-fabricated Ti-6Al-4V parts.

With respect to 3D printing amorphous alloys, the previous literature focused only on the possibility of 3D printing amorphous alloys [30,31], the effect of laser processing parameters on microstructure evolution [12,32,33], thermal stress induced micro-cracks [13,34,35] and possible mechanical properties [13,15,16]. The investigation of residual stress in 3D printed amorphous alloys is actually scarce, and the detailed distribution remains unclear, which is deleterious for controlling the dimensional accuracy and preventing premature fatigue failure of the 3D printed components. Therefore in this work, the finite element method combined with experiments and theoretical analysis was introduced to estimate the residual stress in selective laser melting of a Zr-based amorphous alloy. The results reveal that the XY cross scanning strategy exhibits relatively low residual stress by comparison with X and Y strategies, and the residual stress becomes serious with increasing bar thickness. The residual stress, on the other hand, can be tuning by annealing or preheating the substrate. These results provide a new route to improve the forming quality of 3D printed amorphous alloy components with a large scale and complex structure.

2. Constitutive Model and Finite Element Simulation

A thermal-mechanical coupled analysis model consisting of a cantilever and substrate was established by ABAQUS software (ABAQUS 6.10, Dassault, France), as shown in Figure 1a. Gaussian distributed moving heat flux was applied via an ABAQUS subroutine DFLUX. The moving direction and velocity of the heat flux were controlled by a subroutine, and elements were activated step by step sequentially as the heat flux moved ("birth and death" technique [36]). In order to reduce computational costs and predict the results well, Alvarez et al. [27] suggested a ratio between real manufacturing layers and model layer activation steps of 8 (48 real layers, activated in 6 steps). Because the distortions are mainly caused by the bars, only bars are activated in this model. The dimensions of the cantilever are designed as 3.75 mm (length of Y-axis) × 0.5 mm (width of X-axis) × 1.12 mm (height of Z-axis). Fine hexahedral elements with dimensions of 0.025 mm (width) × 0.025 mm (length) × 0.025 mm (thickness), similar to particle dimension, were used in experiment. In addition, the mesh size of substrate increased gradually from 0.025 mm to 0.25 mm along the X and Y-axes, but was maintained as 0.25 mm along the Z-axis. C3D8T elements were used from the ABAQUS element library for analysis. As a result, an acceptable finite-element model with a compromise in calculation accuracy and computation time is successfully established.

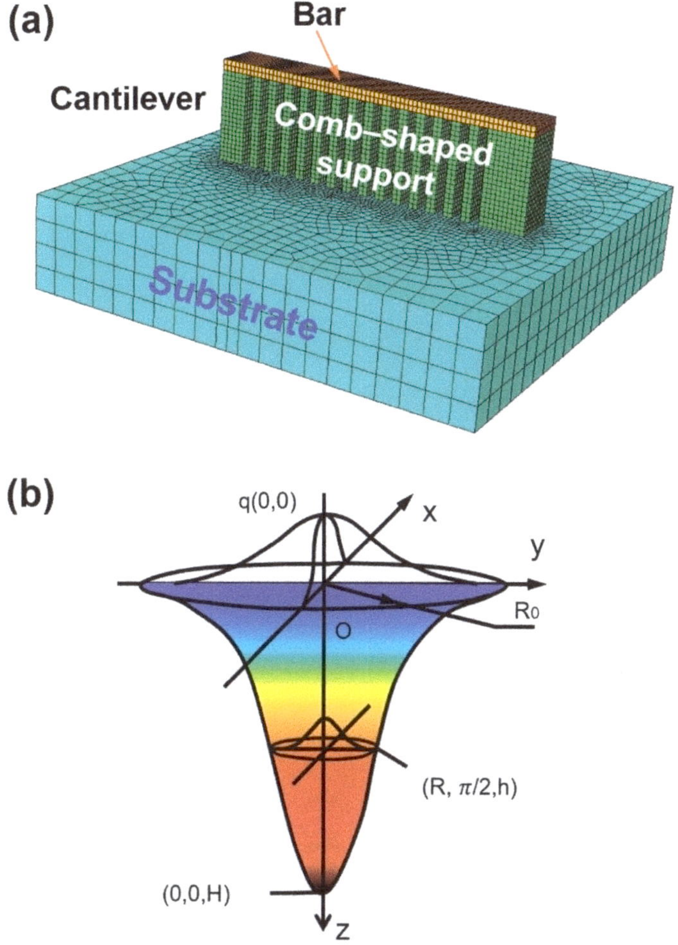

Figure 1. (**a**) Three-dimensional coupled thermo-mechanical finite element model and (**b**) representative heat input modeling with Gaussian distribution.

2.1. Heat Input Modeling

Considering the penetration depth in the powder bed during laser processing, a moving volumetric heat source assumed to obey a Gaussian distribution is designed to simulate the interaction between the laser and material, as illustrated in Figure 1b. The thermal flux density was applied to the powder layer using an ABAQUS subroutine DFLUX, expressed as [37]:

$$q(x,y,z) = \frac{3c_S QA}{\pi H\left(1 - \frac{1}{e^3}\right)} \exp\left[\frac{-3c_S}{\log\left(\frac{H}{z}\right)}\left(x^2 + y^2\right)\right] \tag{1}$$

where A is the laser absorptivity of materials affected by the wavelength, Q is the laser power, H is the height of heat source, and $c_S = \frac{3}{R_0^2}$ is the concentric coefficient of heat-flux distribution of the cross section, in which R_0 is the effective radius of the heat source. The parameters used for finite element simulation are listed in Table 1. Furthermore, the latent heat of phase transformation should be considered, due to the process of melting and solidification of the material during SLM. The enthalpy (H) is defined as a function of temperature [38]:

$$H = \int \rho c dT \tag{2}$$

Table 1. Parameters used for finite element simulation.

Parameter	Value
Laser power, P (W)	240
Scanning speed, V (mm/s)	1200
Density of Zr-based MG powder, ρ (kg m^{-3})	4129
Latent heat of fusion (J Kg^{-1})	2.638×10^5
Melting point of Zr-based BMG (K)	1165
Coefficient for the heat convection, h (W m^{-2} K^{-1})	20
The Stefan-Boltzmann constant, δ (W m^{-2} K^{-4})	5.67×10^{-8}
Radiation emissivity, ε	0.77

2.2. Heat Transfer Modeling

In the SLM process, heat losses to the environment from the molten pool are determined through radiation, convection acting on the surface, and conduction into the support structure and substrate. Therefore, boundary conditions for heat transfer should be considered, described as [13]:

$$q_{cond} = q(x,y,z) - q_c - q_r \tag{3}$$

where the heat losses through conduction q_{cond}, convection q_c and radiation q_r in the laser scan model are expressed as [39]:

$$q_{cond} = -k\Delta T \tag{4}$$

$$q_c = h_c(T - T_0) \tag{5}$$

$$q_r = \sigma \in \left(T^4 - T_0^4\right) \tag{6}$$

in which k is thermal conductivity, h_c is the heat convection coefficient, σ is the Stefan-Boltzman constant, $\in$ is the emissivity of the powder bed, T is the temperature of the molten pool, and T_0 is the ambient temperature.

The transient spatial distribution of the temperature field during SLM can be depicted as follows [40]:

$$\rho \frac{\partial(c_p T)}{\partial t} = \frac{\partial}{\partial x}\left(k\frac{\partial T}{\partial x}\right) + \frac{\partial}{\partial y}\left(k\frac{\partial T}{\partial y}\right) + \frac{\partial}{\partial z}\left(k\frac{\partial T}{\partial z}\right) + q(x,y,z) \tag{7}$$

in which ρ, k are the density and thermal conductivity, respectively, of the material, c_p is the specific heat capacity, and T is the temperature. The temperatures of the powder bed and substrate are both set to ambient temperature as the initial condition and can be defined as [41]:

$$T(x,y,z,t)\big|_{t_0=0,x,y,z\in D} = T_0 = 296K \tag{8}$$

2.3. Residual Stress Model

The moving laser inevitably generates a non-linear thermal gradient, and thermal-mechanical change promotes residual stresses and distortions. Therefore, based on the transient spatial distribution of the temperature field given by Equation (7), the temperature gradients that induced residual stresses can be described as:

$$\sigma_v = \delta_v \lambda \varepsilon_{kk} + 2\mu\varepsilon_v - \delta_v(3\lambda + 2\mu)\alpha T \tag{9}$$

where λ and μ are the Lame constants related to elasticity modulus E and Poisson's ratio v of the material and represent the deformation caused by the temperature gradient. The deformation dependent ε_v can be expressed as:

$$\varepsilon_v = \frac{1}{2}\left(\frac{\partial \mu_i}{\partial x_j} + \frac{\partial \mu_j}{\partial x_i}\right) \tag{10}$$

The elastoplastic model was adopted for residual stress calculation, and relevant data are acquired from the strains generated in the SLM process. The strains comprise elastic, plastic and thermal natures, defined as [29]:

$$\varepsilon_{ij} = \varepsilon_{ij}^e + \varepsilon_{ij}^p + \varepsilon_{ij}^{th} \tag{11}$$

where ε_{ij}, ε_{ij}^e, ε_{ij}^p, ε_{ij}^{th} are the respective total strain, elastic strain, plastic strain and thermal strain, determined by [42]:

$$\varepsilon_{kl}^e = \sigma_{ij}^e \cdot E(T)^{-1} \tag{12}$$

$$d\varepsilon_{kl}^p = d\lambda \frac{\partial f}{\partial \sigma_{ij}} \tag{13}$$

$$\varepsilon_{kl}^{th} = \alpha_{ij}(T - T_\infty) \tag{14}$$

in which f is the flow area capability, α_{ij} is the coefficient of thermal expansion, T_∞ is the reference temperature, λ is a constant that depends on the properties of the material; for a perfect elastic-plastic material, λ can be expressed as:

$$\lambda = \frac{3GS_{ij}S_{kl}}{\sigma e^2} \tag{15}$$

2.4. Inherent Shrinkage Model

Here, we also consider the inherent shrinkage that is the main driving force for distortion during the cooling process [27]. The equivalent thermal strain should be accommodated by the part, which would lead to the redistribution of strains and stresses, described as:

$$\varepsilon^{th} = \alpha \cdot \Delta T \tag{16}$$

where ε^{th} is the equivalent thermal strain, α is the thermal expansion coefficient, ΔT is the temperature gradient.

3. Experimental Procedures

The $Zr_{55}Cu_{30}Ni_5Al_{10}$ (at %) amorphous alloy powders used in this work were produced through high-pressure inert gas atomization. The physical properties of amorphous alloy are listed in Table 2. The SLM experiment was conducted using a commercial SLM machine (FORWEDO LM-120, Forwedo, Harbin, China) equipped with a Nd:YAG fiber laser. The SLM device generates a laser beam with focus diameter of 80 μm, maximum power of 500 W, and wavelength of 1060 nm. An inert high purity argon gas atmosphere with an oxygen content below 100 ppm after vacuum was used during the entire experiment. Here, optimized processing parameters with a laser power of 240 W, scanning speed of 1200 mm/s, powder layer thickness fixed at 60 μm and scanning space of 100 μm were adopted based on our previous experiments [35]. The cantilever structure was designed according to the FEM model. At the two ends of the cantilever arms, the additional supporting body was adopted to stand deformation. The bars were printed with different scanning strategies, such as X, Y and XY cross scanning strategies (scanning direction of 90° alternated among layers), as shown in Figure 2a. The other group of cantilever specimens was fabricated with the same XY cross scanning strategy, but for various bar thicknesses (Figure 2b).

Table 2. Thermal physics parameters of the Zr-based amorphous alloy.

Temperature T (°C)	Specific Heat Capacity c (J Kg^{-1} °C^{-1})	Thermal Conductivity k (W m^{-1} °C^{-1})
20	326	4.9
100	346	6.2
200	342	7.4
300	271	6.8
400	233	6.6
500	−323	7.6
600	112	7.7
700	362	10.1
800	375	9.6

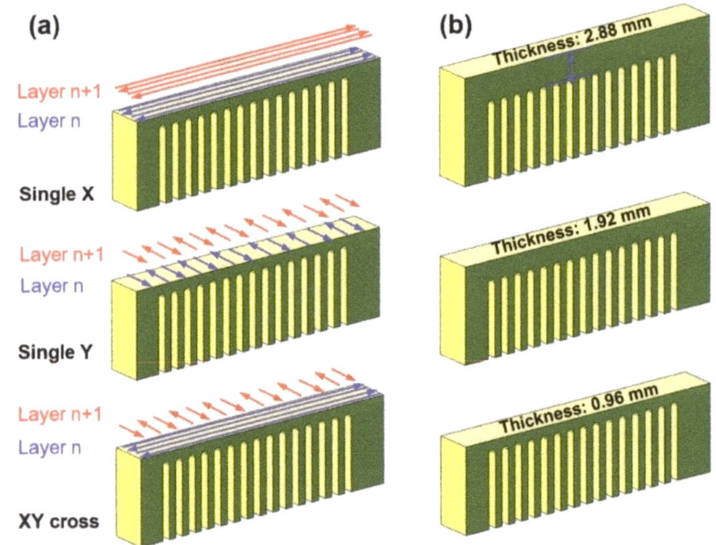

Figure 2. Schematic diagram of different (**a**) laser scanning strategies and (**b**) bar thicknesses.

4. Results

4.1. FEM-Simulated Residual Stress Field

To illustrate the residual stress distribution under different scanning strategies, the simulated contours of residual stress in cantilever parts are described in Figure 3. In general, the stress distribution (Figure 3a) and shrinkage (Figure 3b) in the cantilever components varied with scanning strategies, and the largest residual stresses are obtained in the connection region between the cantilever and substrate (as shown by the cycle in Figure 3a). It is worth noting that the measured shrinkage of the bar under the X scanning strategy is 21.54 µm, corresponding to the maximum residual stress concentration of 1.40 GPa. As for the Y scanning strategy, the shrinkage decreases to 13.79 µm with maximum residual stress concentration of 0.99 GPa. For the XY cross scanning strategy, the shrinkage of the cantilever is 13.80 µm, and the maximum residual stress concentration is 1.02 GPa, slightly different from the Y scanning strategy. Actually, the X and Y scanning strategies are identical; the significant difference in stress concentration and shrinkage between X and Y scanning strategies is caused by the cantilever structure. In order to build the cantilever part with lower residual stress accumulation, the XY cross scanning strategy is adopted in this work.

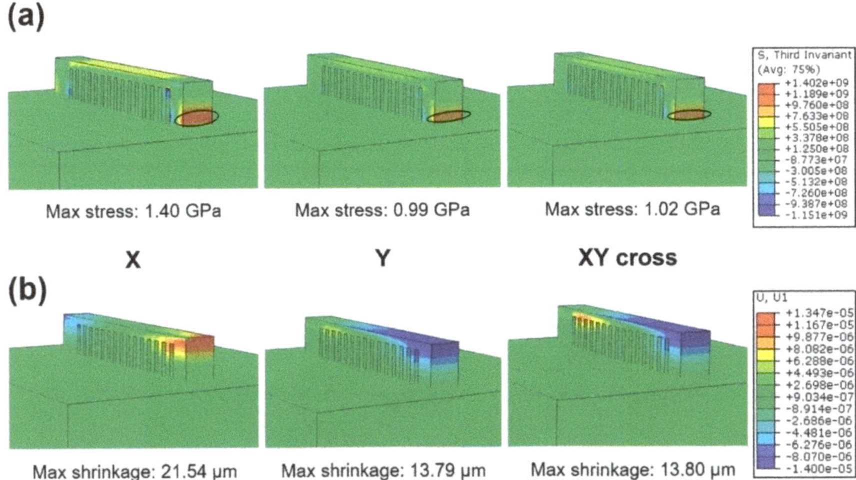

Figure 3. (**a**) Residual stress field and (**b**) longitudinal shrinkage under different laser scanning strategies.

On the other hand, the shrinkage and the maximum stress concentration for bars with different thickness were also investigated, as described in Figure 4, from which the residual stresses are concentrated in similar locations (as shown by the cycle), and shrinkage varies with the thickness of the bars (Figure 4b). For the thin bar (thickness of 0.96 mm), the maximum shrinkage is 33.97 μm with a relatively low residual stress of 592.06 MPa. When the bar thickness increases to 2.88 mm, the maximum shrinkage rises to 42.06 μm, and the residual stress accumulates to 710.42 MPa. These results indicate residual stress is dependent on bar thickness.

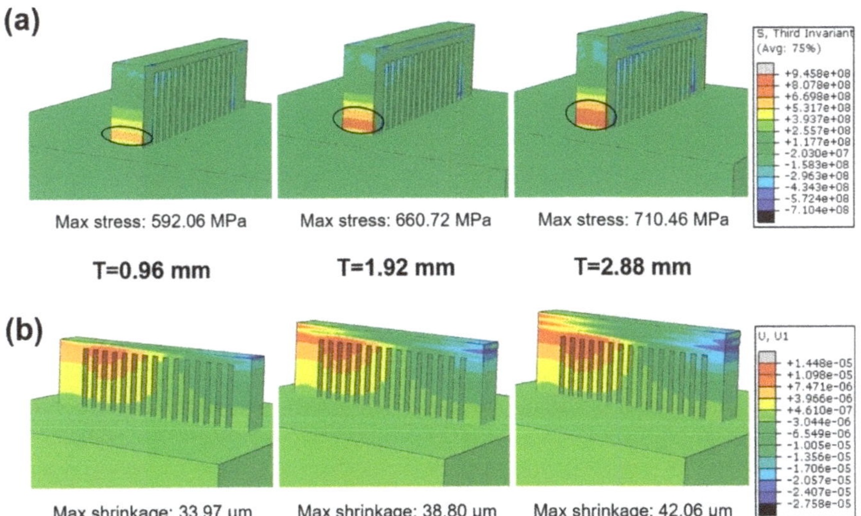

Figure 4. (**a**) Residual stress field and (**b**) longitudinal shrinkage of three typical bar thicknesses under the XY cross scanning strategy.

4.2. Measured Residual Stress Distribution

On the basis of the previous research [35] and the above simulation, Zr-based amorphous alloy cantilever specimens with various bar thicknesses were 3D printed by SLM, as depicted in Figure 5a. The thickness ranging from 0.48 mm to 2.88 mm possibly indicates various residual stresses that will induce distinct distortion. To detect this difference, the comb-shaped supports were cut off from the substrate, and the cantilever bent towards the Z direction (building direction) due to the residual stress release (Figure 5b).

Figure 5. (a) 3D printed Zr-based cantilevers with various bar thicknesses and (b) spreading of cantilevers after separating the support.

To further probe this disparity, the distortion of 7 typical points on the bar surface along the longitudinal direction was measured, and the results are summarized in Figure 6, wherein the distortion of cantilever became serious along the longitudinal direction (such as from point 0 to point 6). However, the degree of lifting varies with the thickness of bars. For instance, the maximum tilted height is 1.36 mm for a bar with thickness 0.96 mm; while when the bar thickness increases to 2.88 mm, the maximum deformation decreases to 0.58 mm.

Owing to the fact that the heat affected zone (HAZ) contains a mixed structure of the amorphous phase and some nanocrystals with dimensions of 100–200 nm as mentioned above, the nanoindentation test [43] is not suitable to measure the residual stress. In order to demonstrate different residual stresses induced by various processing parameters, Vickers micro-indentation was introduced here to measure the hardness along the lateral face with a mirror finish, as shown in Figure 7a. Vickers indentations had an interval of 100 μm, peak load of 2.94 N and dwell time of 10 s. Figure 7b–d illustrates the measured hardness of SLM-fabricated samples with various bar thicknesses. When the bar thickness is 0.96 mm (Figure 7b) and 1.92 mm (Figure 7c), the approximate hardness is about 850 HV, but it decreases to about 600 HV in the region (as shown by the rectangle in Figure 7a) of residual stress concentration (decrease of about 29.4%), identical to the FEM simulated results (Figure 4a). On the other hand, the hardness continuously decreased to 550 HV similar to the SLM-fabricated sample with a bar thickness of 2.88 mm, possibly owing to the residual stress that increased continuously along the OA direction in the relatively large affected zone.

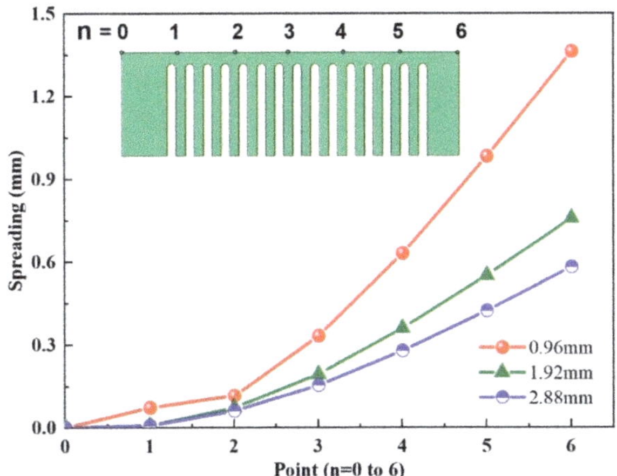

Figure 6. Spreading of the cantilevers depending on bar thickness and measurement point (n).

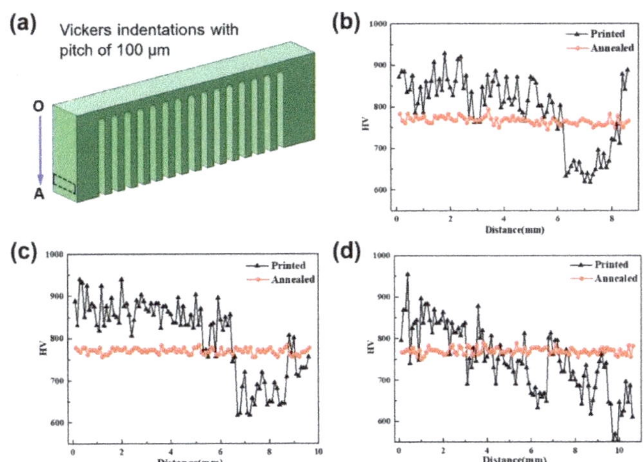

Figure 7. The Vickers hardness before and after annealing at 250 °C for 10 h along the OA direction (**a**) with pitch of 100 μm under the XY cross scanning strategy with bar thicknesses of (**b**) 0.96 mm; (**c**) 1.92 mm; (**d**) 2.88 mm.

4.3. Residual Stress Released after Annealing and Preheating

To probe the possible methods to release residual stress as mentioned above, low-temperature annealing as a relatively economical method was first considered here. For comparison, micro-hardness was also tested after vacuum-annealing at 250 °C for 10 h; in this case the amorphous structure remains [44]. The measured hardness of all samples with different bar thicknesses are almost 770 HV, as shown in Figure 7b–d, demonstrating that annealing is an effective method to alleviate the residual stress. Furthermore, preheating of the substrate was also introduced here, shown in Figure 8. The boundary condition of a constant temperature field of 250 °C was installed for the manufactured cantilever with a bar thickness of 0.96 mm. It is worth noting that by preheating, the maximum residual stress concentration decreases from 592.06 MPa to 279.12 MPa, accompanied by shrinkage of bars from 33.97 µm to 15.57 µm.

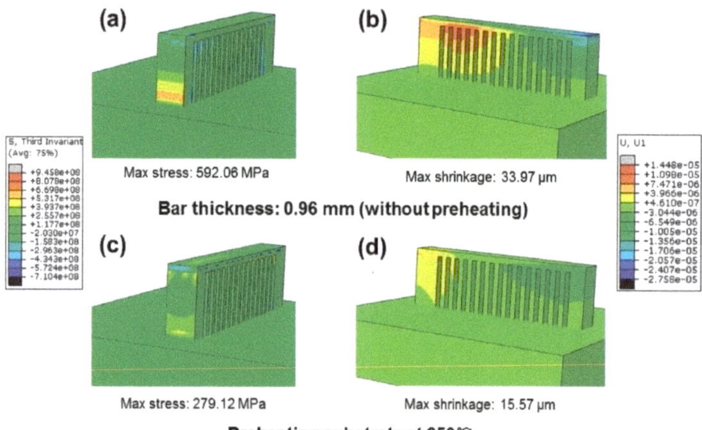

Figure 8. (**a**) Residual stress field and (**b**) longitudinal shrinkage of the bar with a thickness of 0.96 mm under the XY cross scanning strategy compared with preheating at 250 °C (**c**) corresponding residual stress field and (**d**) corresponding longitudinal shrinkage of the bar.

5. Discussion

During the SLM process, steep thermal gradients generate around the molten pool and result in inhomogeneous shrinkage during fast cooling, which inevitably triggers high residual stress. To thoroughly understand the effect of processing parameters on residual stress, the shrinkage of the printed material should be determined, as illustrated in Figure 9. Owing to the temperature rise in flash laser heating, the bar material (assumed as the first layer) expands and then shrinks in all sections in the subsequent cooling process. When a cantilever is built with increasing bar thickness, the second layer cannot freely shrink due to the constraint of the previously solidified layer. Therefore, the shrinkage of the upper layer is relative small. In this case, the lower layer is compressed while the upper layer is lengthened [23], which generates a large tensile residual stress in the upper layer, causing distortion of the fabricated components, as illustrated in Figure 9. On the other hand, the constraint of the comb–shaped support structure causes a stress concentration nearby the connection region between the cantilever and substrate (as shown in Figure 4a). Therefore, the residual stress increases with bar thickness, based on the FEM simulation and shown in Figure 4a, which should induce relative serious distortion of the sample. While from Figure 6, the distortion of the cantilever became moderate with increasing bar thickness, attributed to the considerable section modulus [23].

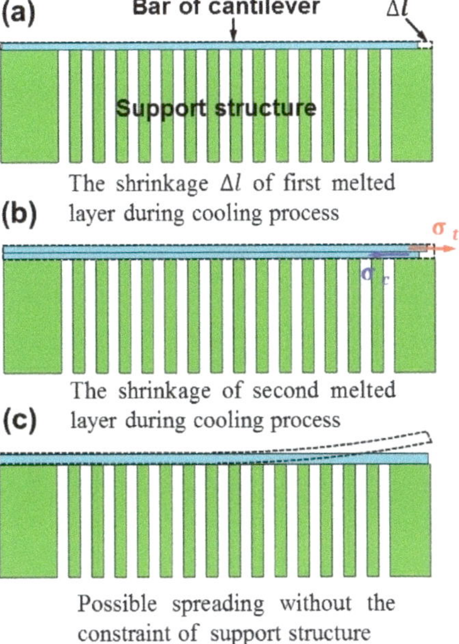

The shrinkage Δl of first melted
layer during cooling process

The shrinkage of second melted
layer during cooling process

Possible spreading without the
constraint of support structure

Figure 9. Schematic description of shrinkage of the (a) first and (b) second melted layer during the cooling process of the bar in the cantilever structure; (c) constraint of the support structure during deformation.

The above phenomenon can also be understood in detail, as shown in Figure 10a, the heated zone is assumed with width and depth of d (zone 1), wherein zones 2 and 3 are normal-temperature zones. During the laser processing, the instantaneous temperature of zone 1 is T, while the temperature in zones 2 and 3 remains T_0 (such as ambient temperature). In this case, the free thermal expansion of zone 1 can be defined as $\alpha_{th}d\Delta T$ (α_{th} is the coefficient of thermal expansion), which indicates that there is compressive stress in zone 1, but tensile stress in zones 2 and 3. Considering the constraint of zones 2 and 3, the shortening value $\sigma_{1a}d/E$ of zone 1 should be considered; therefore, the total expansion value of zone 1, $\Delta\mu = \alpha_{th}d\Delta T + \sigma_{1a}d/E$ ($\Delta T = T - T_0$ is the temperature gradient between zone 1 and zones 2 and 3; σ_{1a} is the compressive stress of zone 1), and the total expansion value of zones 2 & 3 is $\sigma_{2a}d/E$ (σ_{2a} is the tensile stress of zones 2 and 3); therefore,

$$\alpha_{th}d\Delta T + \sigma_{1a}d/E = \sigma_{2a}d/E \tag{17}$$

In addition, the compressive force of zone 1 is expressed as [45]:

$$\sigma_{1a}S_1 = -\sigma_{2a}(S_2 + S_3) \tag{18}$$

where S_1, S_2 and S_3 are the cross-sectional areas of zones 1, 2 and 3, respectively, and $S_2 = S_3$. According to Equations (17) and (18), $\sigma_{1a} = -2\alpha_{th}E\Delta TS_2/S_1 + 2S_2$, $\sigma_{2a} = \alpha_{th}E\Delta TS_1/S_1 + 2S_2$, and $\Delta\mu = \sigma_{2a}d/E = \alpha_{th}d\Delta TS_1/S_1 + 2S_2$.

After cooling, zone 1 is under tensile strain, while zones 2 and 3 are compressed, as described in Figure 10b. The tensile strain of zone 1 is $\sigma_{1b}d/E$ (σ_{1b} is the tensile stress of zone 1), and the final compressive strain of zone 1 is $\Delta v = \Delta\mu - \sigma_{1b}d/E$. The final tensile strain of zones 2 and 3 is $\sigma_{2b}d/E$ (σ_{2b} is the compressive stress of zones 2 and 3), accordingly,

$$\Delta\mu - \sigma_{1b}d/E = -\sigma_{2b}d/E \qquad (19)$$

The tensile force of zone 1 should be equal to the compressive force of zones 2 and 3:

$$\sigma_{1b}S_1 = -\sigma_{2b}(S_2 + S_3) \qquad (20)$$

According to Equations (19) and (20), $\sigma_{1b} = \frac{2\Delta\mu ES_2}{d(S_1+2S_2)}$, $\sigma_{2b} = -\frac{\Delta\mu ES_1}{d(S_1+2S_2)}$, and $\Delta v = \frac{\alpha_{th}d\Delta TS_1^2}{(S_1+2S_2)^2}$. Therefore, the dependence of the compressive value of the heated zone on the temperature gradient with a closed region can be correlated through $\Delta v \propto \Delta T$.

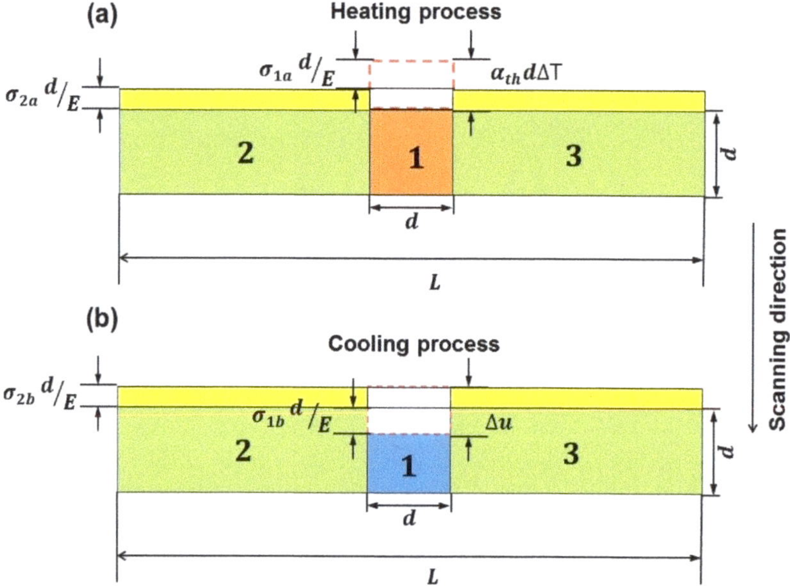

Figure 10. Schematic diagram of heating deformation of the previous layer during laser scanning of the current layer.

When the sample is SLM fabricated with a single X scanning strategy, the heat flux direction is similar because of the parallel laser tracks. Therefore, a large temperature gradient along the building direction (Z direction) was generated in the molten pool by the substrate heat sink and the Marangoni flow [46]. Since the scanning velocity reaches 1.2 m/s, the other temperature gradient along the X direction should also be considered. These steep temperature gradients trigger high residual stress parallel to the X and Z directions. Similar residual stress fields overlapped layer by layer, finally accumulate severe residual stress. On the contrary, when the sample is fabricated by the XY cross scanning strategy, the heat flux direction rotates with the rotation of the laser tracks among successive layers. The rotation of laser tracks changes the residual stress field layer by layer. In this case, partial residual stress is counteracted, resulting in the reduction of the integrated residual stress (Figure 3). According to the Temperature Gradient Mechanism (TGM) [47,48], substrate preheating can reduce the temperature gradient in the molten pool and heat-affected zone, which alleviates the shrinkage of the bar and the residual stress.

The above results also present a micro-hardness dependent residual stress as shown in Figure 7b,c. The FEM results reveal that these local regions suffered high residual stress concentration exhibit tensile stress on the surface, which can be well understood according to the stress-induced dilatation or softening in amorphous alloys [49].

In general, when the applied stress is sufficiently high, dilatation or softening happens owing to the fast free volume creation rate than the annihilation rate [50,51]. Flores et al. [52] investigated the effect of the stress state on strain localization and proposed a relationship between the stress and the average free volume, namely, residual stresses induced volume dilation as free volume change. Under a tensile mean stress, all dilatations are attributed to the change in free volume, and the hard sphere atomic volume is held constant: $v_f^T = v_0 + \Omega \frac{\sigma}{B}$, where, v_0 is the average initial volume assigned to each atom without a superimposed stress, Ω is the atomic volume and B is the bulk modulus. In this work, the maximum tensile stress is 592 MPa (bar thickness of 0.96 mm); therefore, the reduction of hardness (ΔH_T) can be illustrated by [53]:

$$\frac{\Delta H_T}{H_0} = \frac{v_f^T - v_0}{v_0} = \frac{\sigma \Omega}{B v_0} \approx 31.2\% \tag{21}$$

It is interesting that the calculated result (31.2%) is comparable to the experimentally measured micro-hardness reduction (29.4%), demonstrating the validity of the above analysis. Accordingly, annealing treatment would reduce the concentration of defects through free volume annihilation, which causes relative dense packing and results in the increase of hardness. Furthermore, the tensile stress and compressive stress balance each other, inducing homogeneous distribution of hardness as shown in Figure 7.

6. Conclusions

In summary, the residual stress during selective laser melting of a Zr-based amorphous alloy was investigated, based on both experimental and theoretical results, and the following conclusions can be drawn.

(1) Experiments combined with finite element simulation revealed that the XY cross scanning strategy exhibits relatively low residual stress by comparison with the X and Y strategies, and the residual stress becomes serious with increasing bar thickness. The residual stress, on the other hand, could be tuned by annealing or preheating the substrate.

(2) The theoretical analysis suggested that reducing the thermal gradient by the XY cross scanning strategy and preheating the substrate could enhance homogeneous shrinkage during fast cooling, which moderates the residual stress.

Author Contributions: W.X., N.L., D.O. and L.L. designed the experiments; W.X. performed the FEM simulations and the experiments; W.X. and N.L. analyzed the data and wrote the paper.

Funding: This work was financially supported by the National Natural Science Foundation of China under grant (Nos. 51531003 and 51671090) and the Hubei Provincial Natural Science Foundation of China (No. 2018CFA003).

Acknowledgments: The authors are grateful to the Analytical and Testing Center, Huazhong University of Science and Technology, for technical assistance.

Conflicts of Interest: The authors declare no conflict of interest.

References

1. Ghidelli, M.; Idrissi, H.; Gravier, S.; Blandin, J.J.; Raskin, J.P.; Schryvers, D.; Pardoen, T. Homogeneous flow and size dependent mechanical behavior in highly ductile Zr65Ni35metallic glass films. *Acta Mater.* **2017**, *131*, 246–259. [CrossRef]
2. Li, N.; Xu, X.; Zheng, Z.; Liu, L. Enhanced formability of a Zr-based bulk metallic glass in a supercooled liquid state by vibrational loading. *Acta Mater.* **2014**, *65*, 400–411. [CrossRef]
3. Li, N.; Chen, Y.; Jiang, M.Q.; Li, D.J.; He, J.J.; Wu, Y.; Liu, L. A thermoplastic forming map of a Zr-based bulk metallic glass. *Acta Mater.* **2013**, *61*, 1921–1931. [CrossRef]

4. Wang, X.; Dai, W.; Zhang, M.; Gong, P.; Li, N. Thermoplastic micro-formability of TiZrHfNiCuBe high entropy metallic glass. *J. Mater. Sci. Technol.* **2018**, *25*, 4–11. [CrossRef]

5. Li, N.; Liu, Z.; Wang, X.; Zhang, M. Vibration-accelerated activation of flow units in a Pd-based bulk metallic glass. *Mater. Sci. Eng. A* **2017**, *692*, 62–66. [CrossRef]

6. Li, N.; Chen, W.; Liu, L. Thermoplastic Micro-Forming of Bulk Metallic Glasses: A Review. *JOM* **2016**, *68*, 1246–1261. [CrossRef]

7. Li, N.; Chen, Q.; Liu, L. Size dependent plasticity of a Zr-based bulk metallic glass during room temperature compression. *J. Alloys Compd.* **2010**, *493*, 142–147. [CrossRef]

8. Li, N.; Xu, E.; Liu, Z.; Wang, X.; Liu, L. Tuning apparent friction coefficient by controlled patterning bulk metallic glasses surfaces. *Sci. Rep.* **2016**, *6*, 1–9. [CrossRef] [PubMed]

9. Shi, X.; Ma, S.; Liu, C.; Chen, C.; Wu, Q.; Chen, X.; Lu, J. Performance of High Layer Thickness in Selective Laser Melting of Ti6Al4V. *Materials* **2016**, *9*, 975. [CrossRef] [PubMed]

10. Wang, S. Research on High Layer Thickness Fabricated of 316L by Selective Laser Melting. *Materials* **2017**, *10*, 1. [CrossRef] [PubMed]

11. Cai, C.; Gao, X.; Teng, Q.; Li, M.; Pan, K.; Song, B.; Yan, C.; Wei, Q.; Shi, Y. A novel hybrid selective laser melting/hot isostatic pressing of near-net shaped Ti-6Al-4V alloy using an in-situ tooling: Interfacial microstructure evolution and enhanced mechanical properties. *Mater. Sci. Eng. A* **2018**, *717*, 95–104. [CrossRef]

12. Ouyang, D.; Li, N.; Liu, L. Structural heterogeneity in 3D printed Zr-based bulk metallic glass by selective laser melting. *J. Alloys Compd.* **2018**, *740*, 603–609. [CrossRef]

13. Li, N.; Zhang, J.; Xing, W.; Ouyang, D.; Liu, L. 3D printing of Fe-based bulk metallic glass composites with combined high strength and fracture toughness. *Mater. Des.* **2018**, *143*, 285–296. [CrossRef]

14. Jung, H.Y.; Choi, S.J.; Prashanth, K.G.; Stoica, M.; Scudino, S.; Yi, S.; Kühn, U.; Kim, D.H.; Kim, K.B.; Eckert, J. Fabrication of Fe-based bulk metallic glass by selective laser melting: A parameter study. *Mater. Des.* **2015**, *86*, 703–708. [CrossRef]

15. Li, X.P.; Roberts, M.P.; O'Keeffe, S.; Sercombe, T.B. Selective laser melting of Zr-based bulk metallic glasses: Processing, microstructure and mechanical properties. *Mater. Des.* **2016**, *112*, 217–226. [CrossRef]

16. Pauly, S.; Schricker, C.; Scudino, S.; Deng, L.; Kühn, U. Processing a glass-forming Zr-based alloy by selective laser melting. *Mater. Des.* **2017**, *135*, 133–141. [CrossRef]

17. Mishurova, T.; Cabeza, S.; Artzt, K.; Haubrich, J.; Klaus, M.; Genzel, C.; Requena, G.; Bruno, G. An Assessment of Subsurface Residual Stress Analysis in SLM Ti-6Al-4V. *Materials* **2017**, *10*, 348. [CrossRef] [PubMed]

18. Rickhey, F.; Marimuthu, K.P.; Lee, H. Investigation on indentation cracking-based approaches for residual stress evaluation. *Materials* **2017**, *10*, 404. [CrossRef] [PubMed]

19. Le, S.; Salem, M.; Hor, A. Improvement of the bridge curvature method to assess residual stresses in selective laser melting. *Addit. Manuf.* **2018**, *22*, 320–329. [CrossRef]

20. Simulation, N.; Analysis, N.D. Investigation on the residual stress state of drawn tubes by numerical simulation and neutron diffraction analysis. *Materials* **2013**, *6*, 5118–5130. [CrossRef]

21. Method, M.C.; Liu, C. Experimental investigation on the residual stresses in a thick joint with a partial repair weld using. *Materials* **2018**, *11*, 633. [CrossRef]

22. Kruth, J.P.; Froyen, L.; Van Vaerenbergh, J.; Mercelis, P.; Rombouts, M.; Lauwers, B. Selective laser melting of iron-based powder. *J. Mater. Process. Technol.* **2004**, *149*, 616–622. [CrossRef]

23. Buchbinder, D.; Meiners, W.; Pirch, N.; Wissenbach, K.; Schrage, J. Investigation on reducing distortion by preheating during manufacture of aluminum components using selective laser melting. *J. Laser Appl.* **2014**, *26*, 012004. [CrossRef]

24. Kolossov, S.; Boillat, E.; Glardon, R.; Fischer, P.; Locher, M. 3D FE simulation for temperature evolution in the selective laser sintering process. *Int. J. Mach. Tool. Manuf.* **2004**, *44*, 117–123. [CrossRef]

25. Dai, K.; Shaw, L. Thermal and mechanical finite element modeling of laser forming from metal and ceramic powders. *Acta Mater.* **2004**, *52*, 69–80. [CrossRef]

26. Matsumoto, M.; Shiomi, M.; Osakada, K.; Abe, F. Finite element analysis of single layer forming on metallic powder bed in rapid prototyping by selective laser processing. *Int. J. Mach. Tool. Manuf.* **2002**, *42*, 61–67. [CrossRef]

27. Alvarez, P.; Ecenarro, J.; Setien, I.; Sebastian, M.S.; Echeverria, A.; Eciolaza, L. Computationally efficient distortion prediction in Powder Bed Fusion Additive Manufacturing. *Int. J. Eng. Res. Sci.* **2016**, *2*, 2395–6992.

28. Li, C.; Liu, J.F.; Fang, X.Y.; Guo, Y.B. Efficient predictive model of part distortion and residual stress in selective laser melting. *Addit. Manuf.* **2017**, *17*, 157–168. [CrossRef]

29. Parry, L.; Ashcroft, I.A.; Wildman, R.D. Understanding the effect of laser scan strategy on residual stress in selective laser melting through thermo-mechanical simulation. *Addit. Manuf.* **2016**, *12*, 1–15. [CrossRef]

30. Shen, Y.; Li, Y.; Chen, C.; Tsai, H. 3D printing of large, complex metallic glass structures. *Mater. Des.* **2017**, *117*, 213–222. [CrossRef]

31. Pauly, S.; Löber, L.; Petters, R.; Stoica, M.; Scudino, S.; Kühn, U.; Eckert, J. Processing metallic glasses by selective laser melting. *Mater. Today* **2013**, *16*, 37–41. [CrossRef]

32. Hofmann, D.C.; Bordeenithikasem, P.; Pate, A.; Roberts, S.N.; Vogli, E. Developing processing parameters and characterizing microstructure and properties of an additively manufactured FeCrMoBC metallic glass forming alloy. *Adv. Eng. Mater.* **2018**. [CrossRef]

33. Shen, Y.; Li, Y.; Tsai, H. Evolution of crystalline phase during laser processing of Zr-based metallic glass. *J. Non-Cryst. Solids* **2018**, *481*, 299–305. [CrossRef]

34. Li, X.P.; Kang, C.W.; Huang, H.; Sercombe, T.B. The role of a low-energy-density re-scan in fabricating crack-free Al85Ni5Y6Co2Fe2 bulk metallic glass composites via selective laser melting. *Mater. Des.* **2014**, *63*, 407–411. [CrossRef]

35. Ouyang, D.; Li, N.; Xing, W.; Zhang, J.; Liu, L. 3D printing of crack-free high strength Zr-based bulk metallic glass composite by selective laser melting. *Intermetallics* **2017**, *90*, 128–134. [CrossRef]

36. Chand, R.; Kim, I.-S.; Lee, J.-H.; Kim, J.-S. Numerical and Experiment Study of Residual Stress and Strain in Multi-Pass GMA Welding. *Adv. Mater. Res.* **2013**, *717*, 403–409. [CrossRef]

37. Yu, G.; Gu, D.; Dai, D.; Xia, M. Influence of processing parameters on laser penetration depth and melting/re-melting densification during selective laser melting of aluminum alloy. *Appl. Phys. A* **2016**, *122*, 1–12. [CrossRef]

38. Li, Y.; Gu, D. Thermal behavior during selective laser melting of commercially pure titanium powder: Numerical simulation and experimental study. *Addit. Manuf.* **2014**, *1*, 99–109. [CrossRef]

39. Li, Z.; Li, B.; Bai, P.; Liu, B.; Wang, Y. Research on the Thermal Behaviour of a Selectively Laser Melted Aluminium Alloy: Simulation. *Materials* **2018**, *11*, 1172. [CrossRef] [PubMed]

40. Tolochko, N.K.; Khlopkov, Y.V.; Mozzharov, S.E.; Michail, B.; Laoui, T.; Titov, V.I.; Ignatiev, M.B. Emerald Article: Absorptance of powder materials suitable for laser sintering Absorptance of powder materials suitable for laser sintering. *Rapid Prototyp. J.* **2005**, *6*, 155–161. [CrossRef]

41. Gu, D.; He, B. Finite element simulation and experimental investigation of residual stresses in selective laser melted Ti-Ni shape memory alloy. *Comput. Mater. Sci.* **2016**, *117*, 221–232. [CrossRef]

42. Akbari, D. International Journal of Pressure Vessels and Piping Effect of the welding heat input on residual stresses in butt-welds of dissimilar pipe joints. *Int. J. Press. Vessels Pip.* **2009**, *86*, 769–776. [CrossRef]

43. Ghidelli, M.; Sebastiani, M.; Collet, C.; Guillemet, R. Determination of the elastic moduli and residual stresses of freestanding Au-TiW bilayer thin films by nanoindentation. *Mater. Des.* **2016**, *106*, 436–445. [CrossRef]

44. Huang, Y.; Ning, Z.; Shen, Z.; Liang, W.; Sun, H. Bending behavior of as-cast and annealed ZrCuNiAl bulk metallic glass. *J. Mater. Sci. Technol.* **2017**, *33*, 1153–1158. [CrossRef]

45. Shi, Y.; Shen, H.; Yao, Z.; Hu, J. Temperature gradient mechanism in laser forming of thin plates. *Opt. Laser Technol.* **2007**, *39*, 858–863. [CrossRef]

46. Wan, H.Y.; Zhou, Z.J.; Li, C.P.; Chen, G.F.; Zhang, G.P. Effect of scanning strategy on grain structure and crystallographic texture of Inconel 718 processed by selective laser melting. *J. Mater. Sci. Technol.* **2018**, *34*, 1799–1804. [CrossRef]

47. Ali, H.; Ma, L.; Ghadbeigi, H.; Mumtaz, K. In-situ residual stress reduction, martensitic decomposition and mechanical properties enhancement through high temperature powder bed pre-heating of Selective Laser Melted Ti6Al4V. *Mater. Sci. Eng. A* **2017**, *695*, 211–220. [CrossRef]

48. Xiong, J.; Lei, Y.; Li, R. Finite element analysis and experimental validation of thermal behavior for thin-walled parts in GMAW-based additive manufacturing with various substrate preheating temperatures. *Appl. Therm. Eng.* **2017**, *126*, 43–52. [CrossRef]

49. Pan, J.; Wang, Y.X.; Guo, Q.; Zhang, D.; Greer, A.L.; Li, Y. Extreme rejuvenation and softening in a bulk metallic glass. *Nat. Commun.* **2018**. [CrossRef] [PubMed]

Materials **2018**, *11*, 1480

50. Li, N.; Liu, L.; Chen, Q.; Pan, J.; Chan, K.C. The effect of free volume on the deformation behaviour of a Zr-based metallic glass under nanoindentation. *J. Phys. D Appl. Phys.* **2007**, *40*, 6055–6059. [CrossRef]

51. Li, N.; Chan, K.C.; Liu, L. The indentation size effect in Pd40Cu30Ni10P20 bulk metallic glass. *J. Phys. D Appl. Phys.* **2008**, *41*, 5. [CrossRef]

52. Flores, K.M.; Dauskardt, R.H. Mean stress effects on flow localization and failure in a bulk metallic glass. *Acta Mater.* **2001**, *49*, 2527–2537. [CrossRef]

53. Wang, L.; Bei, H.; Gao, Y.F.; Lu, Z.P.; Nieh, T.G. Effect of residual stresses on the hardness of bulk metallic glasses. *Acta Mater.* **2011**, *59*, 2858–2864. [CrossRef]

Article

Investigation of SLM Process in Terms of Temperature Distribution and Melting Pool Size: Modeling and Experimental Approaches

Md Jonaet Ansari, Dinh-Son Nguyen and Hong Seok Park *

Department of Mechanical Engineering, University of Ulsan, Ulsan 44601, Korea;
jonaetansari@gmail.com (M.J.A.); ngdison@gmail.com (D.-S.N.)
* Correspondence: phosk@ulsan.ac.kr; Tel.: +82-(0)52-259-1458

Received: 5 March 2019; Accepted: 16 April 2019; Published: 18 April 2019

Abstract: Selective laser melting (SLM) is an additive manufacturing (AM) technique that has the potential to produce almost any three-dimensional (3D) metallic part, even those with complicated shapes. Throughout the SLM process, the heat transfer characteristics of the metal powder plays a significant role in maintaining the product quality during 3D printing. Thus, it is crucial for 3D-printing manufacturers to determine the thermal behavior over the SLM process. However, it is a significant challenge to accurately determine the large temperature gradient and the melt pool size using only experiments. Therefore, the use of both experimental investigations and numerical analysis can assist in characterizing the temperature evaluation and the melt pool size in a more effective manner. In this study, 3D finite element analysis applying a moving volumetric Gaussian laser heat source was used to analyze the temperature profile on the powder bed and the resultant melt pool size throughout the SLM process. In the experiments, a TELOPS FAST-IR (M350) thermal imager was applied to determine the temperature profile of the melting pool and powder bed along the scanning direction during the SLM fabrication using Ti_6Al_4V powder. The numerically calculated results were compared with the experimentally determined temperature distribution. The comparison showed that the calculated peak temperature for single- and multi-track by the developed thermal model was in good agreement with the experiment results. Secondly, the developed model was verified by comparing the melting pool size for various laser powers and scanning speeds with the experimentally measured melting pool size from the published literature. The developed model could predict the melt pool width (with 2–5% error) and melt pool depth (with 5–6% error).

Keywords: Additive manufacturing; selective laser melting; volumetric heat source; thermal capillary effects; melt pool size

1. Introduction

Additive manufacturing (AM), widely familiar as 3D printing, is rapidly emerging as a new and disruptive manufacturing technology that offers the opportunity to manufacture complex, freeform three-dimensional (3D) metal parts with a 3D computer-aided design (CAD) design, and an AM printing machine [1]. Currently, AM has received increased attention from the aerospace, automotive, biomedical, and energy industries due to its benefits compared with conventional forming techniques. This technology acts an essential role in Industry 4.0 as the dominant technology that permits the cost-effective realization of the Internet of Things across various industrial applications. In recent years, various alloys such as stainless steel, titanium, aluminum, and nickel-based alloys are preferable for use in additive manufacturing technology. Ti_6Al_4V is the most popular alloy due to its lower density, elevated temperature properties, high strength-to-weight ratio, and good weldability characteristics [2–5].

Throughout the SLM process, the metal powder is continuously melted and fused by a high-intensity laser beam to fabricate high density, complicated metallic parts from a CAD model [6]. As a result, cross-sections of a certain zone are melted in layers, which are built up continuously to fabricate entire 3D products. Due to its unique process behaviors, SLM can be used to fabricate any intricate and high precision shape for metal parts, including elements with complicated porous shapes, which would be hard to produce with traditional manufacturing technologies [7,8].

Due to the continuous flow of energy through heat transfer in successive layers of the part, large temperature gradients are generated in the fabricated part. Consequently, it is essential to understand the inherent characteristics related to SLM processes (such as the temperature gradient and the melting pool size) before performing the actual printing process. This is because these characteristics influence the mechanical and physical properties as well as the overall quality of printed products. Moreover, the quality of the final SLM parts significantly depends on the appropriate selection of process parameters including laser power, scan speed, laser spot size, hatch distance, scanning pattern, and layer thickness. For that reason, it is always crucial to find out the optimal process parameters [9]. It is essential to develop a tool that can determine the temperature profile and melt pool size from both industrial and technical perspectives for optimizing the manufacturing processes and to control the consistent quality of the printed products.

The finite element (FE) technique is the most usually applied numerical approach for investigating the temperature profile and melt pool size in the SLM process. Shuai et al. [10] presented a mathematical model and established an FE method for determining the dynamic temperature distributions during the SLS process. Song et al. [4] performed a numerical analysis to predict the temperature field as well as experimental work to determine the relationship between the density, porosity, and laser scanning speed to identify better process parameters for the SLM process. Ma and Bin [11] proposed an FE analysis to determine the effect of two different laser moving strategies on the temperature distribution, residual stresses, and deformation during the selective laser sintering (SLS) process. Hussein et al. [9] developed an FE model that employed thermo-mechanical analysis for evaluating the development of temperature fields, melt pool size, and residual stress at various points on a single layer fabricated by the SLM process. Ali et al. [12] analyzed the influence of layer thickness on the thermal stress and mechanical properties of the SLM process for Ti_6Al_4V products by changing the applied laser power. Fischer et al. [13] conducted an experimental investigation to measure the peak temperature for the continuous wave and pulsed sintering process using an infrared camera. Dai and Shaw [14] presented an FE model to determine the effect of volume shrinkage associated with phase change during laser forming, and experimental work was performed to verify the presented model. Yadroitsev et al. [15] used a CCD camera for determining the highest temperature and the melt pool size of Ti_6Al_4V and also investigated the microstructure of SLM printed parts. Huang et al. [8] provided a detailed model of the temperature distribution and melt pool size that considered the volume shrinkage and linear energy density; this model was verified by comparing with published experimental results. Dilip et al. [3] experimentally measured the melt pool size and identified the relationship between porosity and energy density by changing the SLM process parameters. Lee and Zhang [16] presented a model of the macroscopic fluid and heat transfer incorporating Marangoni effects. They observed that Marangoni convection dramatically influenced the thermal behavior in the developed melt pool at the time of the SLM process. Ali et al. [17] presented a model that can predict the temperature distribution, melt pool width (with 14.5% error), melt pool depth (with 3% error), and the developed residual stress within the fabricated parts for various SLM process parameters. As noticed from this overview, 3D finite element models have been broadly accepted in cases of the SLM process. However, there have been few studies that used this technique to inspect the temperature evaluation and the developed melt pool size throughout the SLM process.

In this study, a 3D finite element (FE-Equations) model was initiated to analyze the thermal behavior and the developed melt pool size in a multi-track pattern during the SLM process. A moving volumetric Gaussian heat source was used with consideration of the temperature-dependent material properties and the phase transition of the Ti_6Al_4V alloy. Moreover, the variation of thermal behavior, and the developed melt pool size were investigated by applying various SLM process parameters. Finally, the corresponding experimental works were performed to validate the predicted results.

2. 3D Finite Element Modeling

A 3D thermal analysis model was developed using COMSOL Multiphysics Modeling Software (Version 5.4, COMSOL Ltd., Cambridge, UK) to predict the global temperature fields generated during laser irradiation. The 3D numerical model, mesh structure, and the SLM scanning patterns are shown in Figure 1. The dimensions of the modeled powder bed were a length of 10 mm, a width of 6.8 mm, and a thickness of 0.2 mm. A block with dimensions of 10 mm × 6.8 mm × 0.2 mm (length × width × thickness) was considered to be the substrate. To acquire the finest calculating efficiency within the smallest computational time, an extremely fine mesh was used for the powder bed, while a comparatively fine mesh has been used for the substrate. The numerical model consists of 30762 total domain elements, 9962 boundary elements, and 432 edge elements. The total computational time required for this presented simulation is about 9–10 h when using a workstation with an Intel Xeon CPU E5-2620 v2 @2.10GHz, RAM 32 GB (Intel, Santa Clara, CA, USA). The heat source moved in a bidirectional scanning direction. The process parameters that were applied in the FE analysis are listed in Table 1. The entire Ti_6Al_4V powder bed was presumed as a continuous and homogenous ambience. The coefficient of convective heat transfer among the powder bed and the circumstances were presumed to be a constant.

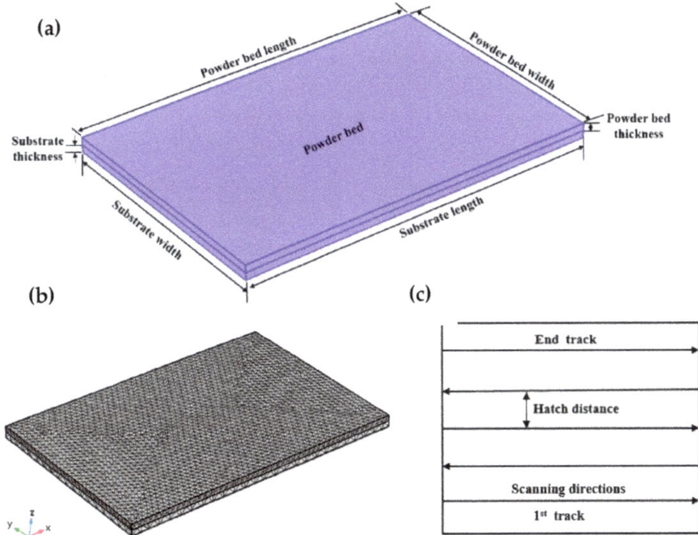

Figure 1. 3D FE model of multi-track laser scanning throughout the SLM process. (**a**) Schematic of the designed geometry containing the substrate and powder bed, (**b**) 3D finite element mesh, and (**c**) scanning strategy.

2.1. Governing Equation and Volumetric Heat Source

The thermal equilibrium is achieved based on the following transient 3D heat conduction equation, which can be expressed as [18]

$$\rho C_p \frac{\partial x}{\partial y} + \rho C_p u \nabla T = \nabla (k \nabla T) + Q \tag{1}$$

in which T is the temperature, ρ is the density, Cp is the specific heat capacity, k is the thermal conductivity, Q is the absorbed heat, and u is the laser scanning speed.

Throughout the SLM process, the heat from the laser beam experiences numerous absorption and reflections across the Ti_6Al_4V powders. The applied laser energy is separated into three portions, including reflection, absorption, and transmission of power. Only the absorbed energy was used to melt the powders. The laser energy can travel a certain depth through the powder bed. Therefore, the heat transfer through the depth direction on the powder bed was also considered in this presented model. As presented by Li et al. [19], the Beer-Lambert attenuation law can be used to define the laser penetration in the depth direction, which is given as

$$Q(x,y,z) = \frac{Q_0(x,y)}{\delta} \exp(-\frac{|z|}{\delta}) \tag{2}$$

Here, Q_0 is the heat flux on the upper surface (W/m^2), δ is the optical penetration depth for used material [20], $|z|$ is the absolute value of the z-coordinate.

The distribution of surface heat flux Q_0 across the powder bed is presumed to be a Gaussian relationship, which can be mathematically represented as [13]

$$Q_0(x,y) = \frac{2AP}{\pi R^2} \exp(-\frac{2((x-ut)^2 + y^2)}{R^2}) \tag{3}$$

where P is the laser power, A is the laser energy absorption coefficient, and R is the resulting heat source radius at which the energy density is minimized to $1/e^2$ at the center of the laser spot. The laser scanning direction is included by replacing x with $(x-ut)$. By replacing Equation (3) into Equation (2), the volumetric heat source is given as:

$$Q(x,y,z) = \frac{2AP}{\pi \delta R^2} \exp(-\frac{2((x-ut)^2 + y^2)}{R^2}) \exp(-\frac{|z|}{\delta}) \tag{4}$$

Temperature-dependent thermal properties (ρ, Cp, k) of solid and liquid Ti_6Al_4V alloy were used in this study and were obtained from the published literature by Boivineau et al. [21], Heigel et al. [22], and Parry et al. [23]. The phase change behavior among the solid and liquid stages of a material can be input into Equation (1) with the following relationships [24]:

$$\rho = \theta \rho_{solid} + (1-\theta)\rho_{liquid} \tag{5}$$

$$C_p = \frac{1}{\rho}(\theta \rho_{solid} C_{p,solid} + (1-\theta)\rho_{liquid} C_{p,liquid}) + L\frac{da}{dT} \tag{6}$$

$$k = \theta k_{solid} + (1-\theta)k_{phase2} \tag{7}$$

$$a = \frac{(1-\theta)\rho_{liquid} - \theta \rho_{solid}}{\theta \rho_{solid} + (1-\theta)\rho_{liquid}} \tag{8}$$

$$\theta = \begin{cases} 0, & if\ T \leq T_S \\ \frac{T-T_L}{T_L-T_S}, & if\ T_S < T < T_L \\ 1, & if\ T \geq T_L \end{cases} \tag{9}$$

where θ is the phase fraction, L is the latent heat of the phase transfer from solid to liquid and a is the mass fraction, T_L is the liquidus temperature and T_S is the solidus temperature.

2.2. Fluid Flow Modeling

The large temperature gradient in the melting pool will cause a significant surface tension gradient, which creates Marangoni effects. These effects are also considered in this work and are given as [16]

$$F^{Marangoni} = \nabla_s \gamma \quad where \quad \gamma = \gamma_0 + \frac{d\gamma}{dT} \Delta T \tag{10}$$

where ∇_s denotes the surface gradient, γ is the surface tension, γ_0 is the surface tension at the liquidous point, $d\gamma/dT$ is the surface tension gradient, and ΔT is the temperature difference.

To consider the flow behavior in the molten pool, the Navier-Stokes equations were used to model the laminar flow in the melt pool and are given as [16]

$$\rho \frac{\partial u}{\partial t} + \rho(u.\nabla)u = \nabla.[-pI + \mu(\nabla u + (\nabla u)^T)] + \rho g + F \tag{11}$$

$$\rho \nabla.(u) = 0 \tag{12}$$

where p is the pressure, μ is the dynamic viscosity, I is the three-dimensional unity tensor, ρg is the gravity force, and F represents the other body forces, which are surface tension gradient-driven Marangoni forces already described in this work.

2.3. Initial and Boundary Conditions

The initial conditions of the finite element model include a uniform temperature field all over the powder bed before applying the heat source, which can be described as,

$$T(x, y, z, 0) = T_0(x, y, z) \text{ for the whole domain at } t = 0 \tag{13}$$

where $T_0(x, y, z)$ is the surrounding temperature and generally assumed as 293.15 K.

At the top surface of the developed model, heat transfer occurs among the powder bed, substrate, and their surroundings.

$$-k\frac{\partial T}{\partial n} = h_c(T - T_0) + \varepsilon\sigma(T^4 - T_0^4) \tag{14}$$

Here, the terms on the right side of the equation denote the heat loss owing to convection and radiation, successively. Furthermore, n denotes the normal direction of the surface, h_c is the convective heat transfer coefficient, ε is the surface emissivity, and σ is the Stefan–Boltzmann constant.

Table 1. Process parameters and thermo-physical properties of Ti_6Al_4V [20,21] used in this finite element simulation.

Name	Description	Value
P	Laser power (W)	120, 150
u	Laser scanning speed (mm/s)	750, 1000
R	Laser spot radius (μm)	50
A	Absorption coefficient	0.3
δ	Optical penetration depth (μm)	65
h	Hatch distance (μm)	30
ε	Emissivity	0.35
σ	Stefan–Boltzmann constant (W/ (mm²·K))	5.67×10^{-14}
μ	Dynamic viscosity (Pa·s)	0.002
L_f	Melting latent heat ($J{\cdot}kg^{-1}$)	3.5×10^5
dγ/dT	Surface tension gradient ($N{\cdot}m^{-1}{\cdot}K^{-1}$)	-2.7×10^{-4}
T_L	Liquidus temperature (K)	1928
T_S	Solidus temperature (K)	1878

To solve the mathematically derived problem, the FE method was developed by means of COMSOL Multiphysics in which two-ways coupling between heat transfer and fluid flow were established.

3. Materials and Experimental Methods

SLM experiments were performed using a SLM printer (MetalSys150, Winforsys co., Ltd., Yongin-si, South Korea) with a YLP-200-AC-Y11 IPG Ytterbium Fiber Laser (Winforsys co., Ltd., Yongin-si, South Korea) (highest laser power of 200 W). The chamber was occupied with argon protection gas to keep the oxygen level below 0.1%.

In this study, a commercial Ti-alloy (Ti_6Al_4V) powder provided by the (SLM Solutions Group AG, Lübeck, Germany), was used as a raw material with the resulting nominal chemical composition (wt.%): Ti-balance, Al-(5.5–6.50), V-(3.50–4.50), Fe-0.25, C-0.08, N-0.03, O-0.13, H-0.0125. The average powder size was about 23–60 μm.

A TELOPS FAST-IR (M350) thermal camera (TELOPS, Quebec City, Canada) with a spatial resolution of 640 pixels × 512 pixels, and a maximum frame rate of 4980 Hz was employed to determine the temperature profile on the melt pool and the powder bed. The built part has dimensions of $10 \times 6 \times 0.03$ mm (length × width × thickness) e.g., one layer. The camera was mounted on a tripod and fitted near to the view window, straightly focusing on the laser scanning area as presented in Figure 2. The process parameters for experiments were chosen as follows: laser powers (P): 120 W, 150 W; scan speeds (u): 750 mm/s, 1000 mm/s.

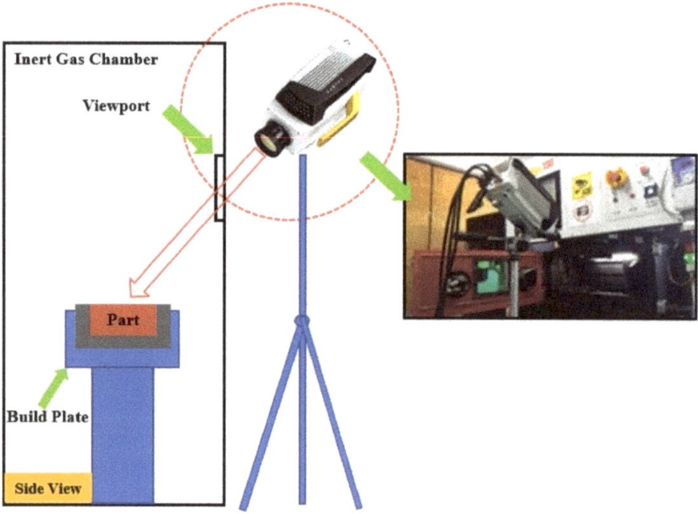

Figure 2. Schematic plot for experimental set up.

4. Numerical Model Validation

To confirm the thermal model and numerical approaches presented in this research, the developed model was initially compared with published experimental results.

Yadroitsev et al. [15] experimentally measured the brightness temperature of the melt pool by using a laser power of 50 W and a scanning speed of 100 mm/s for Ti_6Al_4V alloy. Figure 3a illustrates the contrast of the calculated temperature field in the xy-plane across the laser moving path (presented in Figure 3b) with the experimentally measured peak temperature for the SLM of Ti_6Al_4V. Based on this comparison, the calculated peak temperature across the laser scanning direction concedes well with the trends in the experimentally determined temperature profile.

Fischer et al. [13] used a Raytheon infrared camera for determining the temperature distribution by applying a laser power of 3 W and a scanning speed of 1 mm/s.

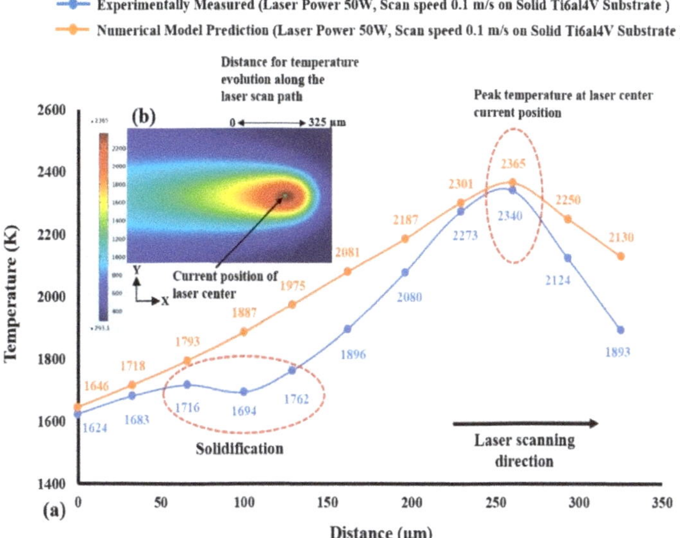

Figure 3. (**a**) Comparison of the FE model calculated temperature in the *xy*-plane along the laser moving path with the experimentally measured peak temperature along the laser scanning direction adapted from reference [15] and (**b**) Numerically predicted temperature distribution along the *xy*-direction considering a 325 μm laser scan path.

Figure 4 demonstrates the temperature profile measured by Fischer et al. [13] and the peak temperature scale was in the range of 2500 K to 3000 K. Numerical analysis was conducted applying process parameters similar to those used by Fischer et al. [13] during their experiments. As demonstrated in Figure 5, the highest temperature after 0.75 s is about 2640.9 K which falls between the measured experimental values.

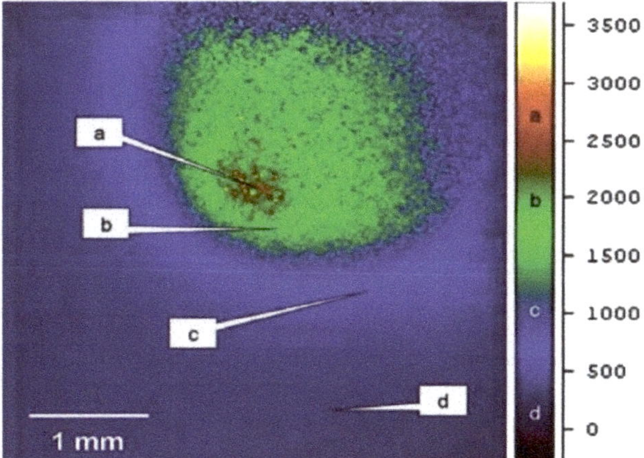

Figure 4. Experimentally measured average temperature profile throughout the laser sintering process at P = 3 W and u = 1 mm/s.

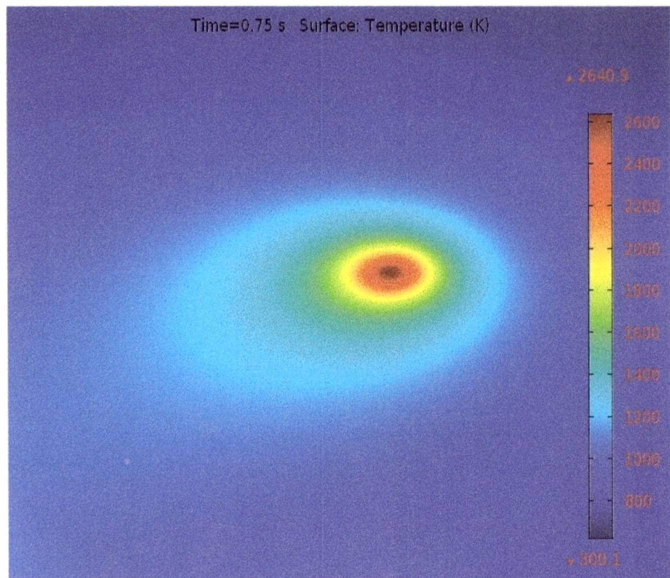

Figure 5. Numerically predicted surface temperature contours at $P = 3$ W and $u = 1$ mm/s.

5. Results and Discussion

5.1. Temperature Distribution

During the SLM process, the temperature field on the powder bed deviates quickly with time and locations; these are critical issues for the printed product quality. The melting zone undergoes solidification in the wake of the heat source because of the moving laser beam and fast heat transmission from the melting zone to the surroundings. The solidification process starts once the temperature of the melt pool falls under the liquidus temperature. Afterwards it cools down to the ambient temperature for fabricating the complete products.

Figure 6 illustrates the temperature contours at the start of laser moving, from which the peak temperature gradient in the laser spot region can be distinctly seen to the used 3D Gaussian heat source. The temperature of the powder bed rises quickly owing to the absorption of high energy irradiated from the heat source, initiating a melting pool in the powder bed once the temperature surpasses the liquidus point of Ti_6Al_4V (1928 K). Figure 7 illustrates the predicted temperature profiles and melt pool formation as the heat source reached various locations throughout the SLM process for $P = 120$ W and $u = 1000$ mm/s. Figure 7a demonstrates the temperature profiles at the final point of the first scanning track (at $t = 0.01$ s). At this point, the predicted highest temperature on the melt pool was about 2108.2 K, which exceeded the liquidus temperature of Ti_6Al_4V. Besides, the lowest temperature was only 293.15 K in most of the area of the powder bed and the substrate. The corresponding temperature contours and the melt pool region is presented in Figure 7b starting with the isothermal contours at 1900 K and going to the highest temperature of the melt pool. The melt pool size, as presented by the isothermal contours, provides a visual idea of the spatial energy distribution for a consistent heat source. The calculated temperature in the isothermal contours was larger than the liquidus temperature of Ti_6Al_4V. As a result, a little melt pool developed within this region. The length and width of the melt pool were approximately 237.6 μm and 90.7 μm, respectively. As the heat source arrived at the middle of the powder bed at $t = 1.015$ s, the maximum temperature of the melt pool increased to 2193.6 K in the center of the melt pool, as presented in Figure 7c. The melt pool length and width improved by approximately 356.4 μm and 110.2 μm, at a time of 1.015 s, as shown in Figure 7d. The heat source moved to its final position, and after that the heat source was no longer specified. Thus, only heat loss happens at this end position. At the ending of the final scanning track at $t = 2.01$ s, the predicted lowest and highest temperature of the powder bed raised to approximately 528.7 K and 2315.5 K, respectively, as presented in Figure 7e. Figure 7f represents the resulting melt pool length and width (approximately 512.2 μm and 137.8 μm) at a time of $t = 2.01$ s, which are larger than those at 0.01 s and 1.015 s. Therefore, the length of the melting pool increased more than the width of the melting pool as the laser irradiating time increased on the powder bed. For the specified numerical circumstances, the width of the melt pool at various positions was larger than the hatch distance (30 μm), which led to smooth melt tracks due to the development of a large enough melt pool between the adjacent tracks.

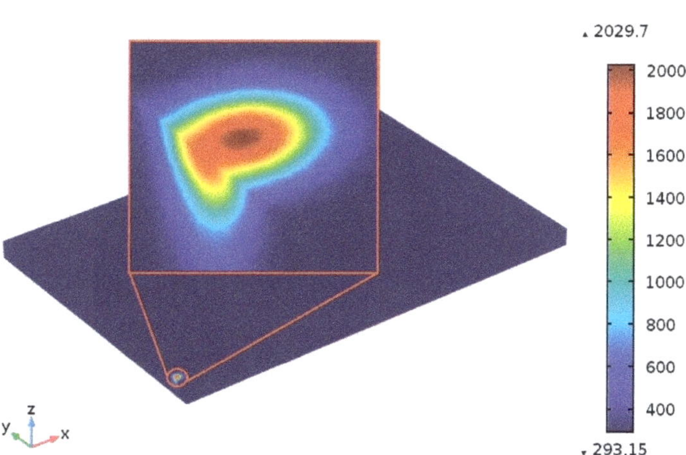

Figure 6. Temperature distribution at the start of laser scanning.

5.2. Variation of Temperature Distribution with Different Process Parameter

The development of peak temperature with respect to time for different process parameters during the SLM process is presented in Figures 8 and 9. Once the laser scan speed decreased from 1000 mm/s to 750 mm/s (at $P = 120$ W), a maximum temperature of 2306.4 K was predicted at a time of 0.0133 s, which is larger than the liquidus temperature of Ti_6Al_4V, as presented in Figure 8a. When the heat source reached at the final scanning track (at $t = 2.6793$ s), the observed temperature was 2624.7 K (Figure 8b), which was also over the liquidus temperature of Ti_6Al_4V. Once the laser power was further raised to 150 W (at $u = 1000$ mm/s), the predicted peak temperature was 2271.2 K (at $t = 0.01$ s), and it increased to 2530.1 K at the ending of the final scanning track (at $t = 2.01$ s), as demonstrated in Figure 9a,b, respectively.

Throughout the SLM process, an elevated temperature gradient can be found among the melt pool when applying higher laser power (150 W) with a comparative lower scanning speed (750 mm/s) and this phenomenon happens owing to absorption of adequate laser energy by the supplied material powders. Furthermore, an excessive heat growth phenomenon can happen which can remelt the previously built scanning path. In this situation, a higher temperature gradient of 2509.6 K (at $t = 0.01333$ s) was obtained in the melt pool which was further raised to 2891.2 K at the final point of the scanning process (at $t = 2.6793$ s), as presented in Figure 9c,d, respectively.

The numerical findings indicated that the applied laser power directly impacted the temperature fields of the powder bed throughout the SLM process; however, the laser scanning speed changed the temperature distribution by varying the laser exposure time among the applied heat source and the powder bed.

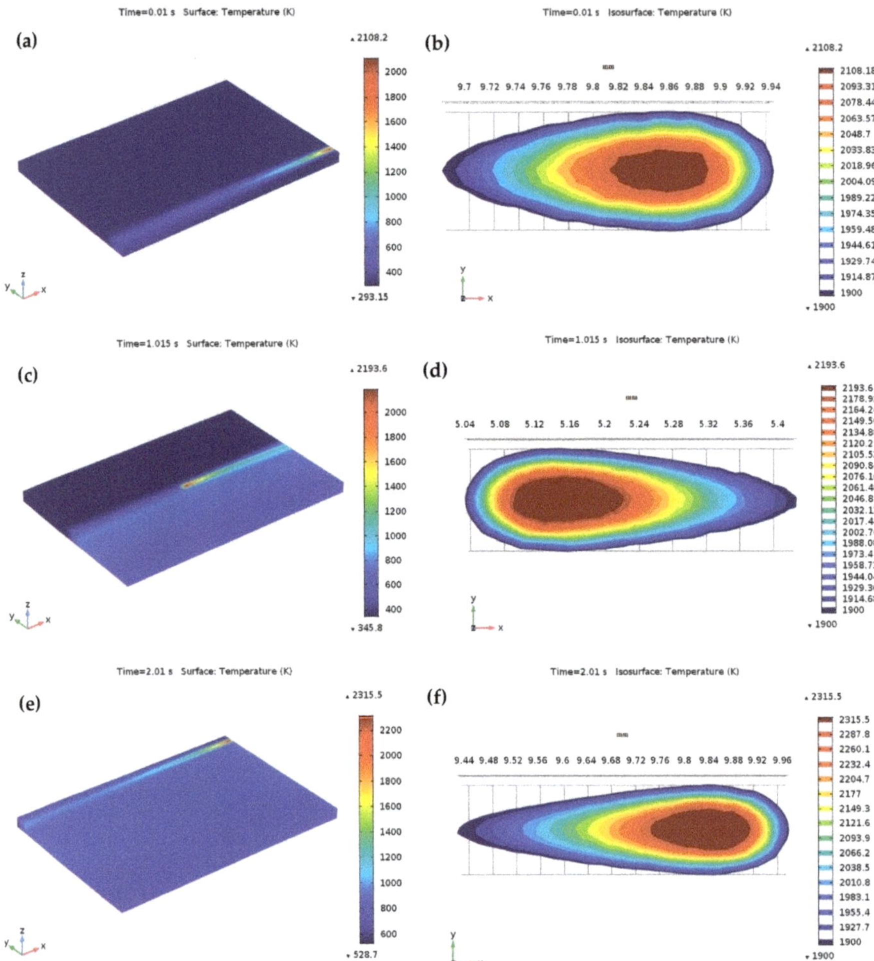

Figure 7. Temperature contours during the SLM process at $P = 120$ W and $u = 1000$ mm/s: (**a**) on the Ti$_6$Al$_4$V powder bed at the ending of the first scanning track at $t = 0.01$s and (**b**) isothermal contours around the melt pool at $t = 0.01$ s; (**c**) on the middle of Ti$_6$Al$_4$V powder bed at $t = 1.015$ s and (**d**) isothermal contours around the melt pool at $t = 1.015$ s; (**e**) on the Ti$_6$Al$_4$V powder bed at the ending of the last scanning track (after scanning a total of 201 tracks) at $t = 2.01$ s and (**f**) isothermal contours around the melt pool at $t = 2.01$ s.

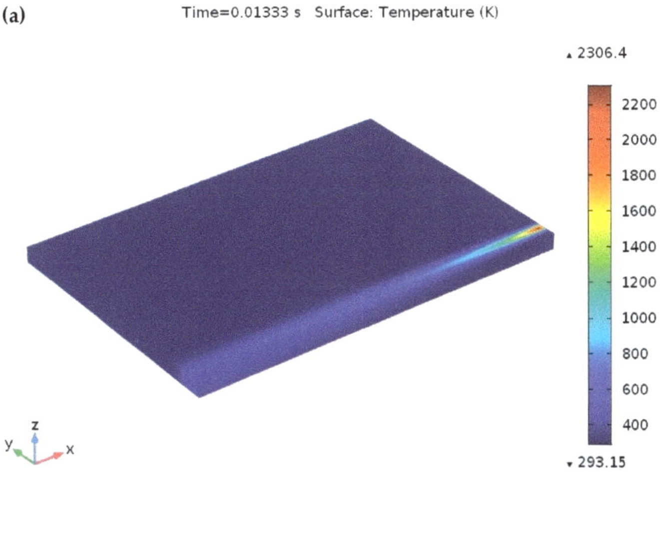

Figure 8. Temperature profiles throughout the SLM process at P =120 W and u = 750 mm/s. On the Ti_6Al_4V powder bed (**a**) at the ending of the first scanning track at t = 0.01333s and (**b**) at the ending of the final scanning track (after scanning a total of 201 tracks) at t = 2.6793s.

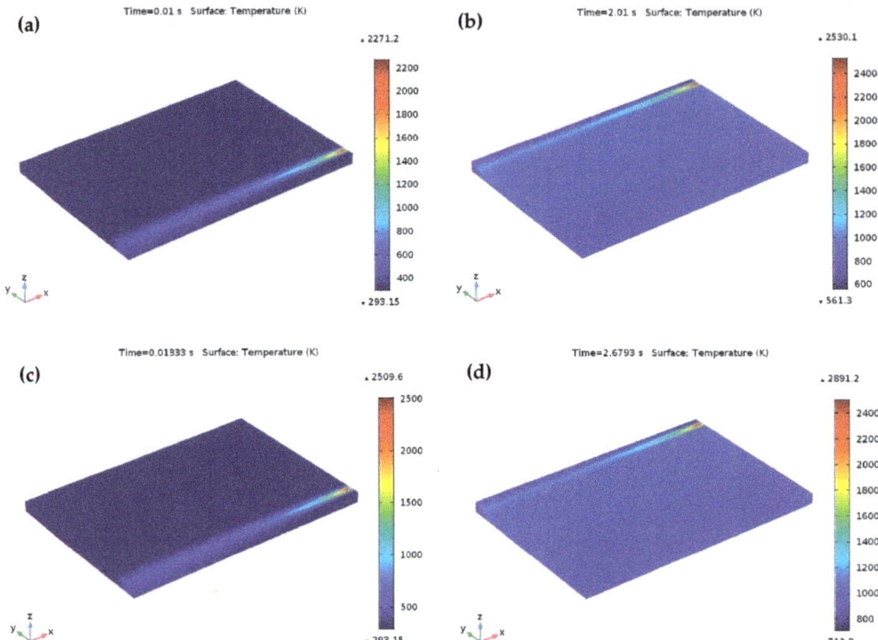

Figure 9. Temperature profiles throughout the SLM process at P =150 W and u = 1000 mm/s, 750 mm/s. On the Ti$_6$Al$_4$V powder bed (**a**) at the ending of the first scanning track at t = 0.01 s, (**b**) at the ending of the final scanning track (after scanning a total of 201 tracks) at t = 2.01 s, (**c**) at the ending of the first scanning track at t = 0.01333s, (**d**) at the ending of the final scanning track (after scanning a total of 201 tracks) at t = 2.6793s.

5.3. Molten Pool Dimensions

Throughout the SLM process, it is a significant challenge to analyze the melting pool length and depth using experiments. Dilip et al. [3] experimentally determined the single-track melt pool width and depth by varying the SLM process parameters. It was reported that single-track with the process parameter sets of 150 W-750 mm/s, and 150 W-1000 mm/s had a modest energy input and produced a regular melt pool size with enough depth of penetration. The experimentally measured melt pool width and depth were approximately 134 μm and 72 μm, respectively, for the combination of 150 W-750 mm/s, and approximately 116 μm and 54 μm, respectively, for the combination of 150 W-1000 mm/s. Therefore, the calculated melt pool width and depth are comparable with the experimental findings for the single-track result. Figure 10 describes the calculated melt pool dimensions using laser powers of 120 W, and 150 W, and scanning speeds of 750 mm/s, and 1000 mm/s for the single-track laser scanning. The melting pool size (length, width, and depth) were observed from the single-track temperature distribution results considered from the melting point (1928 K) to the peak temperature along the scanning direction. At a low scanning speed, the heat source can melt the irradiated zone for a longer time rather than the high scanning speed, resulting in a large melt pool size for the high temperature in the melting zone. However, the heat source can melt the irradiated zone for a shorter time at a high scanning speed, resulting in a small melt pool for the low temperature gradient in the melting zone. Figure 10a,b clearly illustrates the calculated melt pool size for the combination of 120 W-1000 mm/s and 120 W-750 mm/s. At the same laser power of 120 W, the calculated melt pool width and depth was approximately 90.7 μm and 34.2 μm at u = 1000 mm/s, while the calculated melt pool depth and width was approximately 114.6 μm and 53.1 μm at u = 750 mm/s. Figure 10c represents the calculated melt pool size for the combination of 150 W-1000 mm/s. The calculated melt pool width was approximately 110.8 µm, which agrees with the reported experimental results with 5% error. The predicted melt pool depth was approximately 50.9 µm, which coincides with the reported experimental results with 6% error. The melt pool size is also predicted by decreasing the scanning speed to 750 mm/s at the same laser power, which is demonstrated in Figure 10d. The predicted melt pool width was approximately 130.9 µm, which was 2% less than the experimental findings, and the predicted melt pool depth was approximately 68.1 µm, which was 5% less than the reported experimental results.

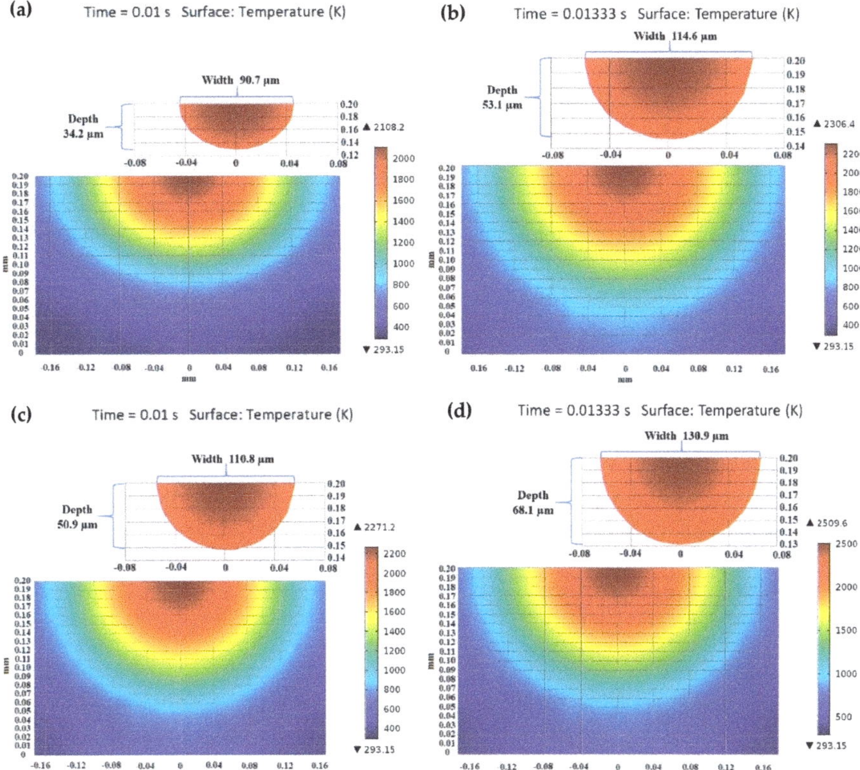

Figure 10. Variation in the melt pool geometry at various laser powers and scan speeds. Predicted melt pool width and depth at (**a**) 120 W-1000 mm/s, (**b**) 120 W-750 mm/s, (**c**) 150 W-1000 mm/s, (**d**) 150 W-750 mm/s, respectively.

Figure 11 was presented to analyze the variation of melt pool length by applying the different SLM process parameters. For each specific laser power, the melting pool length was predicted by varying the scanning speed. At a laser power of 120 W, a clear increasing tendency was determined for the melt pool length from 237.5 µm (at $u = 1000$ mm/s) to 350.7 µm (at $u = 750$ mm/s). Once the applied laser power was raised from 120 W to 150 W, the length of the melt pool improved to 450.7 µm (at $u = 1000$ mm/s) and 520.2 µm (at $u = 750$ mm/s). Therefore, the simulation findings exhibited that the melt pool size (length, width, and depth) raised linearly with the applied laser power. As can be seen from the evaluations made based on the literature, the presented numerical model can determine the melting pool width and depth in an acceptable range.

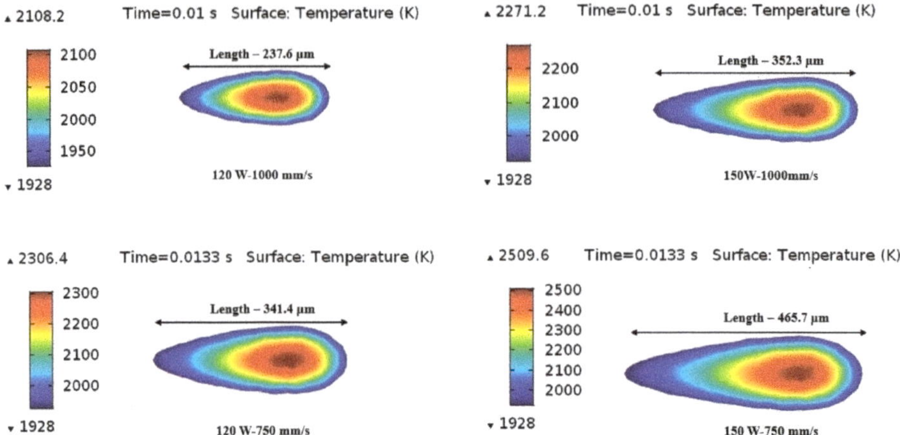

Figure 11. Variation of the melt pool length at different process parameters in the scanning direction at the end of the first scanning track.

5.4. Experimental Validation

To confirm the reliability of the presented simulation model, experiments were carried out to determine the temperature distributing characteristics throughout the SLM process. By using the thermal imager, the developed temperature profiles in the melt pool were monitored, and the images were captured in real time. Figure 12 displays the characteristics of temperature distributions and melt pools at various laser powers and scanning speeds. To measure the temperature, thermal camera emissivity was fixed to 0.35 with a transmission rate of 1.0. These thermal pictures represent the temperature profiles obtained by moving the heat source. A high-temperature region was found close to the laser spot center and progressively cooled down behind the melting pool as the heat source moved away. Figure 12a presents the temperature profiles for the first scanning track at $P = 120$ W and $u = 1000$ mm/s. By means of the heat source moved through the powder bed in the first scanning track, the temperature increased rapidly and exceeded the melting point, leading to powder melting. The temperature progressively decreased as the laser beam moved away. Taking the emissivity as $\varepsilon{\sim}0.35$, the highest temperature of the melt pool at the time of 0.001s was about 2132.3 K (1859.2 °C) (white color) which surpassed the melting temperature of Ti_6Al_4V. For multi-track scanning, the hatch distance influences the highest temperature in the present scanning track due to the enduring heat provided by the previous tracks. Since the radius of the laser beam is larger than the hatch distance of 30 µm, some areas of the former track remelted, which led to the rise of the peak temperature at the ending of the SLM process. As a result, the temperature increased to 2353.7 K (2080.6 °C) at the end point of the final track, and the total scanning time was 2.01 s, as presented in Figure 12b. In case of low scanning speed of 750 mm/s (at the same $P = 120$ W), the temperature increased to 2329.5 K (2056.4 °C) at a time of 0.01333 s, and 2657.3 K (2384.2 °C) at a time of 2.6793 s, as presented in Figure 12c,d, respectively.

Another set of experiments was carried out at the process parameter sets of 150 W-1000 mm/s and 150 W-750 mm/s. Figure 12e describes the temperature contours at the end of the first scanning track for $P = 150$ W and $u = 1000$ mm/s. The peak temperature in the melt pool was determined approximately 2296.8 K (2023.7 °C) at 0.01 s. The temperature increases in the second track leading to the end track were attributed to reheating induced by the heat source due to hatch spacing, and the peak temperature was reached to 2571.7 K (2298.6 °C) at the end of final scanning track, as shown in Figure 12f. Figure 12g further demonstrates that the highest temperature of the melt pool increased to 2527.4 K (2257.8 °C) at the time of 0.01333 s, as the scanning speed decreases to 750 mm/s (at the same $P = 150$ W). The reason behind this peak temperature is the tremendous energy density provided by the heat source. Furthermore, the highest temperature measured by the thermal imager was about 2891.2 K (2648.5 °C) at a time of 2.6793 s.

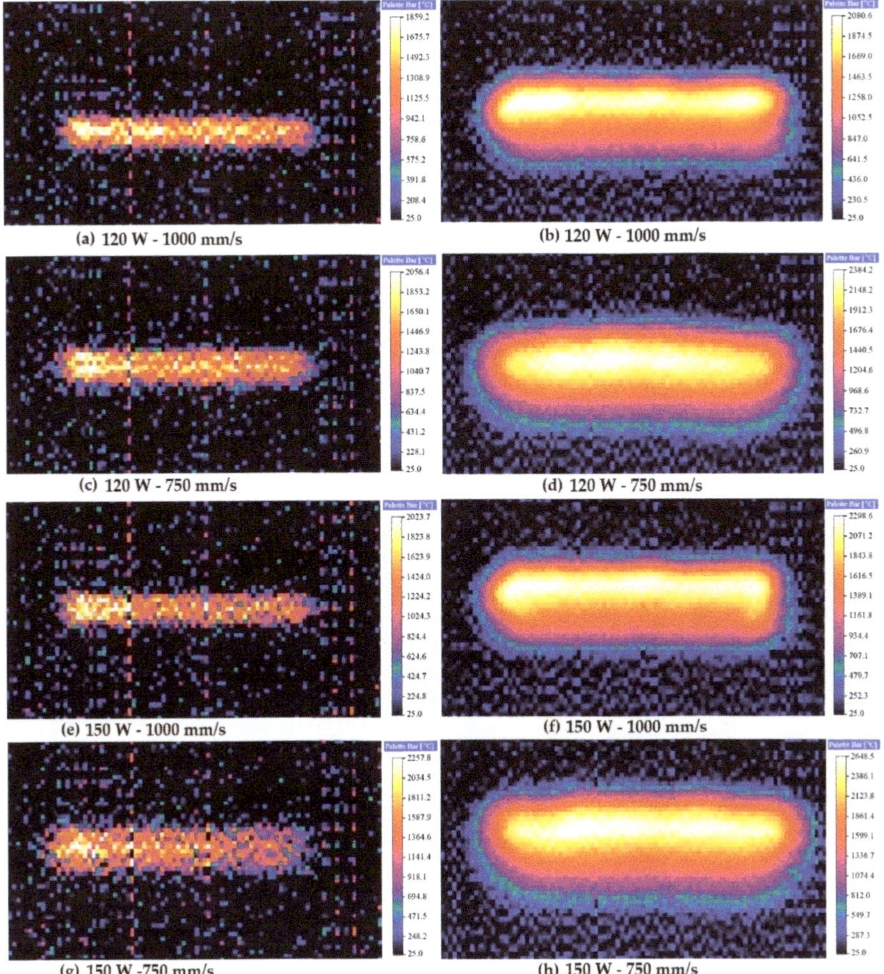

Figure 12. Typical thermal images for different laser powers and scanning speeds along the scanning direction. (**a**) Temperature gradient at a time of 0.01s for 120 W-1000 mm/s. (**b**) Temperature profiles at a time of 2.01s (after scanning a total of 201 tracks) for 120 W-1000 mm/s. (**c**) Temperature profiles at the time of 0.01333s for 120 W-750 mm/s. (**d**) Temperature profiles at a time of 2.6793s (after scanning a total of 201 tracks) for 120 W-750 mm/s. (**e**) Temperature profiles at a time of 0.01s for 150 W-1000 mm/s. (**f**) Temperature profiles at a time of 2.01s (after scanning a total of 201 tracks) for 150 W-1000 mm/s. (**g**) Temperature profiles at a time of 0.01333s for 150 W-750 mm/s. (**h**) Temperature profiles at a time of 2.6793s (after scanning a total of 201 tracks) for 150 W-750 mm/s.

Figure 13 demonstrates a comparison of the numerically predicted and experimentally measured peak temperature distribution results. The developed numerical model presented a reasonably precise prediction based on a comparison with the experimentally measured peak temperature. For the temperature measurement, the predicted errors among the simulation and experimental results are in a range of 21–40 K. These errors can occur because of the scattering in the experimental results and probable differences between the thermo-physical properties used for the numerical analysis and real properties in the experiments. Overall, the developed model is appropriate to employ in the thermal modeling and numerical analysis for the SLM process of Ti$_6$Al$_4$V powder.

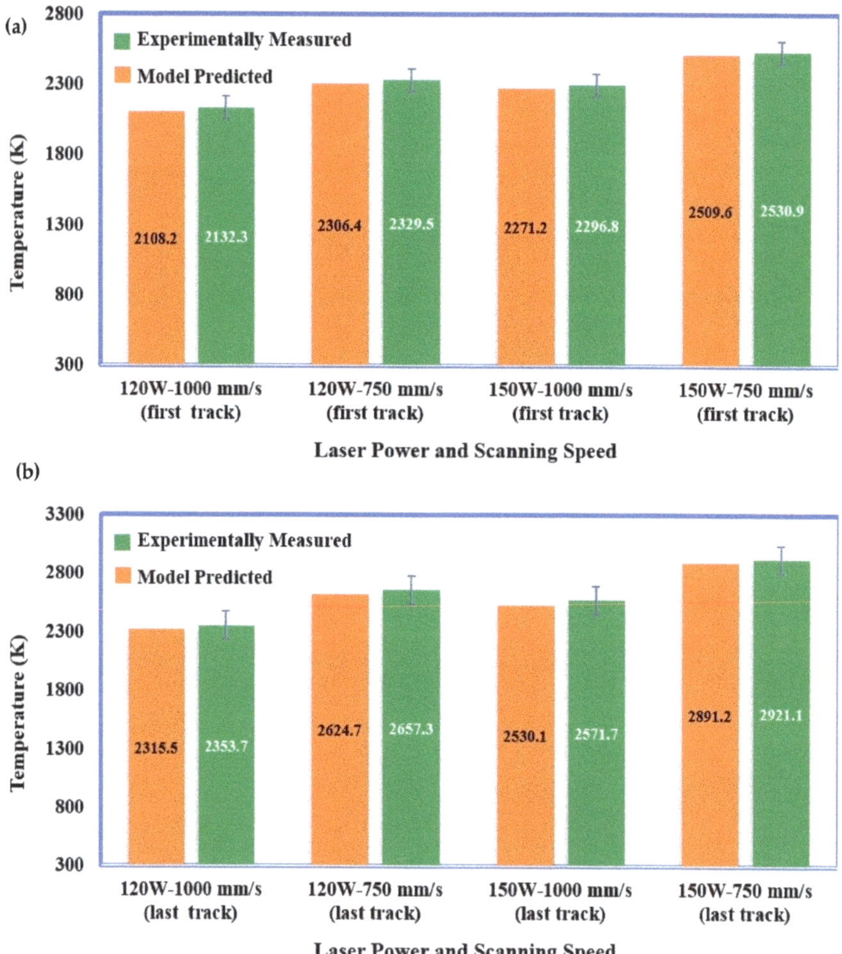

Figure 13. Comparison of experimental and model-predicted peak temperature distribution results: (**a**) at the ending of the first track and (**b**) at the ending of the final scanning track.

6. Conclusions

In this study, a 3D FE model was established to evaluate and predict the temperature distribution and the melt pool size during the SLM process. Furthermore, a thermal imager was used to determine the temperature gradients of the Ti$_6$Al$_4$V material employing the same process parameters as those used in the numerical investigation. The major conclusions of this study can be summarized as follows:

1. The thermal imager was able to capture the temperature profiles at various laser powers and scanning speeds under the same camera setting, e.g., emissivity = 0.35 and transmission rate = 1.0. Besides, the developed model correctly determined the temperature distribution along the laser scanning direction with good correlation to the experimentally measured temperature for both

the single-track and multi-track scanning. The predicting error of the established model is in the range of 21–40 K.

2. At a laser power of 150 W, the predicted melt pool width and depth were approximately 130.9 μm and 68.1 μm, respectively for the 750 mm/s scanning speed; while the length and width would be 114.6 μm and 53.1 μm, respectively for a scanning speed of 1000 mm/s. The developed model can predict the melt pool width (with 2–5% error) and melt pool depth (with 5–6% error).

3. Therefore, the presented fluid flow model that includes the heat flow behavior among the melt pool owing to Marangoni convection is an effective technique for modeling theTi_6Al_4V powder melting behavior. Furthermore, the peak temperature, and the length, width, depth of the melt pool all rises as the laser power rises and the scanning speed decreases.

4. The calculation of peak temperature and melt pool size for the single-track depositions would be an important method for exploring the optimal process parameters for the SLM process. From the above investigation, SLM process parameters with the sets of 150 W-750 mm/s result in enough temperature to melt the powder, provide a well-defined melt pool size and create a uniform single track. Therefore, this set of process parameters would be suggested for manufacturing 3D printed parts by using the SLM process for the Ti_6Al_4V alloy.

Furthermore, the established thermal model can be used to explore the temperature fields and melt pool size for other materials. It could also be expanded to 3D freeform complex geometries manufactured with the SLM process.

Author Contributions: M.J.A. and H.S.P. conceived of the idea of this study. M.J.A. developed and validated the FE method. M.J.A. and D.-S.N. performed the experimental works. M.J.A. written the manuscript and H.S.P. revised it. All authors discussed about the results and commented on the manuscript.

Funding: This research received no external funding.

Acknowledgments: This work has been supported by the ICT R&D program of MSIP/IITP [B0101-14-1081].

Conflicts of Interest: The authors declare no conflict of interest.

References

1. Levy, G.N.; Schindel, R.; Kruth, J.P. Rapid manufacturing and rapid tooling with layer manufacturing (lm) technologies, state of the art and future perspectives. *CIRP Ann. Manuf. Technol.* **2003**, *52*, 589–609. [CrossRef]

2. Masoomi, M.; Thompson, S.M.; Shamsaei, N. Laser powder bed fusion of Ti-6Al-4V parts: Thermal modeling and mechanical implications. *Int. J. Mach. Tools Manuf.* **2017**, *118–119*, 73–90. [CrossRef]

3. Dilip, J.J.S.; Zhang, S.; Teng, C.; Zeng, K.; Robinson, C.; Pal, D.; Stucker, B. Influence of processing parameters on the evolution of melt pool, porosity, and microstructures in Ti-6Al-4V alloy parts fabricated by selective laser melting. *Prog. Addit. Manuf.* **2017**, *2*, 157–167. [CrossRef]

4. Song, B.; Dong, S.; Liao, H.; Coddet, C. Process parameter selection for selective laser melting of Ti_6Al_4V based on temperature distribution simulation and experimental sintering. *Int. J. Adv. Manuf. Technol.* **2012**, *61*, 967–974. [CrossRef]

5. Li, Z.; Li, B.-Q.; Bai, P.; Liu, B.; Wang, Y. Research on the Thermal Behaviour of a Selectively Laser Melted Aluminium Alloy: Simulation and Experiment. *Materials* **2018**, *11*, 1172. [CrossRef]

6. Li, R.; Shi, Y.; Wang, Z.; Wang, L.; Liu, J.; Jiang, W. Densification behavior of gas and water atomized 316L stainless steel powder during selective laser melting. *Appl. Surf. Sci.* **2010**, *256*, 4350–4356. [CrossRef]

7. Bertrand, P.; Yadroitsev, I.; Yadroitsava, I.; Smurov, I. Factor analysis of selective laser melting process parameters and geometrical characteristics of synthesized single tracks. *Rapid Prototyp. J.* **2012**, *18*, 201–208.

8. Huang, Y.; Yang, L.J.; Du, X.Z.; Yang, Y.P. Finite element analysis of thermal behavior of metal powder during selective laser melting. *Int. J. Therm. Sci.* **2016**, *C*, 146–157. [CrossRef]

9. Hussein, A.; Hao, L.; Yan, C.; Everson, R. Finite element simulation of the temperature and stress fields in single layers built without-support in selective laser melting. *Mater. Des. 1980–2015* **2013**, *52*, 638–647. [CrossRef]

10. Shuai, C.; Feng, P.; Gao, C.; Zhou, Y.; Peng, S. Simulation of dynamic temperature field during selective laser sintering of ceramic powder. *Math. Comput. Model. Dyn. Syst.* **2013**, *19*, 1–11. [CrossRef]

11. Ma, L.; Bin, H. Temperature and stress analysis and simulation in fractal scanning-based laser sintering. *Int. J. Adv. Manuf. Technol.* **2007**, *34*, 898–903. [CrossRef]

12. Ali, H.; Ghadbeigi, H.; Mumtaz, K. Processing Parameter Effects on Residual Stress and Mechanical Properties of Selective Laser Melted Ti_6Al_4V. *J. Mater. Eng. Perform.* **2018**, *27*, 4059–4068. [CrossRef]

13. Fischer, P.; Locher, M.; Romano, V.; Weber, H.P.; Kolossov, S.; Glardon, R. Temperature measurements during selective laser sintering of titanium powder. *Int. J. Mach. Tools Manuf.* **2004**, *12–13*, 1293–1296. [CrossRef]

14. Dai, K.; Shaw, L. Finite element analysis of the effect of volume shrinkage during laser densification. *Acta Mater.* **2005**, *53*, 4743–4754. [CrossRef]

15. Yadroitsev, I.; Krakhmalev, P.; Yadroitsava, I. Selective laser melting of Ti_6Al_4V alloy for biomedical applications: Temperature monitoring and microstructural evolution. *J. Alloys Compd.* **2014**, *583*, 404–409. [CrossRef]

16. Lee, Y.S.; Zhang, W. Modeling of heat transfer, fluid flow and solidification microstructure of nickel-base superalloy fabricated by laser powder bed fusion. *Addit. Manuf.* **2016**, *12*, 178–188. [CrossRef]

17. Ali, H.; Ghadbeigi, H.; Mumtaz, K. Residual stress development in selective laser-melted Ti_6Al_4V: A parametric thermal modelling approach. *Int. J. Adv. Manuf. Technol.* **2018**, *97*, 2621–2633. [CrossRef]

18. Gürtler, F.-J.; Karg, M.; Leitz, K.-H.; Schmidt, M. Simulation of Laser Beam Melting of Steel Powders using the Three-Dimensional Volume of Fluid Method. *Phys. Procedia* **2013**, *41*, 881–886. [CrossRef]

19. Li, Y.; Gu, D. Thermal behavior during selective laser melting of commercially pure titanium powder: Numerical simulation and experimental study. *Addit. Manuf.* **2014**, *1–4*, 99–109. [CrossRef]

20. Fischer, P.; Romano, V.; Weber, H.P.; Karapatis, N.P.; Boillat, E.; Glardon, R. Sintering of commercially pure titanium powder with a Nd:YAG laser source. *Acta Mater.* **2003**, *51*, 1651–1662. [CrossRef]

21. Boivineau, M.; Cagran, C.; Doytier, D.; Eyraud, V.; Nadal, M.-H.; Wilthan, B.; Pottlacher, G. Thermophysical Properties of Solid and Liquid Ti-6Al-4V (TA6V) Alloy. *Int. J. Thermophys.* **2006**, *27*, 507–529. [CrossRef]

22. Heigel, J.C.; Michaleris, P.; Reutzel, E.W. Thermo-mechanical model development and validation of directed energy deposition additive manufacturing of Ti–6Al–4V. *Addit. Manuf.* **2015**, *5*, 9–19. [CrossRef]

23. Parry, L.; Ashcroft, I.A.; Wildman, R.D. Understanding the effect of laser scan strategy on residual stress in selective laser melting through thermo-mechanical simulation. *Addit. Manuf.* **2016**, *12*, 1–15. [CrossRef]

24. Kundakcıoğlu, E.; Lazoglu, I.; Poyraz, Ö.; Yasa, E.; Cizicioğlu, N. Thermal and molten pool model in selective laser melting process of Inconel 625. *Int. J. Adv. Manuf. Technol.* **2018**, *95*, 3977–3984. [CrossRef]

Article

Study on the Numerical Simulation of the SLM Molten Pool Dynamic Behavior of a Nickel-Based Superalloy on the Workpiece Scale

Liu Cao * and Xuefeng Yuan

Advanced Institute of Engineering Science for Intelligent Manufacturing, Guangzhou University, Guangzhou 510006, China
* Correspondence: caoliu@gzhu.edu.cn; Tel.: +86-20-39341527

Received: 14 June 2019; Accepted: 9 July 2019; Published: 15 July 2019

Abstract: Nickel-based superalloys are one of the most industrially important families of metallic alloys at present. Selective Laser Melting (SLM), as one of the additive manufacturing technologies for directly forming complex metal parts, has been applied in the production of Inconel 718 components. Based on the more reasonable and comprehensive equivalent processing models (vaporization heat loss, equivalent physical parameters) for the nickel-based superalloy SLM process, an SLM molten pool dynamic behavior prediction model on the workpiece scale was established. Related equivalent processing models were customized by secondary development with the software Fluent. In order to verify the feasibility of the SLM molten pool dynamics model, the SLM single-pass employed to form the Inconel 718 alloy process was calculated. The simulated and experimental solidified track dimensions were in good agreement. Then, the influences of different process parameters (laser power, scanning speed) on the SLM formation of the Inconel 718 alloy were calculated and analyzed. The simulation and experimental solidified track widths were well-matched, and the result showed that, as a rule, the solidified track width increased linearly with the laser power and decreased linearly with the scanning speed. This paper will help lay the foundation for a subsequent numerical simulation study of the thermal-melt-stress evolution process of an SLM workpiece.

Keywords: selective laser melting; molten pool dynamic behavior; equivalent processing model; workpiece scale; nickel-based superalloy; numerical simulation

---

1. Introduction

Superalloys are suitable for long-term operation in high-temperature environments and meet corrosion and abrasion requirements. They are the key metal structural materials in today's aerospace, power, and defense fields [1,2]. Among them, the Inconel 718 nickel-based superalloy is one of the most industrially important families of metallic alloys at present, due to its excellent comprehensive properties, and is widely used in many products, including aircraft engine turbine disks, fasteners, and blades [3,4]. However, with the continuous industrial demand for improvement, traditional Inconel 718 alloy smelting, forging, and reduced material processing methods have gradually made it difficult to meet the growing processing requirements for complex parts. Based on this idea, "layer by layer" additive manufacturing technology can directly and precisely manufacture digital models into three-dimensional solid parts with a high flexibility, no mold, and no restrictions on the part structure [5]. Selective Laser Melting (SLM), as one of the additive manufacturing technologies for directly forming complex metal parts, has been applied in aerospace, automotive, medical, and other fields [6].

At present, the research on SLM formed metal parts mainly relies on experimental means. The research directions include the SLM formation mechanism, the influence of process parameters on the quality of parts, and generating the formation process in situ [7,8]. Kruth et al. [9] found, through experiments, that the effects of the laser on the SLM process were mainly reflected in three aspects: laser wavelength, energy density, and laser mode. Strano et al. [10] presented an investigation of the surface roughness and morphology for SLM parts, and the surface analysis showed an increasing density of spare particles positioned along the step edges as the surface sloping angle increased. A new mathematical model was developed to include the presence of particles on top surfaces, in addition to the stair step effect, for the accurate prediction of surface roughness. Liu et al. [11] investigated the influences of scanning speed, powder thickness, and laser power on the formation of a nickel-based superalloy by SLM, and the results showed that the synergistic effects of laser power and laser scanning speed affected the formation quality. However, the complex thermophysical interactions that existed during the SLM process often occurred on a very short, microsecond time scale. Among them, the thermodynamics and dynamics evolution mechanisms made it difficult to achieve good analytical results through engineering experiments, which restricted the essential understanding of the problems of microstructure control, internal defect formation, deformation, and cracking of the workpiece during the current SLM engineering application process. The method of numerical simulation has been widely used in industrial production for its forward-thinking nature and has been applied in studying physical processes and preventing defects in mechanical manufacturing [12,13].

In the past ten years, numerical simulation studies on the SLM forming process have gradually emerged [14,15], and these theoretical research works can be roughly divided into two directions: based on the particle scale [16,17] and based on the workpiece scale [18,19].

1.1. Numerical Simulation of SLM Molten Pool Dynamic Behavior on the Particle Scale

The so-called particle scale refers to the modeling based on the actual particle morphology, directly calculating the heating and melting effects of the laser on the metal particles, and then describing the complex flow behavior of the metal liquid between the particles on the order of micrometers. Voisin et al. [20] used the multi-physics code ALE3D to study the dynamic behavior of the SLM molten pool based on the particle scale, and directly calculated the distribution of pore defects at different scanning speeds. Lee et al. [21] used the open source discrete element method (DEM) code Yade to obtain the initial distribution of laminated particles, and used the commercial software Flow-3D to calculate the SLM single pass process to study the formation of a ball defect through the simulation results. Panwisawas et al. [22] carried out a numerical simulation of the dynamic behavior of the SLM molten pool based on the open source computational fluid dynamics (CFD) code OpenFOAM, and compared the effects of different lamination thicknesses on the formation effects. This kind of simulation method can directly describe the SLM formation process and directly predict the formation and evolution of defects, such as pores and balls, but the calculation requirements are often huge (the number of elements is tens of millions, and the required computing resources reach the order of 10^5 cpu.hrs). It needs to be implemented with a supercomputer, and the calculation size is often limited to a few hundred microns.

1.2. Numerical Simulation of SLM Molten Pool Dynamic Behavior on the Workpiece Scale

The so-called workpiece scale refers to the powder layer (including metal particles and pores) as a special material, indirectly describing the temperature and flow field evolution in the SLM forming process by setting equivalent physical parameters and flow behavior models, where the mesh size is often a few hundred microns, or even a few millimeters. The reason for the higher computational efficiency of this method is that there is no need to describe the movement of the pores inside the powder bed. Xiao et al. [23] used the idea of a continuous medium (single phase with a uniform material distribution) to calculate the shape of the molten pool during the SLM process and considered the influence of buoyancy and the Marangoni effect on the internal flow behavior of the molten pool, but did not consider vaporization heat loss. Gusarov et al. [24] proposed utilizing equivalent thermal conductivity to characterize the thermal conduction of the powder layer. The equivalent radiation heat transfer model was used to calculate the heating effect of the laser beam on the powder layer, and the influence of the laser beam mode on the SLM process was studied. Yuan et al. [25] carried out a numerical simulation of the SLM process using Fluent, analyzed the internal flow of the molten pool caused by the Marangoni effect, and compared the influence of different process parameters on the size of the molten pool. This kind of simulation method cannot describe the SLM formation process intuitively, but the advantage is that the temperature, flow, and stress field evolution in the SLM process can be described by equivalent processing methods, and then the temperature, the shape of the molten pool, and the deformation of the workpiece during the entire formation process can be obtained. Due to the unusual complexity of the SLM formation process, the accuracy of the simulation method based on the workpiece scale mainly depends on the rationality of the equivalent processing models.

In summary, the calculation efficiency of the research method based on the particle size was too low, which makes it difficult to quickly predict and analyze the SLM process. Therefore, the research method based on the workpiece scale was selected. However, due to the incompleteness of the equivalent processing methods currently used in the research based on the workpiece scale, the calculation accuracy was low. In this paper, by introducing more reasonable and comprehensive equivalent processing models (vaporization heat loss, equivalent physical parameters), a dynamic behavior prediction model of an SLM molten pool based on the workpiece scale has been established for the nickel-based superalloy SLM process. The secondary development method was used to customize the relevant equivalent processing models based on Fluent, and a numerical simulation of the SLM formation process of a nickel-based superalloy was carried out. To verify the feasibility of the SLM molten pool dynamics model, the SLM single-pass formation of the Inconel 718 alloy was calculated and compared to the experimentally obtained solidified track size. Then, the influences of different process parameters (laser power, scanning speed) on the SLM formation of the Inconel 718 alloy were analyzed, and the calculation results were verified with the experimental results. This study can be expected to help lay the foundation for a subsequent numerical simulation study of the thermal-melt-stress evolution process of SLM parts.

2. Mathematical and Numerical Modeling

2.1. Dynamic Behavior Control Equations of the SLM Molten Pool Based on the Workpiece Scale

In a study of the dynamic behavior of the SLM molten pool based on the workpiece scale, the calculation area consists of four parts: the powder bed, the solidified portion, the metal base plate, and the protective atmosphere chamber. In the calculation process, the powder bed is gradually transformed into the solidified portion by the equivalent treatment, and energy and momentum interactions occur between the parts. In addition, in order to ensure the efficiency of the numerical calculation, several appropriate assumptions have been made, including: not considering the mass loss caused by the vaporization of molten metal; not considering the influence of the change in metal density on the volume; and considering that the fluids involved in the calculation are all incompressible, Newtonian fluids. These assumptions mean that the mass of the metal phase in the calculation was constant, the influence of the volume change of the metal phase on the flow behavior was not considered, and the compressibility of the gas phase and the liquid metal was not considered. Next, the three types of conservation equations used in this study will be introduced.

2.1.1. Momentum Conservation Equation

When metal particles are melted by laser radiation, factors affecting the flow behavior of the liquid metal include: surface tension between the liquid metal and the substrate and particles, the Marangoni effect (surface tension gradient caused by the temperature difference on the liquid metal surface), the vaporization recoil force of the liquid metal, buoyancy, the internal pressure of the liquid metal, the internal viscous force of the liquid metal, gravity, and the difference in fluidity between the liquid and solid metal during solidification. Among them, the first three influencing factors are surface forces and the last five influencing factors are volumetric forces. Since the calculation model used here is a single-phase flow model (the coupling between gas phase and liquid metal was not calculated) and considering that vaporization recoil force mainly affects the liquid surface fluctuation of liquid metal, this study does not model these factors. The obtained momentum conservation equation is as below.

$$\frac{\partial \rho u}{\partial t} + \nabla \cdot (\rho u \otimes u) = -\nabla p + \nabla \cdot \overline{\tau} + \rho g + F_{buoyancy} + F_{mushy} \tag{1}$$

where

$$\overline{\tau} = 2\mu\left[\left(\frac{1}{2}\nabla u + \frac{1}{2}(\nabla u)^T\right) - \frac{1}{3}(\nabla \cdot u)\mathbf{I}\right] \tag{2}$$

$$F_{buoyancy} = \rho g \beta\left(T - T_{ref}\right) \tag{3}$$

$$F_{mushy} = -\rho K_C\left[\frac{(1-f_l)^2}{f_l^3 + C_K}\right]u \tag{4}$$

Here, ρ is the density, kg/m^3; u is the velocity, m/s; t is the time, s; $\otimes$ is the tensor product; p is the pressure, Pa; $\overline{\tau}$ is the stress tensor; g is the gravity acceleration, m/s^2; $F_{buoyancy}$ is the buoyancy, N/m^3; F_{mushy} is the mushy zone drag force, which can be used to characterize the difference in fluidity caused by the liquid-solid transition [26], N/m^3; μ is the dynamic viscosity, Pa·s; $\mathbf{I}$ is the unit matrix; β is the thermal expansion coefficient, 1/K; T is the temperature, K; T_{ref} is the thermal expansion reference temperature, K; K_C is the porous media permeability coefficient, 1/s; C_K is a custom smaller value, which is used to avoid the drag force of the mushy zone during the calculation to infinity; and f_l is the liquid fraction of the metal phase.

The right-end terms in Equation (1) characterize the five-volume forces (internal pressure, internal viscous force, gravity, buoyancy, and mushy zone drag force) experienced by the liquid metal, respectively. Because the laser energy density is Gaussian in the horizontal plane, the liquid metal surface temperature shows a central high and a peripheral low, and since the surface tension is related to temperature, the tangential flow on the liquid surface occurs under the influence of the surface tension gradient, so the Marangoni effect needs to be characterized by defining the corresponding surface force. The boundary condition used to describe the Marangoni effect [27] here is

$$-\mu\frac{\partial u_x}{\partial z} = \frac{d\sigma}{dT}\frac{\partial T}{\partial x} \tag{5}$$

$$-\mu\frac{\partial u_y}{\partial z} = \frac{d\sigma}{dT}\frac{\partial T}{\partial y} \tag{6}$$

Here, $\frac{d\sigma}{dT}$ is the surface tension coefficient with the rate of change in temperature, N/(m·K); x, y are the coordinates of the horizontal plane, m; z is the coordinate in the vertical direction, m; and u_x, u_y are the components of the tangential velocity on the liquid metal surface, m/s.

2.1.2. Energy Conservation Equation

The factors to be considered in the calculation of the temperature field of the SLM process include the absorption of laser energy, melting of the solid metal, vaporization of the liquid metal, convection diffusion inside the metal phase, and heat exchange between the metal phase and the surroundings (convection and radiation). The adopted energy conservation equation is

$$\frac{\partial \rho c_e T}{\partial t} + \nabla \cdot (\rho u c_e T) = \nabla \cdot (k\nabla T) + Q_{laser} \tag{7}$$

where

$$c_e = \begin{cases} c + \frac{L_f}{T_l - T_s} & T_l < T < T_s \\ c & T \geq T_l \, or \, T \leq T_s \end{cases} \tag{8}$$

Here, c, c_e represent the specific heat capacity of the metal phase and the equivalent specific heat capacity [28], respectively, J/(kg·K); k is the thermal conductivity, W/(m·K); Q_{laser} is the laser energy density, W/m^3; L_f is the metal melting latent heat, J/kg; and T_l, T_s are the metal liquidus and solidus temperatures, respectively, K.

Since the vaporization heat loss of the liquid metal and the heat exchange between the metal phase and the surroundings are carried out through the surface, the heat transfer boundary condition used is

$$q_{transfer} = -q_{con} - q_{rad} - q_{vap} \tag{9}$$

where

$$q_{con} = h_c(T - T_{con}) \tag{10}$$

$$q_{rad} = \sigma_s \varepsilon \left(T^4 - T_{rad}^4\right) \tag{11}$$

Here, $q_{transfer}$, q_{con}, q_{rad}, q_{vap} are the total heat exchange, convective heat transfer, radiation heat transfer, and vaporization heat loss, respectively, W/m^2; h_c is the convective heat transfer coefficient, W/(m^2·K); T_{con} is the convection temperature of the surroundings, K; σ_s is the Stefan–Boltzmann constant, W/(m^2·K^4); ε is the emissivity; and T_{rad} is the radiative temperature of the surroundings, K.

In addition, the equivalent physical property parameters (to describe the transition of the powder layer to the solidified portion), the laser energy density Q_{laser}, and the vaporization heat loss q_{vap} will be separately described later.

2.1.3. Mass Conservation Equation

Since the fluids involved are considered incompressible fluids in the calculation process, the mass conservation equation is

$$\nabla \cdot \boldsymbol{u} = 0 \tag{12}$$

2.2. Gaussian Body Heat Source Considering Laser Reflection between Particles

Unlike the heat source in the welding process, a laser beam will be reflected multiple times between particles during the SLM formation process [29], so the laser can be considered to heat the particles at different positions (especially in the height direction) almost simultaneously. Therefore, the heat model needs to describe the reflection process of the laser beam between the particles. However, due to the simulation study being based on the workpiece scale, the powder layer is regarded as a special material, so the surface heat source or the body heat source can only be used to characterize the energy propagation of the laser.

The laser energy model used here is a Gaussian body heat source [30]. The energy density in the cross-section of the heat source model is Gaussian, and the energy density in the height direction considers the difference in energy density between the upper and lower end faces caused by laser reflection. Figure 1 is the schematic of the energy density distribution of the body heat source. The mathematical expression of the heat source model is

$$Q_{laser} = \frac{\xi \eta W_{laser}}{\pi (1 - e^{-3})(E + F)} \left(\frac{1 - \chi}{z_e - z_i} z + \frac{\chi z_e - z_i}{z_e - z_i} \right) \exp\left(-\frac{3r^2}{r_0^2} \right) \tag{13}$$

where

$$r_0 = \frac{z^2}{w} + s \tag{14}$$

$$w = \frac{z_e^2 - z_i^2}{r_e - r_i} \tag{15}$$

$$s = \frac{r_i z_e^2 - r_e z_i^2}{z_e^2 - z_i^2} \tag{16}$$

$$E = \frac{1-\chi}{z_e - z_i}\left\{\left(\frac{1}{w^2}\frac{z_e^6}{6} + \frac{s}{w}\frac{z_e^4}{2} + \frac{s^2}{2}z_e^2\right) - \left(\frac{1}{w^2}\frac{z_i^6}{6} + \frac{s}{w}\frac{z_i^4}{2} + \frac{s^2}{2}z_i^2\right)\right\} \tag{17}$$

$$F = \frac{\chi z_e - z_i}{z_e - z_i}\left\{\left(\frac{1}{w^2}\frac{z_e^5}{5} + 2\frac{s}{w}\frac{z_e^3}{3} + s^2 z_e\right) - \left(\frac{1}{w^2}\frac{z_i^5}{5} + 2\frac{s}{w}\frac{z_i^3}{3} + s^2 z_i\right)\right\} \tag{18}$$

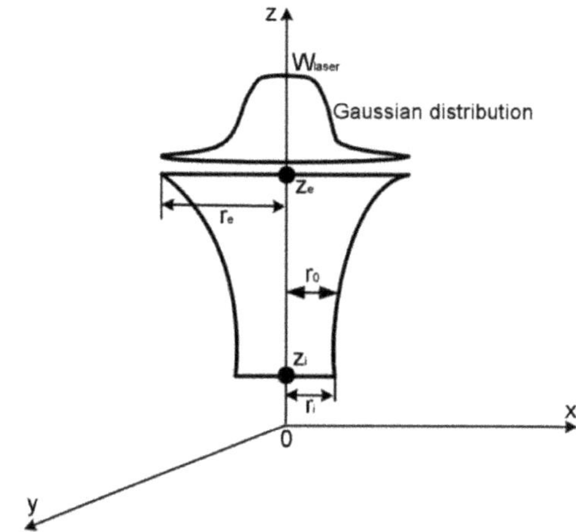

Figure 1. Energy density distribution of the Gaussian body heat source.

Here, W_{laser} is the laser power, W; ξ is the energy distribution factor; η is the effective absorption factor; χ is the ratio of the central energy density of the lower end face to the upper end face; z_e, z_i are the height coordinates of the upper and lower end faces of the laser energy distribution area, respectively, m; r_e, r_i are the radii of the upper and lower end faces of the laser energy distribution area, respectively, m; r_0 is the laser distribution cross-section radius when the height coordinate is z, m; and w, s, E, F are calculated intermediates.

2.3. Vaporization Heat Loss Model

For general metals, the vaporization temperature is around 3000 K. The laser beam has a very high energy density during the SLM process, and it is often able to vaporize the metal in a very short time. Therefore, an accurate SLM numerical simulation needs to consider the effects of vaporization heat loss and the vaporization recoil force. The vaporization heat loss model [31] used here is

$$q_{vap} = \hat{m}_{vap}\Delta H_{vap} \tag{19}$$

where

$$\hat{m}_{vap} = \left(p_{vap} - p_{amb}\right)\sqrt{\frac{m}{2\pi k_B T}} \tag{20}$$

Here, $\hat{m}_{vap}$ is the mass of the vaporized liquid metal on a unit of liquid surface per unit time, kg/(m²·s); ΔH_{vap} is the metal vaporization latent heat, J/kg; m is the metal molecular mass of the metal, kg; k_B is the Boltzmann constant, J/K; p_{vap} is the vaporization pressure, Pa; and p_{amb} is the protective atmosphere pressure, Pa.

The calculation of p_{vap} uses the vaporization pressure model under different environmental pressures, as proposed by Pang et al. [32]:

$$p_{vap} = \begin{cases} p_{amb} & 0 \le T < T_{left} \\ \frac{1+\beta_r}{2} p_0 \exp\left[\frac{m\Delta H_{vap}}{k_B}\left(\frac{1}{T_v} - \frac{1}{T}\right)\right] & T \ge T_{right} \\ p_{smooth} & T_{left} \le T < T_{right} \end{cases} \quad (21)$$

Here, p_0 is the standard atmospheric pressure, Pa; T_v is the metal vaporization temperature, K; T_{left}, T_{right} are the left and right critical temperatures of the transition zone, respectively, K; p_{smooth} is the transition zone pressure, Pa; and β_r is the recombination rate, and its value depends on the Mach number of the vapor plume. For high gasification rate conditions (such as a vacuum or at a high laser intensity), $\beta_r = 0.18$, and for low gasification rate conditions (such as a high ambient pressure or at a low laser intensity), $\beta_r = 1$. In other cases, the value of β_r is between the two.

The effect of the transition zone pressure, p_{smooth}, is employed to achieve a smooth interfacial pressure over the entire temperature range (Figure 2). The junction temperature, T_{vb}, in Figure 2 can be calculated by the following formula:

$$\frac{1+\beta_r}{2} p_0 \exp\left[\frac{m\Delta H_{vap}}{k_B}\left(\frac{1}{T_v} - \frac{1}{T_{vb}}\right)\right] = p_{amb} \quad (22)$$

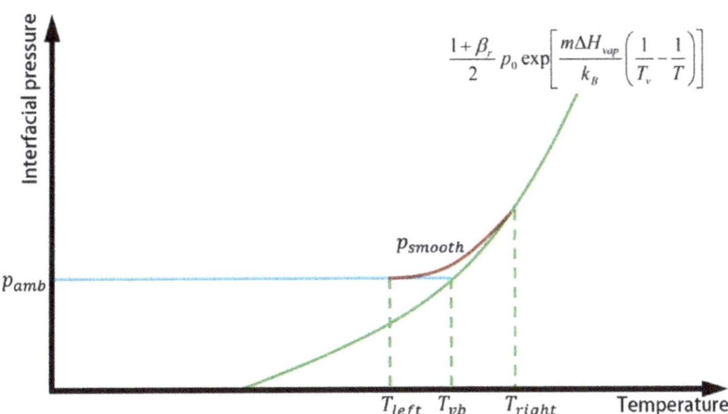

Figure 2. Schematic of the vaporization pressure model.

The left and right critical temperatures of the transition zone, T_{left}, T_{right}, satisfy the following (the coefficient of 0.05 is an artificially set value for smoothing):

$$T_{right} - T_{vb} = T_{vb} - T_{left} = 0.05 T_{vb} \quad (23)$$

The transition zone pressure, p_{smooth}, can be defined as

$$p_{smooth} = aT^3 + bT^2 + cT + d \quad (24)$$

In order to ensure a smooth transition of the interface pressure at T_{left} and T_{right}, the distribution of p_{smooth} can be obtained according to the coordinates of the two ends and the slopes of the tangents.

2.4. Equivalent Thermal Property Parameters Based on the Formation State

The core aim of the numerical simulation based on the workpiece scale is to equate the powder layer to a special material, but, in the actual SLM process, the powder layer will undergo a process of melting into liquid metal and become a dense solid. Therefore, it can be considered that the powder layer undergoes three state transitions: a particle state, a liquid state, and a solid state. In the calculation process, the basis for judging whether the state of the powder layer is changed is: (1) once the temperature of the original particle element exceeds its melting temperature (generally taken as the intermediate value of the liquidus and solidus temperature), the element state is converted to a liquid state; (2) for the elements that were originally in the liquid or solid state, their state will only change between liquid and solid (based on the liquidus-solidus temperature of the metal). The equivalent physical properties (density, specific heat capacity, and thermal conductivity) based on the formation state of the powder layer will be described below.

2.4.1. Equivalent Density and Specific Heat Capacity Based on the Formation State

During the calculation, the density of the powder layer element is

$$\rho = \begin{cases} (1-\phi)\rho_m + \phi\rho_a & particle\ state \\ \rho_m & liquid\ state\ or\ solid\ state \end{cases} \tag{25}$$

Here, ϕ is the initial porosity of the powder layer and ρ_m, ρ_a are the densities of the metal and gas phases, respectively, kg/m^3. It should be noted that ρ_m and ρ_a are temperature-dependent. In addition, the powder layer element was treated equivalently to the specific heat capacity.

2.4.2. Equivalent Thermal Conductivity Based on the Forming State

For the thermal conductivity of the powder layer, a treatment like that in Equation (25) cannot be performed (the equivalent physical property parameter is the weighted average of the physical parameters of the constituent phase). This is because, for the powder layer element in the particle state, the thermal conductivity is mainly determined by the heat conduction of the gas phase between the particles, but is also slightly affected by the thermal conductivity of the particles themselves. The equivalent thermal conductivity model [33] of the powder layer used here is

$$k = \begin{cases} \left(1 - \sqrt{1-\phi}\right)(k_a + \phi k_r) + \sqrt{1-\phi}\left\{ \dfrac{2}{\frac{1}{k_a}-\frac{1}{k_m}}\left[\dfrac{1}{1-\frac{k_a}{k_m}}\ln\left(\frac{k_m}{k_a}\right) - 1\right] + k_r\right\} & particle\ state \\ k_m & liquid\ state\ or\ solid\ state \end{cases} \tag{26}$$

where

$$k_r = 4F_{view}\sigma_s T_P^3 D_P \tag{27}$$

Here, k_m, k_a, k_r are the thermal conductivities of the metal phase, the protective gas, and the internal radiation of the powder layer, respectively, W/(m·K); F_{view} is the internal radiation factor, which is 1/3; T_P is the temperature of the metal particle, K; and D_P is the average particle diameter, m. It should be noted that k_m and k_a are temperature-dependent.

2.5. Numerical Solution of the Dynamic Behavior of the SLM Molten Pool Based on the Workpiece Scale

Based on the commercial CFD software Fluent v19.1, the numerical calculation of the dynamic behavior of the SLM molten pool on the workpiece scale was carried out. Among them, the selected solution models were Energy, Viscous-Laminar, and Solidification & Melting. User Defined Functions (UDFs) included a moving Gaussian body heat source, heat transfer boundary conditions (convection heat dissipation, radiation heat dissipation, and vaporization heat loss), and equivalent physical parameters (density, specific heat capacity, and thermal conductivity). The pressure–velocity coupling algorithm used SIMPLEC, and the time step was 1 ns. Figure 3 is the calculation flow chart for this study.

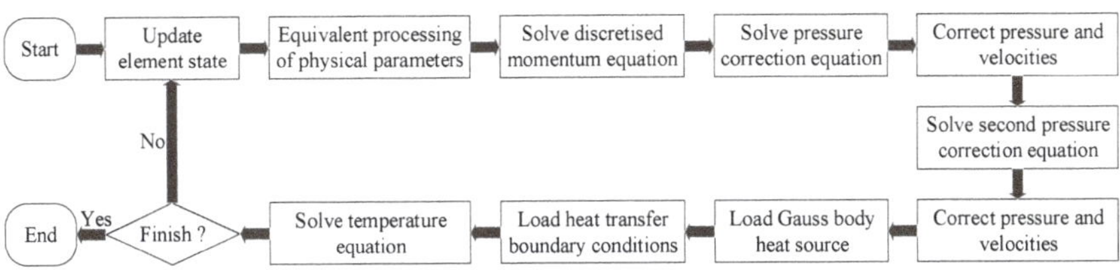

Figure 3. Calculation flow chart.

3. Results and Discussion

According to the physical model and numerical solution described above, the dynamic behavior of the SLM molten pool on the workpiece scale was predicted by using Fluent. First, in order to verify the feasibility of the SLM molten pool dynamics model, the single-pass process was calculated and compared to the experimentally obtained solidified track size, according to the experimental conditions for forming the Inconel 718 alloy by SLM outlined by Zhang et al. [34]. Secondly, the influences of different process parameters (laser power, scanning speed) on the SLM formation of the Inconel 718 alloy were analyzed, and the calculation results were verified with the experimental results according to the SLM experiment done by Wu et al. [35]. The mesh generation tool used ICEM CFD v19.1 and CFD-Post v19.1 was used for post-processing.

3.1. Experimental Verification of the Inconel 718 Nickel-Based Superalloy by the SLM Process

3.1.1. Calculation Parameters and Mesh Model

The composition (mass percentage) of the Inconel 718 alloy is Ni 50.4-Fe 21.86-Cr 18.44-Nb 5.04-Mo 3.02-Ti 0.88-Al 0.33-C 0.03. Table 1 contains the Inconel 718 physical parameters calculated by JMatPro-v7.

Table 1. Physical parameters of the Inconel 718 alloy.

Parameter	Value
Density, kg/m^3	8250 (298 K) $-$ 7488 (1373 K) $-$ 7803 (1638 K) $-$ 7378 (2000 K) $-$ 6470 (2773 K)
Solidus temperature, K	1373
Liquidus temperature, K	1638
Vaporization temperature, K	3000
Latent heat of melting, J/kg	2.19×10^5
Latent heat of vaporization, J/kg	7.34×10^6
Specific heat capacity, J/(kg·K)	760
Surface tension coefficient with temperature change rate, N/(m·K)	-3.24×10^{-4}
Molecular mass, kg	9.9134×10^{-26}
Thermal conductivity, W/(m·K)	11.03 (298 K) $-$ 28.01 (1373 K) $-$ 27.86 (1638 K) $-$ 45.72 (2773 K)
Dynamic viscosity, Pa·s	0.021 (1373 K) $-$ 0.009 (1638 K) $-$ 0.005 (1933 K) $-$ 0.002 (2773 K)

The constant specific heat capacity was chosen to improve the computational efficiency and the temperature dependent values were set using a simple linear interpolation.

The vaporization pressure, p_{vap}, Pa, of the Inconel 718 alloy was calculated according to Equations (21)–(24):

$$p_{vap} = \begin{cases} 1.01325 \times 10^5 & 0 < T < 2935 \\ 3.376372 \times 10^{-3}T^3 - 29.4454291T^2 + 85590.17272T - 8.28202513 \times 10^7 & 2935 \le T < 3244 \\ 60795\exp\left[52724 \times \left(\frac{1}{3000} - \frac{1}{T}\right)\right] & T \ge 3244 \end{cases} \tag{28}$$

The protective atmosphere in the experiment was argon and the other parameters required for the calculation are shown in Table 2.

Table 2. Other required calculation parameters.

Parameter	Value
Initial porosity of the powder layer [36]	0.4
Laser absorption rate [33]	0.36
Laser spot diameter, m	1.0×10^{-4}
Average particle diameter, m	3.0×10^{-5}
Powder bed thickness, m	4.0×10^{-5}
Laser power, W	285
Scanning speed, m/s	0.96
Density of the base plate, kg/m^3	7200
Thermal conductivity of the base plate, W/(m·K)	28
Specific heat capacity of the base plate, J/(kg·K)	640
Density of the gas phase, kg/m^3	1.225
Thermal conductivity of the gas phase, W/(m·K)	0.0242
Specific heat capacity of the gas phase, J/(kg·K)	1006.43
Convective/radiation heat transfer temperature of the surroundings, K	300
Convective heat transfer coefficient of the lower surface of the base plate and the upper surface of the powder layer, W/(m^2·K)	80
Emissivity	0.36
Initial temperature, K	353.15
Stefan–Boltzmann constant, W/(m^2·K^4)	5.67×10^{-8}
Boltzmann constant, J/K	$1.3806505(24) \times 10^{-23}$
Standard atmospheric pressure, Pa	1.01325×10^5

Figure 4 shows the geometry and mesh model used here. The calculation area was divided into three parts: the power layer, the solidified layer, and the base plate. The geometric dimensions of the three parts were $1 \times 0.5 \times 0.04$ mm^3, $1 \times 0.5 \times 0.08$ mm^3, and $1 \times 0.5 \times 0.2$ mm^3, and the mesh sizes were $0.01 \times 0.01 \times 0.0025$ mm^3, $0.01 \times 0.01 \times 0.01$ mm^3, and $0.01 \times 0.01 \times 0.02$ mm^3, respectively. The number of mesh elements obtained in each part was 80,000, 40,000, and 50,000, respectively. Boundary conditions included the top surface of the power layer set to convection, radiation, and vaporization; the bottom surface of the base plate set as convective heat transfer; the contact surface between the powder layer and the solidified layer set as a coupled wall; and the other boundary faces set as heat insulation. In addition, the start position, end position, and scanning direction of the laser in the single pass process are indicated in Figure 4 (the x coordinates of the start and end points are 0.1 mm and 0.9 mm, respectively).

3.1.2. Comparison of Simulation and Experimental Solidified Track Sizes

Figure 5 shows the temperature distributions in the top and middle sections at different times. It can be seen from the calculation results that, since the laser energy density is Gaussian in the horizontal plane, the temperature in the center of the active laser region was high, and the temperature was low around the periphery (Figure 5a–c). From the temperature distribution in the middle section (Figure 5d–f), the highest temperature of the pool was not at the center of the laser beam, indicating that the metal particles at the center of the laser spot were not completely melted. It can also be seen from the figure that, as the laser started to heat the powder layer, the upper surface of the solidified layer was significantly heated, meaning that there was heat exchange between the powder and the solidified layers. This heat came from two sources: part of the laser energy passing through the powder bed, and heat conduction between the powder layer and the solidified layer. Figure 6 shows the molten pool shapes at different times, which were characterized by separately extracting the liquidus temperature isothermal surface of the powder layer and the solidified layer. From the top-view (Figure 6a–c), it can be seen that, when the heat exchange in the formation process reached the quasi-steady state, the shape of the molten pool was in the shape of a teardrop. From the side-view (Figure 6d–f), it can be seen that the solidified layer was partly re-melted, due to indirect heating from the laser, which is typical for the SLM process and required to properly prepare the printed component.

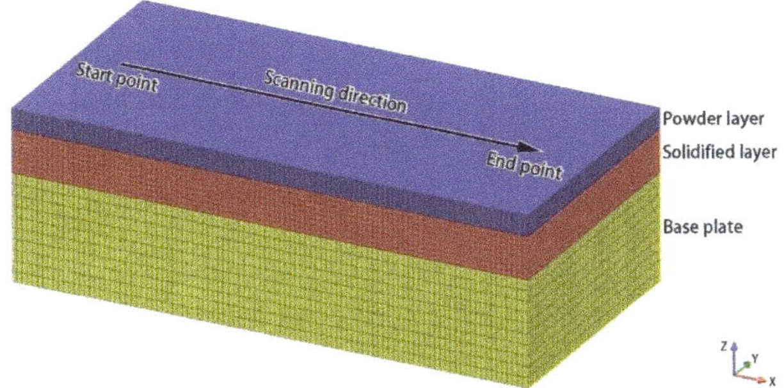

Figure 4. Adopted geometry and mesh models.

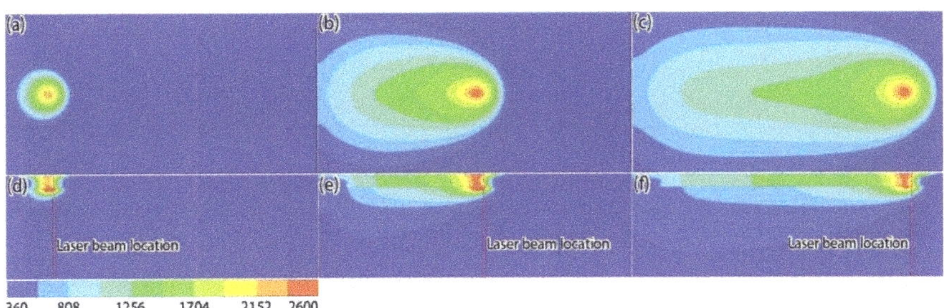

Figure 5. Simulation results of temperature fields in the top (**a–c**) and middle (**d–f**) sections at different times: (**a,d**) 6×10^{-5} s; (**b,e**) 4.5×10^{-4} s; (**c,f**) 8.4×10^{-4} s (unit: K).

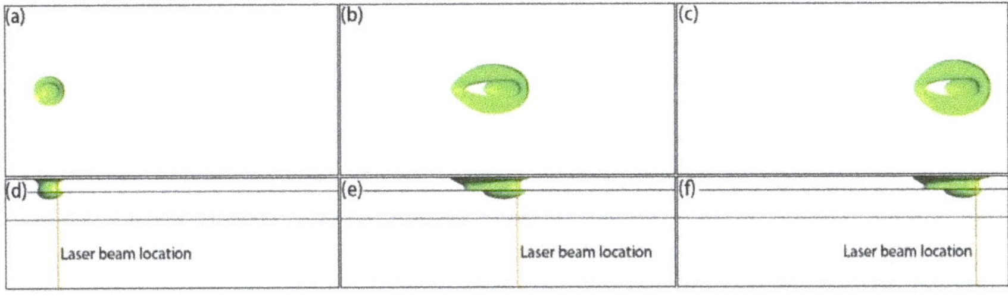

Figure 6. Simulation results of top- (**a–c**) and side- (**d–f**) view molten pool shapes at different times: (**a,d**) 6×10^{-5} s; (**b,e**) 4.5×10^{-4} s; (**c,f**) 8.4×10^{-4} s.

Figure 7 shows the velocity distributions on the top surface at different times. It can be seen from the simulation results that, due to the Gaussian distribution of the temperature on the top surface, the liquid metal flowed from the center of the molten pool to its periphery under the influence of the Marangoni effect (Figure 7d–f). Moreover, the speed distribution results (Figure 7a–c) show that the speed was low in the reverse scanning direction because the temperature of the laser-applied region was high, so the temperature gradient from the center to the activated region was low. Figure 8 shows the velocity distributions in the middle section at different times. Due to the Marangoni flow on the liquid surface, the annular convection phenomenon, centered on the molten pool axis, occurred inside the molten pool, and the tangential flow of the molten pool surface and internal convection affected the morphology of the molten pool. It should be noted that the difference in the flow behavior of the liquid and solid phases during the formation process was achieved by setting different dynamic viscosities, and a continuity condition was required in the calculation process, yielding velocity values outside of the solidified track, but their impact on the simulation results was limited.

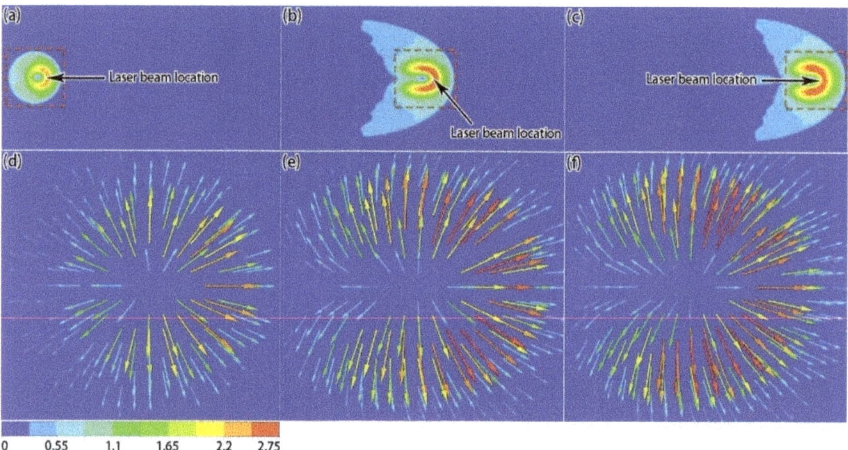

Figure 7. Simulation results of velocity magnitude (**a–c**) and local velocity (**d–f**) distributions on the surface at different times: (**a,d**) 6×10^{-5} s; (**b,e**) 4.5×10^{-4} s; (**c,f**) 8.4×10^{-4} s (unit: m/s).

Figure 9 shows the density distributions of the top surface and the side-view solidified track shapes at different times. It can be seen that the density of the powder layer element changed from the particle state to the liquid or solid state after being melted by heat, and it also reflects that the solidified track sizes were basically stable after the heat exchange reached the quasi-steady state during the formation process. Figure 10 shows the simulation result of the transverse section of the solidified track. By comparing with the experimental result [34], the experimental transverse section of the solidified track was semi-elliptical, and the powder layer and the solidified layer did not show a smooth transition to the solidified track in the simulation result. The influence of the solidified layer re-melting on the internal flow of the powder layer molten pool was not considered here, and the calculation model based on the workpiece scale could not characterize the dynamics, such as the collapse of the particles, so the temperature and velocity fields of the powder layer and the solidified layer were calculated independently. However, the key data of the SLM process was obtained through the simulation results, namely the molten pool width and depth. The simulation solidified track width was 126.08 µm and the depth was 65.26 µm (Figure 10). The experimentally obtained solidified track width was 124.14 µm and the depth was 66.21 µm, which was directly from Ref. [34]. The two agree well. Therefore, the molten pool dynamic behavior model based on the workpiece scale can be used to describe the SLM formation process to a certain extent. The model can feasibly describe the SLM process.

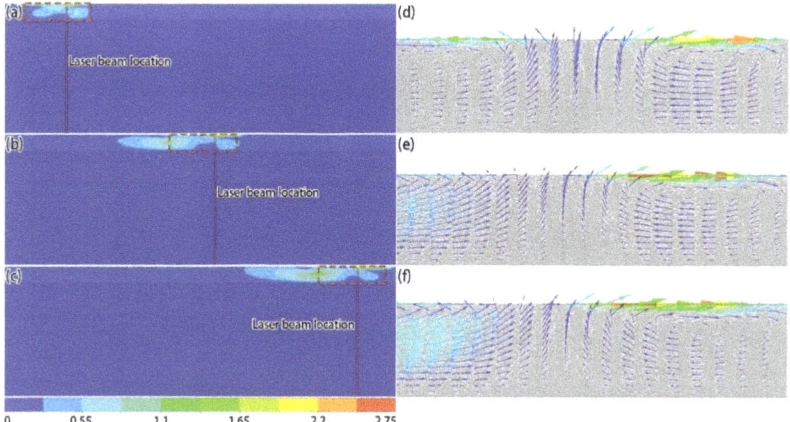

Figure 8. Simulation results of velocity magnitude (**a**–**c**) and local velocity (**d**–**f**) distributions in the middle section at different times: (**a**,**d**) 6×10^{-5} s; (**b**,**e**) 4.5×10^{-4} s; (**c**,**f**) 8.4×10^{-4} s (unit: m/s).

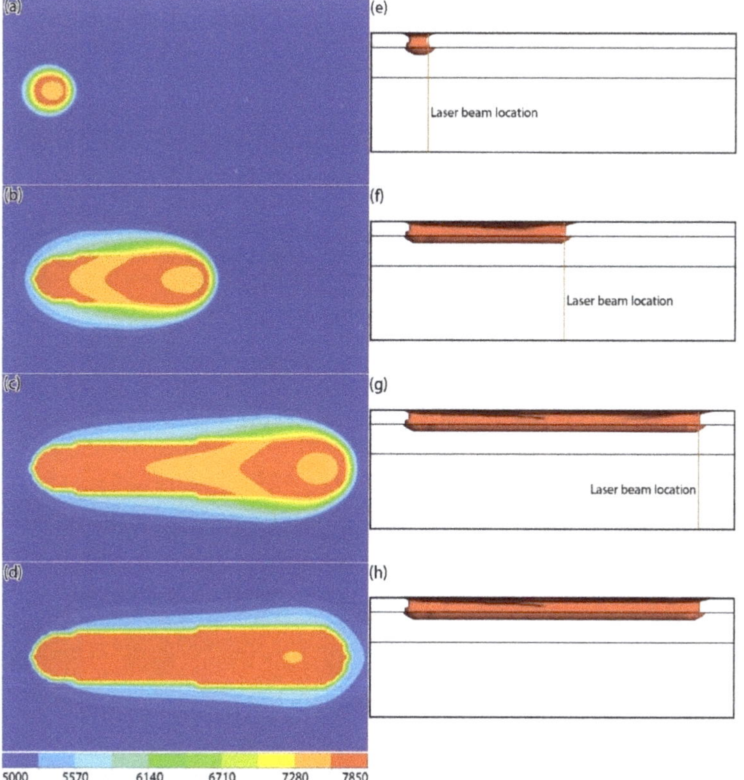

Figure 9. Simulation results of density distributions on the surface (**a**–**d**) and side-view solidified track shapes (**e**–**h**) at different times: (**a**,**e**) 6×10^{-5} s; (**b**,**f**) 4.5×10^{-4} s; (**c**,**g**) 8.4×10^{-4} s; (**d**,**h**) 1.5×10^{-3} s (unit: kg/m^3).

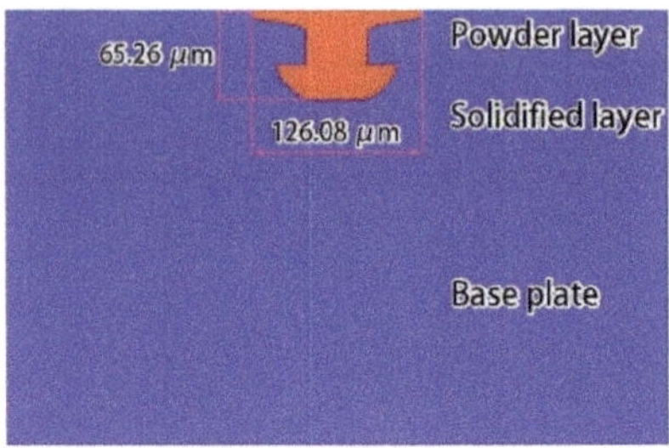

Figure 10. Simulation result of the transverse section of the solidified track.

3.2. Analysis of the SLM Process of the Inconel 718 Nickel-Based Superalloy

3.2.1. Calculation Parameters and Mesh Model

The parameters required to calculate this process were basically the same as those in Section 3.1.1, and Table 3 lists the different parameters. In addition, the geometric models used were nuanced. The geometric dimensions of the calculation area for the powder layer, the solidified layer, and the base plate were $1 \times 0.5 \times 0.03$ mm^3, $1 \times 0.5 \times 0.06$ mm^3, and $1 \times 0.5 \times 0.2$ mm^3, respectively. The corresponding mesh sizes were $0.01 \times 0.01 \times 0.002$ mm^3, $0.01 \times 0.01 \times 0.006$ mm^3, and $0.01 \times 0.01 \times 0.02$ mm^3. The number of mesh elements obtained in each area was 75,000, 50,000, and 50,000, respectively.

Table 3. Related parameters of this experiment.

Parameter	Value
Laser spot diameter, m	7.0×10^{-5}
Powder bed thickness, m	3.0×10^{-5}
Laser power, W	150, 200, 250, 300, 350
Scanning speed, m/s	0.4, 0.5, 0.6, 0.7, 0.8

3.2.2. Influence of the Laser Power on the Solidified Track Width

Figure 11 shows the temperature and local velocity distributions of the top surface under different laser powers when the laser acted on the center of the powder layer, where the scanning speed was set to 0.6 m/s. It can be seen that, as the laser power increased, the temperature of the active laser area increased significantly (Figure 11a–e). From the local velocity distributions (Figure 11f–j), as the laser power increased, the Marangoni effect became more apparent, and the tangential speed of the liquid metal at the surface became larger. Based on the shapes of the molten pools (Figure 12), the sizes of the molten pool also increased as the laser power increased.

Figure 13 shows the simulation results of the final shapes of the solidified tracks under different laser powers. From the simulation results, it can be seen that the solidified track width remained stable after the heat exchange from the SLM process reached a quasi-steady state. As the laser power increased, the width of the solidified track increased. Although the solidified track shapes in the experimental results [35] were not as regular in the simulation results, it is obvious that the solidified track width increased as the laser power increased. From the comparison of the simulation with the experimental solidified track widths under different laser powers (Figure 14, the experimental data was directly from Ref. [35]), the simulation results were in good agreement with the experimental results, and both showed that, as a rule, the solidified track width increased linearly with the laser power. It should be noted that, if the laser power was too large, the solidified track broke down due to balling and liquid instabilities [37].

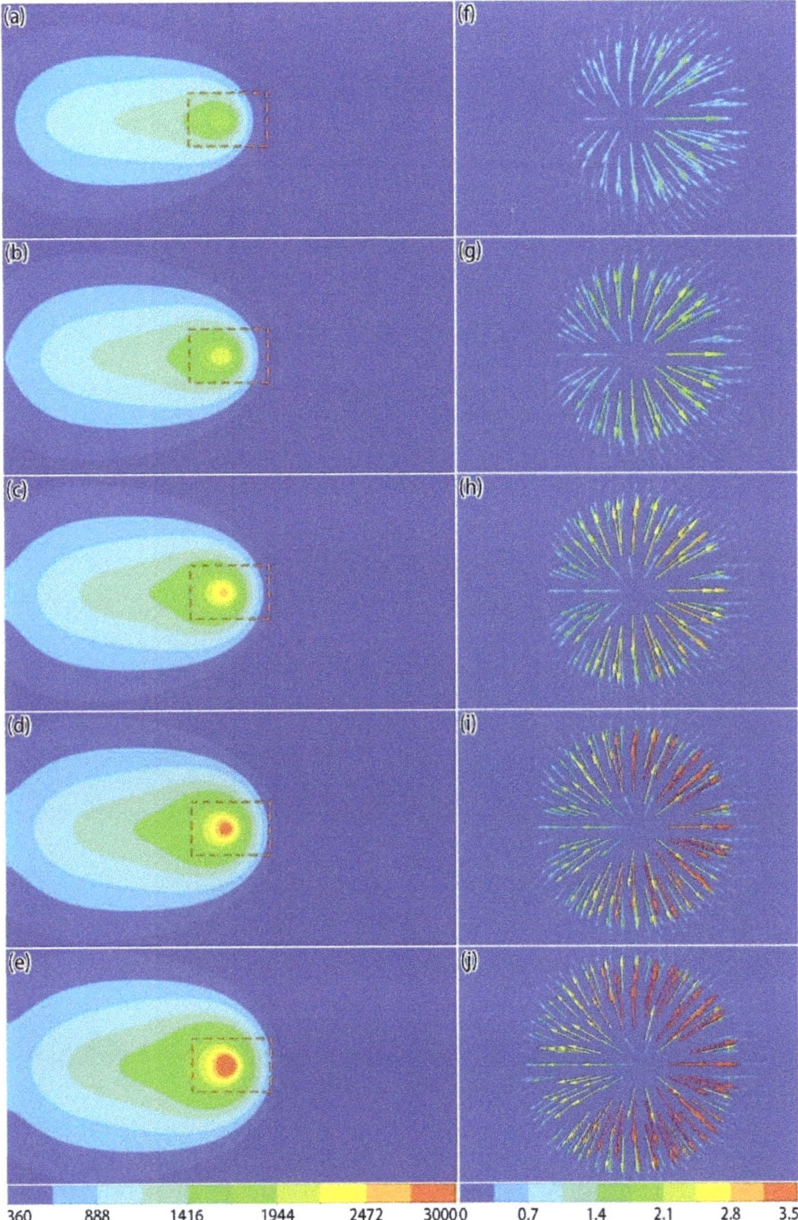

Figure 11. Simulation results of temperature (**a–e**, unit: K) and local velocity (**f–j**, unit: m/s) distributions on the surface under different laser powers when the laser acted on the center of the powder layer: (**a,f**) 150 W; (**b,g**) 200 W; (**c,h**) 250 W; (**d,i**) 300 W; (**e,j**) 350 W.

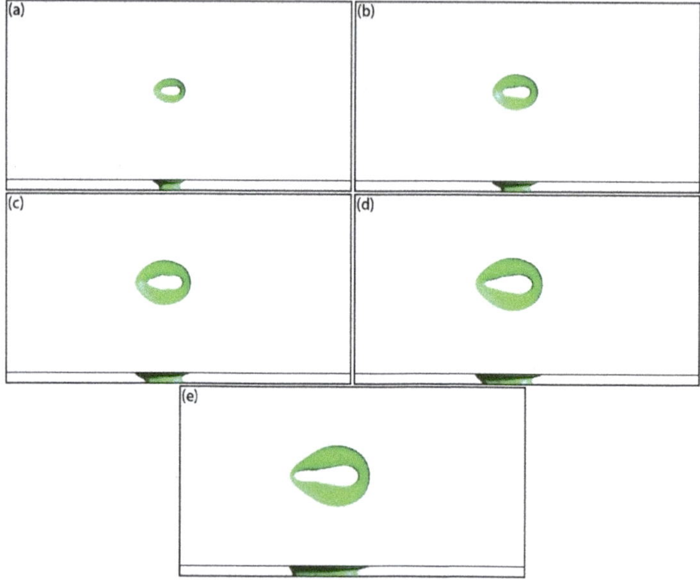

Figure 12. Simulation results of top- and side-view molten pool shapes under different laser powers when the laser acted on the center of the powder layer: (**a**) 150 W; (**b**) 200 W; (**c**) 250 W; (**d**) 300 W; (**e**) 350 W.

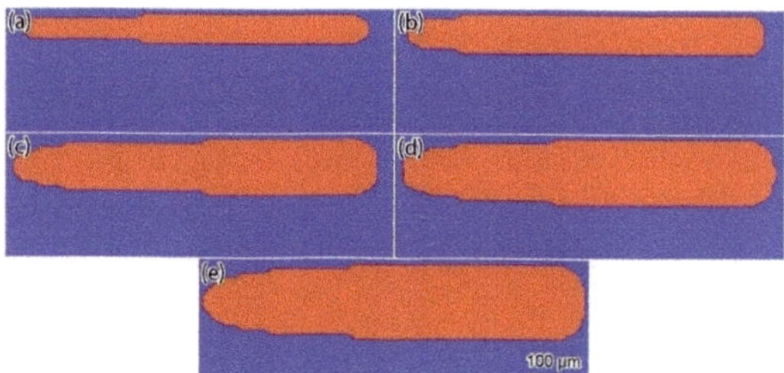

Figure 13. Simulation results of the final shapes of solidified tracks under different laser powers: (**a**) 150 W; (**b**) 200 W; (**c**) 250 W; (**d**) 300 W; (**e**) 350 W.

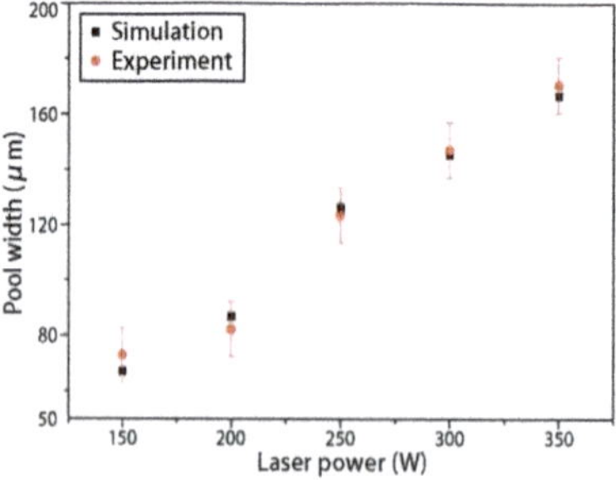

Figure 14. Comparison of simulation and experimental [35] solidified track widths under different laser powers.

3.2.3. Influence of the Scanning Speed on the Solidified Track Width

Figure 15 shows the temperature and local velocity distributions of the top surface under different scanning speeds when the laser acted on the center of the powder layer, where the laser power was set to 250 W. It can be seen that as the scanning speed increased, the temperature of the active laser area was significantly reduced (Figure 15a–e), because the active time of the laser on a fixed position was reduced. From the local velocity distributions (Figure 15f–j), as the scanning speed increased, the Marangoni effect weakened and the tangential speed of the liquid metal at the surface became smaller. From the shape of the molten pools (Figure 16), the sizes of the molten pool decreased as the scanning speed increased.

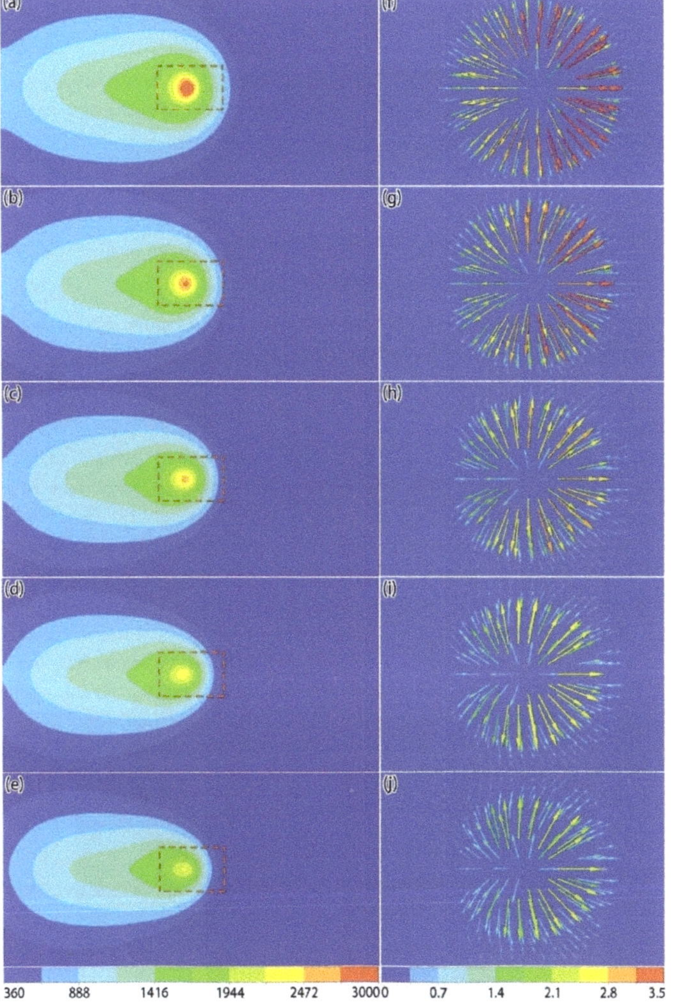

Figure 15. Simulation results of temperature (**a–e**, unit: K) and local velocity (**f–j**, unit: m/s) distributions on the surface under different scanning speeds when the laser acted on the center of the powder layer: (**a,f**) 0.4 m/s; (**b,g**) 0.5 m/s; (**c,h**) 0.6 m/s; (**d,i**) 0.7 m/s; (**e,j**) 0.8 m/s.

Figure 16. Simulation results of top- and side-view molten pool shapes under different scanning speeds when the laser acted on the center of the powder layer: (**a**) 0.4 m/s; (**b**) 0.5 m/s; (**c**) 0.6 m/s; (**d**) 0.7 m/s; (**e**) 0.8 m/s.

Figure 17 displays the simulation results of the final shapes of the solidified tracks under different scanning speeds. From the simulation results, it can be seen that, when the heat exchange of the SLM process reached a quasi-steady state, the width of the solidified track remained stable and, as the scanning speed increased, the width of the solidified track gradually decreased. Although the solidified track shapes in the experimental results [35] were not as regular as the simulation results, it was obvious that the solidified track width decreased as the scanning speed increased. From the comparison of the simulation and the experimental solidified track widths under different scanning speeds (Figure 18, the experimental data was directly from Ref. [35]), the simulation results were in good agreement with the experimental results, and both showed, as a rule, that the solidified track width decreased linearly with the scanning speed. It should be noted that, if the scanning speed was too small, the solidified track broke down due to balling and liquid instabilities [37].

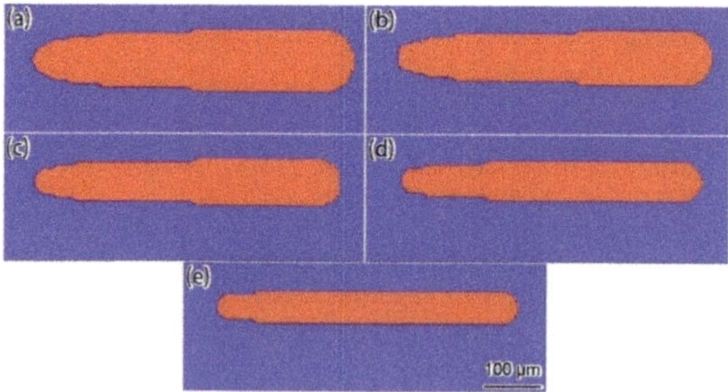

Figure 17. Simulation results of the final shapes of solidified tracks under different scanning speeds: (**a**) 0.4 m/s; (**b**) 0.5 m/s; (**c**) 0.6 m/s; (**d**) 0.7 m/s; (**e**) 0.8 m/s.

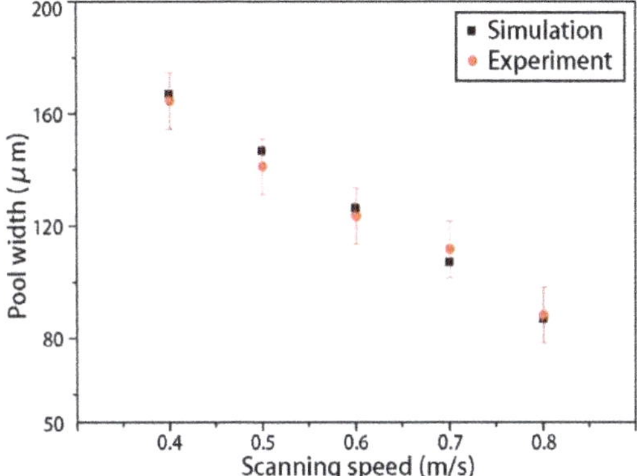

Figure 18. Comparison of simulation and experimental [35] solidified track widths under different scanning speeds.

4. Conclusions

(1) The more reasonable and comprehensive equivalent processing models included the following. Based on the smooth vaporization pressure model, the liquid metal vaporization heat loss models were established. To characterize the transformation of the powder layer state (particle state, liquid state and solid state) in the SLM process, the equivalent density, specific heat capacity, and thermal conductivity models based on the formation state were established.

(2) The SLM single-pass formation of the Inconel 718 alloy process was calculated. The simulation and experimental solidified track sizes were in good agreement, and the feasibility of the SLM molten pool dynamics model was verified.

(3) The influences of different process parameters (laser power, scanning speed) on the SLM formation of the Inconel 718 alloy were calculated and analyzed. Comparing the simulation and the experimentally determined solidified track widths, the two agreed well, and the results showed that, as a rule, the width increased linearly with the laser power and decreased linearly with the scanning speed.

(4) Key to the current metal additive manufacturing process is that the geometry of the workpiece has an important influence on the thermal-melt-stress evolution. To analyze the influence of the "heat transfer process-geometry-stress distribution" on the quality of the workpiece using the molten pool dynamics model, the thermal load under different process parameters must be obtained based on the model discussed here and be introduced into the stress calculation of the workpiece in a reasonable way.

(5) The complex thermophysical interactions existing in the SLM process often occur in a very short period of time and on a microscopic scale, such that the microstructure of the workpiece is greatly affected by the SLM process. Therefore, predicting the evolution behavior of an SLM solidification structure under different process parameters is also an important direction to study the "microstructure-molten pool-performance" of SLM parts.

Author Contributions: Methodology, X.Y.; Project administration, L.C.; Writing—original draft, L.C.

Funding: This research is supported by the Research Platform Construction Funding of Advanced Institute of Engineering Science for Intelligent Manufacturing, Guangzhou University.

Acknowledgments: We thank LetPub (www.letpub.com) for its linguistic assistance during the preparation of this manuscript.

Conflicts of Interest: The authors declare no conflicts of interest.

References

1. Wang, H.Y.; An, Y.Q.; Li, C.Y.; Chao, B.; Ni, Y.; Liu, G.B. Research progress of Ni-based superalloys. *Mater. Rev.* **2011**, *25*, 482–486.
2. Xiao, X.; Xu, H.; Qin, X.Z.; Guo, Y.A.; Guo, J.T.; Zhou, L.Z. Thermal fatigue behaviors of three cast nickel base superalloys. *Acta Metall. Sin.* **2011**, *47*, 1129–1134.
3. Qi, H. Review of Inconel 718 alloy: Its history, properties, processing and developing substitutes. *J. Mater. Eng.* **2012**, *8*, 92–100.
4. Liu, Y.C.; Guo, Q.Y.; Li, C.; Mei, Y.P.; Zhou, X.S.; Huang, Y. Recent progress on evolution of precipitates in Inconel 718 superalloys. *Acta Metall. Sin.* **2016**, *52*, 1259–1266.
5. Yang, Q.; Lu, Z.L.; Huang, F.X.; Li, D.C. Research on status and development trend of laser additive manufacturing. *Aeronaut. Manuf. Technol.* **2016**, *59*, 26–31.
6. Jia, X. Research on the flow field and mass transfer of tiny molten pool during selective laser melting of Inconel 718. Master's Thesis, Harbin Institute of Technology, Harbin, China, 2017.
7. Yadroitsev, I.; Gusarov, A.; Yadroitsava, I.; Smurov, I. Single track formation in selective laser melting of metal powders. *J. Mater. Process. Tech.* **2010**, *210*, 1624–1631. [CrossRef]
8. Huan, J.; Tian, Z.J.; Liang, H.X.; Xie, D.Q.; Shen, L.D.; Lv, F. Study on forming process and surface topography of Titanium alloy by selective laser melting process. *Appl. Laser* **2018**, *38*, 183–189.
9. Kruth, J.P.; Levy, G.; Klocke, F.; Childs, T.H.C. Consolidation phenomena in laser and powder-bed based layered manufacturing. *CIRP Ann.* **2007**, *56*, 730–759. [CrossRef]
10. Strano, G.; Hao, L.; Everson, R.M.; Evans, K.E. Surface roughness analysis, modelling and prediction in selective laser melting. *J. Mater. Process. Technol.* **2013**, *213*, 589–597. [CrossRef]
11. Liu, B.T.; Tian, C.; Zhang, A.P. Forming process of selective laser melting of nickel-based superalloy. *J. Heilongjiang Univ. Sci. Technol.* **2016**, *26*, 138–142. (In Chinese)
12. Cao, L.; Liao, D.M.; Sun, F.; Chen, T. Numerical simulation of cold-lap defects during casting filling process. *Int. J. Adv. Manuf. Technol.* **2018**, *97*, 2419–2430. [CrossRef]
13. Cao, L.; Sun, F.; Chen, T.; Tang, Y.L.; Liao, D.M. Quantitative prediction of oxide inclusion defects inside the casting and on the walls during cast-filling processes. *Int. J. Heat Mass Transf.* **2018**, *119*, 614–623. [CrossRef]
14. Francois, M.M.; Sun, A.; Wayne, K.E.; Henson, N.J.; Tourret, D.; Bronkhorst, C.A. Modeling of additive manufacturing processes for metals: Challenges and opportunities. *Curr. Opin. Solid State Mater. Sci.* **2017**, *21*, 198–206. [CrossRef]
15. Markl, M.; Körner, C. Multiscale modeling of powder bed-based additive manufacturing. *Annu. Rev. Mater. Res.* **2016**, *46*, 93–123. [CrossRef]
16. Gürtler, F.J.; Karg, M.; Leitz, K.H.; Schmidt, M. Simulation of laser beam melting of steel powders using the three-dimensional volume of fluid method. *Phys. Procedia* **2013**, *41*, 874–879. [CrossRef]
17. Khairallah, S.A.; Anderson, A.T.; Rubenchik, A.; King, W.E. Laser powder-bed fusion additive manufacturing: Physics of complex melt flow and formation mechanisms of pores, spatter, and denudation zones. *Acta Mater.* **2016**, *108*, 36–45. [CrossRef]
18. Dai, D.; Gu, D.D. Thermal behavior and densification mechanism during selective laser melting of copper matrix composites: Simulation and experiments. *Mater. Des.* **2014**, *55*, 482–491. [CrossRef]
19. Michaleris, P. Modeling metal deposition in heat transfer analyses of additive manufacturing processes. *Finite Elem. Anal. Des.* **2014**, *86*, 51–60. [CrossRef]
20. Voisin, T.; Calta, N.P.; Khairallah, S.A.; Forien, J.; Balogh, L.; Cunningham, R.W. Defects-dictated tensile properties of selective laser melted Ti-6Al-4V. *Mater. Des.* **2018**, *158*, 113–126. [CrossRef]
21. Lee, Y.S.; Zhang, W. Mesoscopic simulation of heat transfer and fluid flow in laser powder bed additive manufacturing. In Proceedings of the Annual International Solid Freeform Fabrication Symposium, Austin, TX, USA, 10–12 August 2015.
22. Panwisawas, C.; Qiu, C.; Anderson, M.J.; Sovani, Y.; Turner, R.; Attallah, M.M. Mesoscale modelling of selective laser melting: Thermal fluid dynamics and microstructural evolution. *Comp. Mater. Sci.* **2017**, *126*, 479–490. [CrossRef]
23. Xiao, B.; Zhang, Y. Marangoni and buoyancy effects on direct metal laser sintering with a moving laser beam. *Numer. Heat Transf. A-Appl.* **2007**, *51*, 715–733. [CrossRef]

24. Gusarov, A.V.; Smurov, I. Modeling the interaction of laser radiation with powder bed at selective laser melting. *Phys. Procedia* **2010**, *5*, 381–394. [CrossRef]

25. Yuan, P.; Gu, D.D. Molten pool behaviour and its physical mechanism during selective laser melting of TiC/AlSi10Mg nanocomposites: Simulation and experiments. *J. Phys. D-Appl. Phys.* **2015**, *48*, 035303. [CrossRef]

26. Cao, L.; Sun, F.; Chen, T.; Teng, Z.H.; Tang, Y.L.; Liao, D.M. Numerical simulation of liquid-solid conversion affecting flow behavior during casting filling process. *Acta Metall. Sin.* **2017**, *53*, 1521–1531.

27. Yang, J.; Wang, F. 3D finite element temperature field modelling for direct laser fabrication. *Int. J. Adv. Manuf. Tech.* **2009**, *43*, 1060–1068. [CrossRef]

28. Cao, L.; Liao, D.M.; Lu, Y.Z.; Chen, T. Heat transfer model of directional solidification by LMC process for superalloy casting based on finite element method. *Metall. Mater. Trans. A* **2016**, *47*, 4640–4647. [CrossRef]

29. King, W.E.; Anderson, A.T.; Ferencz, R.M.; Hodge, N.E.; Kamath, C.; Khairallah, S.A. Laser powder bed fusion additive manufacturing of metals; physics, computational, and materials challenges. *Appl. Phys. Rev.* **2015**, *2*, 041304. [CrossRef]

30. Qiu, C.; Panwisawas, C.; Ward, M.; Basoalto, H.C.; Brooks, J.W.; Attallah, M.M. On the role of melt flow into the surface structure and porosity development during selective laser melting. *Acta Mater.* **2015**, *96*, 72–79. [CrossRef]

31. Hagelin-Weaver, H. Surface Science: Foundations of Catalysis and Nanoscience. *J. Nanopart. Res.* **2002**, *4*, 575–576. [CrossRef]

32. Pang, S.Y.; Hirano, K.; Fabbro, R.; Jiang, T. Explanation of penetration depth variation during laser welding under variable ambient pressure. *J. Laser Appl.* **2015**, *57*, 022007. [CrossRef]

33. Dai, K.; Shaw, L. Thermal and mechanical finite element modeling of laser forming from metal and ceramic powders. *Acta Mater.* **2004**, *52*, 69–80. [CrossRef]

34. Zhang, L.; Wu, W.H.; Lu, L.; Ni, X.Q.; He, B.B.; Yang, Q.Y. Effect of heat input parameters on temperature field in Inconel 718 alloy during selective laser melting. *J. Mater. Eng.* **2018**, *46*, 29–35.

35. Wu, T.; Liu, B.T.; Liu, J.H. Study on nickel-based superalloy selective laser melting process parameters. *J. Heilongjiang Univ. Sci. Tech.* **2015**, *25*, 361–365.

36. Liu, B.Y. Cyclic aging behavior of TC4 powder and the influence on properties of components built by SLM. Master's Thesis, Shenyang Aerospace University, Shenyang, China, 2018.

37. Gibson, I.; Rosen, D.; Stucker, B. *Additive Manufacturing Technologies: Rapid Prototyping to Direct Digital Manufacturing*; Springer: Heidelberg, Germany, 2010.

MDPI

St. Alban-Anlage 66

4052 Basel

Switzerland

Tel. +41 61 683 77 34

Fax +41 61 302 89 18

www.mdpi.com

Materials Editorial Office

E-mail: materials@mdpi.com

www.mdpi.com/journal/materials